Lecture Notes in Computer Science   8493

Commenced Publication in 1973
Founding and Former Series Editors:
Gerhard Goos, Juris Hartmanis, and Jan van Leeuwen

Arnold Beckmann   Erzsébet Csuhaj-Varjú
Klaus Meer (Eds.)

Language, Life, Limits

10th Conference on Computability in Europe, CiE 2014
Budapest, Hungary, June 23-27, 2014
Proceedings

 Springer

Volume Editors

Arnold Beckmann
Swansea University, Department of Computer Science
Swansea, UK
E-mail: a.beckmann@swansea.ac.uk

Erzsébet Csuhaj-Varjú
Eötvös Loránd University
Department of Algorithms and Their Applications
Budapest, Hungary
E-mail: csuhaj@inf.elte.hu

Klaus Meer
Brandenburg University of Technology Cottbus-Senftenberg
Computer Science Institute
Cottbus, Germany
E-mail: meer@informatik.tu-cottbus.de

ISSN 0302-9743   e-ISSN 1611-3349
ISBN 978-3-319-08018-5   e-ISBN 978-3-319-08019-2
DOI 10.1007/978-3-319-08019-2
Springer Cham Heidelberg New York Dordrecht London

Library of Congress Control Number: 2014940542

LNCS Sublibrary: SL 1 – Theoretical Computer Science and General Issues

Typesetting: Camera-ready by author, data conversion by Scientific Publishing Services, Chennai, India

Printed on acid-free paper

Springer is part of Springer Science+Business Media (www.springer.com)

Preface

CiE 2014: Language, Life, Limits
Budapest, Hungary, June 23–27, 2014

The year 2014 was a very special one for the CiE conference series: Computability in Europe 2014 (CiE 2014) was the tenth meeting since the conference series started in 2005. As emphasized by this year's motto *Language, Life, Limits*, CiE 2014 in particular focused on relations between computational linguistics, natural and biological computing, and more traditional fields of computability theory. This was understood in its broadest sense including computational aspects of problems in linguistics, studying models of computation and algorithms inspired by physical and biological approaches as well as exhibiting limits (and non-limits) of computability when considering different models of computation arising from such approaches.

As with previous CiE conferences, the allover glueing perspective was to strengthen the mutual benefits of analyzing traditional and new computational paradigms in their corresponding frameworks both with respect to practical applications and a deeper theoretical understanding.

The conference series *Computability in Europe* is organized by the Association CiE. The association promotes the development of computability-related science, ranging from mathematics, computer science and applications in various natural and engineering sciences, such as physics and biology, as well as the promotion of related fields, such as philosophy and history of computing. In particular, the conference series successfully brings together the mathematical, logical and computer sciences communities that are interested in developing computability related topics.

The host of CiE 2014 was the Faculty of Informatics, Eötvös Loránd University, Budapest, Hungary, the venue of the conference was Hotel Mercure Budapest Buda in Budapest.

The nine previous CiE conferences were held in Amsterdam (The Netherlands) in 2005, Swansea (Wales) in 2006, Siena (Italy) in 2007, Athens (Greece) in 2008, Heidelberg (Germany) in 2009, Ponta Delgada (Portugal) in 2010, Sofia (Bulgaria) in 2011, Cambridge (England) in 2012, and Milan (Italy) in 2013. The proceedings of all these meetings were published in the Springer series *Lecture Notes in Computer Science*. The annual CiE conference has become a major event and is the largest international meeting focused on computability theoretic issues. CiE 2015 will be held in Bucharest, Romania, and CiE 2016 in Paris, France.

The series is coordinated by the CiE Conference Series Steering Committee consisting of Arnold Beckmann (Swansea, chair), Laurent Bienvenu (Paris), Natasha Jonoska (Tampa FL), Benedikt Löwe (Amsterdam & Hamburg), Mariya Soskova (Sofia & Berkeley CA), Susan Stepney (York), and Peter van Emde Boas (Amsterdam).

The Programme Committee of CiE 2014 was chaired by Erzsébet Csuhaj-Varjú (Budapest) and Klaus Meer (Cottbus). It was responsible for the selection of the invited speakers, the special session organizers and for running the reviewing process of all submitted regular contributions.

Structure and Programme of the Conference

The Programme Committee invited 6 speakers to give plenary lectures: Lev Beklemishev (Moscow), Alessandra Carbone (Paris), Maribel Fernández (London), Przemyslaw Prusinkiewicz (Calgary), Éva Tardos (Cornell), and Albert Visser (Utrecht). Elsevier B.V. sponsored the *Elsevier TCS Lecture* which was given by Maribel Fernández.

In addition to the plenary lectures, the conference had two tutorials by Peter Grünwald (Amsterdam & Leiden) and Wolfgang Thomas (Aachen).

Springer-Verlag generously funded a *Best Student Paper Award* that was given during the CiE 2014 conference. The winner was Ludovic Patey for his paper "*The complexity of satisfaction problems in Reverse Mathematics.*"

The conference CiE 2014 had six special sessions: two sessions, *Computational Linguistics* and *Bio-inspired Computation*, were devoted to the special focus of CiE 2014. In addition to this, new developments in areas frequently covered in the CiE conference series were addressed in the further special sessions on *History and Philosophy of Computing, Computability Theory, Online Algorithms,* and *Complexity in Automata Theory.* Speakers in these special sessions were selected by the special session organizers and were invited to contribute a paper to this volume.

History and Philosophy of Computing.
> *Organizers.* Lisbeth de Mol (Ghent) and Giuseppe Primiero (London).
> *Speakers.* Federico Gobbo (Amsterdam) and Marco Benini (Varese), Jacqueline Léon (Paris), James Power (Maynooth), Graham White (London).

Computational Linguistics.
> *Organizers.* Maria Dolores Jiménez-López (Tarragona) and Gábor Prószéky (Budapest).
> *Speakers.* Leonor Becerra-Bonache (St. Etienne), Henning Christiansen (Odense), Frank Drewes (Umeå), Frédérique Segond (Grenoble).

Computability Theory.
> *Organizers.* Karen Lange (Wellesley) and Barbara Csima (Waterloo).
> *Speakers.* Rachel Epstein (Swarthmore), Andy Lewis (London), Iskander Kalimullin (Kazan), Keng Meng Ng (Nanyang).

Bio-inspired Computation.
  Organizers. Marian Gheorghe (Sheffield) and Florin Manea (Kiel).
  Speakers. Harold Fellermann (Odense), Hendrik Jan Hoogeboom (Leiden),
  Lars Wienbrandt (Kiel), Claudio Zandron (Milan).
Online Algorithms.
  Organizers. Joan Boyar (Odense) and Csanád Imreh (Szeged).
  Speakers. Kim Skak Larsen (Odense), Jiří Sgall (Prague), Rob van Stee (Leicester), Adi Rosen (Paris).
Complexity in Automata Theory.
  Organizers. Markus Lohrey (Siegen) and Giovanni Pighizzini (Milan).
  Speakers. Artur Jez (Wrocław), Christos Kapoutsis (Qatar), Martin Kutrib (Giessen), James Worrell (Oxford).

We received 78 contributed paper submissions which were reviewed by the Programme Committee and many expert referees. In the end, 35% of the submitted papers were accepted for publication in this volume. In addition, this volume contains 15 invited papers. Without the help of our expert referees, the production of the volume would have been impossible. We would like to thank all the sub-reviewers for their excellent work; their names are listed in the organization section of this preface.

All authors who contributed to this conference are encouraged to submit significantly extended versions of their papers with unpublished research content to *Computability. The Journal of the Association CiE.*

The Steering Committee of the conference series CiE is concerned about the representation of female researchers in the field of computability. In order to increase female participation, the series started the *Women in Computability* (WiC) program in 2007, first funded by the *Elsevier Foundation,* then taken over by the publisher *Elsevier.* We were proud to continue this programme with its successful annual WiC workshop and a grant programme for junior female researchers in 2014.

Acknowledgments

The organizers of CiE 2014 would like to acknowledge and thank the following entities for their financial support (in alphabetic order): the *Association for Symbolic Logic* (ASL), *Elsevier B.V.,* the *European Association for Computer Science Logic* (EACSL), the *European Association for Theoretical Computer Science* (EATCS), and *Springer-Verlag.* We would also like to acknowledge the support of our non-financial sponsor, the *Association Computability in Europe* (CiE). Finally, we would like to acknowledge both financial and non-financial support of the following institutions: *Faculty of Informatics, Eötvös Loránd University,* Budapest, Hungary, *Faculty of Informatics and its Department of Computer Science, University of Debrecen,* Hungary.

We thank Andrej Voronkov for his EasyChair system which facilitated the work of the Programme Committee and the editors considerably.

April 2014   Arnold Beckmann
                                               Erzsébet Csuhaj-Varjú
                                                      Klaus Meer

Organization

Programme Committee

Gerard Alberts, Amsterdam
Sandra Alves, Porto
Hajnal Andréka, Budapest
Luís Antunes, Porto
Arnold Beckmann, Swansea
Laurent Bienvenu, Paris
Paola Bonizzoni, Milan
Olivier Bournez, Palaiseau
Vasco Brattka, Munich
Bruno Codenotti, Pisa
S. Barry Cooper, Leeds
Erzsébet Csuhaj-Varjú, Budapest
(co-chair)
Michael J. Dinneen, Auckland
Erich Grädel, Aachen
Marie Hicks, Chicago
Natasha Jonoska, Tampa

Jarkko Kari, Turku
Elham Kashefi, Edinburgh
Viv Kendon, Leeds
Satoshi Kobayashi, Tokyo
András Kornai, Budapest
Marcus Kracht, Bielefeld
Benedikt Löwe, Amsterdam &
   Hamburg
Klaus Meer, Cottbus (co-chair)
Joseph Mileti, Grinnell
Georg Moser, Innsbruck
Benedek Nagy, Debrecen
Sara Negri, Helsinki
Thomas Schwentick, Dortmund
Neil Thapen, Prague
Peter van Emde Boas, Amsterdam
Xizhong Zheng, Glenside

Reviewers

Altenkirch, Thorsten
Angeleska, Angela
Avanzini, Martin
Baartse, Martijn
Battyányi, Péter
Bauwens, Bruno
Besozzi, Daniela
Bordihn, Henning
Bourreau, Pierre
Calude, Cristian
Calvert, Wesley
Carlucci, Lorenzo
Chen, Qingliang
Choffrut, Christian
Csima, Barbara
Dashkov, Evgenij
de Brecht, Matthew

Delaney, Aidan
Della Vedova, Gianluca
Dennunzio, Alberto
Dondi, Riccardo
Downey, Rod
Duncan, Ross
Egri-Nagy, Attila
Eguchi, Naohi
Enayat, Ali
Everitt, Matthew
Finkel, Olivier
Fisseni, Bernhard
Flood, Stephen
Florido, Mário
Frittaion, Emanuele
Gabbay, Murdoch
Gazdag, Zsolt

Genova, Daniela
Ghosh, Sukumar
Gierasimczuk, Nina
Gimenez, Stéphane
Glaßer, Christian
Goncharov, Sergei
Graça, Daniel
Heiner, Monika
Hinze, Thomas
Hirschfeldt, Denis
Hirst, Jeff
Hoyrup, Mathieu
Jeřábek, Emil
Kach, Asher M.
Kara, Ahmet
Kent, Curtis
Kim, Yun-Bum

Kjos-Hanssen, Bjørn	Oliva, Paulo	Seyfferth, Benjamin
Kleijn, Jetty	Palano, Beatrice	Shlapentokh, Alexandra
Kontchakov, Roman	Patey, Ludovic	Souto, André
Kreuzer, Alexander	Păun, Gheorghe	Stannett, Mike
Kuske, Dietrich	Pellegrini, Marco	Stepney, Susan
Leporati, Alberto	Petrocchi, Marinella	Towsner, Henry
Leupold, Peter	Proctor, Timothy	Turán, György
Löding, Christof	Razafindrakoto, Jean	Turetsky, Daniel
Martini, Simone	Rettinger, Robert	Tzevelekos, Nikos
Marx, Dániel	Romero, Daniel	Törmä, Ilkka
Mauri, Giancarlo	Russell, Ben	Vasconcelos, Pedro
Mayordomo, Elvira	Salo, Ville	Vaszil, György
McCartin, Catherine	Savchuk, Dmytro	Viganò, Luca
Meduna, Alexander	Sburlan, Dragos	Vizzari, Giuseppe
Miller, Russell	Schaper, Michael	Vortmeier, Nils
Mitrana, Victor	Schett, Maria Anna	Yokomori, Takashi
Monin, Benoît	Schröder, Matthias	Yokoyama, Keita
Mummert, Carl	Schröder, Lutz	Zandron, Claudio
Németi, István	Seisenberger, Monika	Zeume, Thomas
Normann, Dag	Seki, Shinnosuke	Ziegler, Martin
Okubo, Fumiya	Selivanov, Victor	Zinoviadis, Charalampos

Local Organization

Local organizer of CiE 2014 was the Faculty of Informatics, Eötvös Loránd University, Budapest, Hungary, partly in cooperation with the Department of Computer Science, Faculty of Informatics, University of Debrecen. Some parts of the local arrangements were performed by ELTE-Soft Non-Profit Organization, Budapest, Hungary, and Pannonia Tourist Service, Budapest, Hungary.

Members of the Organizing Committee were Erzsébet Csuhaj-Varjú (chair, Budapest), Zsolt Gazdag (Budapest), Katalin Anna Lázár (Budapest), and Krisztián Tichler (Budapest).

Table of Contents

Computability and Categoricity
of Ultrahomogeneous Structures

Francis Adams and Douglas Cenzer

Department of Mathematics, University of Florida, P.O. Box 118105,
Gainesville, Florida 32611
fsadams@ufl.edu, cenzer@math.ufl.edu

Abstract. This paper investigates the effective categoricity of ultrahomogeneous structures. It is shown that any computable ultrahomogeneous structure is Δ_2^0 categorical. A structure $\mathcal{A}$ is said to be *weakly ultrahomogeneous* if there is a finite (*exceptional*) set of elements $a_1, \ldots, a_n$ such that $\mathcal{A}$ becomes ultrahomogeneous when constants representing these elements are added to the language. Characterizations are obtained for the weakly ultrahomogeneous linear orderings, equivalence structures, and injection structures, and compared with characterizations of the computably categorical and Δ_2^0 categorical structures.

Keywords: computability theory, ultrahomogeneous, injections, effective categoricity, computable model theory.

1 Introduction

Computable model theory studies the algorithmic properties of effective mathematical structures and the relationships among such structures. The effective categoricity of a computable structure $\mathcal{A}$ measures the possible complexity of isomorphisms between $\mathcal{A}$ and computable copies of $\mathcal{A}$, and is an important gauge of the complexity of $\mathcal{A}$.

We say a structure (model) $\mathcal{A}$ is computable if its universe A is computable and all of its functions and relations are uniformly computable. Given two computable structures, we will say they are computably isomorphic if there exists an isomorphism between them that is computable. For a single computable structure $\mathcal{A}$, we will say $\mathcal{A}$ is computably categorical if every computable structure isomorphic to $\mathcal{A}$ is in fact computably isomorphic to $\mathcal{A}$. More generally, we will say two computable structures are Δ_α^0 isomorphic if there exists an isomorphism between them that is Δ_α^0 and we will say a computable structure $\mathcal{A}$ is Δ_α^0 categorical if every computable structure isomorphic to $\mathcal{A}$ is Δ_α^0 isomorphic to $\mathcal{A}$.

A structure $\mathcal{A}$ is said to be *ultrahomogeneous* if every isomorphism between finitely generated substructures extends to an automorphism of $\mathcal{A}$. Ultrahomogeneous structures were first studied by Fraïssé [5], who defined the *age* of a structure to be the family of finitely generated substructures of $\mathcal{A}$ and gave properties which characterized the age of an ultrahomogeneous structure. Csima,

A. Beckmann, E. Csuhaj-Varjú, and K. Meer (Eds.): CiE 2014, LNCS 8493, pp. 1–10, 2014.

Harizanov, R. Miller and A. Montalban [4] studied computable ages and the computability of the canonical ultrahomogeneous structures, called *Fraïssé limits*.

Here are some simple examples of countable ultrahomogeneous structures. See [6] for more details. The linear ordering $(\mathbb{Q}, <)$ of the rationals is the unique ultrahomogeneous countable linear ordering. The age here is just the set of all finite linear orderings. This structure is computably categorical. An equivalence structure (A, E) is ultrahomogeneous if and only if all equivalence classes have the same size k, $1 \leq k \leq \aleph_0$; then the age is the set of all finite equivalence structures with all classes of size $\leq k$. These structures are computably categorical.

An injection structure is a set with a single 1-1 unary function. This function induces a partition of the set into distinct orbits: finite cycles, one-way infinite orbits (ω-orbits), or two-way infinite orbits ($\mathbb{Z}$-orbits). An injection structure is ultrahomogeneous if and only if it has no ω-orbits. For example, there is the injection structure with infinitely many $\mathbb{Z}$-orbits, where the age is the set of structures consisting of finitely many $\mathbb{Z}$-orbits. There is also the injection structure with exactly one orbit of size k for each finite k, where the age is the family of finite injection structures with no more than one orbit of any size k. In these examples, the ultrahomogeneous structure with infinitely many $\mathbb{Z}$-orbits is in fact not computably categorical, but is Δ_2^0 categorical.

We observe that in the first two examples there are computable models of the countable ultrahomogeneous structure. In the third example, one can have an arbitrary number of orbits of various finite sizes and thus a structure which is not computable.

In this paper, we will closely examine the effective categoricity of ultrahomogeneous structures. In section 2, we will prove that any computable ultrahomogeneous structure is Δ_2^0 categorical. In section 3, we introduce the notion of *weakly ultrahomogeneous structures*, where $\mathcal{A}$ is weakly ultrahomogeneous if there is a finite (*exceptional*) set of elements $a_1, \ldots, a_n$ such that $\mathcal{A}$ becomes ultrahomogeneous when constants representing these elements are added to the language. We characterize the weakly ultrahomogeneous linear orderings, equivalence structures, and injection structures. In section 4, we present our conclusions and topics for future research, including nested equivalence structures.

2   Categoricity of Ultrahomogeneous Structures

In this section, we will show that any computable ultrahomogeneous structure is Δ_2^0 categorical. Some lemmas are needed. If $\mathcal{A}$ is an ultrahomogeneous structure and $\mathcal{A} \cong \mathcal{B}$, by composing maps it is easy to see that $\mathcal{B}$ is also ultrahomogeneous. We also have the following stronger fact:

Lemma 1. *Let $\mathcal{A}, \mathcal{B}$ be isomorphic ultrahomogeneous structures. Let X, Y be finitely generated substructures of $\mathcal{A}, \mathcal{B}$ respectively. If $\varphi : X \to Y$ is an isomorphism, then there is an isomorphism of $\mathcal{A}$ and $\mathcal{B}$ extending φ.*

Proof. Let $\theta : \mathcal{A} \to \mathcal{B}$ be an isomorphism. Then $\theta^{-1} \circ \varphi : X \to \mathcal{A}$ is an isomorphism of finitely generated substructures of $\mathcal{A}$. So it extends to an automorphism $\alpha : \mathcal{A} \to \mathcal{A}$. Then $\theta \circ \alpha : \mathcal{A} \to \mathcal{B}$ is an isomorphism that extends φ.   □

Let $\mathcal{A}_s[x_1, \ldots, x_n]$ be the terms of height s starting with the x_i, i.e. the set obtained by starting with the elements of $\{x_1, \ldots, x_n\}$ and applying the functions of the structure up to s-many times. Then $\langle x \rangle = \bigcup_{s \in \omega} \mathcal{A}_s[x]$. While the $\mathcal{A}_s[x]$ aren't structures, we will say that $\mathcal{A}_s[x] \cong \mathcal{A}_s[y]$ if for any terms $t_1, \ldots t_m$ of height s and any relation R, we have $R(t_1[x], \ldots, t_m[x]) \Leftrightarrow R(t_1[y], \ldots, t_m[y])$.

Lemma 2. $\langle x \rangle \cong \langle y \rangle$ with $x_i \to y_i$ iff for all $s \in \omega$ we have $\mathcal{A}_s[x] \cong \mathcal{A}_s[y]$. Thus, asking if two finitely generated substructures are isomorphic is a Π_1^0 question.

Proof. For the left to right direction, it is clear that the restriction of the isomorphism to any height is an instance of the desired map. The reverse implication follows from the fact that given finitely many terms $t_1[x], \ldots t_m[x]$, they occur by some finite height s. So the terms are in $\mathcal{A}_s[x]$, hence for any relation R we have $R(t_1[x], \ldots, t_m[x]) \Leftrightarrow R(t_1[y], \ldots, t_m[y])$ and so the map $t[x] \to t[y]$ is an isomorphism. $\qquad\square$

Theorem 1. *Every computable ultrahomogeneous structure is Δ_2^0-categorical.*

Proof. Let $\mathcal{A}$ be a computable ultrahomogeneous structure and let $\mathcal{B}$ be isomorphic to $\mathcal{A}$ by the isomorphism φ. We want to build a Δ_2^0- isomorphism θ. We do this with a back-and-forth argument, building increasing partial isomorphisms θ_n at each stage and letting $\theta = \bigcup \theta_n$. Let $a_0 \in A$. Since $\langle a_0 \rangle \cong \langle \varphi(a_0) \rangle$, set $\theta_0(a_0) = \varphi(a_0) = b_0$.

Suppose we have defined θ_{2n-1} for $\{a_0, \ldots a_{2n-1}\}$ with $\theta_{2n-1}(a_i) = b_i$. Choose the least $a_{2n} \in A \setminus \{a_0, \ldots a_{2n-1}\}$. There exists a $b \in B$ such that $\langle a_0, \ldots a_{2n} \rangle \cong \langle b_0, \ldots, b_{2n-1}, b \rangle$ and we can choose the isomorphism so it extends θ_{2n-1} by Lemma 1. Now search for this b using a Π_1^0-oracle to check whether $\langle a_0, \ldots a_{2n} \rangle \cong \langle b_0, \ldots, b_{2n-1}, b \rangle$, call it b_{2n}, and define $\theta_{2n}(a_{2n}) = b_{2n}$.

Now suppose we have defined θ_{2n} for $\{a_0, \ldots a_{2n}\}$ with $\theta_{2n}(a_i) = b_i$. Choose the least $b_{2n+1} \in A \setminus \{b_0, \ldots, b_{2n}\}$. There exists an $a \in B$ such that $\langle a_0, \ldots a_{2n}, a \rangle \cong \langle b_0, \ldots, b_{2n+1} \rangle$ and we can choose the isomorphism so it extends θ_{2n} by Lemma 1. Search for this a using a Π_1^0-oracle to check if $\langle a_0, \ldots a_{2n}, a \rangle \cong \langle b_0, \ldots, b_{2n+1} \rangle$, call it a_{2n+1}, and define $\theta_{2n+1}(a_{2n+1}) = b_{2n+1}$.

The map $\theta : \mathcal{A} \to \mathcal{B}$ thus constructed is a bijection, since we took the least a_i and b_i at each stage. It is also clear that it is Δ_2^0, since θ is defined using a Π_1^0-oracle. To show that it is an isomorphism, fix an m-tuple x such that $\theta(a_i) = b_i$ for each a_i in x, and let $y = \theta(x)$. Choose any relation R, and any function f of arity m. Then for some n, $x \subseteq \{a_0, \ldots, a_n\}$, so $y \subseteq \{b_0, \ldots, b_n\}$, and since $\langle a_0, \ldots, a_n \rangle \cong \langle b_0, \ldots, b_n \rangle$, we have $R(x) \Leftrightarrow R(y)$ and $\theta(f(x)) = f(y)$. $\qquad\square$

The complexity of the isomorphism constructed in the theorem is a direct result of the complexity of the problem of determining if two finitely generated substructures are isomorphic. So if this problem is computable for a structure, then that structure is computably categorical. If a structure is relational, then for any finite subset X of the universe we have $\langle X \rangle = X$ and checking if two finite structures are isomorphic is computable. Therefore, all computable relational ultrahomogeneous structures are computably categorical.

More generally, any computable ultrahomogeneous locally finite structure is computably categorical, where locally finite means that every finitely generated substructure is finite. The converse is false: an injection structure consisting of a single $\mathbb{Z}$-orbit is computably categorical, but is clearly not locally finite.

3 Weakly Ultrahomogeneous Structures

In this section, we introduce a weak version of ultrahomogeneity. Whereas ultrahomogeneous structures have the property that 'all points look the same' in a very strong way, the weaker version will allow finitely many elements to look different from the others.

Definition 2. *A structure $\mathcal{A}$ is weakly ultrahomogeneous if there exists a finite set $\{a_1, a_2, \ldots, a_n\} \subseteq A$ such that for all tuples $\boldsymbol{x}, \boldsymbol{y}$ from A with $\langle \boldsymbol{a}, \boldsymbol{x} \rangle \cong \langle \boldsymbol{a}, \boldsymbol{y} \rangle$ where each a_i is fixed, this isomorphism of substructures extends to an automorphism of $\mathcal{A}$. Call such a set $\{a_1, a_2, \ldots, a_n\}$ an exceptional set of $\mathcal{A}$.*

Alternatively, $\mathcal{A}$ is weakly ultrahomogeneous if there is a finite set $a_1, \ldots, a_n$ of elements of $\mathcal{A}$ such that $(\mathcal{A}, a_1, \ldots, a_n)$ is ultrahomogeneous in the extended language with constants for $a_1, \ldots, a_n$. Thus we can prove the following.

Theorem 3. *Every computable weakly ultrahomogeneous structure is Δ_2^0-categorical.*

If $\mathcal{A}$ is a finite structure, it is trivially weakly ultrahomogeneous since the universe can be taken to be an exceptional set. As above, if $\mathcal{A}$ is weakly ultrahomogeneous and locally finite, then $\mathcal{A}$ is computably categorical. Given any exceptional set, we can add finitely many elements to it and obtain another exceptional set. But more interesting are the minimal exceptional sets and the senses in which such sets are unique. To see some instances of this definition, we will look at the weakly ultrahomogeneous analogues of the examples of ultrahomogeneous structures considered in the introduction.

3.1 Linear Orders

We start with a characterization of computably categorical linear orders proved in [3]. Say that an element in a linear order is a successivity if it has an immediate successor, an immediate predecessor, or is an endpoint.

Theorem 4 (Remmel). *A computable linear order $\mathcal{A}$ is computably categorical iff $\mathcal{A}$ has finitely many successivities.*

Next we give a characterization of weakly ultrahomogeneous linear orders.

Theorem 5. *For a countable linear order $\mathcal{A}$, the following are equivalent:*

1. *$\mathcal{A}$ is weakly ultrahomogeneous.*
2. *$\mathcal{A}$ has finitely many successivities.*

3. $\mathcal{A} = L_0 + \mathbb{Q} + L_1 + \mathbb{Q} + \ldots + \mathbb{Q} + L_n$ where the L_i are finite chains, L_0, L_n are possibly empty and $|L_i| \geq 2$ for $1 \leq i \leq n-1$.

Proof. $[1 \Rightarrow 2]$: Suppose $\mathcal{A}$ has infinitely many successivities. Let $\{a_1, \ldots, a_n\}$ be a finite subset of A; we will show it cannot be an exceptional set. The successivities of $\mathcal{A}$ occur in finite chains or in subsets of order type ω, ω^* (the reverse order of ω), or $\mathbb{Z}$. If there is a subset C that has order type ω or $\mathbb{Z}$, choose elements $x_1 < x_2$ and $y_1 < y_2 \in C$ greater than all $a_i \in C$ such that there are more elements between x_1 and x_2 than there are between y_1 and y_2. Then $\langle a, x_1, x_2 \rangle \cong \langle a, y_1, y_2 \rangle$, but the isomorphism can't be extended to an automorphism. If C is of order type ω^*, we can repeat the argument above by choosing the elements below all a_i in C. Finally, if $\mathcal{A}$ has infinitely many successivities in finite chains, choose one of these chains containing none of the a_i and from it choose the first element x and the second element y. Then $\langle a, x \rangle \cong \langle a, y \rangle$, but the isomorphism can't be extended to an automorphism. With infinitely many successivities one of these situations must occur, and in either case we see $\{a_1, \ldots, a_n\}$ isn't exceptional, so $\mathcal{A}$ isn't weakly ultrahomogeneous.

$[2 \Rightarrow 3]$: Suppose $\mathcal{A}$ has finitely many successivities; call this finite set S. Then S can be partitioned into $L_0, \ldots, L_n$ where $x, y \in L_i$ iff there are finitely many elements between x and y, and so $i < j$ implies that for all $x \in L_i$ and for all $y \in L_j$, $x < y$. So between L_i and L_{i+1} are infinitely many elements, none of which are successivities, hence these elements compose a copy of $\mathbb{Q}$.

$[3 \Rightarrow 1]$: Let $S = \bigcup_{i \leq n} L_i = \{a_1, \ldots, a_k\}$. We claim that this set is exceptional. To see this, suppose $\langle a, x \rangle \cong \langle a, y \rangle$ with $a_i \to a_i$ for $i \leq k$ and $x_j \to y_j$ for $j \leq n$. Also assume that $x_j, y_j \notin S$ for $j \leq n$. So for each $j \leq n$, x_j and y_j are in the same copy of $\mathbb{Q}$, since they bear the same relation to all elements of S. If all the x_i, and hence all y_i are in the same copy of $\mathbb{Q}$, by ultrahomogeneity there is an automorphism of that copy of $\mathbb{Q}$ sending x_i to y_i. Then using the identity elsewhere, we have an automorphism of the entire structure. If there are x_i in more than one copy, use ultrahomogeneity in each copy with an x_i and the identity elsewhere to get an automorphism of the whole structure. $\square$

Corollary 1. *A computable linear order $\mathcal{A}$ is computably categorical iff $\mathcal{A}$ is weakly ultrahomogeneous.*

Corollary 2. *For any countable weakly homogeneous linear order $\mathcal{A}$, there is a computable structure isomorphic to $\mathcal{A}$.*

Let us say that a an exceptional set S for a weakly ultrahomogeneous structure is a *minimal exceptional set* if no proper subset of S is exceptional. Such as set must exist since S is finite. We will try to characterize minimal exceptional sets and determine whether they are unique to a structure, or perhaps unique up to automorphism.

For an ultrahomogeneous linear order of the form $L_0 + \mathbb{Q} + L_1 + \mathbb{Q} + \ldots + \mathbb{Q} + L_n$, as in Theorem 5, the set $\bigcup_{i \leq n} L_i$ is an exceptional set, but this set is not necessarily minimal. In general, exceptional sets for this ordering can be described as follows.

Proposition 1. *Let $\mathcal{A} = L_0 + \mathbb{Q} + L_1 + \mathbb{Q} + \ldots + \mathbb{Q} + L_n$ be a countable weakly ultrahomogeneous linear order. The exceptional sets are those subsets S of $\bigcup_{i \le n} L_i$ such that*

(i) $A \setminus S$ doesn't contain two consecutive successivities.
(ii) S contains each last element of $L_0, \ldots L_{n-1}$ and each first element of $L_1, \ldots L_n$.

Proof. If (i) fails for a set $S = \{a_1, \ldots, a_n\}$, let x, y be two consecutive successivities not in S; so both x and y are in some L_i. Then $\langle a, x \rangle \cong \langle a, y \rangle$, but the isomorphism can't extend since they are in different positions in L_i and have a different number of immediate successors.

If (ii) fails for $S = \{a_1, \ldots, a_n\}$, let x witness its failure and let y be an element of the copy of $\mathbb{Q}$ adjacent to x. Then $\langle a, x \rangle \cong \langle a, y \rangle$, but the isomorphism can't extend since x is a successivity and y isn't.

Now suppose $i)$ and $ii)$ hold for $S = \{a_1, \ldots, a_n\}$ and suppose $\langle a, x \rangle \cong \langle a, y \rangle$ by φ. Any x_i which are successivities are uniquely between two a_i, or are the unique element greater than or less than all a_i if they are endpoints, hence they must be fixed. Any non-successivity pairs x_i, y_i must be in the same copy of $\mathbb{Q}$ and by the ultrahomogeneity of each copy, we can extend φ to an automorphism.   □

It follows that the minimal exceptional sets contain each last element of $L_0, \ldots, L_{n-1}$ and each first element of $L_1, \ldots, L)n$ and, for any other elements a and b such that b is the successor of a, S contains exactly one of a, b. As an example, consider the linear order $\mathbb{Q} + L_0 + \mathbb{Q}$ where $L_0 = \{a_1 < a_2 < \ldots < a_5\}$. Then both $\{a_1, a_3, a_5\}$ and $\{a_1, a_2, a_4, a_5\}$ are minimal exceptional sets. This shows that, while we would like the exceptional sets of a weakly ultrahomogeneous structure to be unique in some way, minimal exceptional sets aren't necessarily unique and in fact need not even be isomorphic.

Recall that in a structure $\mathcal{A}$, an element $b \in A$ is definable from a set $S \subseteq A$ if $\{b\}$ is a subset of A definable from S. Define the definable closure of S, $D(S) = \{x \in A : x \text{ is definable from } S\}$. An important fact about definability we will use repeatedly is that if b is definable from S and σ is an automorphism of $\mathcal{A}$ fixing all elements of S, then σ also fixes b. Looking at definable closures reveals a sense in which minimal exceptional sets are unique.

Proposition 2. *Let $\mathcal{A}$ be a weakly ultrahomogeneous linear order and let $M = \{a_1 < \ldots < a_n\}$ be a minimal exceptional set. Then $D(M)$ is the set of successivities of $\mathcal{A}$.*

Proof. Suppose x is in one of the copies of $\mathbb{Q}$ in $\mathcal{A}$. Using the ultrahomogeneity of $\mathbb{Q}$, there is an automorphism of $\mathcal{A}$ moving x while fixing M. So $x \notin D(M)$. Now let x be a successivity of $\mathcal{A}$; we may assume $x \notin M$. Then x is the unique element satisfying $\phi(y) : (y < a_i)$ if it is a left endpoint of $\mathcal{A}$, similarly for right endpoints, or else it is the unique element satisfying $\phi(y) : (a_i < y < a_{i+1})$ for some $i \le n$.   □

3.2 Equivalence Structures

The effective categoricity of equivalence structures was investigated by Calvert, Cenzer, Harizanov and Morozov in [1]. The *character* of an equivalence structure indicates the number of equivalence classes of each size. The structure is said to have *bounded character* if there exists a $k \in \mathbb{N}$ such that all finite classes have size at most k. It is proved in [1] that an equivalence structure $\mathcal{A}$ is computably categorical iff $\mathcal{A}$ has finitely many finite classes, or $\mathcal{A}$ has finitely many infinite classes, bounded character, and there is at most one k such that there are infinitely many classes of size k. This condition is equivalent to saying that all but finitely many classes of $\mathcal{A}$ have the same size.

Theorem 6. *For a countable equivalence structure $\mathcal{A}$, the following are equivalent:*

1. *$\mathcal{A}$ is weakly ultrahomogeneous.*
2. *All but finitely many equivalence classes of $\mathcal{A}$ have the same size.*

In this case, a minimal exceptional set contains exactly one element from each of the exceptional equivalence classes.

Proof. [1 ⇒ 2] Suppose that $\mathcal{A}$ has infinitely many classes of different sizes and let $\{a_1, \ldots, a_n\}$ be a finite subset. Then find elements x, y from classes of different sizes so neither is related to any a_i. Then $\langle a, x \rangle \cong \langle a, y \rangle$, but this can't extend to an automorphism.

[2 ⇒ 1] If all but finitely many equivalence classes of $\mathcal{A}$ are of the same size, let $\{a_1, \ldots, a_n\}$ contain exactly one element from each of these exceptional classes. Then suppose $\langle a, x \rangle \cong \langle a, y \rangle$ via the isomorphism φ so $\varphi(x_i) = y_i$ and $\varphi(a_k) = a_k$. Either each pair x_i, y_i is in an equivalence class with some a_k, or if not they are in classes of the same size. Then the isomorphism extends to an automorphism as follows.

First, we will explain how the equivalence classes are mapped, and then we will describe what happens to the elements of each class. Fix each exceptional class as well as any classes which do not contain any x_i or y_i. If a nonexceptional class has an x_i, then map x_i to y_i and hence the class $[x_i]$ to the class $[y_i]$. If there are nonexceptional classes with a y_i but no x_i, there must be the same number of nonexceptional classes with an x_i but no y_i. Send each class of the first kind to one of the second kind. Thus we may end up with cycles, say $[x_1]$ maps to $[y_1]$ and $y_1 E x_3$, so that $[y_1]$ maps to $[y_3]$, and then $[y_3]$ maps to $[x_1]$.

Within each class, do the following. For classes with no x_i or y_i, the class is fixed and we also fix each element of the class. For the nonexceptional classes containing some of the x_i or y_i, we have mapped the x_i and y_i above, and the remaining elements can be mapped arbitrarily. For the exceptional classes containing some of the x_i or y_i, it follows that $\varphi(x_i) E x_i$, thus we can map those elements respecting φ using a cycle decomposition similar to that described above for the nonexceptional classes. Now the remaining elements can simply be fixed.

The claim about the minimal exceptional sets follows since the proof shows such a set is exceptional, and that a finite set disjoint from two classes of different sizes is not exceptional.   □

Corollary 3. *A computable equivalence structure $\mathcal{A}$ is weakly ultrahomogeneous iff $\mathcal{A}$ is computably categorical.*

Corollary 4. *For any countable weakly homogeneous equivalence structure $\mathcal{A}$, there is a computable structure isomorphic to $\mathcal{A}$.*

With this characterization of weakly ultrahomogeneous structures and their minimal exceptional sets, we can again investigate their uniqueness properties. Given two minimal exceptional sets, there is an automorphism of the structure sending one to the other by interchanging the two elements in each exceptional class and fixing everything else. However, as opposed to linear orders we don't have uniqueness of the definable closures.

Proposition 3. *Let $\mathcal{A} = (A, E)$ be a weakly ultrahomogeneous equivalence structure and let $S = \{a_1, \ldots, a_n\}$ be a minimal exceptional set. Then $x \in A$ is definable from S iff $x \in S$ or x is in an exceptional class of size at most 2.*

Proof. If $x \in A$ isn't in an exceptional class, there is an automorphism fixing all the exceptional classes but moving x by interchanging $[x]$ with another class of the same size. If x is in an exceptional class of size at most 2, then for some $i \leq n$, either $x = a_i$ or $\{x\} = \{y : yEa_i \ \& \ y \neq a_i\}$. If x is in an exceptional class of size greater than 2, then $[x]$ contains a_i for some $i \leq n$ and an element y distinct from x and a_i. In this case, switching x and y and using the identity everywhere else is an automorphism of $\mathcal{A}$ moving x and fixing S.   □

3.3   Injection Structures

The effective categoricity of injection structures was studied by Cenzer, Harizanov and Remmel in [2]. It was shown that an injection structure is computably categorical if and only if it has finitely many infinite orbits, and is Δ_2^0 categorical if and only if it either has only finitely many orbits of type $\mathbb{Z}$ or has only finitely many orbits of type ω.

Proposition 4. *A countable injection structure $\mathcal{A}$ is weakly ultrahomogeneous iff it has finitely many ω-orbits. In this case, a minimal exceptional set contains exactly one member from each ω-orbit.*

Proof. Suppose that $\mathcal{A}$ is an injection structure having only finitely many ω-orbits. Let $\{a_1, \ldots, a_n\}$ contain exactly one element from each of the ω-orbits, and assume that $\langle \boldsymbol{a}, \boldsymbol{x} \rangle \cong \langle \boldsymbol{a}, \boldsymbol{y} \rangle$ via the isomorphism φ. The isomorphism is extended to an automorphism as follows.

First, orbits not containing any x_i or y_h are fixed. If x_i is in a finite orbit of size k, then y_i is also in a finite orbit of size k, and the orbit of x_i is mapped to the orbit of y_i. If there are finite orbits of size k containing some y_j but no x_i, then there must be an equal number of orbits of size k containing some x_i but no y_j, and then we map each class of the first kind to one of the second kind.

If x_i is in a $\mathbb{Z}$-orbit, then φ maps the sequence $(x_i, f(x_i), \dots)$ to the sequence $(y_i, f(y_i), \dots)$ and this can be extended to the entire orbits. Each ω-orbit must be fixed, since it contains one of the a_i, and φ fixes a_i and respects f.

Now assume $\mathcal{A}$ has infinitely many ω-orbits, and let $\{a_1, \dots, a_n\}$ be a finite set. In an ω-orbit containing none of the a_i, let x_0 be the initial element and $x_1 = f(x_0)$. Then $\langle \boldsymbol{a}, x_0 \rangle \cong \langle \boldsymbol{a}, x_1 \rangle$, but the isomorphism can't extend since x_1 is in the range of f while x_0 isn't. Thus $\mathcal{A}$ isn't weakly ultrahomogeneous.

If a finite set S doesn't include an element from each ω-orbit, we may repeat the above argument with the orbit not intersecting S to show the finite set isn't exceptional. The claim about minimal exceptional sets follows.   □

It follows that, for computable injection structures, computable categoricity implies weak ultrahomogeneity which implies Δ_2^0-categoricity. Neither implication can be reversed as witnessed by computable injection structures consisting of only infinitely many $\mathbb{Z}$-orbits, and structures consisting of only infinitely many ω-orbits, respectively.

Proposition 5. *Let $\mathcal{A}$ be a weakly ultrahomogeneous injection structure and let $S = \{a_1, \dots, a_n\}$ be a minimal exceptional set. Then $D(S)$ may be characterized in two cases as follows.*

1. *If $\mathcal{A}$ has exactly one orbit $\{a\}$ of size 1, then $D(S)$ consists of a together with the union of the finitely many ω-orbits of $\mathcal{A}$.*
2. *If $\mathcal{A}$ does not have exactly one orbit of size 1, then $D(S)$ consists of a together with the union of the finitely many ω-orbits of $\mathcal{A}$.*

Proof. Suppose that $b \in \mathcal{A}$ is not in an ω-orbit or in an orbit of size 1. Then the map fixing all ω-orbits and sending x to $f(x)$ otherwise is an automorphism fixing S but moving a. Suppose next that there are two (or more) orbits $\{b\}$ and $\{c\}$ of size 1. Then the map interchanging b and c and fixing every other element shows that b and c are not definable from S. For the other direction, suppose b is in an ω-orbit with a_i for some $i \leq n$. Then for some $n \in \mathbb{N}$ we have $f^{(n)}(a) = a_i$ or $f^{(n)}(a_i) = a$. In either case, a is definable from S. Finally, if $\{a\}$ is the unique orbit of size 1, then this fact provides a definition of a.   □

4   Conclusions and Future Research

In this paper, we have examined countable ultrahomogeneous and weakly ultra-homogeneous structures. We observed that countable ultrahomogeneous linear orderings and also countable ultrahomogeneous equivalence structures all have computable models. We showed that for computable linear orders and computable equivalence structures, the weakly homogeneous structures are exactly the computably categorical structures. For computable injection structures, we observed that there are ultrahomogeneous structures with no computable copy, and there are ultrahomogeneous structures which are not computably categor-ical, although every computably categorical structure is weakly ultrahomoge-neous. We proved that any computable weakly ultrahomogeneous structure is Δ_2^0 categorical.

We introduced the notion of a *minimal exceptional set* for weakly ultraho-mogeneous structures. We gave characterizations of these sets for linear orders, equivalence structures, and injection structures.

Future research topics include the study of other structures such as vector spaces, Boolean algebras, Abelian p-groups, graphs and partial orderings. One goal is to classify the weakly ultrahomogeneous structures. A second goal is to determine the effective categoricity of computable weakly ultrahomogeneous structures. We have also begun to look at nested equivalence structures.

For $n < \omega$, an n-equivalence structure is a structure $\mathcal{A} = (A, E_1, \ldots, E_n)$ where each E_i is an equivalence relation on A. An ω-equivalence structure is a structure $\mathcal{A} = (A, E_1, E_2, \ldots)$ where each E_i for $i < \omega$ is an equivalence relation. For $1 \leq n \leq \omega$, an n-equivalence structure is nested if $i < j \Rightarrow xE_iy \rightarrow xE_jy$, i.e $E_i \subseteq E_j$ as subsets of $A \times A$.

It is easy to find a necessary condition for ultrahomogeneity.

Proposition 6. *Let $\mathcal{A}$ be a n-equivalence structure, with $1 \leq n \leq \omega$. If $\mathcal{A}$ is ultrahomogeneous, then for $1 \leq i \leq n$, (A, E_i) is ultrahomogeneous.*

In general, this condition is not sufficient. For a 2-equivalence structure $\mathcal{A} = (A, E_1, E_2)$, if $[x]_1 \cap [x]_2$ and $[y]_1 \cap [y]_2$ have different cardinalities for some $x, y \in A$, then $\langle x \rangle \cong \langle y \rangle$ but the isomorphism can't extend since such an automorphism would have to be an isomorphism between the intersections. A similar condition on the intersections must hold for any ultrahomogeneous n-equivalence structure with $n \leq \omega$.

In the case of nested n-equivalence structures for finite n however, this condi-tion on the ultrahomogeneity of the individual equivalence relations is enough.

Theorem 7. *Let $\mathcal{A} = (A, E_1, \ldots, E_n)$ be an nested n-equivalence structure. Then $\mathcal{A}$ is ultrahomogeneous iff for $1 \leq i \leq n$, (A, E_i) is ultrahomogeneous.*

References

1. Calvert, W., Cenzer, D., Harizanov, V., Morozov, A.: Effective categoricity of equiv-alence structures. Annals of Pure and Applied Logic 141, 61–78 (2006)
2. Cenzer, D., Harizanov, V., Remmel, J.B.: Computability theoretic properties of injection structures. Algebra and Logic (to appear)
3. Remmel, J.: Recursively Categorical Linear Orders. Proceedings of the AMS 83, 387–391 (1981)
4. Csima, B.F., Harizanov, V.S., Miller, R., Montalban, A.: Computability of Fraisse Limits. Journal of Symbolic Logic 76, 66–93 (2011)
5. Fraisse, R.: Theory of Relations. North-Holland, Amsterdam (1986)
6. Marker, D.: Model Theory: An Introduction. Springer, New York (2002)

Parameterized Inapproximability of Target Set Selection and Generalizations

Cristina Bazgan[1,4], Morgan Chopin[2], André Nichterlein[3], and Florian Sikora[1]

[1] PSL, Université Paris-Dauphine, LAMSADE UMR CNRS 7243, France
{bazgan,florian.sikora}@lamsade.dauphine.fr
[2] Institut für Optimierung und Operations Research, Universität Ulm, Germany
morgan.chopin@uni-ulm.de
[3] Institut für Softwaretechnik und Theoretische Informatik, TU Berlin, Germany
andre.nichterlein@tu-berlin.de
[4] Institut Universitaire de France

Abstract. In this paper, we consider the TARGET SET SELECTION problem: given a graph and a threshold value $\mathrm{thr}(v)$ for each vertex v of the graph, find a minimum size vertex-subset to "activate" s.t. all the vertices of the graph are activated at the end of the propagation process. A vertex v is activated during the propagation process if at least $\mathrm{thr}(v)$ of its neighbors are activated. This problem models several practical issues like faults in distributed networks or word-to-mouth recommendations in social networks. We show that for any functions f and ρ this problem cannot be approximated within a factor of $\rho(k)$ in $f(k) \cdot n^{O(1)}$ time, unless FPT = W[P], even for restricted thresholds (namely constant and majority thresholds). We also study the cardinality constraint maximization and minimization versions of the problem for which we prove similar hardness results.

1 Introduction

Diffusion processes in graphs have been intensively studied [1, 4, 6, 7, 14, 16, 21, 22]. One model to represent them is to define a *propagation rule* and choose a subset of vertices that, according to the given rule, activates all or a fixed fraction of the vertices where initially all but the chosen vertices are inactive. This models problems such as the spread of influence or information in social networks via word-of-mouth recommendations, of diseases in populations, or of faults in distributed computing [14, 16, 21]. One representative problem that appears in this context is the INFLUENCE MAXIMIZATION problem introduced by Kempe et al. [16]. Given a directed graph and an integer k, the task is to choose a vertex subset of size at most k such that the number of activated vertices at the end of the propagation process is maximized. The authors show that the problem is polynomial-time $(\frac{e}{e-1} + \varepsilon)$-approximable for any $\varepsilon > 0$ under some stochastic propagation rules, but NP-hard to approximate within a ratio of $n^{1-\varepsilon}$ for any $\varepsilon > 0$ for general propagation rules.

In this paper, we use the following deterministic propagation model. We are given an undirected graph, a threshold value $\mathrm{thr}(v)$ associated to each vertex

A. Beckmann, E. Csuhaj-Varjú, and K. Meer (Eds.): CiE 2014, LNCS 8493, pp. 11–20, 2014.

v, and the following propagation rule: a vertex becomes active if at least $\text{thr}(v)$ many neighbors of v are active. The propagation process proceeds in several rounds and stops when no further vertex becomes active. Given this model, finding and activating a minimum-size vertex subset such that all the vertices become active is known as the TARGET SET SELECTION problem and was introduced by Chen [7].

TARGET SET SELECTION has been shown NP-hard even for bipartite graphs of bounded degree when all thresholds are at most two [7]. Moreover, the problem was shown to be hard to approximate in polynomial time within a ratio $O(2^{\log^{1-\varepsilon} n})$ for any $\varepsilon > 0$, even for constant degree graphs with thresholds at most two and for general graphs when the threshold of each vertex is half its degree (called *majority* thresholds) [7]. If the threshold of each vertex equals its degree (*unanimity* thresholds), then the problem is equivalent to the vertex cover problem [7] and, thus, admits a 2-approximation and is hard to approximate with a ratio better than 1.36 [11]. Concerning the parameterized complexity, the problem is shown to be W[2]-hard with respect to (w.r.t.) the solution size, even on bipartite graphs of diameter four with majority thresholds or thresholds at most two [19]. Furthermore, it is W[1]-hard w.r.t. each of the parameters "treewidth", "cluster vertex deletion number", and "pathwidth" [4, 9]. On the positive side, the problem becomes fixed-parameter tractable w.r.t. each of the single parameters "vertex cover number", "feedback edge set size", and "bandwidth" [9, 19]. If the input graph is complete, has a bounded cliquewidth, or has a bounded treewidth and bounded thresholds then the problem is polynomial-time solvable [4, 10, 19].

Motivated by the hardness of approximation and parameterized hardness we showed in previous work [3] that the cardinality constraint maximization version of TARGET SET SELECTION, that is to find a fixed number k of vertices to activate such that the number of activated vertices at the end is maximum, is strongly inapproximable in fpt-time w.r.t. the parameter k, even for restricted thresholds. For the special case of unanimity thresholds, we showed that the problem is still inapproximable in polynomial time, but becomes $r(n)$-approximable in fpt-time w.r.t. the parameter k, for any strictly increasing function r.

Continuing this line of research, we study in this paper TARGET SET SELECTION and its variants where the parameter relates to the optimum value. This requires the special definition of "fpt cost approximation" since in parameterized problems the parameter is given which is not the case in optimization problems (see Section 2 for definitions). Fpt approximation algorithms were introduced by Cai and Huang [5], Chen et al. [8], Downey et al. [12], see also the survey of Marx [17]. Besides this technical difference observe that TARGET SET SELECTION can be seen as a special case of the previously considered problem, since activating all vertices is a special case of activating a given number of vertices. Strengthening the known inapproximability results, we first prove in Section 3 that TARGET SET SELECTION is not fpt cost ρ-approximable, for any computable function ρ, unless FPT = W[P], even for majority and constant thresholds. Complementing our previous work, we also study in Section 4 the cardinality constraint maximization and minimization versions of TARGET SET SELECTION. We prove that these two problems are not fpt cost ρ-approximable, for any computable function ρ, unless FPT = W[1]. Due to space limitation, some proofs are deferred to a full version of the paper.

2   Preliminaries and Basic Observations

In this section, we provide basic backgrounds and notation used throughout this paper and define TARGET SET SELECTION. For details on parameterized complexity we refer to the monographs of Downey and Fellows [13], Flum and Grohe [15], Niedermeier [20]. For details on parameterized approximability we refer to the survey of Marx [17].

Graph terminology. Let $G = (V, E)$ be an *undirected graph.* For a subset $S \subseteq V$, $G[S]$ is the subgraph induced by S. The *open neighborhood* of a vertex $v \in V$ in G, denoted by $N_G(v)$, is the set of all neighbors of v in G. The *closed neighborhood* of a vertex v in G, denoted $N_G[v]$, is the set $N_G(v) \cup \{v\}$. The *degree* of a vertex v is denoted by $\deg_G(v)$ and the *maximum degree* of the graph G is denoted by Δ_G. We skip the subscripts if G is clear from the context.

Parameterized complexity. A parameterized problem (I, k) is said *fixed-parameter tractable* (or in the class FPT) w.r.t. parameter k if it can be solved exactly in $f(k) \cdot |I|^c$ time, where f is any computable function and c is a constant. The parameterized complexity hierarchy is composed of the classes FPT $\subseteq$ W[1] $\subseteq$ W[2] $\subseteq \cdots \subseteq$ W[P]. A W[1]-hard problem is not fixed-parameter tractable (unless FPT = W[1]) and one can prove the W[1]-hardness by means of a *parameterized reduction* from a W[1]-hard problem. Such a reduction between two parameterized problems A_1 and A_2 is a mapping of any instance (I, k) of A_1 in $g(k) \cdot |I|^{O(1)}$ time (for some computable function g) into an instance (I', k') for A_2 such that $(I, k) \in A_1 \Leftrightarrow (I', k') \in A_2$ and $k' \leq h(k)$ for some function h.

Parameterized approximation. An NP-optimization problem Q is a tuple $(\mathcal{I}, Sol, val, goal)$, where $\mathcal{I}$ is the set of instances, $Sol(I)$ is the set of feasible solutions for instance I, $val(I, S)$ is the value of a feasible solution S of I, and $goal$ is either max or min. We assume that $val(I, S)$ is computable in polynomial time and that $|S|$ is polynomially bounded by $|I|$ *i.e.* $|S| \leq |I|^{O(1)}$.

Let Q be an optimization problem and $\rho\colon \mathbb{N} \to \mathbb{R}$ be a function such that $\rho(k) \geq 1$ for every $k \geq 1$ and $k \cdot \rho(k)$ is nondecreasing (when $goal$ = min) and $\frac{k}{\rho(k)}$ is unbounded and nondecreasing (when $goal$ = max). The following definition was introduced by Chen et al. [8].

A decision algorithm $\mathcal{A}$ is an *fpt cost ρ-approximation algorithm* for Q (when ρ satisfies the previous conditions) if for every instance I of Q and integer k, with $Sol(I) \neq \emptyset$, its output satisfies the following conditions:

1. If $opt(I) > k$ (when $goal = \min$) or $opt(I) < k$ (when $goal = \max$), then $\mathcal{A}$ rejects (I, k).
2. If $k \geq opt(I) \cdot \rho(opt(I))$ (when $goal = \min$) or $k \leq \frac{opt(I)}{\rho(opt(I))}$ (when $goal = \max$), then $\mathcal{A}$ accepts (I, k).

Moreover the running time of $\mathcal{A}$ on input (I, k) is $f(k) \cdot |I|^{O(1)}$. If such a decision algorithm $\mathcal{A}$ exists then Q is called fpt cost ρ-approximable.

The notion of a gap-reduction was introduced in [2] by Arora and Lund. We use in this paper a variant of this notion, called fpt gap-reduction.

Definition 1 (fpt gap-reduction). *A problem A parameterized by k is called fpt gap-reducible to an optimization problem Q with gap ρ if for any instance (I, k) of A we can construct an instance I' of Q in $f(k) \cdot |I|^{O(1)}$ time while satisfying the following properties: (i) If I is a yes instance then $opt(I') \leq \frac{g(k)}{\rho(opt(I'))}$ (when goal = min) or $opt(I') \geq g(k)\rho(opt(I'))$ (when goal = max), (ii) If I is a no instance then $opt(I') > g(k)$ (when goal = min) or $opt(I') < g(k)$ (when goal = max), for some function g. The function ρ satisfies the aforementioned conditions.*

The interest of the fpt gap-reduction is the following result that immediately follows from the previous definition:

Lemma 1. *If a parameterized problem A is C-hard and fpt gap-reducible to an optimization problem Q with gap ρ then Q is not fpt cost ρ-approximable unless $\mathsf{FPT} = \mathcal{C}$ where $\mathcal{C}$ is any class of the parameterized complexity hierarchy.*

Problem statement. Let $G = (V, E)$ be an undirected graph and let thr: $V \to \mathbb{N}$ be a threshold function such that $1 \leq \mathrm{thr}(v) \leq \deg(v)$, $\forall v \in V$. The definition of TARGET SET SELECTION is based on the notion of "activation". Let $S \subseteq V$. Informally speaking, a vertex $v \in V$ gets activated by S in the ith round if at least $\mathrm{thr}(v)$ of its neighbors are active after the previous round (where S are the vertices active in the 0th round). Formally, for a vertex set S, let $\mathcal{A}^i_{G,\mathrm{thr}}(S)$ denote the set of vertices of G that are *activated by* S at the ith round, with $\mathcal{A}^0_{G,\mathrm{thr}}(S) = S$ and $\mathcal{A}^{i+1}_{G,\mathrm{thr}}(S) = \mathcal{A}^i_{G,\mathrm{thr}}(S) \cup \{v \in V : |N(v) \cap \mathcal{A}^i_{G,\mathrm{thr}}(S)| \geq \mathrm{thr}(v)\}$. For $S \subseteq V$, the unique positive integer r with $\mathcal{A}^{r-1}_{G,\mathrm{thr}}(S) \neq \mathcal{A}^r_{G,\mathrm{thr}}(S) = \mathcal{A}^{r+1}_{G,\mathrm{thr}}(S)$ is called the *number* $r_G(S)$ *of activation rounds*. It is easy to see that $r_G(S) \leq |V(G)|$ for all graphs G. Furthermore, we call $\mathcal{A}_{G,\mathrm{thr}}(S) = \mathcal{A}^{r_G(S)}_{G,\mathrm{thr}}(S)$ the set of vertices that are *activated by* S. If $\mathcal{A}_{G,\mathrm{thr}}(S) = V$, then S is called a *target set for* G. TARGET SET SELECTION is formally defined as follows.

TARGET SET SELECTION
Input: A graph $G = (V, E)$ and a threshold function thr: $V \to \mathbb{N}$.
Output: A target set for G of minimum cardinality.

We also consider the following cardinality constrained version.

MAX CLOSED k-INFLUENCE
Input: A graph $G = (V, E)$, a threshold function thr: $V \to \mathbb{N}$, and an integer k.
Output: A subset $S \subseteq V$ with $|S| \leq k$ maximizing $|\mathcal{A}_{G,\mathrm{thr}}(S)|$.

The MAX OPEN k-INFLUENCE problem asks for a set $S \subseteq V$ with $|S| \leq k$ such that $|\mathcal{A}_{G,\mathrm{thr}}(S) \setminus S|$ is maximum. We remark that this difference in the definition is important when considering the approximability of these problems. Finally, MIN CLOSED k-INFLUENCE (resp. MIN OPEN k-INFLUENCE) is also defined similarly, but one ask for a solution $S \subseteq V$ with $|S| = k$ such that $|\mathcal{A}_{G,\mathrm{thr}}(S)|$ is minimum (resp. $|\mathcal{A}_{G,\mathrm{thr}}(S) \setminus S|$ is minimum).

Directed edge gadget. We will use the *directed edge gadget* as used by Chen [7] throughout our work: A directed edge gadget from a vertex u to another vertex v consists of a 4-cycle $\{a, b, c, d\}$ such that a and u as well as c and v are adjacent. Moreover $\mathrm{thr}(a) = \mathrm{thr}(b) = \mathrm{thr}(d) = 1$ and $\mathrm{thr}(c) = 2$. The idea is that the vertices in the directed edge gadget become active if u is activated but not if v is activated. Hence, the activation process may go from u to v via the gadget but not in the reverse direction.

3   Parameterized Inapproximability of Target Set Selection

Marx [18] showed that the MONOTONE CIRCUIT SATISFIABILITY problem admits no fpt cost ρ-approximation algorithm for any function ρ unless FPT = W[P]. In this section we show that we can transfer this strong inapproximability result from MONOTONE CIRCUIT SATISFIABILITY to TARGET SET SELECTION.

Before defining MONOTONE CIRCUIT SATISFIABILITY, we recall the following notations. A *monotone (boolean) circuit* is a directed acyclic graph. The nodes with in-degree at least two are labeled with *and* or with *or*, the n nodes with in-degree zero are input nodes, and due to the monotonicity there are no nodes with in-degree one (negation nodes in standard circuits). Furthermore, there is one node with out-degree zero, called the *output node*. For an assignment of the input nodes with true/false, the circuit is satisfied if the output node is evaluated (in the natural way) to true. The weight of an assignment is the number of input nodes assigned to true. We denote an assignment as a set $A \subseteq \{1, \ldots, n\}$ where $i \in A$ if and only if the ith input node is assigned to true. The MONOTONE CIRCUIT SATISFIABILITY problem is then defined as follows:

MONOTONE CIRCUIT SATISFIABILITY
Input: A monotone circuit C.
Output: A satisfying assignment of minimum weight, that is, a satisfying assignment with a minimum number of input nodes set to true.

By reducing MONOTONE CIRCUIT SATISFIABILITY to TARGET SET SELECTION in polynomial time such that there is a "one-to-one" correspondence between the solutions, we next show that the inapproximability result transfers to

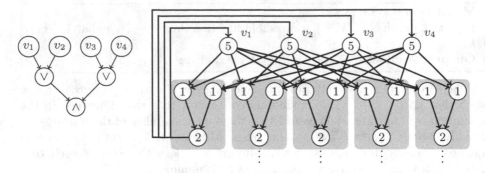

Fig. 1. Illustration of the reduction described in Theorem 1. All arrows of the right graph represent a directed edge gadget. Thresholds are represented inside each vertex.

TARGET SET SELECTION. First, we show one reduction working with general thresholds and then describe, using further gadgets, how to achieve constant or majority thresholds in our constructed instance.

3.1   General Thresholds

As mentioned above, we will reduce from MONOTONE CIRCUIT SATISFIABILITY, and thus derive the same inapproximability result for TARGET SET SELECTION as for MONOTONE CIRCUIT SATISFIABILITY.

Theorem 1. TARGET SET SELECTION *is not fpt cost ρ-approximable, for any computable function ρ, unless* FPT = W[P].

Proof. Let C be an instance of MONOTONE CIRCUIT SATISFIABILITY. We construct an instance of TARGET SET SELECTION as follows. Initialize $G = (V, E)$ as a copy of the directed acyclic graph C where each directed edge is replaced by a directed edge gadget. We call a vertex in G an input vertex (resp. output vertex, and-vertex, or-vertex) if it corresponds to an input node (resp. output node, and-node, or-node). Next, for each and-node in C with in-degree d set the threshold of the corresponding and-vertex in G to d and for each or-vertex in G set the threshold to 1. Set the threshold of each input vertex in G to $n+1$. Next, add n copies to G and "merge" all vertices corresponding to the same input node. This means, that for an input node v with an outgoing edge (v, w) in C the graph G contains $n+1$ vertices $w_1, \ldots, w_{n+1}$ and $n+1$ directed edges from v to w_i, $1 \le i \le n + 1$. Finally, add directed edges from each output vertex to each input vertex. This completes our construction (see Figure 1). To complete the proof, it remains to show that

(i) for every satisfying assignment A for C there exists a target set of size $|A|$ for G, and

(ii) for every target set S for G there exists a satisfying assignment of size $|S|$ for C.

(i) Let $A \subseteq \{1, \ldots, n\}$ be a satisfying assignment for C. We show that the set S of vertices of G that correspond to the input nodes in A form a target set. Clearly, $|S| = |A|$. First, observe that by construction, all the $n+1$ output vertices of G become active. Hence, also all input vertices that are not in S become active. Thus, all remaining vertices in G are activated since $\text{thr}(v) \leq \deg(v)$ for all $v \in V$.

(ii) Let $S \subseteq V$ be a target set for G. First, observe that we can assume that $|S| < n$ since otherwise the satisfying assignment simply sets all input nodes to true. Next, observe that we can assume that S is a subset of the input vertices. Indeed, since G contains $n+1$ copies of the circuit (excluding the input vertices), there is at least one copy without vertices in S and, hence, the output vertex in that copy becomes active solely because of the input vertices in S. Finally, assume by contradiction that the set of input nodes that correspond to the vertices in S do not form a satisfying assignment. Hence, the output node of C is evaluated to false. However, due to the construction, this implies that the vertices corresponding to the output node are not activated, contradicting that S is a target set for G.   □

3.2 Restricted Thresholds

In this subsection, we enhance the inapproximability results to variants of TARGET SET SELECTION with restricted threshold functions. To this end, we use the construction desbribed in Lemma 2 of Nichterlein et al. [19] which transforms in polynomial time any instance $I = (G = (V, E), \text{thr})$ of TARGET SET SELECTION into a new instance $I' = (G' = (V', E'), \text{thr}')$ where thr' is the majority function such that

(i) for every target set S for I there is a target set S' for I' with $|S'| \leq |S| + 1$, and

(ii) for every target set S' for I' there is a target set S for I with $|S| \leq |S'| - 1$.

Hence, the next corollary follows.

Corollary 1. TARGET SET SELECTION *with majority thresholds is not fpt cost ρ-approximable, for any computable function ρ, unless* FPT = W[P].

Next, we show a similar statement for constant thresholds.

Lemma 2. *Let $I = (G = (V, E), \text{thr})$ be an instance of* TARGET SET SELECTION. *Then, we can construct in polynomial time an instance $I' = (G' = (V', E'), \text{thr}')$ of* TARGET SET SELECTION *where $\text{thr}'(v) \leq 2$ for all $v \in V'$ and G' is bipartite such that*

(i) *for every target set S for I there is a target set S' for I' with $|S'| = |S|$, and*

(ii) *for every target set S' for I' there is a target set S for I with $|S| \leq |S'|$.*

Theorem 1 and Lemma 2 imply the following.

Corollary 2. TARGET SET SELECTION *with thresholds at most two is not fpt cost ρ-approximable even on bipartite graphs, for any computable function ρ, unless* FPT = W[P].

4 Parameterized Inapproximability of Max and Min k-Influence

We consider in this section the cardinality constraint maximization and minimization versions of TARGET SET SELECTION.

Theorem 2. MAX CLOSED k-INFLUENCE *and* MAX OPEN k-INFLUENCE *are not fpt cost ρ-approximable even on bipartite graphs, for any computable function ρ, unless* FPT = W[1].

We remark that the proof of Theorem 2 shows a stronger result: Unless FPT = W[1], there is no fpt cost ρ-approximation for MAX CLOSED k-INFLUENCE and MAX OPEN k-INFLUENCE, for any computable function ρ, even if the running time is of the form $f(k, \ell) \cdot n^{O(1)}$. Here ℓ is the cost-parameter passed as an argument to the algorithm, that is, ℓ indicates the number of activated vertices.

As the reductions behind Corollaries 1 and 2 are not fpt gap-reductions, we cannot use them to prove the same cost inapproximability results for MAX CLOSED k-INFLUENCE or MAX OPEN k-INFLUENCE with majority thresholds and thresholds at most two.

Minimization variants. In contrast with the maximization versions, we can show that the problems are polynomial-time solvable for unanimity thresholds.

Proposition 1. MIN OPEN k-INFLUENCE *and* MIN CLOSED k-INFLUENCE *are solvable in polynomial time for unanimity thresholds.*

The next result shows that MIN CLOSED k-INFLUENCE and MIN OPEN k-INFLUENCE are also computationally hard even for thresholds bounded by two. To this end, we consider the decision version of MIN CLOSED k-INFLUENCE (resp. MIN OPEN k-INFLUENCE) denoted by CLOSED k-INFLUENCE$_\leq$ (resp. OPEN k-INFLUENCE$_\leq$) and defined as follows: Given a graph $G = (V, E)$, a threshold function thr : $V \to \mathbb{N}$, and integers k and ℓ, determine whether there is a subset $S \subseteq V$, $|S| = k$ such that $|\mathcal{A}_{G,\mathrm{thr}}(S)| \leq \ell$ (resp. $|\mathcal{A}_{G,\mathrm{thr}}(S) \setminus S| \leq \ell$).

Theorem 3. CLOSED k-INFLUENCE$_\leq$ *is* W[1]-*hard w.r.t. parameter (k, ℓ) even for threshold bounded by two and bipartite graphs.* OPEN k-INFLUENCE$_\leq$ *is* NP-*hard even for threshold bounded by two, bipartite graphs and $\ell = 0$.*

We remark that the previous theorem rules out the possibility of any fixed-parameter algorithm with parameter ℓ for OPEN k-INFLUENCE$_\leq$ assuming P $\neq$ NP. Moreover, due to its NP-hardness when $\ell = 0$, MIN OPEN k-INFLUENCE is not at all fpt cost approximable, unless P = NP.

In the following, we provide a final result regarding fpt cost approximation of MIN CLOSED k-INFLUENCE.

Theorem 4. MIN CLOSED k-INFLUENCE *with thresholds at most two is not fpt cost ρ-approximable even on bipartite graphs, for any computable function ρ, unless* FPT = W[1].

Proof. We provide a fpt gap-reduction with gap ρ from INDEPENDENT SET to MIN CLOSED k-INFLUENCE Given an instance $(G = (V, E), k)$ of INDEPENDENT SET, we construct an instance $(G' = (V', E'), k')$ of MIN CLOSED k-INFLUENCE by considering the incidence graph, that is G' is a bipartite graph with two vertex sets V and E and for each edge $e = uv \in E$, there is $ue, ve \in E'$. We define $\text{thr}(u) = 1, \forall u \in V$ and $\text{thr}(e) = 2, \forall e \in E$. We choose the function h such that $h(k)$ is an integer and $k + h(k) + 1 \geq k\rho(k)$. Then, we add $h(k)$ additional vertices F of threshold 1 in G' and a complete bipartite graph between E and F. Define $k' = k$, $g(k) = k + h(k) + 1$. If G contains an independent set of size at least k then, by activating the same k vertices in G', we obtain a solution that activates no more vertex in G' and thus $opt(I') = k \leq \frac{g(k)}{\rho(k)} = \frac{g(k)}{\rho(opt(I'))}$.

If there is no independent set of size k, if one activate only two vertices from F, it will activate the whole vertex set E on the next step, and then the whole graph. Moreover, activating a vertex from the vertex set E will also activate the whole set F on the next step, and then the whole graph. Finally, activating k vertices of V will activate at least one vertex of E since there is no independent set of size k. Note that activating $k - 1$ vertices from V and 1 from F will result not be better since vertices of F are connected to all vertices of E. Therefore, $opt(I') \geq k + h(k) + 1 = g(k)$.

The result follows from Lemma 1 together with the W[1]-hardness of INDEPENDENT SET [13]. $\square$

5   Conclusion

Despite the variety of our intractability results, some questions remains open. Are MAX CLOSED k-INFLUENCE and MAX OPEN k-INFLUENCE fpt cost approximable for constant or majority thresholds? We believe that these problems remain hard, but the classical gadgets used to simulate these thresholds changes does not work for this type of approximation. Similarly, is MIN CLOSED k-INFLUENCE fpt cost approximable for majority thresholds?

Finally, the dual problem of TARGET SET SELECTION (i.e. find a target set of size at most $|V| - k$) seems unexplored. Using the fact that TARGET SET SELECTION with unanimity thresholds is exactly Vertex Cover, we know that the dual problem is therefore W[1]-hard, even with unanimity thresholds. But it is still the case for constant or majority thresholds? Moreover, is the dual of TARGET SET SELECTION fpt cost approximable?

References

[1] Aazami, A., Stilp, K.: Approximation algorithms and hardness for domination with propagation. SIAM J. Discrete Math. 23(3), 1382–1399 (2009)

[2] Arora, S., Lund, C.: Hardness of approximations. In: Approximation Algorithms for NP-Hard Problems, pp. 399–446. PWS Publishing Company (1996)

[3] Bazgan, C., Chopin, M., Nichterlein, A., Sikora, F.: Parameterized approximability of maximizing the spread of influence in networks. In: Du, D.-Z., Zhang, G. (eds.) COCOON 2013. LNCS, vol. 7936, pp. 543–554. Springer, Heidelberg (2013)

[4] Ben-Zwi, O., Hermelin, D., Lokshtanov, D., Newman, I.: Treewidth governs the complexity of target set selection. Discrete Optim. 8(1), 87–96 (2011)

[5] Cai, L., Huang, X.: Fixed-parameter approximation: Conceptual framework and approximability results. Algorithmica 57(2), 398–412 (2010)

[6] Chang, C.-L., Lyuu, Y.-D.: Spreading messages. Theor. Comput. Sci. 410(27-29), 2714–2724 (2009)

[7] Chen, N.: On the approximability of influence in social networks. SIAM J. Discrete Math. 23(3), 1400–1415 (2009)

[8] Chen, Y., Grohe, M., Grüber, M.: On parameterized approximability. In: Bodlaender, H.L., Langston, M.A. (eds.) IWPEC 2006. LNCS, vol. 4169, pp. 109–120. Springer, Heidelberg (2006)

[9] Chopin, M., Nichterlein, A., Niedermeier, R., Weller, M.: Constant thresholds can make target set selection tractable. In: Even, G., Rawitz, D. (eds.) MedAlg 2012. LNCS, vol. 7659, pp. 120–133. Springer, Heidelberg (2012)

[10] Cicalese, F., Cordasco, G., Gargano, L., Milanič, M., Vaccaro, U.: Latency-bounded target set selection in social networks. In: Bonizzoni, P., Brattka, V., Löwe, B. (eds.) CiE 2013. LNCS, vol. 7921, pp. 65–77. Springer, Heidelberg (2013)

[11] Dinur, I., Safra, S.: The importance of being biased. In: Proc. of STOC 2002, pp. 33–42. ACM (2002)

[12] Downey, R.G., Fellows, M.R., McCartin, C.: Parameterized approximation algorithms. In: Bodlaender, H.L., Langston, M.A. (eds.) IWPEC 2006. LNCS, vol. 4169, pp. 121–129. Springer, Heidelberg (2006)

[13] Downey, R.G., Fellows, M.R.: Fundamentals of Parameterized Complexity. Springer (2013)

[14] Dreyer, P.A., Roberts, F.S.: Irreversible k-threshold processes: Graph-theoretical threshold models of the spread of disease and of opinion. Discrete Appl. Math. 157(7), 1615–1627 (2009)

[15] Flum, J., Grohe, M.: Parameterized Complexity Theory. Springer (2006)

[16] Kempe, D., Kleinberg, J., Tardos, É.: Maximizing the spread of influence through a social network. In: Proc. of KDD 2003, pp. 137–146. ACM (2003)

[17] Marx, D.: Parameterized complexity and approximation algorithms. Comput. J. 51(1), 60–78 (2008)

[18] Marx, D.: Completely inapproximable monotone and antimonotone parameterized problems. J. Comput. Syst. Sci. 79(1), 144–151 (2013)

[19] Nichterlein, A., Niedermeier, R., Uhlmann, J., Weller, M.: On tractable cases of target set selection. Soc. Network Anal. Mining 3(2), 233–256 (1869) ISSN 1869-5450

[20] Niedermeier, R.: Invitation to Fixed-Parameter Algorithms. Oxford University Press (2006)

[21] Peleg, D.: Local majorities, coalitions and monopolies in graphs: a review. Theor. Comput. Sci. 282(2), 231–257 (2002)

[22] Reddy, T.V.T., Rangan, C.P.: Variants of spreading messages. J. Graph Algorithms Appl. 15(5), 683–699 (2011)

How can Grammatical Inference Contribute to Computational Linguistics?

Leonor Becerra-Bonache

Laboratoire Hubert Curien, Université Jean Monnet
18 rue du Professeur Benoit Lauras, 42000, Saint Etienne, France
leonor.becerra@univ-st-etienne.fr

Abstract. Grammatical Inference refers to the process of learning grammars and languages from data. Although there are clear connections between Grammatical Inference and Computational Linguistics, there have been a poor interaction between these two fields. The goals of this article are: i) To introduce Grammatical Inference to computational linguists; ii) To explore how Grammatical Inference can contribute to Computational Linguistics.

1 Introduction

According to the *Association of Computational Linguistics*, Computational Linguistics (CL) is the scientific study of language from a computational perspective. We can mainly distinguish two approaches:

- Theoretical approach: the goal is to provide models that can help us to better understand how humans process natural language.
- Practical approach: the goal is to develop systems that can deal with natural language data in an application context.

Enabling machines to have human-like linguistic abilities, also involves to enable them to learn a grammar. It is here where we find the field of *Grammatical Inference* (GI). GI is a subfield of Machine Learning that studies how grammars can be learnt from data. We can also distinguish two approaches:

- Theoretical approach: the goal is to proof efficient learnability of grammars.
- Practical approach: the goal is to develop systems that learn grammars from real data.

It is worth noting that theoretical results in GI have been used in discussions on children's language acquisition [18]. Moreover, empirical systems that learn grammars have been applied to natural language; some research has been done in syntactic parsing, morphological analysis of words and machine translation.

Although there are some clear connections between CL and GI, the interaction between these two communities is scarce. The goal of this paper is to introduce GI to computational linguists and explore how GI can contribute to CL.

A. Beckmann, E. Csuhaj-Varjú, and K. Meer (Eds.): CiE 2014, LNCS 8493, pp. 21–31, 2014.

2 What is Grammatical Inference?

Machine learning concerns the development of techniques that allow computers *to learn.* GI is a specialized subfield of Machine Learning that deals with the learning of formal languages from a set of data. The learning process involves two parts: a) a learner (or learning algorithm) that has to identify the language; b) a teacher that provides information about the language (e.g., strings that belong to the language) to the learner. For example, imagine that the language to be learnt (i.e., the target language) is $(ab)^+$:

- The teacher could provide to the learner strings that belong to the target language (i.e., positive data), such as $ab, abab, ababab....$
- The learner, from this information, should infer that the target language is $(ab)^+$.

As we can see, there are some similarities with the process of natural language acquisition. Instead of a teacher and a learner, we would have an adult and a child. Like the learner in the above example, a child also learns a language from the data he/she receives; a child in an English environment will learn to speak English, and the same child in a Japanese environment, will learn to speak Japanese.

In 1967, E.M. Gold gave the initial theoretical foundations of this field [23], mainly motivated by the problem of how children acquire their native language. His goal was *"to construct a precise model for the intuitive notion 'able to speak a language' in order to be able to investigate theoretically how it can be achieved artificially."* [23, p. 447–448]. Moreover, he stated that the results and methods would have implications in CL, in particular the construction of discovery procedures, and in Psycholinguistics, in particular the study of child learning. Therefore, from the origins of GI, we can see clear connections between GI and CL.

It is worth noting that GI researchers come from different areas: Machine Learning, Pattern Recognition, CL, etc. Several conferences, workshops and tutorials have brought together GI researchers. The main conference devoted to GI is called ICGI (International Colloquium on Grammatical Inference); it is organized every two years since 1994. Some workshops, schools and tutorials have also been organized in conjunction with EACL, NIPS, IJCAI and ACL, some of them directly related to CL (for ex., CLAGI: Workshop on Computational Linguistic Aspects of Grammatical Inference).

For more information on GI, see [21].

3 Theoretical Approaches to GI

3.1 Learning Paradigms

We can distinguish three main learning paradigms in GI: identification in the limit [23], active learning [3] and PAC learning [40]. Next we review the main ideas of these models.

We assume that the reader is familiar with some basic notions on formal languages and automata theory. For detailed information, see [32].

Identification in the Limit. *Identification in the limit* was proposed by E.M. Gold in [23]. In this model, the learner receives more and more examples from the teacher, and has to produce a hypothesis of the target language. If the learner receives a new example that is not consistent with its current hypothesis, then it has to change it. The learner identifies the target language in the limit if, after a finite number of examples, makes a correct hypothesis and does not change it.

Two kind of presentations are possible in this learning paradigm:

- *Text*: only strings that belong to the target language are given to the learner. In this case we say that the learner learns from only *positive data*. For example, imagine that the target language is $\{a^n b^n : n \in \mathbb{N}\}$. An example of a text presentation would be: λ, $a^2 b^2$, $a^8 b^8$, ...

- *Informant*: strings that belong to the target language and strings that do not belong to it are given to the learner. In this case, we say that the learner learns from *positive* and *negative data*. Following the previous example, an informant presentation would be: $(\lambda, +)$, $(abab, -)$, $(aabb, +)$, ...

Active Learning. *Active learning*, also known as *Query learning*, was introduced by D. Angluin in [3]. In this model, instead of passively receive strings of the language (like in the *Identification in the limit* paradigm), the learner *interacts* with the teacher through *queries*. Her motivation to introduce this learning framework was that, in some scenarios (for example, a human expert in cancer diagnosis, trying to communicate his/her method in that domain to an expert system), it is more reasonable to investigate learning methods that are based on helpful examples.

D. Angluin proposed to use a teacher (also called *oracle*) that knows the target language and can correctly answer specific questions asked by the learner. The learner, after asking a finite number of queries, has to return a hypothesis of the language. Its hypothesis has to be the correct one, i.e. exact learning is required.

There exists different type of queries, but just two of them have been established as the standard combination to be used:

- *Membership query* (MQ): the learner asks to the teacher if a string x is in the language, and the teacher answers "yes" or "no".

- *Equivalence query* (EQ): the learner asks to the teacher if its hypothesis is the correct one, and the teacher answers "yes", and otherwise returns a counterexample (i.e., a string in the symmetric difference of the learner's hypothesis and the target language, that is, a string that is in the learner's hypothesis but not in the target language, or a string that is in the target language but not in its hypothesis).

A teacher that is able to answer MQ and EQ is called a Minimally Adequate Teacher (MAT).

PAC Learning. PAC (Probably Approximately Correct) learning was introduced by Valiant in [40]. He proposed a probabilistic model of learning from random examples.

In this model, there exist an unknown distribution over the examples, and the learner receives examples sampled under this distribution. It is required to learn under any distribution, but in contrast to *Identification in the limit* and *Active learning*, exact learning is not required (a small error is permitted). A successful learner is one that finds, with high probability, a grammar whose error is small.

3.2 Main Formal Results

Researchers in GI have been specially focused on obtaining formal results. Many works have been developed in the three learning settings explained in the previous section, specially in *Identification in the limit* and *Active Learning*. Moreover, the classes of languages that have specially attracted their attention are regular (REG) and context-free (CF), which constitute the first two levels of the Chomsky Hierarchy. There are very few studies about identifying classes of languages more powerful than CF. One of the reasons is that it is already too hard to get positive learnability results with these classes.

In *Identification in the limit*, one of the main results obtained is the one given by Gold [23], who proved that *superfinite* classes of languages (i.e., classes that contain all finite languages and at least one infinite language) are not identifiable in the limit from positive data. This implies that none of the classes of languages defined by Chomsky, including REG and CF, are identifiable in the limit from positive data. The problem of learning with only positive data is that overgeneralization can not be controlled; if our hypothesis consists of a grammar whose language is larger than the target language, no positive example will be able to refute our hypothesis.

Despite Gold's results, it is desirable to learn from only positive data, since in the most part of applications the available data is positive. For example, in Natural Language Processing, large sets of positive data may be available for learning, but it is not common to obtain a set of negative data. To overcome this problem, researchers have adopted different solutions: to study subclasses of the languages to be learned [2], to provide structural information [34] or to make also available negative data [30].

In *Active learning*, one of the main results is the one obtanied by D. Angluin, who proved that Deterministic Finite Automata (DFA) are learnable from MQs and EQs in polynomial time [3]. After Angluin's work, researchers developed more efficient versions of the same algorithm, trying to increase the parallelism level, to reduce the number of queries, etc. [7].

Is it possible to extend the polynomial MAT learnability beyond DFA? Angluin and Kharitonov [4] showed that the problem of identifying the class of CF grammars by using MAT is computationally as hard as the cryptographic problems (for which there is currently no known polynomial-time algorithm). In order to avoid this problem, researchers have adopted different solutions: to study subclasses of CF grammars [24], to provide structural information [33] or to reduce the learning problem to some other learning problem whose result is known [35].

What about learning from another kind of queries? Based on studies on children's language acquisition, a new type of query called *correction query* was proposed in [8]. Children, in the early stages of their linguistic development, when produce an erroneous utterance (for example, "milk milk"), they are often corrected by their parents (for example, "you want milk?"); these corrections are made in the form of reformulations and they are mainly used by the parents to be sure that they have understood their children [17]. Taking into account all this, the idea of corrections was applied to GI studies, more concretely to Angluin's model.

In a correction query, the learner asks if a string is in the language, and if it does not belong to the language, the teacher returns a correction. How does the teacher correct the learner? Two different type of corrections have been mainly developed in GI: i) A correction consists of the shortest extension of the queried string; ii) A correction consists of a string of the language closed to the queried string with respect to the edit distance. By using the first type of corrections, it has been showed that it is possible to learn DFA [11], k-reversible languages and pattern languages [39]. By using a correction based on edit distance, it has been showed that it is possible to learn topological balls of strings (i.e., classes of languages defined via edit distance) [10], pattern languages and regular expressions [27].

In *PAC learning*, more negative results have been proved than positive [21]. Even for the case of DFA, most results are negative. The requirement that the learning algorithm must learn under any arbitrary (but fixed) probability distribution seems too strong. To overcome this problem, one of the solutions adopted has been to consider that the data are sampled according to a distribution that is defined by a grammar or an automaton. The goal in this case is to learn this distribution [20].

It is worth noting that a well studied technique in GI for learning automata is *state-merging*. Positive learnability results have been obtained by using this technique, in both probabilistic and non-probabilistic frameworks. Some of the most well known algorithms are: RPNI [30] (for learning DFA from positive and negative data), ALERGIA [14] (for learning probabilistic DFA from only positive data) and OSTIA [31] (for learning subsequential transducers from only positive data).

As we can see, there is a rich literature in GI about learning REG and CF grammars, finite-state automata, transducers, etc. (due to space restrictions, we have just reviewed some of them. For more details about formal results in GI, see [21]). All these formal objects are also important in CL. Hence, CL could explore how to benefit from all these results obtained in the field of GI.

3.3   Which Classes are Relevant to Natural Languages?

Most research in GI has been focused on learning classes of languages with a limited expressive power, i.e. REG and CF. However, a question very interesting from a linguistic point of view is: what classes are relevant to natural languages?

CF grammar is the most popular used to define the syntax of natural languages. However, as it was showed in the 80's, there are some examples of natural language structures that cannot be described using a CF grammar; in particular, examples of multiple agreement, crossed agreement and duplication were found in some natural languages such as Dutch [13], Bambara [19] or Swiss German [36]. These examples showed that natural languages are not CF and that more generative capacity than CF grammar is needed to describe natural languages.

How much power beyond CF is it necessary? CF have not enough expressive power, but a positive aspect is that they are computationally tractable. In contrast, context-sensitive (CS) have enough expressive power to describe natural languages, but they are computationally complex to deal with. Therefore, it would be desirable to find a class that is expressive enough to deal with natural languages and that has good computational properties.

In 1985, A.K. Joshi introduced the notion of Mildly Context-Sensitive (MCS) [25]. His idea was to provide a device capable of generating CF and non-CF structures, keeping under control the generative power. Since then, several formalisms have been introduced and used to fabricate MCS families of languages, such as: tree adjoining grammars, head grammars, combinatory categorial grammars, linear indexed grammars, etc. [26]

It is worth noting that all these formalisms occupy a concentric position in the Chomsky hierarchy, between CF and CS. However, as some researchers have pointed out [29,28], there are some examples of natural languages constructions that are neither REG or CF, and also some REG or CF constructions that do not appear naturally in sentences. Therefore, these researchers suggest that natural languages could occupy an orthogonal position in the Chomsky hierarchy.

If we take into account all these ideas, we can conclude that it would be desirable to find new formalisms that have the following properties: i) They generate MCS languages (i.e., they can generate multiple agreement, crossed agreement and duplication structures, and they are computational feasible); ii) They occupy an orthogonal position in the Chomsky hierarchy (i.e., they contain some REG, some CF, but are included in CS).

Researchers in GI have tried to study classes with such desirable properties. An example is the class of *Simple p-dimensional External Contextual* grammars (SEC). It was proved that SEC can generate MCS languages and it is incomparable with REG and CF, but it is included in CS [8]. Some positive learnability results have also been obtained with this class [9]. Results on the learnability of other classes that also generate MCS languages can be found in [42].

Therefore, GI researchers have studied the learnability of some classes that seem to be appropriate candidates to model some aspects of natural language syntax. It would be very interesting that computational linguists study the relevance of these classes.

4   Practical Approaches to GI

Although research in GI has been specially focused on obtaining formal results, GI algorithms have also been applied to some other domains, such as natural language learning.

According to [22], GI methods for natural language can be classified into: i) *unsupervised methods* (the teacher does not interact with the learner, so the learner does not have any information about the structure of correct sentences in the language); ii) *supervised methods* (the learner can ask to the teacher if its hypothesis about the language is correct. Treebank or structured corpus are used); iii) *semi-supervised methods* (a combination of unstructured data with small structured training sets are used). Most of the GI methods are based on an unsupervised approach and only use positive data during the learning process. Examples of such methods are:

- EMILE [1]: the general idea is to identify substitution classes by means of clustering (expressions that occur in the same context are clustered together and thus, substitutable in similar contexts).
- ABL (Alignment-Based Learning) [41]: the idea is to find possible constituents by aligning sentences and then, select the best constituents by using probabilistic methods.
- ADIOS (Automatic DIstillation Of Structure) [37]: since the information is represented by a graph, the main idea is to find the best paths in the graphs.

Most of the GI methods are evaluated by using a *treebank*. One of the most used is ATIS (Air Traffic Information System); English corpus that contains 577 sentences on air traffic.

It is worth noting that researchers in GI have also developed methods to learn subsequential transducers for language understanding [16] and stochastic transducers for machine translation [15]. These methods have been successfully applied to different non-trivial tasks, such as: *Miniature Language Acquisition task* (this task involves sentences that describe visual scenes), *Traveler task* (it involves human-to-human communication situations in the front-desk of a hotel), etc.

4.1   Learning with Semantics

Most works in GI reduce the learning problem to syntax learning and omit any semantic information. However, as linguistic and cognitive studies suggest, semantic and pragmatic information not only are available to children, but also seem to play an important role in the early stages of children's linguistic development. For instance, let us take the example given in the previous section; if a child produces an incorrect sentence such as "milk milk", thanks to the context in which it is produced, parents can understand the meaning of this sentence, although it is not syntactically correct. Moreover, if parents reformulate the child's sentence to be sure that they have understood their child ("you want milk?"), this correction is going to be based on the meaning that the child intend to express (i.e., the correction *preserves* the intended meaning of the child) [17].

The first attempt to incorporate semantics in the field of GI can be found in [5,6]. Based on all these ideas, it was proposed a simple computational model that takes into account semantics for language learning. Thanks to this model, it was possible to investigate aspects of the roles of semantics and corrections in the process of learning to understand and speak a natural language. In this new approach, the teacher is able to understand a flawed utterance provided by the learner and answer with a correct utterance for that meaning (by using meaning-preserving corrections). Moreover, the learner can recognize that the teacher's utterance has the same meaning but different form. The model was tested with limited sublanguages of ten different natural languages (a simplified version of the *Miniature Language Acquisition task* was used). The empirical results showed that the access to the semantics facilitates language learning, and the presence of corrections by the teacher has an effect on language learning by the learner (even if the learner does not treat corrections specially).

Following this line of research, a work based on pair-Hidden Markov Models was proposed in [12]. It was showed that, by taking into account semantics, it is possible to accelerate the language learning process.

5   Concluding Remarks

GI and CL have many aspects in common. From a theoretical point of view, one of the goals of CL is to go deeper in the understanding of how humans learn language. GI studies were initially motivated by the problem of natural language acquisition. Research in GI have been specially focused on obtaining theoretical results about the learnability of grammars and have provided tools for learning grammars that could be of interest for computational linguists. Moreover, GI researchers have tried to find new classes of languages that are relevant for natural language and have studied their learnability.

GI and CL study the same formal objects: REG grammars, CF grammars, finite-state automata, transducers, etc. GI has specially studied these objects from a theoretical point of view and many relevant results have been obtained. It would be interesting that computational linguists tried to benefit from the results obtained in the field of GI.

From a practical point of view, one of the goals of CL is to develop systems that deal with natural language data. We have seen that some works in this line have also been developed in GI. Instead of proving the learnability of grammars, researchers have focused on providing empirical systems that learn grammars from real data. Moreover, GI methods for other tasks, such as machine translation, have been developed.

CL also aims to develop systems that behaves more and more like humans. In GI there has also been several efforts for taking more natural aspects during the learning process. An example of that is the development of models that take semantics into account during the learning process.

With this paper we hope to have showed the relevance of GI research for the study of natural language, and in particular for CL, and to stimulate further interaction between these two fields.

References

1. Adriaans, P.: Language learning from a categorial perspective. PhD thesis, University of Amsterdam (1992)
2. Angluin, D.: Inference of reversible languages. Journal of the Association for Computing Machinery 29(3), 741–765 (1982)
3. Angluin, D.: Learning regular sets from queries and counterexamples. Information and Computation 75, 87–106 (1987)
4. Angluin, D., Kharitonov, M.: When won't membership queries help? In: STOC 1991, pp. 444–454 (1991)
5. Angluin, D., Becerra-Bonache, L.: Learning meaning before syntax. In: Clark, A., Coste, F., Miclet, L. (eds.) ICGI 2008. LNCS (LNAI), vol. 5278, pp. 1–14. Springer, Heidelberg (2008)
6. Angluin, D., Becerra Bonache, L.: Effects of Meaning-Preserving Corrections on Language Learning. In: CoNLL 2011, pp. 97–105 (2011)
7. Balcázar, J.L., Díaz, J., Gavaldà, R., Watanabe, O.: Algorithms for Learning Finite Automata from Queries: A unified view. In: Du, D.Z., Ko, K.I. (eds.) Advances in Algorithms, Languages, and Complexity, pp. 53–72 (1997)
8. Becerra-Bonache, L.: On the Learnability of Mildly Context-Sensitive Languages using Positive Data and Correction Queries. PhD thesis, Rovira i Virgili Univ. (2006)
9. Becerra-Bonache, L., Case, J., Jain, S., Stephan, F.: Iterative learning of simple external contextual languages. Theoretical Computer Science 411, 2741–2756 (2010)
10. Becerra-Bonache, L., de la Higuera, C., Janodet, J.C., Tantini, F.: Learning balls of strings from edit corrections. JMLR 9, 1841–1870 (2008)
11. Becerra-Bonache, L., Dediu, A.-H., Tîrnăucă, C.: Learning DFA from correction and equivalence queries. In: Sakakibara, Y., Kobayashi, S., Sato, K., Nishino, T., Tomita, E. (eds.) ICGI 2006. LNCS (LNAI), vol. 4201, pp. 281–292. Springer, Heidelberg (2006)
12. Becerra-Bonache, L., Fromont, E., Habrard, A., Perrot, M., Sebban, M.: Speeding up Syntactic Learning Using Contextual Information. In: ICGI 2012, vol. 21, pp. 49–53 (2012)
13. Bresnan, J., Kaplan, R.M., Peters, S., Zaenen, A.: Cross-serial dependencies in dutch. In: Savitch, W.J., Bach, E., Marsh, W., Safran-Naveh, G. (eds.) The Formal Complexity of Natural Language, pp. 286–319 (1987)
14. Carrasco, R.C., Óncina, J.: Learning stochastic regular grammars by means of a state merging method. In: Carrasco, R.C., Oncina, J. (eds.) ICGI 1994. LNCS, vol. 862, pp. 139–152. Springer, Heidelberg (1994)
15. Casacuberta, F., Vidal, E.: Learning finite-state models for machine translation. Machine Learning 66(1), 69–91 (2007)
16. Castellanos, A., Vidal, E., Var, M.A., Oncina, J.: Language understanding and subsequential transducer learning. Computer Speech and Language 12(3), 193–228 (1998)
17. Chouinard, M.M., Clark, E.V.: Adult reformulations of child errors as negative evidence. Journal of Child Language 30, 637–669 (2003)

18. Clark, A.: Grammatical inference and first language acquisition. In: Workshop on Psychocomputational Models of Human Language Acquisition, pp. 25–32 (2004)
19. Culy, C.: The complexity of the vocabulary of bambara. In: Savitch, W.J., Bach, E., Marsh, W., Safran-Naveh, G. (eds.) The Formal Complexity of Natural Language, pp. 349–357 (1987)
20. de la Higuera, C., Thollard, F.: Identification in the limit with probability one of stochastic deterministic finite automata. In: Oliveira, A.L. (ed.) ICGI 2000. LNCS (LNAI), vol. 1891, pp. 15–24. Springer, Heidelberg (2000)
21. de la Higuera, C.: Grammatical Inference: Learning Automata and Grammars. Cambridge University Press, Cambridge (2010)
22. D'Ulizia, A., Ferri, F., Grifoni, P.: A survey of grammatical inference methods for natural language learning. Artificial Intelligence Review 36(1), 1–27 (2011)
23. Gold, E.M.: Language identification in the limit. Information and Control 10, 447–474 (1967)
24. Ishizaka, H.: Polynomial time learnability of simple deterministic languages. Machine Learning 5, 151–164 (1990)
25. Joshi, A.K.: How much context-sensitivity is required to provide reasonable structural descriptions: Tree adjoining grammars. In: Dowty, D., Karttunen, L., Zwicky, A. (eds.) Natural Language Parsing: Psychological, Computational and Theoretical Perspectives, pp. 206–250. Cambridge University Press, New York (1985)
26. Joshi, A.K., Shanker, K.V., Weir, D.: The Convergence of Mildly Context-Sensitive Grammar Formalisms. In: Technical Report, University of Pennsylvania (1990)
27. Kinber, E.: On learning regular expressions and patterns via membership and correction queries. In: Clark, A., Coste, F., Miclet, L. (eds.) ICGI 2008. LNCS (LNAI), vol. 5278, pp. 125–138. Springer, Heidelberg (2008)
28. Kudlek, M., Martín-Vide, C., Mateescu, A., Mitrana, V.: Contexts and the concept of mild context-sensitivity. Linguistics and Philosophy 26(6), 703–725 (2002)
29. Manaster-Ramer, A.: Some uses and abuses of mathematics in linguistics. In: Martín-Vide, C. (ed.) Issues in Mathematical Linguistics, pp. 73–130. John Benjamins, Amsterdam (1999)
30. Oncina, J., García, P.: Identifying regular languages in polynomial time. In: Bunke, H. (ed.) Advances in Structural and Syntactic Pattern Recognition, vol. 5, pp. 99–108 (1992)
31. Oncina, J., García, P., Vidal, E.: Learning subsequential transducers for pattern recognition interpretation tasks. IEEE Transactions on Pattern Analysis and Machine Intelligence 15(5), 448–458 (1993)
32. Rozenberg, G., Salomaa, A.: Handbook of formal languages. Springer (1997)
33. Sakakibara, Y.: Learning context-free grammars from structural data in polynomial time. Theoretical Computer Science 76, 223–242 (1990)
34. Sakakibara, Y.: Efficient learning of context-free grammars from positive structural examples. Information Processing Letters 97, 23–60 (1992)
35. Sempere, J.M., García, P.: A characterization of even linear languages and its application to the learning problem. In: Carrasco, R.C., Oncina, J. (eds.) ICGI 1994. LNCS, vol. 862, pp. 38–44. Springer, Heidelberg (1994)
36. Shieber, S.M.: Evidence against the context-freeness of natural languages. In: Savitch, W.J., Bach, E., Marsh, W., Safran-Naveh, G. (eds.) The Formal Complexity of Natural Language, pp. 320–334. D. Reidel, Dordrecht (1987)
37. Solan, Z., Horn, D., Ruppin, E., Edelman, S.: Unsupervised learning of natural languages. PNAS 102(33), 11629–11634 (2005)
38. Takada, Y.: Grammatical inference for even linear languages based on control sets. Information Processing Letters 28(4), 193–199 (1988)

39. Tîrnăucă, C., Knuutila, T.: Polynomial time algorithms for learning k-reversible languages and pattern languages with correction queries. In: Hutter, M., Servedio, R.A., Takimoto, E. (eds.) ALT 2007. LNCS (LNAI), vol. 4754, pp. 272–284. Springer, Heidelberg (2007)
40. Valiant, L.G.: A theory of the learnable. Communication of the ACM 27, 1134–1142 (1984)
41. van Zaanen, M.M.V.: Bootstrapping structure into language: alignment-based learning. PhD thesis, University of Leeds (2001)
42. Yoshinaka, R.: Learning mildly context-sensitive languages with multidimensional substitutability from positive data. In: Gavaldà, R., Lugosi, G., Zeugmann, T., Zilles, S. (eds.) ALT 2009. LNCS, vol. 5809, pp. 278–292. Springer, Heidelberg (2009)

Algorithms and Their Explanations

Marco Benini[1] and Federico Gobbo[2]

[1] Università degli Studi dell'Insubria, via Mazzini 5, 21100 Varese, Italy
marco.benini@uninsubria.it
[2] Universiteit van Amsterdam, Spuistraat 210, 1012 VT Amsterdam,
The Netherlands
F.Gobbo@uva.nl

Abstract. By analysing the explanation of the classical heapsort algorithm via the method of levels of abstraction mainly due to Floridi, we give a concrete and precise example of how to deal with algorithmic knowledge. To do so, we introduce a concept already implicit in the method, the 'gradient of explanations'. Analogously to the gradient of abstractions, a gradient of explanations is a sequence of discrete levels of explanation each one refining the previous, varying formalisation, and thus providing progressive evidence for hidden information. Because of this sequential and coherent uncovering of the information that explains a level of abstraction—the heapsort algorithm in our guiding example—the notion of gradient of explanations allows to precisely classify purposes in writing software according to the informal criterion of 'depth', and to give a precise meaning to the notion of 'concreteness'.

1 Introduction

We are currently living in the age of the zettabyte (10^{21} bytes), a quantity of information "expected to grow fourfold approximately every three years... every day, enough new data is being generated to fill all US libraries eight times over" Floridi [9], page 5. This quantity of information is mostly produced through digital computers, and therefore it is algorithmic in nature, at least in part. Even from a syntactic point of view, algorithmic information is of a very different character than ordinary information: while the latter relies on the classic theory by Shannon and Weaver, the fundamentals of algorithmic information are in the theory of computation as initiated by Turing—see Chapter 14 in Allo *et al* [2].

Furthermore, we need semantics to upgrade information (the agent in the state of being informed) to knowledge (the agent in the state of being able to perform a conscious informational analysis). Chapter 10 of Floridi [8] solves this problem by giving two different logics at the basis of the states above (see also Allo [3,1]) while Primiero [15] analyses the special case of information locally valid, i.e., when functional information is in charge. Functional information, commonly used in information sciences, particularly in software engineering, entails realisable instructions to obtain reliable data not yet available. The aim of this paper is to analyse knowledge, in the sense above, in the case of algorithms inside this line of research known as Philosophy of Information, see Allo *et al* [2].

A. Beckmann, E. Csuhaj-Varjú, and K. Meer (Eds.): CiE 2014, LNCS 8493, pp. 32–41, 2014.

To do so, we will analyse the heapsort algorithm. The heapsort algorithm has been chosen as the guiding example for two reasons: in the first place, heapsort is a classical algorithm, deeply studied and used, and non elementary; in the second place, heapsort exhibits in a nutshell all the features that appears in larger and more complex software, so it provides an ideal case study to test and to explain ideas about the epistemology of computing and programming.

The paper is organised as follows: in the next section, the terminology and the basic concepts of the method of levels of abstraction are introduced, tailored to our purposes. Section 3 is devoted to illustrate the heapsort algorithm from three different points of view: the ones of a programmer, of a software designer, and of an algorithm designer. Section 4 introduces the notion of gradient of explanations, showing how the analysis conducted in Sect. 3 generates one, and some consequences are drawn. The paper concludes with a brief summary and discussion of possible future developments.

2   The Method of Levels of Abstraction

The method of levels of abstraction comes from modelling, a common practice in science: in its standard presentation, variables model observations of reality, where only necessary details are retained. The method is flexible, as it can be used in qualitative terms without technicalities, as in Floridi [9], where ethical issues are analysed, as well as in the advanced educational settings presented in Gobbo and Benini [10]. Oppositely, the method can be used in a technical sense, as for instance in the case of algorithmic information analysed in this paper.

In fact, algorithmic information presupposes that the informational organisms in charge are computational in nature. In other words, computational informational organisms (c-inforgs), are formed by (at least) a human being and by some kind of computing machine—typically a modern digital computer. As Gobbo and Benini [13] argued information can be hidden to the eyes of the observers according to the growth of complexity of the c-inforg itself, even if it can be revealed if the agent holds the necessary knowledge to cope with the complexity at the given level of abstraction. In fact, the key feature of a c-inforg is being programmable, and not every variable in the given level of abstraction is granted to be completely observable – instead it could be hidden, exactly because of the nature of algorithmic information.

The method distinguishes three kinds of levels: proper *Levels of Abstraction* (LoAs); the *Levels of Organisation* (LoO), the machinery part of the c-inforg, and *Levels of Explanation* (LoE). In general, the LoAs and the LoOs are always strictly connected in every kind of informational organisms. In particular, in the case of c-inforgs, this connection is particularly clear. In fact, each software abstraction (LoA) is run over a correspondent hardware abstraction (LoO): the history of modern computing shows a continuous drift from hardware to software; in our terms, more LoAs are introduced so to abstract over the hardware, see Gobbo and Benini [12] for details. Moreover, for each pair of LoA/LoO it is possible to identify more than one LoE because c-inforgs are programmable, and at least the programmer's and end-user's views are possible for the same pair LoA/LoO, see Gobbo and Benini [11].

If a range of LoAs is made of discrete variables and each level can be nested into another within a sequence, that range is called nested *Gradient of Abstraction* (GoA)—see subsection 3.2.6 in Floridi [8]. As we will see in the next section, the explanation of algorithms needs a new concept which is implicit in the method presented until now, that is a *Gradient of Explanation* (GoE), which holds if a GoA is in charge. We will justify the epistemic need of GoE inductively, via the examination of heapsort algorithm.

3   The Heapsort Algorithm

The heapsort procedure is a classical topic in the study of algorithms, see, e.g., pages 144–148 of Knuth [14]. In the following, the presentation is mainly based on Chapter 6 of Cormen *et al* [5], although we adopt the method of levels of abstraction to make explicit the various hypotheses and building passages, to let the non-technical reader follow our arguments.

Heapsort solves the problem of sorting an array: given an array A with homogeneous elements and a total ordering $\preceq$ among its elements, the problem asks to construct an array B whose elements are ordered by $\preceq$, i.e., $B[1] \preceq B[2] \preceq B[3] \preceq \cdots$, and such that B is a permutation of A.

The reason why heapsort is an interesting solution to the sorting problem is its efficiency: the resulting array B is constructed out of A using only a small and constant amount of additional memory, in contrast with mergesort for example, and the computing process takes a number of steps proportional to $n \log n$, with n the number of elements in A. The time complexity is optimal, because no comparison-based solution to the sorting problem may be computed in a number of steps whose order of magnitude is less than $n \log n$ in the worst case.

3.1   A Programmer's View

When a programmer is asked to implement heapsort, he should consider to be part of a c-inforg P. The LoO P_O and the LoA P_A describing the computing device in P are known to the programmer. For the sake of clarity, let us assume the LoA P_A to be the bare programming language C and the LoO P_O to be the computer memory as seen through the primitives and libraries of the language. The purpose of the programmer is to construct another LoA S_A on the top of P_A, providing a new operation, the sorting algorithm, which becomes observable in the corresponding LoO S_O by the Heapsort syntax.

To perform the implementation, the programmer needs a complex and structured amount of knowledge. First, he knows the syntax of the programming language that will be used to implement heapsort—C, in our illustration; also, he knows how the various instructions and constructions of the language modify the state of the machine, the so-called operational semantics of the language. These pieces of knowledge come from being a programmer. Then, he needs to know the

Heapsort(A : array) $\equiv$
   BuildMaxHeap(A)
   for $i \leftarrow$ len(A) **downto** 2 **do**
      exchange($A[1], A[i]$)
      heapsize(A) $\leftarrow$ heapsize(A) $- 1$
      MaxHeapify($A, 1$)

BuildMaxHeap(A : array) $\equiv$
   heapsize(A) $\leftarrow$ len(A)
   for $i \leftarrow \lfloor$len(A)$/2\rfloor$ **downto** 1 **do**
      MaxHeapify(A, i)

left(i: $\mathbb{N}$) $\equiv 2i$

right(i: $\mathbb{N}$) $\equiv 2i + 1$

MaxHeapify(A: array, i: $\mathbb{N}$) $\equiv$
   $m \leftarrow i$
   $l \leftarrow$ left(i)
   $r \leftarrow$ right(i)
   if $l \leq$ heapsize(A) & $A[l] \succ A[m]$ **then**
      $m \leftarrow l$
   if $r \leq$ heapsize(A) & $A[r] \succ A[m]$ **then**
      $m \leftarrow r$
   if $m \neq i$ **then**
      exchange($A[i], A[m]$)
      MaxHeapify(A, m)

exchange($A[i]$: element, $A[j]$: element) $\equiv$
   $x \leftarrow A[i]$;
   $A[i] \leftarrow A[j]$;
   $A[j] \leftarrow x$

Fig. 1. The heapsort algorithm

description of the heapsort algorithm in some (semi-)formal notation which is the *specification* of his task. For example, we may assume the programmer knows the pseudo-code in Fig. 1. Of course, he needs to understand the notation: specifically, he has to know that $\leftarrow$ means assignment; that **for** $x \leftarrow e$ **downto** n **do** B means a loop; that indentation is used for grouping instructions; and so on.

This knowledge is not yet enough. For example, the presented pseudo-code assumes the array A to be a data structure having two operations: len(A) which tells the number of elements in the array A, and $A[i]$ which gives access to the i-th element of A, provided $1 \leq i \leq$ len(A). A careful inspection of Fig. 1 reveals that the pseudo-code assumes that the ordering relation $\succ$ is embedded into the algorithm rather than a parameter.

It is clear that the amount of information described so far is enough to allow the programmer to fulfil his implementation task. Thus, this amount of information forms an *explanation* of heapsort, the one allowing to implement the algorithm inside the c-inforg P, producing a new pair (S_A, S_0) of LoA and LoO, respectively. This new abstraction allows the programmer to use the computing device in P in a new way, because a new concept is available, heapsort.

The new LoA S_A and LoO S_O are explained by a corresponding LoE $S_{E;P}$ which can be stated in natural language as follows: 'the LoE is all that is needed to fulfill the purpose of encoding in C language the heapsort algorithm'.

The amount of knowledge necessary to fulfil the purpose is the one sketched above, used by the programmer to implement S_A and S_O from P_A and P_O. As we have seen, this kind of knowledge can be adequately represented in what Primiero [15] calls functional information. Also, it should be clear that almost no creativity is involved here, and so at this rather low level of abstraction programmers are not exploiting the artistic possibilities inside the act of programming – see Gobbo and Benini [11].

3.2   A Software Designer's View

The point of view of a prototypical software designer is a step beyond the basic programmer's: in our terms, the LoE $S_{E;P}$ is *nested* in another, broader LoE. In fact, the software designer has to provide a specification to the programmer so that the implementation shall be coherent with the needs of a software which contains many other algorithms. Thus, the designer is aware of the sorting problem and recognises the problem as a node in a larger network of problems, whose overall solution forms the software where the heapsort implementation is only a small component.

Hence, the software designer is part of the *c*-inforg P together with the programmer, and they share the same LoA P_A and LoO P_O. The programmer and the software designer are similar to the carpenter whose LoE is to make a chair (example in Floridi [8], section 3.4): the functional organisation (the blueprint) and the realisation (the chair) are similar to the design of the algorithm and its actual implementation in our example. The 'only' difference between making the chair and the design and development of heapsort is the need for two agents with different degrees of knowledge and specialisation. In fact, the designer and the programmer have distinct LoEs, because their purposes are very different. Precisely, the designer has to choose an algorithm which solves the problem he is examining, the sorting problem in our example. The choice is guided by many issues: how frequently the problem has to be solved inside the complete application; the size of the array to be sorted; how the array data structure has to be organised. All these issues, and maybe others, shape the particular instance of the 'right' algorithm to choose. For example, the designer may choose heapsort because he knows it is efficient—a feature which is relevant when arrays may be large or it is not possible to predict their size in advance—and because *stability*, the relative order of equal elements with respect to the ordering should not necessarily be preserved by the sorting process. Also, he may choose heapsort instead of mergesort or quicksort because a destructive manipulation of the array is acceptable, and because a good performance in the worst case is preferred to a better performance in the average case, which is the difference between heapsort and quicksort, see Cormen *et al* [5].

But heapsort is an algorithm, i.e., an abstract, ideal computation, and before becoming a valid specification, a number of decisions have to be taken. For example, in Fig. 1 it is clear that the length of an array A is considered an attribute of A; alternatively, the length could have been passed to the heapsort procedure as an additional parameter. Moreover, although it is clear that the heap data structure is a sub-type of array, having to obey an additional constraint, and this fact is rendered by adding the additional attribute heapsize to heaps only, there is no explicit 'recasting' of types in the specification, which means that this piece of information in not required to be made explicit, e.g., to the language type-checker. Furthermore, the ordering relation $\succ$, see Fig. 1, is not structurally linked to the array data structure, suggesting that $\succ$ is not a parameter of the heapsort procedure—while it may be, if different orderings are required in distinct contexts of the same program. Finally, the decision to abstract over exchanges of elements and the identification of left and right branches in the heap suggests that the designer may want to leave space for changing the heapsort procedure in future maintenance releases of the program.

All the above descriptions of the possible reasons behind the shape of the presented pseudo-code, and possibly others, form the LoE $S_{E;D}$ which interprets the LoA S_A and the LoO S_O. According to the designer's perspective, the programmer's LoE could be fulfilled by other pairs LoA/LoO.

3.3   An Algorithm Designer's View

Given a problem, usually arising from a concrete application, an algorithm designer is faced with the task to conceive a computable method to calculate its solutions. Often, software designers pick algorithms off the shelf, using and adapting the vast literature Computer Science is continuing to produce. But, ideally, an algorithm designer is another human agent in the same c-inforg P we previously introduced: in fact, each instance of an algorithm has been conceived for a specific computational architecture which has to be shared among the algorithm designer, the software designer and the programmer to be effectively implemented in the context (program) where it is intended to be used—recall the parallel with the carpenter and the chair.

The first step in designing an algorithm is to polish the problem, to abstract over what is not needed to solve it, and to eventually reveal the inner structure which will be used to calculate the solution. This abstraction process is guided by ingenuity and a well-established set of techniques (again Cormen *et al* [5] for a comprehensive introduction). We will follow our example of heapsort to illustrate the design process of an algorithm. At the first stage, the elements in the array A can be considered to be composed by two distinct parts: the *key* and the *datum*. Ordering considers just the key, so this is the only relevant part with respect to the sorting problem. It is not important that elements really have two distinct parts: as far as it is possible to extract the key value from an element, the abstraction is fair, and this possibility amounts to have an effective, computable predicate $\preceq$, representing the ordering relation. It is worth noticing that the above abstraction leaves a trace in the properties of a sorting algorithm: it is exactly the separation between the key and the datum that identifies the stability of an algorithm. In fact, if in the input array A, $A[i] \preceq A[j]$ and the key of the i-th element equals the key of the j-th element, stability says that in the sorted array B, $A[i] \equiv B[k] \preceq B[h] \equiv A[j]$ exactly when $i \leq j$. Thus, elements which are equal with respect to $\preceq$, may be different when considered as elements, revealing how they have been rearranged. In general, the fact that equality may not be identity in Computer Science has a number of consequences, discussed in Gobbo and Benini [10].

Since elements can be identified with their keys when solving the sorting problem, the idea behind heapsort is to define a sub-class of arrays, the *heaps*, whose members have a distinctive property of interest. Posing parent$(i) = \lfloor i/2 \rfloor$, left$(i) = 2i$ and right$(i) = 2i + 1$, we can interpret the elements in an array as

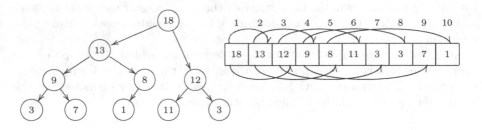

Fig. 2. A heap represented as a tree and as an array

if they were nodes in a tree, as illustrated in Fig. 2. Now, a heap H is an array satisfying, for each valid index $i > 1$:

$$H[i] \preceq H[\text{parent}(i)] \ . \tag{1}$$

By abusing terminology, an array A will be called a heap when its initial segment running from index 1 to a known index heapsize(A) satisfies (1). Hence, the root node $H[1]$ in a heap H is the greatest element in H, see Fig. 2, and moreover, its left and right sub-trees are heaps, too. Thus, we can sort the array by moving the root past the end of the heap, and then we can combine the left and the right sub-trees to obtain a new heap. In fact, this is exactly how heapsort operates, see the Heapsort procedure in Fig. 1.

Therefore, the algorithm designer is left with two sub-problems: building an initial heap out of a given array, and constructing a heap given two heaps. The former problem can be easily reduced to the latter: given an array A, its elements beyond $\lfloor \text{len}(A)/2 \rfloor$ are leaves in the tree representation, so they are trivially heaps. Each node i which is a parent of some other node, must satisfy (1): it is immediate to see that exchanging the parent node with the greatest node among $A[i]$, $A[\text{left}(i)]$, and $A[\text{right}(i)]$ forces (1) to hold, except for the sub-tree whose root has been changed—something which can be recursively restored. Thus, recursively applying this process, coded by the MaxHeapify procedure in Fig. 1, from the sub-trees with lowest depth, eventually the whole array A is rearranged in a heap, as done by BuildMaxHeap in Fig. 1. Some caution should be taken since the tree may not be complete, and so the first conjunct in the **if** statements of MaxHeapify. It is clear that induction on the structure of trees allows to prove that MaxHeapify, BuildMaxHeap, and Heapsort operate correctly, so the algorithm designer can formally derive the first essential property of heapsort, namely that it solves the sorting problem.

The second property the algorithm designer wants to establish is the computational complexity of heapsort: a detailed analysis of the derivation is presented in Chapter 6 of Cormen *et al.* [5]. For our aims, it suffices to notice that recursion plays a fundamental role in the algorithm, and thus computational complexity gets calculated as the solution of a system of recurrence equations.

So, the LoE $S_{E;A}$ of the algorithm designer contains at least the theory of binary trees, because of the induction and recursion principles used to derive the properties of heapsort, the mathematics of recurrence equations, to calculate the complexities, and most of what is needed by the programmer and the software designer, to ensure that heapsort will be implementable in the given architecture.

4 Gradients of Explanations

The heapsort example shows a very simple GoA: the c-inforg P, after the programmer has implemented the sorting procedure, exhibits two LoAs: the pair (P_A, P_O) describing the computing device, and the pair (S_A, S_O) having heapsort as a primitive. The LoA S_A abstracts over P_A by *hiding* the implementation of heapsort, and provides a new action, sorting, along with its specification; at the same time, S_O extends P_O by adding a new observable, the reference to the procedure Heapsort, which reifies the action. The relation between (P_A, P_O) and (S_A, S_O) is given by the implementation, which is the *encoding* of the action 'sorting' of S_A into P_A. It is the presence of an encoding that justifies to consider the sequence $\langle (P_A, P_O), (S_A, S_O) \rangle$ as a GoA: encodings are an essential aspect of c-inforgs, as discussed in Gobbo and Benini [10,11].

The sequence $\langle S_{E;P}, S_{E;D}, S_{E;A} \rangle$ given by the LoEs of the programmer, the software designer, and the algorithm designer as described in Sect. 3, behave similarly, being nested one into the other: in fact, each LoE explains (S_A, S_O) and, in turn, each LoE explains some parts of the preceding LoEs in the sequence. For example, the algorithm designer proves that heapsort solves the sorting problem and this statement together with its proof is in $S_{E;A}$; the software designer knows the statement but has no need for the proof, so just the statement is in $S_{E;D}$; also, the programmer knows the statement because this has to become part of S_A since the observable Heapsort has to be paired with a specification but, again, the programmer has no need for the proof. Hence, the sequence $\langle S_{E;P}, S_{E;D}, S_{E;A} \rangle$ is a gradient, and since it relates LoEs, it is called a *Gradient of Explanations* (GoE).

The relations between the LoEs in the gradient is similar to encodings in GoAs, but subtler. To understand these relations, it is useful to compare them with encodings between pairs of LoAs: instead of considering an encoding as the way a concept of the abstract LoA A gets implemented in the concrete LoA C, we may think that the encoding is the way to construct the concept in A from C. This construction resembles mathematical induction, since new objects are built starting from simpler ones. But there is a dual construction, *coinduction*, see [4]: starting from a given, large universe, coinduction operates by progressively discarding the elements of the universe not satisfying the construction property. In a GoE, the LoEs are linked to each other via a relation that follows the coinductive pattern: starting from a very general, wide and abstract universe of concepts, each LoE in the gradient refines the next one in the sequence in the sense of hiding what is not strictly needed to provide a coherent explanation of the purpose the LoE has to fulfil. The first LoE in the sequence, in turn, is refined by the LoA it is called to explain. This fact suggests that each LoA should be considered a LoE with no content deputed to explain.

In the heapsort example, the LoA S_A is explained by itself in an empty way—a coherent explanation, but not so useful. In turn, S_A is a refined explanation of $S_{E;P}$ where the contribution of the programmer is thrown away. In the opposite sense, which is the usual one when presenting LoEs, $S_{E;P}$ explains S_A by the knowledge a programmer has to provide in order to build the (S_A, S_O) pair from (P_A, P_O). It is important to remark that the process of constructing gradients is not necessarily inductive or coinductive: although, as a rule of thumb, GoAs follow an inductive pattern, while GoEs follow a coinductive one, it is not difficult to imagine counterexamples of any sort. It is the 'direction' of construction that matters, not the instrument, exactly as in the case of the carpenter: a chair can be made using different instruments, but the style and therefore the 'personal touch' of the carpenter is given by the direction of construction.

An immediate consequence of having a GoE is the possibility to give a *measure*—in the sense stated by Gobbo and Benini [10]—of concreteness of the various concepts that explain, in some sense, a piece in a *c*-inforg. Consider the lowest LoE E in a gradient G over a LoA A that uses the concept C under examination: we define the measure of C with respect to A as the distance from A, calculated by counting the number of LoEs separating E from A. This measure is loosely related to Krull dimension in commutative algebra (Eisenbud [6]), but this relation is out of the scope of the present paper. This measure is a direct expression of the level of concreteness of C with respect to the LoA A: by definition, it shows the distance between the 'concrete' basis of the GoE, the LoA A, and the first occurrence of the concept C in the gradient. Also, the same measure can be used to classify purposes. Since a purpose becomes explained in some point of a GoE, the distance d between the concrete realisation of the purpose, which is the explanation behind the LoA, and the first LoE that explains the purpose, classifies the purposes. It is important to remark that the suggested measure is relative to a gradient and based upon a LoA. It does not make sense to use this measure to compare objects not pertaining to the same LoA or not part of the same GoA.

5   Conclusions

By using the methods of levels of abstractions by Floridi [7], our analysis has naturally driven the reader toward a novel concept, extending the aforementioned method, which coherently fits into the epistemological framework. The newly synthesised concept, called Gradient of Explanations (GoE) is analogous to the Gradient of Abstractions (GoA) explained in paragraph 2.2.6 of Floridi [8], but applied to levels of explanation instead of levels of abstractions. Despite this similarity, which justifies why the new concept is conservative with respect to the method, a GoE has a rather different epistemological status.

The consequences of the introduction of the GoE are not yet fully explored and its formalisation is still preliminary. However, throughout this paper differences in the status between GoE and GoA were clarified, permitting to derive some of its consequences, using the heapsort example as a concrete guideline.

Acknowledgements. Marco Benini has been supported by the project *Correctness by Construction*, EU 7th framework programme, grant n. PIRSES-GA-2013-612638, and by the project *Abstract Mathematics for Actual Computation: Hilberts Program in the 21st Century* from the John Templeton Foundation.

F. Gobbo holds the Special Chair in Interlinguistics and Esperanto at the University of Amsterdam (UvA), Faculty of Humanities, on behalf of the Universal Esperanto Association (UEA, Rotterdam).

The content and opinions expressed in this article are the authors' and they do not necessarily reflect the opinions of the institutions supporting them.

References

1. Allo, P.: Information and Logical Discrimination. In: Cooper, S.B., Dawar, A., Löwe, B. (eds.) CiE 2012. LNCS, vol. 7318, pp. 17–28. Springer, Heidelberg (2012)
2. Allo, P., Baumgaertner, B., D'Alfonso, S., Fresco, N., Gobbo, F., Grubaugh, C., Iliadis, A., Illari, P., Kerr, E., Primiero, G., Russo, F., Schulz, C., Taddeo, M., Turilli, M., Vakarelov, O., Zenil, H. (eds.): The Philosophy of Information: An Introduction. Version 1.0. Society for the Philosophy of Information (2013)
3. Allo, P.: The Logic of 'Being Informed' Revisited and Revised. Philosophical Studies 153(3), 417–434 (2011)
4. Barwise, J., Moss, L.: Vicious Circles: on the Mathematics of Non-Wellfounded Phenomena. CSLI Publications (1996)
5. Cormen, T.H., Leiserson, C.E., Rivest, R.L., Stein, C.: Introduction to Algorithms, 2nd edn. MIT Press (2001)
6. Eisenbud, D.: Commutative Algebra with a View Toward Algebraic Geometry. Graduate Texts in Mathematics, vol. 150. Springer (1995)
7. Floridi, L.: The Method of Levels of Abstraction. Minds & Machines 18(3), 303–329 (2008)
8. Floridi, L.: The Philosophy of Information. Oxford University Press (2011)
9. Floridi, L.: The Ethics of Information. Oxford University Press (2013)
10. Gobbo, F., Benini, M.: What Can We Know of Computational Information? The Conceptual Re-Engineering of Measuring, Quantity, and Quality. Topoi (forthcoming)
11. Gobbo, F., Benini, M.: Why Zombies Can't Write Significant Source Code: The Knowledge Game and the Art of Computer Programming. Journal of Experimental & Theoretical Artificial Intelligence (in publication)
12. Gobbo, F., Benini, M.: From Ancient to Modern Computing: A History of Information Hiding. IEEE Annals of the History of Computing 35(3), 33–39 (2013)
13. Gobbo, F., Benini, M.: The Minimal Levels of Abstraction in the History of Modern Computing. Philosophy & Technology (2013)
14. Knuth, D.E.: The Art of Computer Programming, Volume 3, Sorting and Searching, 2nd edn. Addison-Wesley (1998)
15. Primiero, G.: Offline and Online Data: on Upgrading Functional Information to Knowledge. Philosophical Studies (2012)

Gene Tree Correction by Leaf Removal and Modification: Tractability and Approximability

Stefano Beretta[1] and Riccardo Dondi[2]

[1] Inst. for Biomedical Technologies, National Research Council, Segrate, Italy
[2] Dip. di Scienze Umane e Sociali, Università degli Studi di Bergamo, Bergamo, Italy
stefano.beretta@itb.cnr.it, riccardo.dondi@unibg.it

Abstract. The reconciliation of a gene tree and a species tree is a well-known method to understand the evolution of a gene family in order to identify which evolutionary events (speciations, duplications and losses) occurred during gene evolution. Since reconciliation is usually affected by errors in the gene trees, they have to be preprocessed before the reconciliation process. A method to tackle with this problem aims to correct a gene tree by removing the minimum number of leaves (Minimum Leaf Removal). In this paper we show that Minimum Leaf Removal is not approximable within factor $b \log m$, where m is the number of leaves of the species tree and $b > 0$ is a constant. Furthermore, we introduce a new variant of the problem, where the goal is the correction of a gene tree with the minimum number of leaf modifications. We show that this problem, differently from the removal version, is $W[1]$-hard, when parameterized by the number of leaf modifications.

1 Introduction

Genome evolution can be explained by a combination of different events: micro-evolutionary events, such as insertions, deletions and substitutions, and macro-evolutionary events, like duplications and losses. These latter events are fundamental in the evolution of species [8, 11] and can be responsible for the presence of many gene copies inside a genome. Genes originating from duplications of a single gene form a *gene family*.

Given a gene family, a gene tree G representing the evolution of sequences associated with a set of species is usually built by using a micro-evolutionary model. Then, the gene tree is compared with a species tree that represents the speciation history of the genomes of the considered species, with the goal of inferring the macro-evolutionary events that occurred during evolution. Such a method in known as *reconciliation* [2–4, 7, 12, 13, 15, 18].

When no species tree is known, then the problem asks to infer a species tree from a set of possibly discordant gene trees, usually with a parsimonious evolution scenario [1, 4, 10].

It is known that reconciliation is highly sensitive to errors in the gene trees. It has been shown that few errors can produce a completely misleading evolutionary

A. Beckmann, E. Csuhaj-Varjú, and K. Meer (Eds.): CiE 2014, LNCS 8493, pp. 42–52, 2014.
© Springer International Publishing Switzerland 2014

scenario, usually increasing the number of duplications and losses [9, 15]. Hence gene trees have to be preprocessed before the reconciliation process.

Errors in gene trees can be related to a special kind of duplications, called *Non-Apparent Duplications* (NAD) [4]. NAD nodes are considered as potential results of errors in the gene trees, since each NAD node represents a contradiction with the structure of a species tree not explainable by gene duplications. Motivated by this observation, a recent approach aims to correct a gene tree, before the reconciliation, by removing misplaced leaves/labels [16]. In [5, 16], the complexity of two combinatorial problems related to the removal of leaves/labels (Minimum Leaf Removal and Minimum Label Removal) has been investigated. In [5] the two problems have been shown to be APX-hard even if each label has at most two occurrences in the gene tree. Other fixed-parameter results have been presented in [5, 16], in particular showing that both Minimum Leaf Removal and Minimum Label Removal are fixed-parameter tractable when parameterized by the number of leaves/labels removed.

In this paper, we study the approximation complexity of Minimum Leaf Removal, and we show in Section 3 that it is not approximable within factor $b \log m$, for some constant $b > 0$, where m is the number of leaves of the species tree, even if each label has at most two occurrences in the input gene tree. Then, we introduce a new variant of the problem, Minimum Leaf Modification, where the aim is to correct the given gene tree by modifying the minimum number of leaves. We show that this problem, differently from the removal version, is $W[1]$-hard (Section 4), when parameterized by the number of leaf modification.

Due to space limitations some of the proofs are omitted.

2 Preliminaries

In this section, we introduce some preliminary definitions that will be useful in the rest of the paper.

Consider a set $\Lambda = \{1, 2, \ldots, m\}$ of integers, each one representing a different species. Consider a tree U, then we denote by $L(U)$ the set of its leaves, by $\Lambda(U)$ the set of labels associated with $L(U)$, and by $V(U)$ the set of its nodes. Given an internal node x of U, x_l (x_r respectively) denotes the left child (the right child respectively) of x. $U[x]$ denotes the subtree of U rooted at node x, and $\Lambda(U[x])$ denotes the set of labels associated with leaves of $U[x]$. When there is no ambiguity on the tree, we consider $\mathcal{C}(x) = \Lambda(U[x])$ (we call $\mathcal{C}(x)$ the *cluster* of x). Any node on the path from the root of U to a node x is called an *ancestor* of x; the *parent* y of x is the ancestor of x such that (y, x) is an arc of U.

In this paper, we consider two kinds of rooted binary trees leaf-labeled by the elements of Λ: *species trees* and *gene trees*. For a *species tree* T there exists a bijection from $L(S)$ to Λ (hence each element of Λ labels at most one leaf of T). For a *gene tree* G there exists a function from $L(G)$ to Λ (hence each element of Λ may label more than one leaf of G). In the rest of the paper, we denote by m the size of $L(T)$ and by n the size of $L(G)$.

Given a tree U, a *leaf removal* of leaf l consists of: (1) removing l from U, and (2) contracting the resulting node having degree two (that is the parent of l).

A tree U' obtained from a tree U through a sequence of leaf removals, is said to be *included* in U. Given a set $X \subseteq \Lambda(U)$, we denote by $U|X$ the *homomorphic restriction* of subtree U to X, that is the subtree of U obtained by a sequence of leaf removals, one for every leaf with a label in $\Lambda(U) \setminus X$.

We compare a gene tree G and a species tree T both leaf-labeled by Λ by means of the *LCA mapping* (Least Common Ancestor mapping), denoted as $\text{lca}_{G,T}$. $\text{lca}_{G,T}$ maps every node x of G to a node of T. Formally, $\text{lca}_{G,T} = y$, where y is the node of T such that (1) $\mathcal{C}(y) \supseteq \mathcal{C}(x)$, and (2) $\mathcal{C}(y_l) \not\supseteq \mathcal{C}(x)$, $\mathcal{C}(y_r) \not\supseteq \mathcal{C}(x)$. A node x of G is a *duplication node* (or a duplication occurs in x), when x and at least one of its children are mapped by $\text{lca}_{G,T}$ to the same node y of the species tree T. A node of G, which is not a duplication node, is a *speciation* node.

Consider a duplication node x. Then if $\mathcal{C}(x_l) \cap \mathcal{C}(x_r) \neq \emptyset$, x is called an *Apparent Duplication node (AD node)*. It can be easily shown that if x is an AD node, then x is a duplication node for any species tree T. A duplication node x which is not an AD node, that is when $\mathcal{C}(x_l) \cap \mathcal{C}(x_r) = \emptyset$, is called a *Non-Apparent Duplication node (NAD node)*. A gene tree G is said to be *consistent* with a species tree T if and only if each node of G is either a speciation or an AD node.

As observed in [4, 16], NAD nodes are related to errors in the gene tree. Therefore, the following combinatorial problem, Minimum Leaf Removal Problem, has been introduced in [16] for error-correction in gene trees.

Problem 1 *Minimum Leaf Removal Problem[MinLeafRem]*
Input: *A gene tree G and a species tree T, both leaf-labeled by Λ.*
Output: *A tree G^* consistent with T such that G^* is obtained from G by a minimum number of leaf removals.*

Moreover, we introduce a new combinatorial problem, where we modify, instead of removing, leaves of the gene tree so that the resulting tree is consistent with the given species tree. Given a leaf x of G labeled by $\lambda_x \in \Lambda$, a *leaf modification* consists of replacing λ_x with a label in $\Lambda \setminus \{\lambda_x\}$.

Problem 2 *Minimum Leaf Modification Problem[MinLeafMod]*
Input: *A gene tree G and a species tree T, both leaf-labeled by Λ.*
Output: *A tree G^* consistent with T such that G^* is obtained from G by a minimum number of leaf modifications.*

3   Inapproximability of MinLeafRem

In this section, we consider the approximation complexity of the MinLeafRem problem, even if each label has at most two occurrences in the gene tree. We denote this restriction of MinLeafRem as MinLeafRem(2). We show that MinLeafRem(2) is not approximable within factor $c \log m$, for some constant $c > 0$, by giving a factor preserving the reduction from the Minimum Set Cover (MinSC) problem. We refer the reader to [17] for details on gap-preserving reduction. We recall that MinSC, given a collection $\mathcal{C} = \{S_1, \ldots, S_p\}$ of sets over a finite set $U = \{u_1, \ldots u_q\}$, asks for a minimum subcollection $\mathcal{C}'$ of $\mathcal{C}$ such that each $u_x \in U$ belongs to at least one set of $\mathcal{C}'$. Notice that MinSC is known to be not approximable in polynomial time within factor $b \log q$, for some constant $b > 0$ [14].

Let $(\mathcal{C}, U)$ be an instance of MinSC. In the following, we define an instance of MinLeafRem(2) associated with $(\mathcal{C}, U)$, consisting of a gene tree G and a species tree T, both leaf-labeled by a set Λ.

First, we define the set Λ of labels. For each element $u_i \in U$, let $d(u_i) = |\{S_j : u_i \in S_j, 1 \leq j \leq p\}|$. Set $k = p^2 q^2$, $t = qk + 2pq + 1$. The set Λ is defined as:

$$\Lambda = \left(\bigcup_{j=1}^{p} A_j \cup B_j \right) \cup \left(\bigcup_{i=1}^{q} U_i \right) \cup Z \cup \{\alpha\}.$$

where the sets A_j, B_j, $1 \leq j \leq p$, U_i, $1 \leq i \leq q$, and Z are defined as follows:
- $A_j = \{a_{j,l} : 1 \leq l \leq k\}$, with $1 \leq j \leq p$;
- $B_j = \{b_{j,l} : u_l \in S_j\} \cup \{b'_{j,l} : 1 \leq l \leq q - |S_j|\}$, with $1 \leq j \leq p$;
- $U_i = \{u_{i,l} : 1 \leq l \leq t\} \cup \{u'_{i,l} : 1 \leq l \leq p - d(u(i))\}$, with $1 \leq i \leq q$;
- $Z = \{z_i : 1 \leq i \leq t\}$.

Let U be a tree, which is either the gene G, the species tree T, or a tree included in G with a leaf labeled by α. The *spine* of U is the unique path that connects the root of U to the unique leaf of U labeled by α.

The gene tree G is shown in Fig. 1. Informally, it consists of the following subtrees connected to the spine of G:

1. a subtree $G(S_i)$, for each set S_i in $\mathcal{C}$, where $\Lambda(S_i) = A_i \cup B_i$;
2. a collection of t subtrees $G_1(u_i), \ldots, G_t(u_i)$, for each $u_i \in U$, where $\Lambda(U_i) = U_i \cup \{b_{l,i} : u_i \subset S_l\}$. Subtree $G_1(u_i)$ is leaf labeled by the set $\{u_{i,1}\} \cup \{u'_{i,l} : 1 \leq l \leq p - d(u(i))\} \cup \{b_{j,i} : u_i \in S_j\}$ and subtree $G_l(u_i)$, $2 \leq l \leq t$, is leaf labeled by the set $\{u_{i,l-1}, u_{i,l}\}$;
3. t leaves, each one labeled by a distinct z_i, $1 \leq i \leq t$.

Similarly, T is defined as in Fig. 2 and, informally, consists of the following subtrees connected to the spine of T:

1. a subtree $T(S_i)$, for each set $S_i \in \mathcal{C}$, where $\Lambda(T(S_i)) = A_i \cup B_i$;
2. t leaves, each one associated with a distinct label in U_i;
3. t leaves, each one labeled by a distinct z_i, $1 \leq i \leq t$.

It is easy to see that T is a species tree uniquely leaf-labeled by Λ. The gene G is leaf-labeled by Λ, and each label in Λ is associated with at most two leaves of G. Indeed the set of labels associated with more than one leaf consists of the set $\{b_{j,i} : u_i \in S_j\}$ ($b_{j,i}$ labels one leaf of the subtree $G(S_j)$ and one leaf of the subtree $G(u_i)$), and the set $\{u_{i,l} : 1 \leq l \leq t\}$ ($u_{i,l}$ labels one leaf of the subtree $G_l(u_i)$ and one leaf of the subtree $G_{l+1}(u_i)$).

Before giving the details of the proof, we present an outline of the reduction. First, we prove some local properties of the subtrees $G(S_j)$, with $S_j \in \mathcal{C}$: in Remark 1 and in Lemma 1, we show that a solution of MinLeafRem(2) over

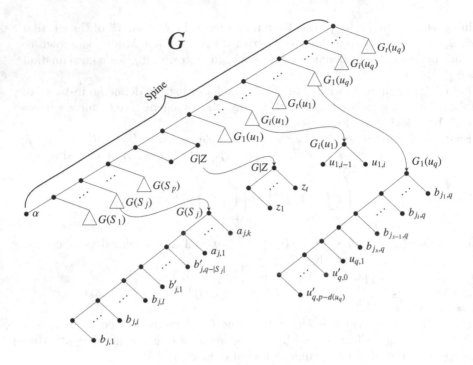

Fig. 1. The gene tree G and the subtrees $G|Z$, $G(S_j)$, $G(u_i)$. Notice that in $G_1(u_q)$ the leaves $b_{j_x,q}, b_{j_{x-1},q}, \ldots, b_{j_i,q}, \ldots, b_{j_1,q}$ refer to the sets $S_{j_x}, S_{j_{x-1}}, \ldots, S_{j_i}, \ldots, S_{j_1}$ respectively, containing u_q (with $j_x > j_{x-1} > \cdots > j_i > \cdots > j_1$).

instance (G, T) can be computed by removing leaves from $G(S_j)$, in (essentially) two possible ways: the set of leaves labeled by A_j or the set of leaves labeled by B_j. Then, exploiting some properties of the subtrees $G_l(u_i)$, with $u_i \in U$ and $1 \leq l \leq t$, and by Lemma 2 and Lemma 4, we are able to relate the former case (the removal of leaves labeled by A_j) to a set S_j in a set cover (see Lemma 5), and the latter (the removal of leaves labeled by B_j) to a set S_j not in a set cover (see Lemma 6). First, we introduce two preliminary properties of G and T.

Remark 1. *Let S_j be a set of $\mathcal{C}$, and let $G(S_j)$ ($T(S_j)$ respectively) be the subtree of G (of T respectively) associated with S_j. Then (1) the subtree of $G(S_j)$ obtained by removing the leaves with labels in A_j is consistent with $T(S_j)$; (2) the subtree of $G(S_j)$ obtained by removing the leaves with labels in B_j is consistent with $T(S_j)$.*

Remark 2. *Consider the subtree $G|Z$ of G. Each node v of G such that $\mathcal{C}(v) \supseteq Z$ is mapped to the root of S.*

A consequence of Remark 2 is that every ancestor of $G|Z$ in G is a duplication node, either a NAD-node or an AD-node.

Next, we introduce a property of the subtrees $G(S_i)$ of G, with $S_i \in \mathcal{C}$.

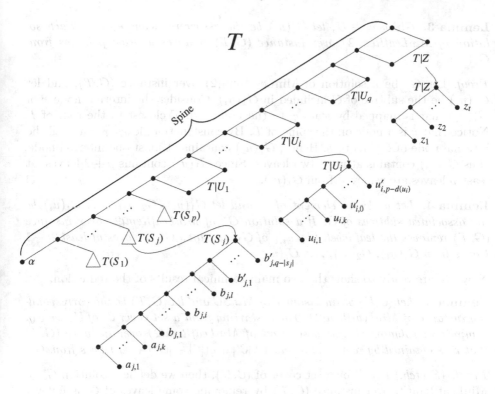

Fig. 2. The species tree T and the subtrees $T|Z$, $T|U_i$, $T(S_j)$ of T

Lemma 1. *Let S_j be a set of C, and consider the corresponding subtrees $G(S_j)$ of G and $T(S_j)$ of T. Then: (1) a solution of MinLeafRem(2) over instance (G, T) removes at least q leaves from $G(S_j)$; (2) a solution of MinLeafRem(2) over instance (G, T) that contains a leaf of $G(S_j)$ with a label in B_j removes at least k leaves from $G(S_j)$.*

Now, we show that we can assume that a solution of MinLeafRem(2) over instance (G, T) contains all the leaves of G with a label in Z.

Lemma 2. *Given a solution G^* of MinLeafRem(2) over instance (G, T) that removes less than t leaves from G and removes a leaf with a label in Z, we can compute in polynomial time a solution of MinLeafRem(2) over instance (G, T) that removes less leaves than G^* and contains all the leaves with labels in Z.*

Hence, in what follows we assume that all the leaves with a label in Z belong to G. Now, we introduce some properties of the subtree $G_1(u_i)$, $1 \leq i \leq q$ (Lemma 3), and of the subtrees $G_l(u_i)$, $1 \leq l \leq t$ (Lemma 4). The two lemmata imply that a solution contains all leaves labeled by $u_{i,l}$, with $1 \leq l \leq t$, for each $u_i \in U$.

Lemma 3. *Given $u_i \in U$, let $G_1(u_i)$ be the associated subtree of G. Each solution of MinLeafRem(2) over instance (G, T) removes at least p leaves from $G_1(u_i)$.*

Proof. Let G^* be a solution of MinLeafRem(2) over instance (G, T), and let $G_1^*(u_i)$ be the subtree of G included in $G_1(u_i)$. Consider the internal node x of $G_1^*(u_i)$ that is mapped by $\mathrm{lca}_{G^*, T}$ to the node y of T closest to the root of T. Notice that y is a node on the spine of T. By construction, $\mathrm{lca}_{G^*, T}$ maps all the internal node of $G_1^*(u_i)$ to y. Hence, $G_1^*(u_i)$ contains at most one internal node, thus $G_1^*(u_i)$ contains at most two leaves. Since $G_1(u_i)$ contains $p + 2$ leaves, at least p leaves are removed from $G_1(u_i)$.   □

Lemma 4. *Let u_i be an element of U and let $G_1(u_i)$, $G_2(u_i)$, ..., $G_t(u_i)$ be the associated subtrees of G. If a solution G^* of MinLeafRem(2) over instance (G, T) removes the leaf labeled by $u_{i,1}$ of $G_1(u_i)$, then G^* removes at least $2t - 1$ leaves from $G_1(u_i)$, $G_2(u_i)$, ..., $G_t(u_i)$.*

Now, we are ready to show the two main technical results of the reduction.

Lemma 5. *Let $(\mathcal{C}, U)$ be an instance of MinSC and let (G, T) be the corresponding instance of MinLeafRem(2). Then, starting from a set cover $\mathcal{C}'$ of U, we can compute in polynomial time a solution of MinLeafRem(2) over instance (G, T) that it is obtained by removing at most $k|\mathcal{C}'| + q(|\mathcal{C}| - |\mathcal{C}'|) + pq$ leaves from G.*

Proof. (Sketch.) Let $\mathcal{C}'$ be a set cover of $(\mathcal{C}, U)$, then we define a solution G^* of MinLeafRem(2) over instance (G, T) by removing some leaves of G as follows: (1) for each S_i in $\mathcal{C}'$, remove from the subtree $G(S_i)$ the set of leaves labeled by A_i (hence $G^*(S_i)$ is obtained with k leaf removals); (2) for each S_i not in $\mathcal{C}'$, remove from the subtree $G(S_i)$ the set of leaves labeled by B_i (hence $G^*(S_i)$ is obtained with q leaf removals); (3) for each $u_i \in U$, remove from $G_1(u_i)$ all the leaves, except for the leaf labeled by $u_{i,1}$ and a leaf labeled by $b_{j,i}$, where $u_i \in S_j$ and S_j is in $\mathcal{C}'$ (hence $G_1^*(u_i)$, with $1 \leq i \leq q$, is obtained with p leaf removals). It follows that G^* is obtained by removing $k|\mathcal{C}'| + q(|\mathcal{C}| - |\mathcal{C}'|) + pq$ leaves from G.   □

Lemma 6. *Let $(\mathcal{C}, U)$ be an instance of MinSC and let (G, T) be the corresponding instance of MinLeafRem(2). Then, from a solution of MinLeafRem(2) over instance (G, T) that is obtained by removing at most $kh + q(|\mathcal{C}| - h) + pq$ leaves, we can compute in polynomial time a solution of MinSC over instance $(\mathcal{C}, U)$ that consists of at most h sets.*

Proof. (Sketch.) Let G^* be a solution of MinLeafRem(2) over instance (G, T), such that G^* is obtained by removing at most $kh + q(|\mathcal{C}| - h) + pq$ leaves. By Lemmata 1-4, we can assume that: (i) all the leaves with label in Z belong to G^*; (ii) all the leaves with a label $u_{i,w}$, $1 \leq i \leq q$ and $1 \leq w \leq t$, belongs to G^*; (iii) either $G^*(S_j)$ has leafset B_j or it has leafset A_j. As a consequence we can define a cover $\mathcal{C}'$ of U as follows: $\mathcal{C}' = \{S_j : \Lambda(G^*(S_j)) = B_i\}$. Since at most $kh + q(|\mathcal{C}| - h) + pq$ leaves are removed from G^*, it follows that G^* contains at most h subtrees $G^*(S_j)$, with $\Lambda(G^*(S_j)) = B_i$, hence $|\mathcal{C}'| = h$.   □

The inapproximability of MinLeafRem(2) follows from Lemma 5 and Lemma 6.

Theorem 1. *MinLeafRem(2) is not approximable within factor $c \log m$, for some constant $c > 0$.*

4 W[1]-hardness of MinLeafMod

In this section, we investigate the parameterized complexity of MinLeafMod and we show that the problem is W[1]-hard when parameterized by the number of modified leaves, by giving a parameterized reduction from the Maximum Independent Set (MaxIS) problem. We recall that MaxIS, given a graph $\mathcal{G} = (V, E)$, asks for a subset $V' \subseteq V$ of maximum cardinality such that for each $u, v \in V'$ it holds $\{u, v\} \notin E$. Notice that the parameterized version of MaxIS asks whether there exists an independent set of $\mathcal{G}$ of size at least h. Hence, in what follows h will denote the size of an independent set of $\mathcal{G}$. We recall that MaxIS is known to be W[1]-hard [6].

Now, consider an instance $\mathcal{G}$ of MaxIS. Then we will show how to construct (in polynomial time) a corresponding instance (G, T) of MinLeafMod. First, we introduce the leafset Λ that labels the leaves of the two trees:
$$\Lambda = \{x_i, \alpha_i : 0 \le i \le |V|\} \cup \{y_i : 0 \le i \le h + 1\} \cup \{\beta\}.$$

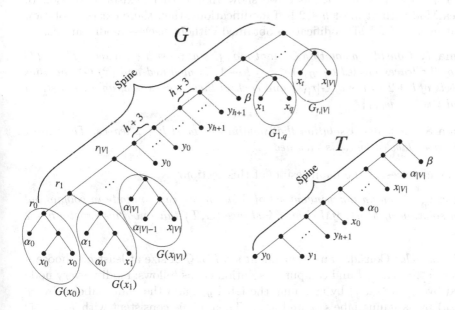

Fig. 3. The gene tree G and the species tree T. Notice that the subtrees $G_{1,q}$, $G_{t,|V|}$ encode the edges $\{v_1, v_q\}$, $\{v_t, v_{|V|}\}$ of $\mathcal{G}$. These subtrees are connected to the spine of G following the lexicographic order of the corresponding edges.

Now, we describe the two trees (see Fig. 3). Similarly to the previous reduction, the *spine* of G is the unique path that connects the root of G to the internal node of G denoted as r_0, while the *spine* of T, is the unique path that connects the root of T to the unique leaf of T labeled by y_0.

The species tree T is a caterpillar over leafset Λ. Informally, G is built by connecting the following subtrees to the spine of G:

- a subtree $G(x_i)$, $0 \le i \le |V|$; $G(x_0)$ is a caterpillar over three leaves labeled by α_0, x_0, x_0 respectively; $G(x_i)$, $1 \le i \le |V|$, is a caterpillar over three leaves labeled by x_i, α_{i-1} and α_i respectively. The nodes of the spine connected to $G(x_i)$, $0 \le i \le |V|$ are denoted as r_i;
- $h + 3$ leaves each one labeled by y_i, for each i, with $0 \le i \le h + 1$;
- a leaf labeled by β;
- for each edge $\{v_i, v_j\} \in E$, a subtree $G_{i,j}$ having two leaves labeled by x_i and x_j respectively.

First, we state a property of the instance (G, T).

Remark 3. *Consider the instance (G, T). Then each node that connects the first leaf labeled by y_i, $0 \le i \le h + 1$, to the spine of the gene tree G is a NAD node.*

We call a leaf modification *useless* if it does not change the label of a leaf to a label y_i, $0 \le i \le h + 1$. Next, we show that if there exists a solution of MinLeafMod with at most $h + 2$ leaf modifications, then there exists a solution with at most $h + 2$ leaf modification obtained without useless modifications.

Lemma 7. *Consider a solution G^* that modifies at most $h+2$ leaves. Then: (1) none of the leaves labeled by y_i, $0 \le i \le h + 1$, is modified and (2) G^* modifies the labels of $h + 2$ leaves of $G[r_{|V|}]$ and each of these leaves is assigned a distinct label of $\{y_0, \ldots, y_{h+1}\}$.*

Lemma 8. *Consider a solution that modifies at most $h + 2$ leaves. Then none of the leaves labeled by α_i is modified.*

Now, we can present the main results of this section.

Lemma 9. *Given an independent set of $\mathcal{G}$ size h, we can compute in polynomial time a solution of MinLeafMod over instance (G, T) that modifies exactly $h + 2$ leaves.*

Proof. (Sketch.) Consider an independent set I of $\mathcal{G}$ of size at least h. Choose h vertices, $v_{i_1}, \ldots v_{i_h} \in I$ and compute a solution G^* as follows: modify every node labeled by x_{i_j} of $G(x_{i_j})$ by assigning the label y_{i_j}, and the nodes labeled by x_0 of $G(x_0)$ by assigning labels y_0 and y_{h+1}. Then, G^* is consistent with T. $\square$

Lemma 10. *Given a solution of MinLeafMod over instance (G, T) that modifies exactly $h + 2$ leaves, we can compute in polynomial time an independent set of $\mathcal{G}$ consisting of at least h vertices.*

Proof. Consider a solution G^* of MinLeafMod that modifies exactly $h+2$ leaves. By Lemma 7, it follows that the solution must modify exactly $h + 2$ leaves of

$G[r_{|V|}]$. Then, for each tree $G_{i,j}$ having leaves labeled by x_i, x_j, at least one of the leaves in $G[r_{|V|}]$ labeled by x_i, x_j is not modified. If this is not the case, the node on the spine of G^* connected to the root of $G_{i,j}$ is a NAD node. Moreover, by Lemma 7 and by Lemma 8, it follows that the modified leaves of $G^*[r_{|V|}]$ are associated with labels in $\{x_0, \ldots, x_{|V|}\}$. Hence, define an independent set of $\mathcal{G}$ as follows: $V' = \{v_i \in V : x_i$ is a label associated with a modified leaf of $G^*[r_{|V|}]\}$. Then, since for each $G_{i,j}$, with leaves labeled by x_i, x_j, at least one of the leaves in $G[r_{|V|}]$ is labeled by x_i, x_j, it follows that for each $v_i, v_j \in V'$, it holds that $\{v_i, v_j\} \notin E$. Then, it follows that V' is an independent set of $\mathcal{G}$ of size h.   □

As a consequence of Lemma 8, of Lemma 9, and of the $W[1]$-hardness of MaxIS [6], we have the following result.

Theorem 2. *MinLeafMod is $W[1]$-hard.*

References

1. Blin, G., Bonizzoni, P., Dondi, R., Rizzi, R., Sikora, F.: Complexity insights of the minimum duplication problem. Theoretical Computer Science (to appear, 2014)
2. Bonizzoni, P., Della Vedova, G., Dondi, R.: Reconciling a gene tree to a species tree under the duplication cost model. Theoretical Computer Science 347, 36–53 (2005)
3. Chang, W., Eulenstein, O.: Reconciling gene trees with apparent polytomies. In: Chen, D.Z., Lee, D.T. (eds.) COCOON 2006. LNCS, vol. 4112, pp. 235–244. Springer, Heidelberg (2006)
4. Chauve, C., El-Mabrouk, N.: New perspectives on gene family evolution: losses in reconciliation and a link with supertrees. In: Batzoglou, S. (ed.) RECOMB 2009. LNCS, vol. 5541, pp. 46–58. Springer, Heidelberg (2009)
5. Dondi, R., El-Mabrouk, N., Swenson, K.M.: Gene tree correction for reconciliation and species tree inference: Complexity and algorithms. Journal of Discrete Algorithms 25, 51–65 (2014)
6. Downey, R.G., Fellows, M.R.: Fixed-parameter tractability and completeness ii: On completeness for w[1]. Theor. Comput. Sci. 141(1&2), 109–131 (1995)
7. Durand, D., Haldórsson, B., Vernot, B.: A hybrid micro-macroevolutionary approach to gene tree reconstruction. Journal of Computational Biology 13, 320–335 (2006)
8. Eichler, E., Sankoff, D.: Structural dynamics of eukaryotic chromosome evolution. Science 301, 793–797 (2003)
9. Hahn, M.: Bias in phylogenetic tree reconciliation methods: implications for vertebrate genome evolution. Genome Biology 8(R141) (2007)
10. Ma, B., Li, M., Zhang, L.: From gene trees to species trees. SIAM J. on Comput. 30, 729–752 (2000)
11. Ohno, S.: Evolution by gene duplication. Springer, Berlin (1970)
12. Page, R.: Genetree: comparing gene and species phylogenies using reconciled trees. Bioinformatics 14, 819–820 (1998)
13. Page, R., Cotton, J.: Vertebrate phylogenomics: reconciled trees and gene duplications. In: Pacific Symposium on Biocomputing, pp. 536–547 (2002)

14. Raz, R., Safra, S.: A sub-constant error-probability low-degree test, and a sub-constant error-probability pcp characterization of np. In: Leighton, F.T., Shor, P.W. (eds.) STOC, pp. 475–484. ACM (1997)
15. Sanderson, M., McMahon, M.: Inferring angiosperm phylogeny from EST data with widespread gene duplication. BMC Evolutionary Biology 7, S3 (2007)
16. Swenson, K.M., Doroftei, A., El-Mabrouk, N.: Gene tree correction for reconciliation and species tree inference. Algorithms for Molecular Biology 7, 31 (2012)
17. Vazirani, V.V.: Approximation algorithms. Springer (2001)
18. Vernot, B., Stolzer, M., Goldman, A., Durand, D.: Reconciliation with non-binary species trees. Journal of Computational Biology 15, 981–1006 (2008)

Uniform Schemata for Proof Rules

Ulrich Berger and Tie Hou

Department of Computer Science, Swansea University, UK
{u.berger,cshou}@swansea.ac.uk

Abstract. Motivated by the desire to facilitate the implementation of interactive proof systems with rich sets of proof rules, we present a uniform system of rule schemata to generate proof rules for different styles of logical calculi. The system requires only one schema for each logical operator to generate introduction and elimination rules in natural deduction and sequent calculus style. In addition, the system supports program extraction from proofs by generating realizers for the proof rules automatically.

Keywords: Proof calculi, Semantics and logic of computation, Realizability.

1 Introduction

In mathematical logic, specifically in proof-theory, one usually tries to be minimalistic regarding the design of a logical calculus, that is, one tries to find a minimal number of complete proof rules. The reason is that when reasoning *about* a logical calculus one often argues by induction on the construction of proofs, which generates a proof case for each proof rule. However, if one reasons *with* a logical calculus, for example within an interactive theorem prover, one is interested in a calculus that provides a rich set of rules in order to allow convenient and fast proof development. In fact, in current interactive proof systems one usually finds proof rules that correspond to natural deduction style, sequent style, or combinations and variants thereof.

The main motivation for this work is to provide a systematic approach to a concise and efficient implementation of logical calculi with such rich sets of proof rules. We introduce a uniform system of *rule schemata*, which directly express the meaning of logical operators and which, in a uniform way, allow to derive the rules of different styles of proof calculi, such as sequent calculus and natural deduction, but also further rules that are used in interactive proof assistants. Surprisingly, the approach requires only one schema for each logical operator. The introduction and elimination rules of natural deduction as well as left and right rules in sequent calculus are derived automatically. Moreover, our system is able to automatically derive realizers of intuitionistic proof rules, thus facilitating the implementation of proof systems that support program extraction from proofs, such as Coq [4] and Minlog [7]. We are currently developing a prototype of such a proof system using rule schemata as a basis of the implementation.

A. Beckmann, E. Csuhaj-Varjú, and K. Meer (Eds.): CiE 2014, LNCS 8493, pp. 53–62, 2014.

An additional advantage of rule schemata is the fact that they are built on a data structure of finitary sets, a generalization of finite sets. Finitary sets have the structure of a monad and can therefore be very conveniently implemented and manipulated in a programming language that supports monads and provides a special syntax for them.

2 Rule Schemata and Their Associated Generating Rules

Briefly, the global strategy is as follows. First we introduce rule schemata, from which we derive generating rules. These generating rules are different rules that correspond to different styles of proving e.g. sequent calculus, or natural deduction or the mixture of these two. Then from generating rules we obtain the real rules in the proof system by instantiation and adding side formulas.

2.1 Finitary Sets

The premise and conclusion of a rule schema will be a set of sets of sequents. For propositional logic finite sets would suffice, but in order to deal with quantifiers the notion of finiteness needs to be slightly extended. Let us assume we are given a class of objects e, called 'expressions', for which the notions of free variable and substitution $e[\boldsymbol{x}/\boldsymbol{t}]$ are defined, where $\boldsymbol{t}$ is a tuple of objects called 'terms'. A *finitary set* of expressions, *f-set* for short, is of the form

$$E_{\boldsymbol{x}}.$$

where E is a finite set and $\boldsymbol{x}$ is finite tuple of variables called *abstractions*. The intended meaning of $E_{\boldsymbol{x}}$ is the set $\{e[\boldsymbol{t}/\boldsymbol{x}] \mid e \in E, \boldsymbol{t} \text{ terms }\}$. In $E_{\boldsymbol{x}}$ all free occurrences of the variables $\boldsymbol{x}$ in E are bound. In fact, regarding free and bound variables $E_{\boldsymbol{x}}$ is analogous to the lambda abstraction $\lambda \boldsymbol{x}.E$. Using this analogy, we can define a notion of substitution for f-sets, hence f-sets can be regarded as expressions again, and the notion of an 'f-set of f-sets' makes sense.

The passage from expressions to f-sets of expression gives rise to a functor which has the additional structure of a monad [8]. The monadic structure greatly facilitates the implementation of f-sets in functional languages, such as Haskell, that support monads and provide a concise and intuitive syntax [11] for them. We took advantage of this syntax in our prototype implementation, but will not use it here, because there is no space to explain it, and we wish to keep the paper accessible to readers unfamiliar with it.

The union of two f-sets is defined as

$$E_{\boldsymbol{x}} \cup F_{\boldsymbol{y}} = (E \cup F)_{\boldsymbol{xy}}$$

where w.l.o.g. it is assumed that the tuples $\boldsymbol{x}$ and $\boldsymbol{y}$ are disjoint and don't create undesired bindings. Note that any finite set of expressions can be viewed as an f-set of expression (with an empty tuple of abstractions).

2.2 Rule Schemata

We consider first-order formulas $\top, \bot, P(t), A \wedge B, A \vee B, A \to B, \forall x\, A, \exists x\, A$, where $\top$ and $\bot$ are symbols for truth and falsity, P ranges of predicate symbols of fixed arities, and t ranges over finite vectors of first-order terms built from variables, constants and function symbols. An *atomic proposition* is a predicate symbol of arity 0. We identify an atomic proposition P with the formula $P()$. An *instance* of a formula is obtained by substituting each constant by a term, each function symbol f by a function abstraction $\lambda x\,.\,s$ (that is, replacing each occurrence of a subterm $f(t)$ by $s[t/x]$), and each predicate symbol P by a comprehension term $\{x \mid A\}$ (that is, replacing each occurrence of a subformula $P(t)$ by $A[t/x]$).

A *sequent*, $S = \Delta \vdash A$, consists of a finite set of formulas Δ called *antecedent*, and a formula A called *succedent* of S. We write a sequent $\{A_1, \ldots, A_n\} \vdash B$ as $A_1, \ldots, A_n \vdash B$ and identify a formula A with the sequent $\vdash A$. A sequent $S = A_1, \ldots, A_n \vdash B$ represents the formula $[S] = A_1 \wedge \ldots \wedge A_n \to B$. For an f-set of sequents $X = \{S_1, \ldots, S_n\}_x$ we define the formulas

$$\bigwedge X = \forall x([S_1] \wedge \ldots \wedge [S_n]), \quad \bigvee X = \exists x([S_1] \vee \ldots \vee [S_n]).$$

For an f-set of f-sets of sequents, $\mathcal{X} = \{X_1, \ldots, X_n\}_x$, we define the formulas

$$\bigwedge \mathcal{X} = \forall x(\bigvee X_1 \wedge \ldots \wedge \bigvee X_n), \quad \bigvee \mathcal{X} = \exists x(\bigwedge X_1 \vee \ldots \vee \bigwedge X_n).$$

The general form of a *rule schema* (*schema* for short) is

$$\frac{\mathcal{X}}{\mathcal{Y}} \tag{1}$$

where $\mathcal{X}$ and $\mathcal{Y}$ arc f-sets of f-sets of sequents. The schema (1) represents the formula

$$\bigvee \mathcal{X} \to \bigwedge \mathcal{Y}.$$

2.3 Schemata for Intuitionistic Logic

The rule schemata for intuitionistic logic consist of a defining schema for each logical operator plus a structural schema Ax that corresponds to an axiom or assumption rule,

$$
\begin{array}{cccccc}
\wedge & \dfrac{\{\{A, B\}\}}{\{\{A \wedge B\}\}} & \vee & \dfrac{\{\{A\}, \{B\}\}}{\{\{A \vee B\}\}} & \to & \dfrac{\{\{A \vdash B\}\}}{\{\{A \to B\}\}} \\[3ex]
\forall & \dfrac{\{\{P(x)\}_x\}}{\{\{\forall x\, P(x)\}\}} & \exists & \dfrac{\{\{P(x)\}\}_x}{\{\{\exists x\, P(x)\}\}} & & \\[3ex]
\top & \dfrac{\{\{\}\}}{\{\{\top\}\}} & \bot & \dfrac{\{\}}{\{\{\bot\}\}} & \text{Ax} & \dfrac{\{\{\}\}}{\{\}}
\end{array} \tag{2}
$$

where A, B are different atomic propositions and P is a unary predicate symbol.

Theorem 1 (Soundness of Rule Schemata for Intuitionistic Logic). *The schemata for intuitionistic logic (2) are logically valid. The formulas represented by defining schemata are of the form $C \to C$ where C is the formula in the schema's conclusion. The schema Ax represents the formula $\top \to \top$.*

Remark. If we regard schemata as formulas in a meta-logic, the defining schemata in (2) can be viewed as *definitions* of the logical operators in a meta-logic (where in this paper we refrain from distinguishing between operators from the meta-logic and the object-logic). This is similar to categorical logic [5, 6] where one defines the logical operators through appropriate adjunctions. In categorical logic one can use the categorical laws to derive logical proof rules. Similarly, we will use the laws of an intuitionistic meta-logic to derive (in Sect. 2.5 and 2.6) proof rules of the object-logic. What we gain is the fact that the meta-logic can be formalized with a minimal set of rules, but the resulting proof rules of the object-logic will have a rich set of rules.

2.4   Invertible Rule Schemata

We call an f-set of f-sets of sequents $\mathcal{X}$ *dualizable* and define its *dual* $\delta\mathcal{X}$ if one of the following two conditions holds:

(1) $\mathcal{X} = \{\{S_1, \ldots, S_n\}_x\}$, with $\delta\mathcal{X} = \{\{S_1\}, \ldots, \{S_n\}\}_x$.
(2) $\mathcal{X} = \{\{S_1\}, \ldots, \{S_n\}\}_x$, with $\delta\mathcal{X} = \{\{S_1, \ldots, S_n\}_x\}$.

Clearly, if $\mathcal{X}$ is dualizable, then $\delta\mathcal{X}$ is dualizable, and $\delta\delta\mathcal{X}$ is the same as $\mathcal{X}$. A rule schema

$$\frac{\mathcal{X}}{\mathcal{Y}}$$

is *invertible* if the sets $\mathcal{X}$ and $\mathcal{Y}$ are both dualizable. In this case the *inverse* is defined as

$$\frac{\delta\mathcal{Y}}{\delta\mathcal{X}}$$

Theorem 2. *If $\mathcal{X}$ is an invertible f-set of f-sets of sequents, then $\bigvee \mathcal{X}$ is equivalent to $\bigwedge \delta\mathcal{X}$, and $\bigwedge \mathcal{X}$ is equivalent to $\bigvee \delta\mathcal{X}$. Hence the inverse of an invertible scheme represents the converse implication represented by the original scheme.*

Clearly, the schemata for intuitionistic logic (2) are invertible, with inverses

$$
\begin{array}{ccc}
\wedge^- \ \dfrac{\{\{A \wedge B\}\}}{\{\{A\}, \{B\}\}} & \vee^- \ \dfrac{\{\{A \vee B\}\}}{\{\{A, B\}\}} & \to^- \ \dfrac{\{\{A \to B\}\}}{\{\{A \vdash B\}\}} \\[3ex]
\forall^- \ \dfrac{\{\{\forall x\, P(x)\}\}}{\{\{P(x)\}\}_x} & \exists^- \ \dfrac{\{\{\exists x P(x)\}\}}{\{\{P(x)\}_x\}} & \\[3ex]
\top \ \dfrac{\{\{\top\}\}}{\{\}} & \bot^- \ \dfrac{\{\{\bot\}\}}{\{\{\}\}} & \mathrm{Ax}^- \ \dfrac{\{\{\}\}}{\{\}}
\end{array}
\tag{3}
$$

Note that the schema Ax is identical to its inverse Ax^-.

2.5   Generating Rules

We describe two ways (**Rules 1, 2**) of associating with a schema a *generating rule*. There will also be **Rules v1, v2** which produce variants of generating rules.

The general form of a generating rule is

$$\frac{X}{S} \tag{4}$$

where X is a f-set of sequents and S is a sequent. (4) represents the formula

$$\bigwedge X \to S$$

If $X = \{S_1, \ldots, S_n\}_x$ then we will display the generating rule (4) usually as

$$\frac{S_1 \quad \cdots \quad S_n}{S} \; x$$

Below, C denotes an atomic proposition not occurring in X, Y. We associate with a schema

$$\frac{\mathcal{E}_x}{\mathcal{F}_y}$$

where $\mathcal{E}$ and $\mathcal{F}$ are finite sets of f-sets of sequents, generating rules according to the following Rules 1, 2:

Rule 1 associates with any f-set $X \in \mathcal{E}$ and f-sets $(F_1)_{u_1}, \ldots, (F_n)_{u_n} \in \mathcal{F}$, where the F_i are finite sets of formulas (i.e. sequents with empty antecedents), the generating rule

$$\frac{X \cup \{A_1, \ldots, A_n \vdash C \mid A_1 \in F_1, \ldots, A_n \in F_n\}_{u_1, \ldots, u_n}}{C}.$$

Note that the abstractions x and y are discarded.

Rule 2 associates with $X \in \mathcal{E}$ and $\{\Delta \vdash A\} \in \mathcal{F}$ the generating rule

$$\frac{X \cup \Delta}{A}$$

Rules v1 allows to produce variants of a given generating rule by moving formulas from the premise to the antecedent of the conclusion. More precisely, let a generating rule of the form

$$\frac{X \cup \Delta'}{\Delta \vdash A}$$

be given, where Δ' is a finite (not just finitary) set of formulas, i.e the elements of Δ' are sequents with empty antecedent. We transform this into

$$\frac{X}{\Delta \cup \Delta' \vdash A}$$

Rule v2 transforms a generating rule

$$\frac{X}{\Delta \vdash A}$$

into the variant

$$\frac{X \cup \Delta \cup \{A \vdash C\}}{C}$$

From the construction of the generating rules associated with schemata one immediately sees:

Theorem 3 (Soundness of the Rules 1, 2, v1, v2). *The generating rule associated with a schema by rules 1 and 2 are intuitionistically implied by the schema. Rules v1, v2 produce equivalent generating rules in the sense that the sets of instances of formulas they represent are equivalent in intuitionistic logic.*

2.6   Proof Rules

A *proof rule* is a set of rules of the form

$$\frac{\Delta_1 \vdash A_1 \quad \ldots \quad \Delta_n \vdash A_n}{\Delta \vdash A}.$$

A generating rule generates a proof rule by adding side formulas and instantiating predicate symbols. More precisely, a generating rule

$$\frac{\Delta_1 \vdash A_1 \quad \ldots \quad \Delta_n \vdash A_n}{\Delta \vdash A}\ \boldsymbol{x}$$

generates the proof rule consisting of the rules

$$\frac{\Gamma \cup \Delta_1' \vdash A_1' \quad \ldots \quad \Gamma \cup \Delta_n' \vdash A_n'}{\Gamma \cup \Delta' \vdash A'}$$

where Γ is a finite set of formulas (the *side formulas*) not containing $\boldsymbol{x}$ free, and the primed Δs and As are instances of the Δs and As leaving $\boldsymbol{x}$ unchanged.

3   Deriving the Rules of Natural Deduction and Sequent Calculus

We now discuss the generating rules associated with the schemata (2) and their inverses (3) and show that all logical rules of intuitionistic natural deduction and sequent calculus are generated. We omit the defining schema for $\top$ and its inverse since their generating rules are less interesting and are largely subsumed by the schema Ax. Note also that the defining schema for $\bot$ has no associated generating rule (but the inverse of this schema does have generating rules).

$\wedge\ \dfrac{\{\{A, B\}\}}{\{\{A \wedge B\}\}}$. By Rules 1, 2, v1, v2 we have the generating rules

$$1 \quad \frac{A \quad B \quad A \wedge B \vdash C}{C} \qquad 2 \quad \frac{A \quad B}{A \wedge B}$$

and the variants

$$1.1 \quad \frac{B \quad A \wedge B \vdash C}{A \vdash C} \qquad 1.2 \quad \frac{A \quad A \wedge B \vdash C}{B \vdash C} \qquad 1.3 \quad \frac{A \wedge B \vdash C}{A, B \vdash C}$$

$$2.1 \quad \frac{B}{A \vdash A \wedge B} \qquad 2.2 \quad \frac{A}{B \vdash A \wedge B} \qquad 2.3 \quad \frac{}{A, B \vdash A \wedge B}$$

2 corresponds to the $\wedge$-introduction rule of natural deduction, which is the same as the $\wedge$-right rule of sequent calculus. 1.3 corresponds to the inverse of the $\wedge$-left rule in sequent calculus. 2.3 is the axiom of $\wedge$-introduction. To give an example of a generated proof rule, the proof rules corresponding to 2 consists of all rules of the form

$$\frac{\Gamma \vdash A \quad \Gamma \vdash B}{\Gamma \vdash A \wedge B} \quad .$$

where Γ ranges over an arbitrary finite set of formulas and A, B range over arbitrary formulas.

In the following we only show a selection of generating rules concentrating on those that correspond to proof rules in natural deduction and sequent calculus.

$\wedge^{-} \quad \dfrac{\{\{A \wedge B\}\}}{\{\{A\}, \{B\}\}}$. We have the generating rules

$$\frac{A, B \vdash C}{A \wedge B \vdash C} \qquad \frac{A \wedge B}{A} \qquad \frac{A \wedge B}{B}$$

which are the $\wedge$-left rule of sequent calculus and the $\wedge$-elimination rules of natural deduction.

$\vee \quad \dfrac{\{\{A\}, \{B\}\}}{\{\{A \vee B\}\}}$. Only the generating rules derived from Rule 2 are of interest:

$$\frac{A}{A \vee B} \qquad \frac{B}{A \vee B}$$

These are the $\vee$-introduction rules of natural deduction, which are the same as the $\vee$-right rules of sequent calculus.

$\vee^{-} \quad \dfrac{\{\{A \vee B\}\}}{\{\{A, B\}\}}$. We have the generating rules

$$\frac{A \vee B \quad A \vdash C \quad B \vdash C}{C} \qquad \frac{A \vdash C \quad B \vdash C}{A \vee B \vdash C}$$

which are the $\vee$-elimination rule in natural deduction and the $\vee$-left rule in sequent calculus.

$\rightarrow \quad \dfrac{\{\{A \vdash B\}\}}{\{\{A \rightarrow B\}\}}$. Only the generating rule from Rule 2, $\dfrac{A \vdash B}{A \rightarrow B}$, is interesting. It corresponds to $\rightarrow$-introduction in natural deduction which is the same as the $\rightarrow$-right rule in the sequent calculus.

$\rightarrow^{-} \quad \dfrac{\{\{A \rightarrow B\}\}}{\{\{A \vdash B\}\}}$. The generating rules of interest are

$$\frac{A \to B \quad A}{B} \qquad \frac{A \to B}{A \vdash B} \qquad \frac{A \quad B \vdash C}{A \to B \vdash C}$$

which are $\to$-elimination in natural deduction a.k.a. modus ponens, the inverse of $\to$-introduction, and (obtained from the former generating rule by applying rules v2 and then v1) the $\to$-left rule of sequent calculus.

$\forall \; \dfrac{\{\{P(x)\}_x\}}{\{\{\forall x \, P(x)\}\}}$. By Rule 2, we have $\dfrac{P(x)}{\forall x \, P(x)} \; x$, which is the $\forall$-introduction rule of natural deduction and the $\forall$-right rule of the sequent calculus. The corresponding proof rule is $\dfrac{\Gamma \vdash A(x)}{\Gamma \vdash \forall x \, A(x)}$ where $A(x)$ is an arbitrary formula and Γ is a finite set of formulas not containing x free.

$\forall^- \; \dfrac{\{\{\forall x \, P(x)\}\}}{\{\{P(x)\}\}_x}$. By Rules 1 and v1, we have $\dfrac{P(x) \vdash C}{\forall x \, P(x) \vdash C} \; x$, which is the $\forall$-left rule of the sequent calculus. By Rule 2, we have $\dfrac{\forall x \, P(x)}{P(x)}$, which is the $\forall$-elimination rule of natural deduction. The corresponding proof rules are

$$\frac{\Gamma, A(t) \vdash C}{\Gamma, \forall x \, A(x) \vdash C} \qquad \frac{\Gamma \vdash \forall x \, A(x)}{\Gamma \vdash A(t)} \; .$$

$\exists \; \dfrac{\{\{P(x)\}\}_x}{\{\{\exists x \, P(x)\}\}}$. By Rules 1 and v1, we have $\dfrac{\exists x \, P(x) \vdash C}{P(x) \vdash C}$. By Rule 2, we have $\dfrac{P(x)}{\exists x \, P(x)}$, which is the $\exists$-introduction rule of natural deduction and the $\exists$-right rule of the sequent calculus. The corresponding proof rules are

$$\frac{\Gamma, \exists x \, A(x) \vdash C}{\Gamma, A(t) \vdash C} \qquad \frac{\Gamma \vdash A(t)}{\Gamma \vdash \exists x \, A(x)} \; .$$

$\exists^- \; \dfrac{\{\{\exists x \, P(x)\}\}}{\{\{P(x)\}_x\}}$. By Rule 1, we have $\dfrac{\exists x \, P(x) \quad P(x) \vdash C}{C} \; x$, which is the $\exists$-elimination rule of natural deduction. By Rule v1, we have $\dfrac{P(x) \vdash C}{\exists x \, P(x) \vdash C} \; x$ which is the $\exists$-left rule of sequent calculus. The corresponding proof rules are

$$\frac{\Gamma \vdash \exists x \, A(x) \quad \Gamma, A(x) \vdash C}{C} \qquad \frac{\Gamma, A(x) \vdash C}{\Gamma, \exists x \, A(x) \vdash C} \; ,$$

where Γ and C must not contain x free.

$\bot^- \; \dfrac{\{\{\bot\}\}}{\{\{\}\}}$. By Rule 1, we have $\dfrac{\bot}{C}$, which is the efq rule.

$\text{Ax} \; \dfrac{\{\{\}\}}{\{\}}$. By Rules 1 and v1, we have $\overline{C \vdash C}$, which is the axiom or assumption rule. By Rules 1 and v2, we have $\dfrac{C \quad C \vdash C'}{C'}$, which is the cut rule.

Theorem 4 (Completeness of Schemata for Minimal Logic). *The propositional rule schemata are complete for minimal Natural Deduction and Sequent Calculus in the sense that every axiom or rule of these calculi is an instance of a generating rule derived from one of the schemata (2) or their inverses (3) by the Rule 1, 2, v1, v2.*

4 Realizability

In addition to the generation of proof rules, schemata allow to automatically generate realizers for proof rules. Regarding the notion of realizability we refer to [10] and [2]. From a programming perspective, the most interesting aspect of realizability is the Soundness Theorem stating that from an intuitionistic proof of a formula A one can extract a realizer of A which can be viewed as solution to the computational problem specified by A. The Soundness Theorem is based on the fact that the proof rules of intuitionistic logic, more precisely, the formulas they represent, are realizable. Therefore, the main building blocks of an implementation of program extraction based on realizability are (implementations of) realizers of proof rules. Since the proof rules of intuitionistic logic and their inverses all represent formulas of the form $A \to A$, they are trivially realized by the identity function. The Rules 1, 2, v1, v2 correspond to simple transformations of realizers (one can view them as the programs extracted from the proof of the Soundness Theorem for these rules (Thm 3)). Rule 1, which is based on the fact that $\land$ distributes over $\lor$ is realized by a cascade of case analyses. The remaining rules are realized by simple combinations of projections, currying and uncurrying. Applying these transformations to the identity function one obtains realizers of the derived generating rules and hence realizers of the corresponding proof rules.

5 Conclusion

We presented a uniform system of rule schemata for intuitionistic first-order logic and showed how to derive generating rules corresponding to the usual proof rules of natural deduction and sequent calculus as well as realizers thereof. The main motivation for this work is to obtain a framework facilitating the implementation of logic and program extraction from proofs.

Rule schemata are not restricted to first-order logic. In our prototype we apply them to an intuitionistic version of Church's Simple Theory of Types [3] extended by inductive and coinductive definitions (to be detailed in a forthcoming publication). In Church's Simple Theory of Types (which essentially is higher-order logic) one can view predicate constants as higher-type variables and write the collection of schemata 2 as a single f-set with A, B, P as abstracted variables. Something similar can be done for the derived generating rules and proof rules. This has the advantage that the process of instantiation of formulas and rules is covered by f-sets as well. One can also give a defining schema for Leibniz equality

$$\frac{\{\{P(x) \vdash P(y)\}_P\}}{\{\{x = y\}\}}$$

from which the expected proof rules can be derived.

Our rule schemata have some resemblance with the hypersequent calculus [1, 9]. A hypersequent is a finite set of sequents, hence the premise of a rule in hypersequent calculus can be viewed as a set of sets of sequents. Note however, that a hypersequent is always interpreted disjunctively, while in the context of schemata the interpretation of an f-set of sequents depends on whether it appears in the premise or conclusion of a schema. Note also that the hypersequent calculus is a proof calculus where sequents are replaced by hypersequents while rule schemata are seeds for proof rules based on ordinary sequents. It is conceivable though that rule schemata based on hypersequents can be developed leading to a compact representation of the rules of the hypersequent calculus.

References

1. Avron, A.: A constructive analysis of RM. Journal of Symbolic Logic 52(4), 939–951 (1987)
2. Berger, U., Seisenberger, M.: Proofs, programs, processes. Theory of Computing Systems 51(3), 313–329 (2012)
3. Church, A.: A Formulation of the Simple Theory of Types. The Journal of Symbolic Logic 5(2), 56–68 (1940)
4. The Coq Proof Assistant, http://coq.inria.fr/
5. Jacobs, B.: Categorical logic and type theory. Studies in Logic and the Foundations of Mathematics, vol. 141. North Holland, Elsevier (1999)
6. Lawvere, W.: Functorial semantics of algebraic theories and some algebraic problems in the context of functorial semantics of algebraic theories. Ph.D. thesis, Columbia University (1963); Republished in: Reprints in Theory and Applications of Categories 5, 1–121 (2004)
7. The Minlog System, http://www.minlog-system.de
8. Moggi, E.: Notions of Computation and Monads. Information and Computation 93(1), 55–92 (1991)
9. Pottinger, G.: Uniform cut-free formulations of T, S4 and S5 (abstract). Journal of Symbolic Logic 48, 900 (1983)
10. Schwichtenberg, H., Wainer, S.S.: Proofs and Computations. Cambridge University Press (2012)
11. Wadler, P.: Comprehending monads. Mathematical Structures in Computer Science 2, 461–493 (1992)

Graph Polynomials Motivated by Gene Rearrangements in Ciliates

Robert Brijder[1,*] and Hendrik Jan Hoogeboom[2]

[1] Hasselt University and Transnational University of Limburg, Belgium
robert.brijder@uhasselt.be
[2] Leiden Institute of Advanced Computer Science,
Leiden University, The Netherlands
h.j.hoogeboom@liacs.leidenuniv.nl

Abstract. Gene rearrangements within the process of gene assembly in ciliates can be represented using a 4-regular graph. Based on this observation, Burns et al. [Discrete Appl. Math., 2013] propose a graph polynomial abstracting basic features of the assembly process, like the number of segments excised. We show that this *assembly polynomial* is essentially *(i)* a single variable case of the *transition polynomial* by Jaeger and *(ii)* a special case of the *bracket polynomial* introduced for simple graphs by Traldi and Zulli.

1 Introduction

Ciliates are an ancient group of unicellular organisms. They have the remarkable property that their DNA is stored in two vastly different types of nuclei. During conjugation a germline nucleus called the *micronucleus* (MIC) is transformed into a somatic nucleus called the *macronucleus* (MAC). In this way, each MIC gene is transformed into its corresponding MAC gene, in a process that we call *gene assembly*. Various formal models for this gene transformation process are presented in [3].

One of these formal models is string based, with letters representing "pointers" (special DNA sequences in a MIC gene) together with their relative orientation in the corresponding MAC gene [10]. The model postulates that three operations called *loop excision*, *hairpin recombination*, and *double loop recombination* accomplish the transformation of a MIC gene to its corresponding MAC gene. This model has been significantly generalized using the notion of circle graph, see, e.g., [3, 10].

This string-based formal model can be very naturally fitted within the well-developed theory of transformations of Eulerian circuits in 4-regular graphs [4]. An example using the Actin I gene of Sterkiella nova, taken from Prescott [18], is recalled below. It is known that these Eulerian circuit transformations may in turn be viewed both as a special case of a matrix operation called *principal pivot transform* (see [22]) and as a set system operation called *twist*. It turns out that the interplay of the successive operations can be much better understood in these more general settings compared to the string or graph settings. These generalizations for gene assembly are outlined in [4].

---

* R.B. is a postdoctoral fellow of the Research Foundation - Flanders (FWO).

A. Beckmann, E. Csuhaj-Varjú, and K. Meer (Eds.): CiE 2014, LNCS 8493, pp. 63–72, 2014.
© Springer International Publishing Switzerland 2014

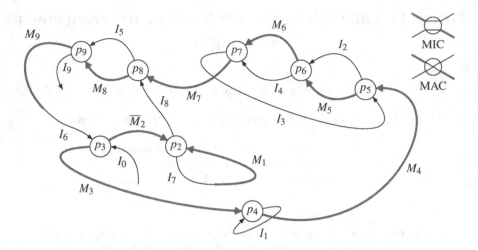

Fig. 1. Actin I gene of Sterkiella nova. Schematic diagram, based on [18]

Here we recall the 4-regular graph representation that represents both the MIC and MAC form of a gene. We recall how the MIC and MAC form of a gene are two different sets of circuits in a 4-regular graph, and may be obtained from one another by making different decisions where to continue a path at each vertex. Possible intermediate results of this recombination can also be read from this 4-regular graph. Based on this observation, Burns et al. [7] have proposed the *assembly polynomial*, a graph polynomial intended to abstract basic features of the assembly process, like the number of molecules excised during this process.

In this paper we show that the assembly polynomial is closely related to the following graph polynomials: *(i)* the *transition polynomial* by Jaeger [13, 15] and *(ii)* the *bracket polynomial* introduced for simple graphs by Traldi and Zulli [21]. We show how notions and results related to these polynomials may be carried over to the assembly polynomial. We note that the transfer of notions and results between other (related) graph polynomials have been accomplished, as can be seen in [11, 12].

Gene Assembly as 4-Regular Graphs

Prescott and Greslin [19] have unraveled the different representations of genetic material in the MIC and MAC form of genes in ciliates. We illustrate this using the example of the Actin I gene of Sterkiella nova, the presentation of which is based on [18]. In MIC form this gene consists of macronuclear destined sequences (MDSs for short), which end up in the MAC form of this gene, that are scrambled (their order may be permuted and some MDSs may be inverted). The segments in between the MDSs are called internal eliminated sequences (IESs for short) that are excised and do not appear in the MAC form of this gene.

The MIC of the Actin I gene can be written as $I_0 M_3 I_1 M_4 I_2 M_6 I_3 M_5 I_4 M_7 I_5 M_9$ $I_6 \overline{M}_2 I_7 M_1 I_8 M_8 I_9$, where M_i and I_j represent MDSs and IESs, respectively. The final MAC can be written as $M_1 M_2 \ldots M_8 M_9$. Both the MIC and MAC form of this gene can

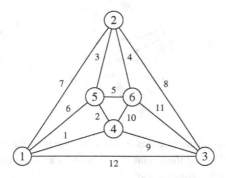

Fig. 2. Graph G_w defined by string $w = 145265123463$, see Example 1. The edges trace the Eulerian circuit C_w as defined by w according to the edge numbers given.

be read from the 4-regular graph G in Fig. 1. The vertices of G represent the *pointers*, i.e., the places where recombination takes place, and the edges represent the MDSs and IESs. Each vertex has two incoming and two outgoing edges. The MIC and MAC form both trace paths in the diagram. For the MIC form successive incoming and outgoing edges are chosen such that the two visits of a vertex do not cross, whereas for the MAC form the successive visits are connected in a crossing fashion (see top right illustration).

Without loss of generality, in the abstract context considered in the next sections, the initial edge and final edge are joined to form a single edge, in order to obtain an actual 4-regular graph.

2   Eulerian Circuits in 4-regular Graphs

The example in the introduction illustrates that 4-regular graphs are one of the key tools for describing the transformation of a MIC gene into its MAC gene. The MIC gene corresponds to an Eulerian circuit, while the MAC gene corresponds to a number of circuits (each edge belongs to one such circuit), such that one of those contains the MDSs in the proper order and orientation, while the other circuits contain only IESs that are excised during the gene assembly process. We abstract from genes and describe circuits in 4-regular graphs over an abstract alphabet (the elements of which, in effect, denote pointers).

A *double-occurrence string* over an alphabet V contains each letter from V exactly twice. Each double-occurrence string describes a 4-regular graph G_w together with an Eulerian circuit C_w for G_w, choosing vertex set V and edges that follow consecutive letters in w (in a cyclic fashion).

Example 1. Let $V = \{1, \ldots, 6\}$. For $w = 145265123463$, the 4-regular graph G_w is given in Fig. 2, where the edges are numbered according to the Eulerian circuit C_w. Thus, first introduce a vertex for each element in V. Then continue by reading w and adding (undirected) edges between consecutive letters (in a cyclic fashion): $1 \overset{1}{-} 4 \overset{2}{-} 5 \overset{3}{-} 2 \overset{4}{-} 6 \overset{5}{-} 5 \overset{6}{-} 1 \overset{7}{-} 2$, etc.   □

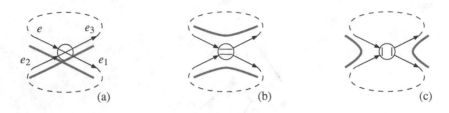

Fig. 3. Three ways to connect pairs of edges in a 4-regular graph relative to an (oriented) Eulerian circuit: (a) $\overset{e}{\to} v \overset{e_1}{\to}$ following the circuit, (b) $\overset{e}{\to} v \overset{e_3}{\to}$ in an orientation-consistent way, and (c) $\overset{e}{\to} v \overset{e_2}{\leftarrow}$ in an orientation-inconsistent way

We discuss here the basic theory of splitting and joining Eulerian circuits in 4-regular graphs, see [16] and also [14, Chapter 17]. Let G be a undirected 4-regular graph and C an Eulerian circuit of G. We assume here that circuits are not oriented.

Consider a *circuit partition* P of G, i.e., a set of circuits of G that together contain all the edges exactly once. We can describe P relative to C by considering at each vertex how the circuits in P follow the edges relative to C. In fact, when during a walk of P, P enters a vertex v, then P can leave in one of exactly three directions.[1] They can be classified as follows, cf. Fig. 3. Let $\dots \overset{e}{\to} v \overset{e_1}{\to} \dots \overset{e_2}{\to} v \overset{e_3}{\to} \dots$ be an arbitrary orientation of C. (*i*) If, by entering v via e in P, P leaves v via e_1, then we say that P *follows* C at v. (*ii*) If P leaves v via e_3, then we say that P is *orientation-consistent* with C at v, and finally (*iii*) if P leaves v via e_2, then we say that P is *orientation-inconsistent* with C at v. In [15], the first two are called *coherent* and the last one is called *anticoherent*. Moreover, in [7] the last two are called (parallel) *p-smoothing* and *n-smoothing*, respectively.

We let $D(C,P) = (D_1,D_2,D_3)$ be the ordered partition (we allow some D_i to be empty) of V such that the vertices of P that (*i*) follow C are in D_1, (*ii*) are orientation-consistent with C are in D_2, and (*iii*) are orientation-inconsistent with C are in D_3. (We return to this notion later, see Example 6 and Theorem 5.)

A *transition* at a vertex v is a partition in (unordered) pairs of the edges incident to v. A *transition system* of G is a set of transitions, one for each vertex of G. Note that a circuit partition uniquely determines a transition system. Indeed, if a circuit of P visits vertex v, entering and leaving at v via a pair of edges, then another visit of v (which may occur either in the same circuit or in another circuit of P) uses the remaining pair of edges. In the same way, a transition system uniquely determines a circuit partition.

Remark 2. In the process of gene assembly in ciliates, the MIC form of a gene traces an Eulerian circuit, while its corresponding MAC form has a different transition at each vertex (i.e., pointer). For each vertex v, this transition is uniquely determined by the relative orientation in the MIC form of the two MDSs with edges incident to v. When the MDSs have the same orientation, the orientation-consistent transition is taken; otherwise, the orientation-inconsistent transition is taken. In this way, an intermediate result

---

[1] In case G has loops one has to consider "half edges" to obtain three directions, but these technicalities are left to the reader.

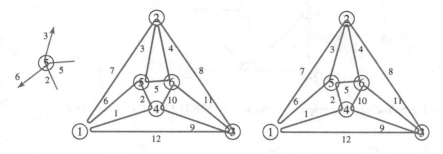

Fig. 4. Changing the Eulerian circuit of Fig. 2 at vertex 5 in orientation-consistent and orientation-inconsistent ways, respectively. The transition that follows C_w at vertex 5 is indicated on the left.

of the gene assembly process either follows the MIC or the MAC transition at each vertex. Thus in modelling the rearrangements in gene assembly in ciliates one has to keep track of the relative orientation of the MDS-segments.

In Fig. 3, we see that changing an Eulerian circuit at a single vertex by choosing the transition in an orientation-consistent manner, splits the circuit into two circuits. Choosing the orientation-inconsistent transition instead "inverts" part of the circuit.

Example 3. The Eulerian circuit C_w is given by $w = 145265123463$, cf. Example 1. By changing C_w at vertex 5 in an orientation-consistent way, we have the transitions $\frac{2}{}5\frac{6}{}$ and $\frac{5}{}5\frac{3}{}$, so we obtain two circuits 145123463 and 526. By recombining in an orientation-inconsistent way we have the transitions $\frac{2}{}5\frac{5}{}$ and $\frac{3}{}5\frac{6}{}$, so we obtain a single circuit 145625123463 with segment inverted, see Fig. 4. □

We noted that changing the transition of vertex v in an Eulerian circuit C to an orientation-consistent way splits the circuit in two. However, when we in parallel change transitions in an Eulerian circuit at distinct vertices u and v, both in an orientation-consistent way, then again an Eulerian circuit is formed, provided that u and v are "interlaced" in C, i.e., they occur in the order $\cdots u \cdots v \cdots u \cdots v \cdots$ (again, strings are regarded cyclic). In that case, two segments between the occurrences u and v are swapped. This is an important observation in the context of intra-molecular models of gene assembly. The interlacement of pairs of vertices is captured in an *interlace graph* (also called a *circle graph* as it can be defined using a collection of chords in a circle with edges denoting their intersection, see, e.g., [14, Chapter 17]).

The *interlace matrix* $I(C)$ of C is the $V \times V$-matrix (i.e., the rows and columns are not ordered, but indexed by V) over $GF(2)$ that has 1 at position (u, v) if vertices u and v are interlaced in C, i.e., occur in the order $\cdots u \cdots v \cdots u \cdots v \cdots$, and 0 otherwise. Since $I(C)$ is symmetric, $I(C)$ may be viewed as the adjacency matrix of a simple graph G, called the *interlace graph* of C. We are sloppy, and also use $I(C)$ as a notation for G.

Example 4. For C_w defined by $w = 145265123463$, the interlace graph $I(C_w)$ is given in Fig. 5. □

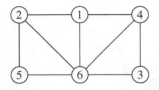

Fig. 5. The interlace graph $I(C_w)$ for $w = 145265123463$, cf. Example 4

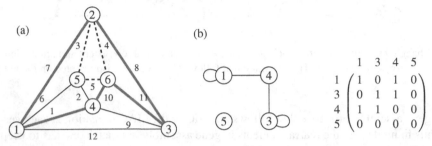

Fig. 6. Cf. Example 6. **(a)** Partition P of the edges of the 4-regular graph G_w into three closed walks; edge numbers according to C_w. **(b)** Both the graphical and matrix representation of the graph $(I(C_w) + \Delta(\{1,3\})) \setminus \{2,6\}$.

Cohn and Lempel [8] proved a surprisingly simple formula for the number of circuits that result from an Eulerian circuit C in a 4-regular graph by changing C in an orientation-consistent way at several vertices. The formula is in terms of the nullity (dimension of the null space) of the interlace matrix. A more general treatment of this result is given by Traldi [20], allowing one to change transitions in both orientation-consistent and orientation-inconsistent ways. Above we have borrowed the terminology of [20].

Traldi has shown that the number of circuits in P can be elegantly expressed in terms of the interlace matrix $I(C)$ and the partition $D(C,P)$. Let A be a $V \times V$-matrix. If $D \subseteq V$, then $A \setminus D$ denotes the restriction of A to the principal submatrix of A indexed by $V \setminus D$, and $\Delta(D)$ is the $V \times V$-matrix that has 1 only on diagonal elements (u,u) with $u \in D$, and 0 elsewhere. Also, ν and ρ denote the nullity and rank of matrices computed over GF(2), respectively.

Theorem 5 (Traldi-Cohn-Lempel). *Let G be an undirected, connected 4-regular graph with Euler cycle C, and let P be a circuit partition of $E(G)$, such that $D(C,P) = (D_1, D_2, D_3)$. Then $|P| = \nu((I(C) + \Delta(D_3)) \setminus D_1) + 1$.*

Example 6. We continue the running example. Fig. 6(a) shows a circuit partition P of G_w in three parts, with circuits 14632, 1345, and 265. At vertex 5, the circuits trace the transitions $\underline{}2\,5\,\underline{6}$ and $\underline{}5\,5\,\underline{3}$, which is orientation-consistent w.r.t. the circuit C_w. Hence 5 belongs to the second component of the partition $D(C_w,P)$. Considering all vertices, we find $D(C_w,P) = (\{2,6\},\{4,5\},\{1,3\})$. Graph $(I(C_w) + \Delta(\{1,3\})) \setminus \{2,6\}$ is obtained from the interlace graph $I(C_w)$ by deleting vertices 2,6 and adding loops to 1,3. It is given in Fig. 6(b). The corresponding matrix has (dimension 4×4, rank 2, and) nullity 2. Indeed $|P| = 2 + 1$ satisfies Theorem 5.   □

3 Graph Polynomials Motivated by Gene Rearrangement

One of the first graph polynomials is defined by Martin [17] for 4-regular graphs. Given such a graph G, the coefficient of y^k in its polynomial $M_G(y)$ equals the number of circuit partitions P of G with $|P| = k$. Currently, there is an impressive body of results on graph polynomials, see, e.g., the overview papers by Ellis-Monaghan and Merino [11, 12]. Typical topics that are studied are algebraic and combinatorial in nature, and include recursive formulations of the polynomials, and the interpretation of evaluations at specific values. Generalizations have been obtained for structures like knots and matroids [15]. Also, a multimatroid polynomial is proposed as a unified framework to several polynomials for graphs and matroids [5].

The study of gene rearrangements has motivated the introduction of new polynomials, most notably the *interlace* [1, 2] and *assembly* [7] polynomials, as, e.g., a feature that could measure and compare the complexity of the rearrangement process. Both these polynomials fit in the corpus of existing graph polynomials, and thus techniques and results can be carried over. For the interlace polynomial it has been shown that it is tied to the well-known Tutte polynomial [1, 5]. Here we discuss how the assembly polynomial is related to other known polynomials.

The Transition Polynomial and Its Relatives

It was observed by Jaeger [15] that several polynomials for 4-regular graphs are special cases of the *transition polynomial*, which is a multivariate/weighted polynomial. Similar as the Martin polynomial, the transition polynomial counts circuit partitions P w.r.t. an arbitrary Eulerian circuit C. However, the circuit partitions P have weights in the transition polynomial that depend on $D(C, P)$. Unfortunately, it seems that [15] is not widely distributed.

We show that the assembly polynomial is closely related to the transition polynomial, essentially by embedding its second variable into the weights. Secondly, we observe a less obvious relation to the bracket polynomial for graphs in terms of the nullity of the adjacency matrix of the circle graph using Theorem 5.

Burns et al. [7, Section 6] (see also [9]) define the assembly polynomial for a 4-regular graph G_w together with an Eulerian circuit C_w belonging to a double-occurrence string w over alphabet V. The *assembly polynomial* of G_w w.r.t. Eulerian circuit C_w is

$$S(G_w)(p,t) = \sum_s p^{\pi(s)} t^{c(s)-1},$$

where the sum is taken over all $2^{|V|}$ transition systems s that differ at each vertex from the transition system corresponding to C_w (a transition that differs from the transition of C_w is called a smoothing in [7]), $\pi(s)$ equals the number of orientation-consistent transitions of s w.r.t. C_w (called p-smoothings in [7]), and $c(s)$ equals the number of circuits in the circuit partition corresponding to s.

Example 7. Consider the double-occurrence string $w = 112323$. In Fig. 7 we show that its corresponding Eulerian circuit C_w in the 4-regular graph G_w (top-left) and its eight

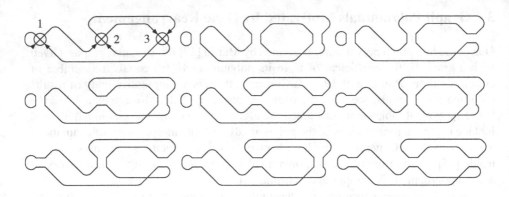

Fig. 7. Construction of G_w for $w = 112323$, and the eight smoothings that determine its assembly polynomial, see Example 7

possible smoothings (at each vertex, vertical connections correspond to a consistent-orientation change, i.e., a p-smoothing). Note that G_w has both a loop and parallel edges. The assembly polynomial equals $S(G_w) = p^3t + 2p^2t + p^2 + pt^2 + 2p + t$ (as can be verified using the Assembly Words online tool [6]).

We discuss the transition polynomial from [15]. Let $P(G)$ be the set of all transition systems of G. Note that $|P(G)| = 3^{|V|}$, where V is the set of vertices of G. A weight function W assigns a weight to each of the three possible transitions at each vertex. The weight $\omega(s)$ of the transition system s is the product of the weights of s at each vertex. The (weighted) *transition polynomial* of G is now defined as

$$q(G,W;x) = \sum_{s \in P(G)} \omega(s)x^{c(s)-1}.$$

We obtain the assembly polynomial as a special case of the transition polynomial by fixing the weights $W(C)$ relative to the Eulerian circuit C. Transitions that follow C have weight 0 (so are not counted at all), orientation-consistent and and orientation-inconsistent transitions have weight p and 1 respectively. We have $\omega(s) = p^{\pi(s)}$ if s contains no transitions that follow C, and $\omega(s) = 0$ otherwise. Hence $S(G_w)(p,t) = q(G_w, W(C_w);t)$.

The interlace polynomial on the other hand, is equal to a suitable generalisation of the transition polynomial where zero weight is assigned to the orientation-inconsistent transitions.

Various results are known for the transition polynomial. For example, let P be a circuit partition of G and $W(P)$ be such that the transitions belonging to P have weight 0, and the other transitions have weight 1. Then $q(G,W(P);-2) = (-1)^{|V|}(-2)^{c(P)-1}$, see [15, Proposition 11]. Consequently, $S(G_w)(1,-2) = (-1)^{|V|}$ since P is the Euler circuit C_w and $c(C_w) = 1$.

We now move to a second interpretation of the assembly polynomial. Consider a transition system s that never follows circuit C, as is relevant in this context. It determines a set of circuits P_s, and the number of circuits $c(s) = |P_s|$ is given by the Traldi-Cohn-Lempel formula, and equals $v(I(C_w) + \Delta(D_3)) + 1$, where $D(C, P_s) = (\varnothing, D_2, D_3)$. Recall that s determines $D(C, P_s)$ but also vice versa. Thus we reformulate the assembly polynomial of G_w w.r.t. C_w replacing summation over transition systems s by summation over partitions $(\varnothing, D_2, D_3)$. The notation $\dot\cup$ represents disjoint set union.

$$S(G_w)(p, t) = \sum_{D_2 \dot\cup D_3 = V} p^{|D_2|} t^{v(I(C_w) + \Delta(D_3))}.$$

In this formulation we recognize a related graph polynomial defined by Traldi and Zulli [21]. Let $\mathscr{A}(G)$ be the adjacency matrix of graph G. The *bracket polynomial* of a graph G is

$$[G](A, B, d) = \sum_{\Delta} A^{v(\Delta)} B^{\rho(\Delta)} d^{v(\mathscr{A}(G) + \Delta)}$$

with a summand for each $n \times n$ diagonal matrix $\Delta = \Delta(D)$ for some $D \subseteq V$.

Now let $D \subseteq V$, and set $D_3 = D$ and $D_2 = V \setminus D$. Then $v(\Delta(D)) = |V| - |D_3| = |D_2|$. Taking $A = p$, $B = 1$, and $d = t$ we see that the terms for the bracket and assembly polynomials match. To be precise, the assembly polynomial $S(G_w)$ of the 4-regular graph G_w defined by the double-occurrence string w is equal to the bracket polynomial $[I(C_w)](p, 1, t)$ of the interlace graph $I(C_w)$. As a consequence, when two double-occurrence strings have the same interlace graph, their 4-regular graphs have the same assembly polynomial, cf. [9, Proposition 3].

The transition polynomial allows for a straightforward recursive relation [15, Proposition 4], cf. [7, Lemma 6.4] for the case of the assembly polynomial. In fact, this recursive relation *characterizes* the transition polynomial. In a similar way, the bracket polynomial allows for a characteristic recursive relation [21], using a generalization of Euler circuit transformations called local complementation and edge complementation (which in turn is a special case of the general matrix operation of principal pivot transform [22]).

We hope these connections may be a starting point for transferring notions and results from one of these fields to another.

References

1. Aigner, M., van der Holst, H.: Interlace polynomials. Linear Algebra and its Applications 377, 11–30 (2004)
2. Arratia, R., Bollobás, B., Sorkin, G.B.: The interlace polynomial of a graph. Journal of Combinatorial Theory, Series B 92(2), 199–233 (2004)
3. Brijder, R., Daley, M., Harju, T., Jonoska, N., Petre, I., Rozenberg, G.: Computational nature of gene assembly in ciliates. In: Rozenberg, G., Bäck, T., Kok, J. (eds.) Handbook of Natural Computing, vol. 3, pp. 1233–1280. Springer (2012)
4. Brijder, R., Hoogeboom, H.J.: The algebra of gene assembly in ciliates. In: Jonoska, N., Saito, M. (eds.) Discrete and Topological Models in Molecular Biology. Natural Computing Series, pp. 289–307. Springer, Heidelberg (2014)
5. Brijder, R., Hoogeboom, H.J.: Interlace polynomials for multimatroids and delta-matroids. European Journal of Combinatorics [arXiv:1010.4678] (to appear, 2014)

6. Burns, J., Dolzhenko, E.: Assembly words (properties), http://knot.math.usf.edu/assembly/properties.html (visited March 2014)
7. Burns, J., Dolzhenko, E., Jonoska, N., Muche, T., Saito, M.: Four-regular graphs with rigid vertices associated to DNA recombination. Discrete Applied Mathematics 161(10-11), 1378–1394 (2013)
8. Cohn, M., Lempel, A.: Cycle decomposition by disjoint transpositions. Journal of Combinatorial Theory, Series A 13(1), 83–89 (1972)
9. Dolzhenko, E., Valencia, K.: Invariants of graphs modeling nucleotide rearrangements. In: Jonoska, N., Saito, M. (eds.) Discrete and Topological Models in Molecular Biology. Natural Computing Series, pp. 309–323. Springer, Heidelberg (2014)
10. Ehrenfeucht, A., Harju, T., Petre, I., Prescott, D.M., Rozenberg, G.: Computation in Living Cells – Gene Assembly in Ciliates. Springer (2004)
11. Ellis-Monaghan, J.A., Merino, C.: Graph polynomials and their applications I: The Tutte polynomial. In: Dehmer, M. (ed.) Structural Analysis of Complex Networks, pp. 219–255. Birkhäuser, Boston (2011)
12. Ellis-Monaghan, J.A., Merino, C.: Graph polynomials and their applications II: Interrelations and interpretations. In: Dehmer, M. (ed.) Structural Analysis of Complex Networks, pp. 257–292. Birkhäuser, Boston (2011)
13. Ellis-Monaghan, J.A., Sarmiento, I.: Generalized transition polynomials. Congressus Numerantium 155, 57–69 (2002)
14. Godsil, C., Royle, G.: Algebraic Graph Theory. Springer (2001)
15. Jaeger, F.: On transition polynomials of 4-regular graphs. In: Hahn, G., Sabidussi, G., Woodrow, R. (eds.) Cycles and Rays. NATO ASI Series, vol. 301, pp. 123–150. Kluwer (1990)
16. Kotzig, A.: Eulerian lines in finite 4-valent graphs and their transformations. In: Theory of graphs, Proceedings of the Colloquium, Tihany, Hungary, pp. 219–230. Academic Press, New York (1968)
17. Martin, P.: Énumérations eulériennes dans les multigraphes et invariants de Tutte-Grothendieck. PhD thesis, Institut d'Informatique et de Mathématiques Appliquées de Grenoble (IMAG) (1977),
http://tel.archives-ouvertes.fr/tel-00287330_v1/
18. Prescott, D.M.: Genome gymnastics: Unique modes of DNA evolution and processing in ciliates. Nature Reviews 1, 191–199 (2000)
19. Prescott, D.M., Greslin, A.F.: Scrambled actin I gene in the micronucleus of Oxytricha nova. Developmental Genetics 13, 66–74 (1992)
20. Traldi, L.: Binary nullity, Euler circuits and interlace polynomials. European Journal of Combinatorics 32(6), 944–950 (2011)
21. Traldi, L., Zulli, L.: A bracket polynomial for graphs. I. Journal of Knot Theory and Its Ramifications 18(12), 1681–1709 (2009)
22. Tsatsomeros, M.J.: Principal pivot transforms: properties and applications. Linear Algebra and its Applications 307(1-3), 151–165 (2000)

On the Equivalence of Automata
for KAT-expressions*

Sabine Broda, António Machiavelo, Nelma Moreira, and Rogério Reis

CMUP & DM-DCC, Faculdade de Ciências da Universidade do Porto,
Porto, Portugal
{sbb,nam,rvr}@dcc.fc.up.pt, ajmachia@fc.up.pt

Abstract. Kleene algebra with tests (KAT) is a decidable equational
system for program verification that uses both Kleene and Boolean algebras. In spite of KAT 's elegance and success in providing theoretical
solutions for several problems, not many efforts have been made towards
obtaining tractable decision procedures that could be used in practical
software verification tools. The main drawback of the existing methods
relies on the explicit use of all possible assignments to boolean variables.
Recently, Silva introduced an automata model that extends Glushkov's
construction for regular expressions. Broda et al. extended also Mirkin's
equation automata to KAT expressions and studied the state complexity of both algorithms. Contrary to other automata constructions from
KAT expressions, these two constructions enjoy the same descriptional
complexity behaviour as their counterparts for regular expressions, both
in the worst case as well as in the average case. In this paper, we generalize, for these automata, the classical methods of subset construction for
nondeterministic finite automata, and the Hopcroft and Karp algorithm
for testing deterministic finite automata equivalence. As a result, we obtain a decision procedure for KAT equivalence where the extra burden
of dealing with boolean expressions avoids the explicit use of all possible
assignments to the boolean variables. Finally, we specialize the decision
procedure for testing KAT expressions equivalence without explicitly constructing the automata, by introducing a new notion of derivative and a
new method of constructing the equation automaton.

Keywords: Kleene algebra with tests, automata, equivalence, derivative.

1 Introduction

Kleene algebra with tests (KAT) [12] is an equational system that extends Kleene
algebra (KA), the algebra of regular expressions, and that is specially suited to
capture and verify properties of simple imperative programs. In particular, it
subsumes the propositional Hoare logic which is a formal system for the specification and verification of programs, and that is, currently, the base of most

* This work was funded by the European Regional Development Fund through the
programme COMPETE and by the FCT under projects PEst-C/MAT/UI0144/2013
and FCOMP-01-0124-FEDER-020486.

A. Beckmann, E. Csuhaj-Varjú, and K. Meer (Eds.): CiE 2014, LNCS 8493, pp. 73–83, 2014.

of the tools for checking program correctness. The equational theory of KAT is PSPACE-complete and can be reduced to the equational theory of KA, with an exponential cost [9, 15]. Regular sets of guarded strings are standard models for KAT (as regular languages are for KA). The decidability, conciseness and expressiveness of KAT motivated its recent automatization within several theorem provers [4, 17, 18] and functional languages [3]. Those implementations use (variants of) the coalgebraic automaton on guarded strings developed by Kozen [14]. In this approach, derivatives are considered over symbols of the from αp, where p is an alphabetic symbol (program) and α a valuation of boolean variables (the *guard*, normally called atom). This induces an exponential blow-up on the number of states or transitions of the automata and an accentuated exponential complexity when testing the equivalence of two KAT expressions. Recently, Silva [19] introduced an automata model for KAT expressions that extends Glushkov's construction for regular expressions. In this automaton, transitions are labeled with KAT expressions of the form bp, where b is a boolean expression (and not an atom) and p an alphabetic symbol. Using similar ideas, Broda et al. [6] extended the Mirkin's equation automata to KAT expressions and studied the state complexity of both algorithms. Contrary to other automata constructions for KAT expressions, these two constructions enjoy the same descriptional complexity behaviour as their counterparts for regular expressions, both in the worst-case as well as in the average-case. In this paper, we generalize, for these automata, the classical methods of subset construction for nondeterministic finite automata, and the Hopcroft and Karp algorithm for testing deterministic finite automata equivalence. As a result, we obtain a decision procedure for KAT equivalence where the extra burden of dealing with boolean expressions avoids the explicit use of all possible assignments to the boolean variables. Finally, we specialize the decision procedure for testing KAT expressions equivalence without explicitly constructing the automata, by introducing a new notion of derivative and a new method of constructing the equation automaton.

Due to limited number of pages, proofs of lemmas and propositions can be found in the extended version of this paper [7], which is available online.

2   KAT Expressions, Automata, and Guarded Strings

Let $P = \{p_1, \ldots, p_k\}$ be a non-empty set, usually referred to as the set of *program* symbols, and $T = \{t_0, \ldots, t_{l-1}\}$ be a non-empty set of *test* symbols. The set of boolean expressions over T together with negation, disjunction and conjunction, is denoted by BExp, and the set of KAT expressions with disjunction, concatenation, and Kleene star, by Exp. The abstract syntax of KAT expressions, over an alphabet $P \cup T$, is given by the following grammar, where $p \in P$ and $t \in T$,

$$\text{BExp}: \quad b \rightarrow 0 \mid 1 \mid t \mid \neg b \mid b + b \mid b \cdot b$$
$$\text{Exp}: \quad e \rightarrow p \mid b \mid e + e \mid e \cdot e \mid e^\star.$$

As usual, we will omit the operator $\cdot$ whenever it does not give rise to any ambiguity. For the negation of test symbols we frequently use $\bar{t}$ instead of $\neg t$.

The set At, of *atoms* over T, is the set of all boolean assignments to all elements of T, $At = \{x_0 \cdots x_{l-1} \mid x_i \in \{t_i, \bar{t}_i\}, \ t_i \in T\}$. Each atom $\alpha \in At$ has associated a binary word of l bits $(w_0 \cdots w_{l-1})$ where $w_i = 0$ if $\bar{t}_i \in \alpha$, and $w_i = 1$ if $t_i \in \alpha$.

Now, the set of *guarded strings* over P and T is $GS = (At \cdot P)^* \cdot At$. Regular sets of guarded strings form the standard language-theoretic model for KAT [13]. For $x = \alpha_1 p_1 \cdots p_{m-1} \alpha_m, y = \beta_1 q_1 \cdots q_{n-1} \beta_n \in GS$, where $m, n \geq 1$, $\alpha_i, \beta_j \in At$ and $p_i, q_j \in P$, we define the *fusion product* $x \diamond y = \alpha_1 p_1 \cdots p_{m-1} \alpha_m q_1 \cdots q_{n-1} \beta_n$, if $\alpha_m = \beta_1$, leaving it undefined, otherwise. For sets $X, Y \subseteq GS$, $X \diamond Y$ is the set of all $x \diamond y$ such that $x \in X$ and $y \in Y$. Let $X^0 = At$ and $X^{n+1} = X \diamond X^n$, for $n \geq 0$. Given a KAT expression e, we define $GS(e) \subseteq GS$ inductively as follows:

$$GS(p) = \{\alpha p \beta \mid \alpha, \beta \in At\} \qquad\qquad GS(e_1 + e_2) = GS(e_1) \cup GS(e_2)$$
$$GS(b) = \{\alpha \mid \alpha \in At \wedge \alpha \leq b\} \qquad GS(e_1 \cdot e_2) = GS(e_1) \diamond GS(e_2)$$
$$GS(e_1^*) = \cup_{n \geq 0} GS(e_1)^n,$$

where $\alpha \leq b$ if $\alpha \to b$ is a propositional tautology. For $E \subseteq Exp$, let $GS(E) = \bigcup_{e \in E} GS(e)$. Given two KAT expressions e_1 and e_2, we say that they are *equivalent*, and write $e_1 = e_2$, if $GS(e_1) = GS(e_2)$.

A *(nondeterministic) automaton with tests* (NTA) over the alphabets P and T is a tuple $\mathcal{A} = \langle S, s_0, o, \delta \rangle$, where S is a finite set of states, $s_0 \in S$ is the initial state, $o : S \to BExp$ is the output function, and $\delta \subseteq 2^{S \times (BExp \times P) \times S}$ is the transition relation. We denote by $BExp_{\mathcal{A}}$ the set of boolean expressions that occur in δ. In general, we assume that there are no transitions $(s, (b, p), s') \in \delta$ such that b is not satisfiable.

A guarded string $\alpha_1 p_1 \ldots p_{n-1} \alpha_n$, with $n \geq 1$, is accepted by the automaton $\mathcal{A}$ if and only if there is a sequence of states $s_0, s_1, \ldots, s_{n-1} \in S$, where s_0 is the initial state, and, for $i = 1, \ldots, n-1$, one has $\alpha_i \leq b_i$ for some $(s_{i-1}, (b_i, p_i), s_i) \in \delta$, and $\alpha_n \leq o(s_{n-1})$. The set of all guarded strings accepted by $\mathcal{A}$ is denoted by $GS(\mathcal{A})$. Formally, given an NTA $\mathcal{A} = \langle S, s_0, o, \delta \rangle$, one can naturally associate to the transition relation $\delta \subseteq 2^{S \times (BExp \times P) \times S}$ a function $\delta' : S \times (At \cdot P) \longrightarrow 2^S$, defined by $\delta'(s, \alpha p) = \{s' \mid (s, (b, p), s') \in \delta, \ \alpha \leq b\}$. Moreover, one can define a function $\hat{\delta} : S \times GS \longrightarrow \{0, 1\}$ over pairs of states and guarded strings as follows

$$\hat{\delta}(s, \alpha) = \begin{cases} 1 \text{ if } \alpha \leq o(s), \\ 0 \text{ otherwise,} \end{cases} \qquad \hat{\delta}(s, \alpha p x) = \sum_{s' \in \delta'(s, \alpha p)} \hat{\delta}(s', x).$$

Given a state s, $GS(s) = \{x \in GS \mid \hat{\delta}(s, x) = 1\}$ is the set of guarded strings accepted by s, and $GS(\mathcal{A}) = GS(s_0)$. We say that a KAT expression $e \in Exp$ is *equivalent* to an automaton $\mathcal{A}$, and write $e = \mathcal{A}$, if $GS(\mathcal{A}) = GS(e)$.

Example 1. Given the KAT expression $e = t_1 p(pq^* t_2 + t_3 q)^*$, an equivalent NTA $\mathcal{A}$, obtained by the equation algorithm (see [6]), is the following, where $e_0 = e$, $e_1 = (pq^* t_2 + t_3 q)^*$ and $e_2 = q^* t_2 (pq^* t_2 + t_3 q)^*$. Both objects accept, for instance, the guarded string $t_1 t_2 t_3 p t_1 t_2 t_3 p t_1 t_2 t_3 q t_1 t_2 t_3$.

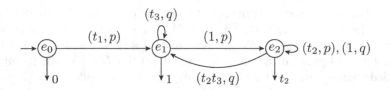

An NTA is called *deterministic* (DTA) if and only if for every pair $(\alpha, p) \in$ At $\times$ P and every state $s \in S$, there is at most one transition $(s, (b, p), s') \in \delta$ such that $\alpha \leq b$, i.e. $\delta'(s, \alpha p)$ is either empty or a singleton.

3   Determinization

The standard subset construction for converting a nondeterministic finite automaton (NFA) into an equivalent deterministic finite automaton (DFA) may be adapted as follows. Given a set of states $X \subseteq S$, whenever there are transitions $(s_1, (b_1, p), s_1'), \ldots, (s_m, (b_m, p), s_m')$, with $s_1, \ldots, s_m \in X$, in the NTA, in the equivalent DTA we consider "disjoint" transitions to subsets $\{s_{i_1}', \ldots, s_{i_k}'\} \subseteq \{s_1', \ldots, s_m'\}$, labeled by $(b_{i_1} \cdots b_{i_k} \neg b_{i_{k+1}} \cdots \neg b_{i_m})p$, where $\{s_{i_{k+1}}', \ldots, s_{i_m}'\} = \{s_1', \ldots, s_m'\} \setminus \{s_{i_1}', \ldots, s_{i_k}'\}$.

Consider the set of atoms At $= \{\alpha_0, \ldots, \alpha_{2^l-1}\}$, with the natural order induced by their binary representation. We define the function

$$V : \mathsf{BExp} \longrightarrow 2^{\{0, \ldots, 2^l-1\}}$$
$$b \longmapsto V_b = \{\, i \mid \alpha_i \leq b, 0 \leq i \leq 2^l - 1 \,\}.$$

This representation of boolean expressions is such that $V_b = V_{b'}$ if and only if b and b' are logically equivalent expressions. We consider V_b as a canonical representation of b and write $\alpha_i \leq V_b$ if and only if $i \in V_b$. Conversely, to each $U \subseteq \{0, \ldots, 2^l - 1\}$ we associate a unique boolean expression $B(U)$, where

$$B : 2^{\{0, \ldots, 2^l-1\}} \longrightarrow \mathsf{BExp}$$
$$U \longmapsto \sum_{i \in U} \alpha_i.$$

For $b, b' \in \mathsf{BExp}$ we have $V_{\neg b} = \overline{V_b}$, $V_{b+b'} = V_b \cup V_{b'}$ and $V_{b \cdot b'} = V_b \cap V_{b'}$, where $\overline{U} = \{0, \ldots, 2^l - 1\} \setminus U$ for any $U \subseteq \{0, \ldots, 2^l - 1\}$.

Example 2. For T $= \{t_1, t_2\}$ and At $= \{\bar{t}_1 \bar{t}_2, \bar{t}_1 t_2, t_1 \bar{t}_2, t_1 t_2\}$, we have $V_{t_2} = \{1, 3\}$ and $V_{t_1 + \neg t_2} = \{0, 1, 3\}$. Also, $B(\{1, 3\}) = t_1 t_2 + \bar{t}_1 t_2$.

We now describe the subset construction that, given an NTA, $\mathcal{A} = \langle S, s_0, o, \delta \rangle$ over the alphabets P and T, produces an DTA, $\mathcal{A}_{\mathsf{det}} = \langle 2^S, \{s_0\}, o_{\mathsf{det}}, \delta_{\mathsf{det}} \rangle$ over P and T, such that $GS(\mathcal{A}) = GS(\mathcal{A}_{\mathsf{det}})$. First, we define two functions

$$\tilde{\delta}_{\mathsf{det}} : 2^S \times (2^{\{0, \ldots, 2^l-1\}} \times P) \longrightarrow 2^S \quad \text{and} \quad \tilde{o}_{\mathsf{det}} : 2^S \longrightarrow 2^{\{0, \ldots, 2^l-1\}}.$$

Then, we take $\delta_{\mathsf{det}} = \{\, (X, (\mathsf{B}(V), p), Y) \mid (X, (V, p), Y) \in \tilde{\delta}_{\mathsf{det}} \,\}$ as well as $o_{\mathsf{det}} = \mathsf{B} \circ \tilde{o}_{\mathsf{det}}$. For $X \subseteq S$, we define $\tilde{o}_{\mathsf{det}}(X) = \bigcup_{s \in X} V_{o(s)}$. To define $\tilde{\delta}_{\mathsf{det}}$, we consider the following sets. Given $X \subseteq S$ and $p \in \mathsf{P}$, Let

$$\Gamma(X, p) = \{\, (b, s') \mid (s, (b, p), s') \in \delta, \; s \in X \,\},$$
$$\Delta(X, p) = \{\, s' \mid (b, s') \in \Gamma(X, p) \,\}.$$

For $s' \in \Delta(X, p)$, we define $\mathsf{V}_{X,p,s'} = \bigcup \{\, \mathsf{V}_b \mid (b, s') \in \Gamma(X, p) \,\}$, And for each $Y \subseteq \Delta(X, p)$ the set

$$\mathsf{V}_{X,p,Y} = \bigcap (\{\, \mathsf{V}_{X,p,s'} \mid s' \in Y \,\} \cup \{\, \overline{\mathsf{V}}_{X,p,s'} \mid s' \in \Delta(X, p) \setminus Y \,\}).$$

Finally, we have

$$\tilde{\delta}_{\mathsf{det}} = \{\, (X, (\mathsf{V}_{X,p,Y}, p), Y) \mid X \subseteq S, \; Y \subseteq \Delta(X, p), \; p \in \mathsf{P}, \; \mathsf{V}_{X,p,Y} \neq \emptyset \,\}.$$

Proposition 1. *For every NTA $\mathcal{A} = \langle S, s_0, o, \delta \rangle$, the automaton $\mathcal{A}_{\mathsf{det}}$ is deterministic and $\mathsf{GS}(\mathcal{A}) = \mathsf{GS}(\mathcal{A}_{\mathsf{det}})$.*

Example 3. Applying the construction above to the NTA from Example 1, we obtain the following DTA:

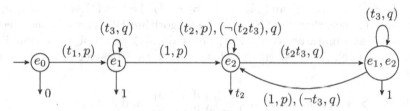

3.1   Implementation and Complexity

It is important to notice that in the determinization algorithm, the construction of all the $2^{|S|}$ subsets X of the set of states S can be avoided by considering only reachable states from the initial state. In order to efficiently deal with boolean operations it is essential to have an adequate representation for the boolean expressions b as well as the sets V_b. A possible choice is to use OBDDs (ordered binary decision diagrams), for which there are several software packages available. The sets $\mathsf{V}_{X,p,Y}$ may also be constructed using a (variant) of the standard Quine-McCluskey algorithm.

In the worst case, the determinization algorithm exhibits an extra exponential complexity to compute the sets $\mathsf{V}_{X,p,Y}$. The deterministic automaton $\mathcal{A}_{\mathsf{det}}$ has at most 2^n states and $k \cdot 2^{n+l}$ transitions where $n = |S|$, $l = |\mathsf{T}|$, and $k = |\mathsf{P}|$. Contrary to what happens with other KAT automata where the set At is explicitly used, in practice and with adequate data structures, we can expect that the number of sets $X \subseteq S$ and of sets $\mathsf{V}_{X,p,Y}$ is kept within tractable limits. It is an open problem to theoretically obtain the average-case complexity of both the power set construction and the sets $\mathsf{V}_{X,p,Y}$.

4   Equivalence of Deterministic Automata

Hopcroft and Karp [11] presented an almost linear algorithm (HK) for testing the equivalence of two DFAs that avoids their minimization. Considering the merge of the two DFAs as a single one, the algorithm computes the finest right-invariant relation, on the set of states, that makes the initial states equivalent. Recently this algorithm was analyzed and extended to NFAs in [2, 5]. In this section, we extend it, again, for testing equivalence of deterministic automata for guarded strings. We will only consider DTAs, $\mathcal{A} = \langle S, s_0, o, \delta \rangle$, where all states are useful, i.e. for every state $s \in S$, $GS(s) \neq \emptyset$.

Given a DTA, $\mathcal{A} = \langle S, s_0, o, \delta \rangle$, over the alphabets P and T, and $s, t \in S$, we say that s and t are *equivalent*, and write $s \approx t$, if $GS(s) = GS(t)$. A binary relation R on S is *right invariant* if for all $s, t \in S$ if sRt then the following conditions hold:

- $\forall \alpha \in At, \ \alpha \leq o(s) \Leftrightarrow \alpha \leq o(t)$;
- $\forall \alpha p \in At \cdot P, \ (\delta'(s, \alpha p) = \delta'(t, \alpha p) = \emptyset)$ or $(\delta'(s, \alpha p) = \{s'\}, \delta'(t, \alpha p) = \{t'\}$ and $s'Rt')$.

It is easy to see that the relation $\approx$ is right invariant. Furthermore, whenever R is a right-invariant relation on S and sRt, for $s, t \in S$, one has $s \approx t$.

Let $\mathcal{A}_1 = \langle S_1, s_0, o_1, \delta_1 \rangle$ and $\mathcal{A}_2 = \langle S_2, r_0, o_2, \delta_2 \rangle$ be two DTAs over the alphabets P and T, such that $S_1 \cap S_2 = \emptyset$. The algorithm HK, given below, decides if these two automata are equivalent, i.e. if $GS(\mathcal{A}_1) = GS(\mathcal{A}_2)$, by building a right-invariant relation that checks whether $s_0 \approx r_0$. Consider $\mathcal{A} = \langle S, s_0, o, \delta \rangle$, where

$$S = S_1 \cup S_2, \quad o(s) = \begin{cases} o_1(s) \text{ if } s \in S_1 \\ o_2(s) \text{ if } s \in S_2 \end{cases} \text{ and } \delta = \delta_1 \cup \delta_2.$$

Lemma 1. *Given two DTAs, $\mathcal{A}_1$ and $\mathcal{A}_2$, let $\mathcal{A}$ be defined as above. Then, $s_0 \approx r_0$ (in $\mathcal{A}$) if and only if $GS(\mathcal{A}_1) = GS(\mathcal{A}_2)$.*

The algorithm uses an initially empty stack H and a set partition $\mathcal{P}$ of S, which are both updated during the computation. The set partition $\mathcal{P}$ is built using the UNION-FIND data structure [10]. Within this structure, three functions are defined:

- MAKE(i): creates a new set (singleton) for one element i (the identifier);
- FIND(i): returns the identifier S_i of the set which contains i;
- UNION(i, j, k): combines the sets identified by i and j into a new set $S_k = S_i \cup S_j$; S_i and S_j are destroyed.

An arbitrary sequence of i operations MAKE, UNION, and FIND, j of which are MAKE operations, necessary to create the required sets can, in worst-case, be performed in $O(i\alpha(j))$ time, where $\alpha(j)$ is related to a functional inverse of the Ackermann function, and, as such, grows very slowly, and for practical uses can be considered a constant. In the whole we assume that whenever FIND(i) fails, MAKE(i) is called.

Algorithm 1.1. HK algorithm for deterministic automata.

```
1   def HK(A₁, A₂):
2     MAKE(s₀);  MAKE(r₀)
3     H = ∅
4     UNION(s₀, r₀, r₀);  PUSH(H, (s₀, r₀))
5     while (s, t) = POP(H):
6       if V_o(s) ≠ V_o(t):  return False
7       for p ∈ P:
8         B₁ = Γ({s}, p)
9         B₂ = Γ({t}, p)
10        if  ⋃      V_b₁  =  ⋃      V_b₂:
              (b₁,_)∈B₁        (b₂,_)∈B₂
11            for (b₁, s') ∈ B₁:
12              for (b₂, t') ∈ B₂:
13                if V_b₁ ∩ V_b₂ ≠ ∅:
14                  s' = FIND(s')
15                  t' = FIND(t')
16                  if s' ≠ t':
17                    UNION(s', t', t')
18                    PUSH(H, (s', t'))
19        else: return False
20    return True
```

The algorithm terminates because every time it pushes a pair onto the stack it performs a union of two disjoint sets in the partition, and this can be done at most $|S| - 1$ times. Given that set operations introduce what can be considered a constant time factor, the worst-case running time of the algorithm HK is $O(m^2 kn)$, where $n = |S|$, $k = |P|$, and $m = |\mathsf{BExp}_\mathcal{A}|$. The correctness of this version of algorithm HK is given by the proposition below, whose proof follows closely the one for DFAs.

Proposition 2. *The algorithm returns* True *if and only if* $s_0 \approx r_0$.

5   Equivalence of Nondeterministic Automata

We can embed the determinization process directly into the HK algorithm, extending it, so that it can be used to test the equivalence of NTAs. As before, given two NTAs with disjoint sets of states, $\mathcal{A}_1$ and $\mathcal{A}_2$, we consider them as a single NTA, $\mathcal{A} = \langle S, s_0, o, \delta \rangle$. The worst-case complexity of this combined algorithm, which we denote by HKN, is consequently $O(m^2 k 2^n)$, where $n = |S|$, $k = |P|$ and $m = |\mathsf{BExp}_\mathcal{A}|$.

Algorithm 1.2. HKN algorithm for nondeterministic automata.

```
1   def HKN(A₁, A₂):
2       MAKE({s₀});  MAKE({r₀})
3       H = ∅
4       UNION({s₀}, {r₀}, {r₀});  PUSH(H, ({s₀}, {r₀}))
5       while (X, Y) = POP(H):
6           if õ_det(X) ≠ õ_det(Y): return False
7           for p ∈ P:
8               B₁ = Γ(X, p)
9               B₂ = Γ(Y, p)
10              if  ⋃      V_{b₁} =  ⋃      V_{b₂}:
                  (b₁,_)∈B₁       (b₂,_)∈B₂
11                  for X' ⊆ Δ(X, p):
12                      for Y' ⊆ Δ(Y, p):
13                          if V_{X,p,X'} ∩ V_{Y,p,Y'} ≠ ∅:
14                              X' = FIND(X')
15                              Y' = FIND(Y')
16                              if X' ≠ Y':
17                                  UNION(X', Y', Y')
18                                  PUSH(H, (X', Y'))
19              else: return False
20      return True
```

6   Equivalence of **KAT** Expressions

Given two KAT expressions, e_1 and e_2, their equivalence can be tested by first converting each expression to an equivalent NTA and then, either by determinizing both and applying the HK algorithm (Section 4), or by directly using the resulting NTAs in algorithm HKN (Section 5). In particular, we could use the equation construction given in [6] to obtain NTAs equivalent to the given KAT expressions and then apply the HKN algorithm. The equation automaton for KAT expressions is an adaptation of Mirkin's construction [16] for regular expressions. Given $e_0 \equiv e \in$ Exp, a set of KAT expressions $\pi(e) = \{e_1, \ldots, e_n\}$ is defined inductively by $\pi(p) = \{1\}$, $\pi(b) = \emptyset$, $\pi(e + f) = \pi(e) \cup \pi(f)$, $\pi(e \cdot f) = \pi(e)f \cup \pi(f)$, and $\pi(e^*) = \pi(e)e^*$. This set satisfies the following system of equations $e_i = \sum_{j=1}^{n} b_{ij1}p_1e_j + \cdots + \sum_{j=1}^{n} b_{ijk}p_ke_j +$ out(e_i) for $i = 0, \ldots, n$, $p_r \in$ P, $k = |$P$|$, some $b_{ijr} \in$ BExp, and where function out is defined below. The equation automaton is $\mathcal{A}_{eq}(e) = \langle \{e\} \cup \pi(e), e, \text{out}, \delta_{eq} \rangle$ with $\delta_{eq} = \{ (e_i, (b_{ijr}, p_r), e_j) \mid$ if $b_{ijr}p_re_j$ is a component of the equation for $e_i \}$.

In this section, we will use this construction to define an algorithm for testing equivalence of KAT expressions by a recursive computation of their derivatives without explicitly building any automaton. However, the correctness of this procedure is justified by the correctness of the equation automaton. We first present a slightly different formalization of this automaton construction, which is more adequate for our purposes. The construction resembles the partial derivative automaton for regular expressions (that is known to be identical to the Mirkin automaton [8]). The resulting decision procedure for KAT equivalence is also similar to the ones recently presented for regular expressions (see [1, 2]) and can be seen as a *syntactic* (and more *compact*) version of the one presented by Kozen [14].

For $e \in \mathsf{Exp}$ and a program symbol $p \in \mathsf{P}$, the set $\partial_p(e)$ of *partial derivatives* of e w.r.t. p is inductively defined as follows:

$$\partial_p : \mathsf{Exp} \longrightarrow \mathsf{BExp} \times \mathsf{Exp}$$

$$\partial_p(p') = \begin{cases} \{(1,1)\} & \text{if } p' \equiv p \\ \emptyset & \text{otherwise} \end{cases} \qquad \begin{aligned} \partial_p(e + e') &= \partial_p(e) \cup \partial_p(e') \\ \partial_p(ee') &= \partial_p(e)e' \cup \mathsf{out}(e)\partial_p(e') \\ \partial_p(e^\star) &= \partial_p(e)e^\star, \end{aligned}$$

$$\partial_p(b) = \emptyset$$

where $\mathsf{out} : \mathsf{Exp} \longrightarrow \mathsf{BExp}$ is defined by

$$\begin{aligned} \mathsf{out}(p) &= 0 & \mathsf{out}(e_1 + e_2) &= \mathsf{out}(e_1) + \mathsf{out}(e_2) \\ \mathsf{out}(b) &= b & \mathsf{out}(e_1 \cdot e_2) &= \mathsf{out}(e_1) \cdot \mathsf{out}(e_2) \end{aligned} \qquad \mathsf{out}(e^\star) = 1,$$

and for $R \subseteq \mathsf{BExp} \times \mathsf{Exp}$, $e \in \mathsf{Exp}$, and $b \in \mathsf{BExp}$, $Re = \{ (b', e'e) \mid (b', e') \in R \}$ and $bR = \{ (bb', e') \mid (b', e') \in R \}$. We also define $\Delta_p(e) = \{ e' \mid (b, e') \in \partial_p(e) \}$. The functions ∂_p, out, and Δ_p are naturally extended to sets $X \subseteq \mathsf{Exp}$. Moreover we define the KAT expression $\sum \partial_p(e) \equiv \sum_{(b_i,e_i) \in \partial_p(e)} b_i p e_i$.

Example 4. The states of the equation automaton in Example 1 satisfy the following system of equations:

$$\begin{aligned} e_0 &= t_1 p e_1 \\ e_1 &= 1 p e_2 + t_3 q e_1 & + 1 \\ e_2 &= t_2 p e_2 + t_2 t_3 q e_1 + 1 q e_2 + t_2. \end{aligned}$$

For instance, note that $\partial_p(e_2) = \{(t_2, e_2)\}$, $\partial_q(e_2) = \{(t_2 t_3, e_1), (1, e_2)\}$, $\mathsf{out}(e_2) = t_2$, and $e_2 = \sum \partial_p(e_2) + \sum \partial_q(e_2) + \mathsf{out}(e_2)$.

Given $e \in \mathsf{Exp}$, we define the *partial derivative* automaton $\mathcal{A}_{\mathsf{pd}}(e) = \langle \{e\} \cup \pi(e), e, \delta_{\mathsf{pd}}, \mathsf{out}(e) \rangle$ where $\delta = \{ (e_1, (b, p), e_2) \mid p \in \mathsf{P}, (b, e_2) \in \partial_p(e_1) \}$.

Lemma 2. *For* $e \in \mathsf{Exp}$, $\mathcal{A}_{\mathsf{pd}}(e)$ *and* $\mathcal{A}_{\mathsf{eq}}(e)$ *are identical.*

Proposition 3. $\mathsf{GS}(\mathcal{A}_{\mathsf{pd}}(e)) = \mathsf{GS}(e)$.

Now, it is easy to see that we can define a procedure, HKK, that directly tests the equivalence of any given two KAT expressions. For that, it is enough to modify HKN by taking $(\{e_1\}, \{e_2\})$ as the initial pair and, for $X \subseteq \mathsf{Exp}$, $\tilde{o}_{\mathsf{det}}(X) = \mathsf{out}(X)$, $\Gamma(X, p) = \partial_p(X)$ and $\Delta(X, p) = \Delta_p(X)$.

7 Conclusions

We considered an automata model for KAT expressions where each transition is labeled by a program symbol and, instead of an atom, a boolean expression. Each transition can, thus, be seen as labeled, in a compact way, by a set of atoms, the ones that satisfy the appropriate boolean expression. Recently, symbolic finite automata (SFA) where transitions are labeled with sets of alphabetic symbols were introduced in order to deal with large alphabets [20]. Although the extension of classical finite automata algorithms to SFAs bears similarities with the ones here presented, SFAs are interpreted over sets of finite words and not over sets of guarded strings. Experiments with the algorithms presented in this paper must be carried out in order to validate their practical applicability and also to suggest goals for a theoretical study of their average-case complexity.

References

1. Almeida, M., Moreira, N., Reis, R.: Antimirov and Mosses's rewrite system revisited. International Journal of Foundations of Computer Science 20(4), 669–684 (2009)
2. Almeida, M., Moreira, N., Reis, R.: Testing regular languages equivalence. Journal of Automata, Languages and Combinatorics 15(1/2), 7–25 (2010)
3. Almeida, R., Broda, S., Moreira, N.: Deciding KAT and Hoare logic with derivatives (96). In: Faella, M., Murano, A. (eds.) Proc. 3rd GANDALF. EPTCS, vol. 96, pp. 127–140 (2012)
4. Armstrong, A., Struth, G., Weber, T.: Program analysis and verification based on Kleene algebra in Isabelle/HOL. In: Blazy, S., Paulin-Mohring, C., Pichardie, D. (eds.) ITP 2013. LNCS, vol. 7998, pp. 197–212. Springer, Heidelberg (2013)
5. Bonchi, F., Pous, D.: Checking NFA equivalence with bisimulations up to congruence. In: Giacobazzi, R., Cousot, R. (eds.) The 40th Annual ACM SIGPLAN-SIGACT Symposium POPL 2013, pp. 457–468. ACM (2013)
6. Broda, S., Machiavelo, A., Moreira, N., Reis, R.: On the Average Size of Glushkov and Equation Automata for KAT Expressions. In: Gąsieniec, L., Wolter, F. (eds.) FCT 2013. LNCS, vol. 8070, pp. 72–83. Springer, Heidelberg (2013)
7. Broda, S., Machiavelo, A., Moreira, N., Reis, R.: Automata for KAT expressions. Tech. Rep. DCC-2014-01, FCUP & CMUP, Universidade do Porto (January 2014), http://www.dcc.fc.up.pt/dcc/Pubs/TReports/
8. Champarnaud, J.M., Ziadi, D.: From Mirkin's prebases to Antimirov's word partial derivatives. Fundam. Inform. 45(3), 195–205 (2001)
9. Cohen, E., Kozen, D., Smith, F.: The complexity of Kleene algebra with tests. Tech. Rep. TR96-1598, Computer Science Department, Cornell University (July 1996)
10. Cormen, T.H., Leiserson, C.E., Rivest, R.L., Stein, C.: Introduction to Algorithms, 2nd edn. The MIT Press (2003)
11. Hopcroft, J., Karp, R.M.: A linear algorithm for testing equivalence of finite automata. Tech. Rep. TR 71 -114, University of California, Berkeley, California (1971)
12. Kozen, D.: Kleene algebra with tests. Trans. on Prog. Lang. and Systems 19(3), 427–443 (1997)
13. Kozen, D.: Automata on guarded strings and applications. Matématica Contemporânea 24, 117–139 (2003)
14. Kozen, D.: On the coalgebraic theory of Kleene algebra with tests. Tech. Rep. Cornell University (May 2008), http://hdl.handle.net/1813/10173
15. Kozen, D., Smith, F.: Kleene algebra with tests: Completeness and decidability. In: van Dalen, D., Bezem, M. (eds.) CSL 1996. LNCS, vol. 1258, pp. 244–259. Springer, Heidelberg (1997)

16. Mirkin, B.G.: An algorithm for constructing a base in a language of regular expressions. Engineering Cybernetics 5, 51–57 (1966)
17. Pereira, D.: Towards Certified Program Logics for the Verification of Imperative Programs. Ph.D. thesis, University of Porto (2013)
18. Pous, D.: Kleene algebra with tests and Coq tools for While programs. CoRR abs/1302.1737 (2013)
19. Silva, A.: Position automata for Kleene algebra with tests. Sci. Ann. Comp. Sci. 22(2), 367–394 (2012)
20. Veanes, M., de Halleux, P., Tillmann, N.: Rex: Symbolic regular expression explorer. In: 3rd Inter. Conference on Software Testing, Verification and Validation, ICST 2010, Paris, France, April 7-9, pp. 498–507. IEEE Computer Society (2010)

Algorithmic Randomness for Infinite Time Register Machines

Merlin Carl

Abstract. A concept of randomness for infinite time register machines ($ITRMs$), resembling Martin-Löf-randomness, is defined and studied. In particular, we show that for this notion of randomness, computability from mutually random reals implies computability and that an analogue of van Lambalgen's theorem holds.

1 Introduction

Martin-Löf-randomness (ML-randomness, see e.g. [5]) provides an intuitive and conceptually stable clarification of the informal notion of a random sequence over a finite alphabet. Since its introduction, several strengthenings and variants of ML-randomness have been considered; a recent example is the work of Hjorth and Nies on Π_1^1-randomness, which led to interesting connections with descriptive set theory ([7]).

We are interested in obtaining a similar notion based on machine models of transfinite computations. In this paper, we will exemplarily consider infinite time register machines. Infinite Time Register Machines ($ITRMs$), introduced in [8] and further studied in [9], work similar to the classical unlimited register machines ($URMs$) described in [4]. In particular, they use finitely many registers each of which can store a single natural number. The difference is that $ITRMs$ use transfinite ordinal running time: The state of an $ITRM$ at a successor ordinal is obtained as for $URMs$. At limit times, the program line is the inferior limit of the earlier program lines and there is a similar limit rule for the register contents. If the limit inferior of the earlier register contents is infinite, the register is reset to 0.

The leading idea of ML-randomness is that a sequence of 0 and 1 is random iff it has no special properties, where a special property should be a small (e.g. measure 0) set of reals that is in some way accessible to a Turing machine. Classical Turing machines, due to the finiteness of their running time, have the handicap that the only decidable null set of reals is the empty set: If a real x is accepted by a classical Turing machine M within n steps, then M will also accept every y agreeing with x on the first n bits. In the definition of ML-randomness, this difficulty is overcome by merely demanding the set X in question to be effectively approximated by a recursively enumerable sequence of sets of intervals with controlled convergence behaviour. For models of transfinite computations, this trick is unnecessary: The decidable sets of reals form a rich

A. Beckmann, E. Csuhaj-Varjú, and K. Meer (Eds.): CiE 2014, LNCS 8493, pp. 84–92, 2014.

class (including all ML-tests and, by [8], all Π_1^1-sets). This is still a plausible notion of randomness, since elements of an $ITRM$-decidable meager set can still be reasonably said to have a special property. In fact, some quite natural properties like coding a well-ordering can be treated very conveniently with our approach. Hence, we define:

Definition 1. $X \subseteq \mathfrak{P}(\omega)$ is called $ITRM$-decidable iff there is an $ITRM$-program P such that $P^x \downarrow = 1$ iff $x \in X$ and $P^x \downarrow = 0$, otherwise. In this case we say that P decides X. P is called deciding iff there is some X such that P decides X. We say that X is decided by P in the oracle y iff $X = \{x \mid P^{x \oplus y} \downarrow = 1\}$ and $\mathfrak{P}(\omega) - X = \{x \mid P^{x \oplus y} \downarrow = 0\}$. The other notions relativize in the obvious way.

Definition 2. Recall that a set $X \subseteq \mathfrak{P}(\omega)$ is meager iff it is a countable union of nowhere dense sets. $X \subseteq \mathfrak{P}(\omega)$ is an $ITRM$-test iff X is $ITRM$-decidable and meager. $x \subseteq \omega$ is $ITRM$-c-random iff there is no $ITRM$-test X such that $x \in X$.

Remark: This obviously deviates from the definition of ML-randomness in that we use meager sets rather than null sets as our underlying notion of 'small'. The reason is simply that this variant turned out to be much more convenient to handle for technical reasons. The use of category rather than measure gives this definition a closer resemblance to what is, in the classical setting, refered to as genericity (see e.g. section 2.24 of [5]). We still decided to use the term '$ITRM$-c-randomness' to avoid confusion with the frequently used concept of Cohen genericity, hence reserving the term '$ITRM$-random' for reals that do not lie in any $ITRM$-decidable null set. We are pursuing the notion of $ITRM$-randomness in ongoing work. In contrast to strong Π_1^1-randomness ([7], [15]), it can be shown that there is no universal $ITRM$-test.

We will now summarize some key notions and results on $ITRM$s that will be used in the paper.

Definition 3. For P a program, $x, y \in \mathfrak{P}(\omega)$, $P^x \downarrow = y$ means that the program P, when run with oracle x, halts on every input $i \in \omega$ and outputs 1 iff $i \in y$ and 0, otherwise. $x \subseteq \omega$ is $ITRM$-computable in the oracle $y \subseteq \omega$ iff there is an $ITRM$-program P such that $P^y \downarrow = x$, in which case we occasionally write $x \leq_{ITRM} y$. If y can be taken to be $\emptyset$, x is $ITRM$-computable. We denote the set of $ITRM$-computable reals by $COMP$.

Remark: We occasionally drop the $ITRM$-prefix as notions like 'computable' always refer to $ITRM$s in this paper.

Theorem 4. Let $x, y \subseteq \omega$. Then x is $ITRM$-computable in the oracle y iff $x \in L_{\omega_\omega^{CK,y}}[y]$, where $\omega_i^{CK,y}$ denotes the ith x-admissible ordinal.

Proof. This is a straightforward relativization of Theorem 5 of [13], due to P. Koepke. □

Theorem 5. Let $\mathbb{P}_n$ denote the set of $ITRM$-programs using at most n registers, and let $(P_{i,n}|i \in \omega)$ enumerate $\mathbb{P}_n$ in some natural way. Then the bounded halting problem $H_n^x := \{i \in \omega | P_{i,n}^x \downarrow\}$ is computable uniformly in the oracle x by an $ITRM$-program (using more than n registers).

Furthermore, if $P \in \mathbb{P}_n$ and $P^x \downarrow$, then P^x halts in less than $\omega_{n+1}^{CK,x}$ many steps. Consequently, if P is a halting $ITRM$-program, then P^x stops in less than $\omega_\omega^{CK,x}$ many steps.

Proof. The corresponding results from [8] (Theorem 4) and [13] (Theorem 9) easily relativize. □

We will freely use the following standard proposition:

Proposition 6. Let $X \subseteq [0,1] \times [0,1]$ and $\tilde{X} := \{x \oplus y \mid (x,y) \in X\}$. Then X is meager/comeager/non-meager iff $\tilde{X}$ is meager/comeager/non-meager.

Most of our notation is standard. By a real, we mean an element of $^\omega 2$. $L_\alpha[x]$ denotes the αth level of Gödel's constructible hierarchy relativized to x. For $a, b \subseteq \omega$, $a \oplus b$ denotes $\{p(i,j) \mid i \in a \wedge j \in b\}$, where $p : \omega \times \omega \to \omega$ is Cantor's pairing function.

2 Computability from Random Oracles

In this section, we consider the question which reals can be computed by an $ITRM$ with an $ITRM$-c-random oracle. We start by recalling the following theorem from [3]. The intuition behind it is that, given a certain non-$ITRM$-computable real x, one has no chance of computing x from some randomly chosen real y.

Theorem 7. Let x be a real, Y be a set of reals such that x is $ITRM$-computable from every $y \in Y$.

Then, if Y has positive Lebesgue measure or is Borel and non-meager, x is $ITRM$-computable.

Corollary 8. Let x be $ITRM$-c-random. Then, for all $i \in \omega$, $\omega_i^{CK,x} = \omega_i^{CK}$.

Proof. Lemma 46 of [3] shows that $\omega_i^{CK,x} = \omega_i^{CK}$ for all $i \in \omega$ whenever x is Cohen-generic over $L_{\omega_\omega^{CK}}$ (see e.g. [3] or [14]) and that the set of Cohen-generics over $L_{\omega_\omega^{CK}}$ is comeager. Hence $\{x|\omega_i^{CK,x} > \omega_i^{CK}\}$ is meager. For each program P, the set of reals x such that P^x computes a code for the ith x-admissible which is greater than ω_i^{CK} is decidable using the techniques developed in [1] and [2]. (The idea is to uniformly in the oracle x compute a real c coding $L_{\omega_{i+1}^{CK,x}}[x]$ in which the natural numbers m and n coding ω_i^{CK} and $\omega_i^{CK,x}$ can be identified in the oracle x, then check - using a halting problem solver for P, see Theorem 5 - whether P^x computes a well-ordering of the same order type as the element of $L_{\omega_{i+1}^{CK,x}}[x]$ coded by n and finally whether the element coded by m is an element of that coded by n.) Hence, if x is $ITRM$-c-random, then there can be no $ITRM$-program P computing such a code in the oracle x. But a code for $\omega_i^{CK,x}$ is $ITRM$-computable in the oracle x for every real x and every $i \in \omega$. Hence, we must have $\omega_i^{CK,x} = \omega_i^{CK}$ for every $i \in \omega$, as desired. $\square$

Lemma 9. Let $a \subseteq \omega$ and suppose that z is Cohen-generic over $L_{\omega_\omega^{CK,a}+1}[a]$. Then $a \leq_{ITRM} z$ iff a is $ITRM$-computable. Consequently (as the set $C_a := \{z \subseteq \omega \mid z$ is Cohen-generic over $L_{\omega_\omega^{CK,a}+1}[a]\}$ is comeager), $S_a := \{z \subseteq \omega \mid a \leq_{ITRM} z\}$ is meager whenever a is not $ITRM$-computable.

Proof. Assume that z is Cohen-generic over $L_{\omega_\omega^{CK,a}+1}[a]$ and $a \leq_{ITRM} z$. By the forcing theorem for provident sets (see e.g. Lemma 32 of [3]), there is an $ITRM$-program P and a forcing condition p such that $p \Vdash P^{\dot{G}} \downarrow = \check{a}$, where $\dot{G}$ is the canonical name for the generic filter and $\check{a}$ is the canonical name of a. Now, let y and z be mutually Cohen-generic over $L_{\omega_\omega^{CK,a}+1}[a]$ both extending p. Again by the forcing theorem and by absoluteness of computations, we must have $P^x \downarrow = a = P^y \downarrow$, so $a \in L_{\omega_\omega^{CK,x}}[x] \cap L_{\omega_\omega^{CK,y}}[y]$. By Corollary 8, $\omega_\omega^{CK,x} = \omega_\omega^{CK,y} = \omega_\omega^{CK}$. By Lemma 28 of [3], we have $L_\alpha[x] \cap L_\alpha[y] = L_\alpha$ whenever x and y are mutually Cohen-generic over L_α and α is provident (see [14]). Consequently, we have $a \in L_{\omega_\omega^{CK,x}}[x] \cap L_{\omega_\omega^{CK,y}}[y] = L_{\omega_\omega^{CK}}[x] \cap L_{\omega_\omega^{CK}}[y] = L_{\omega_\omega^{CK}}$, so a is $ITRM$-computable.

The comeagerness of C_a is standard (see e.g. Lemma 29 of [3]). To see that S_a is meager for non-$ITRM$-computable a, observe that the Cohen-generic reals over $L_{\omega_\omega^{CK,a}+1}[a]$ form a comeager set of reals to non of which a is reducible. $\square$

Definition 10. Let $x, y \subseteq \omega$. Then x is $ITRM$-c-random relative to y iff there is no meager set X such that $x \in X$ and X is $ITRM$-decidable in the oracle y. If x is $ITRM$-c-random relative to y and y is $ITRM$-c-random relative to x, we say that x and y are mutually $ITRM$-c-random.

Intuitively, we should expect that mutually random reals have no non-trivial information in common. This is expressed by the following theorem:

Theorem 11. If z is $ITRM$-computable from two mutually $ITRM$-c-random reals x and y, then z is $ITRM$-computable.

Proof. Assume otherwise, and suppose that z, x and y constitute a counterexample. By assumption, z is computable from x. Also, by Theorem 5, let P be a program such that $P^a(i) \downarrow$ for every $a \subseteq \omega$, $i \in \omega$ and such that P computes z in the oracle y. In the oracle z, the set $A_z := \{a \mid \forall i \in \omega P^a(i) \downarrow = z(i)\}$ is decidable by simply computing $P^a(i)$ for all $i \in \omega$ and comparing the result to the ith bit of z. Clearly, we have $A_z \subseteq \{a \mid z \leq_{ITRM} a\}$. Hence, by our Lemma 9 above, A_z is meager as z is not $ITRM$-computable by assumption. Since A_z is decidable in the oracle z and z is computable from x, A_z is also decidable in the oracle x. Now, x and y are mutually $ITRM$-c-random, so that $y \notin A_z$. But P computes z in the oracle y, so $y \in A_z$ by definition, a contradiction. $\square$

While, naturally, there are non-computable reals that are reducible to a c-random real x (such as x itself), intuitively, it should not be possible to compute a non-arbitrary real from a random real. We approximate this intuition by taking 'non-arbitrary' to mean '$ITRM$-recognizable' (see [9], [1] or [2] for more information on $ITRM$-recognizability). It turns out that, in accordance with this intuition, recognizables that are $ITRM$-computable from $ITRM$-c-random reals are already $ITRM$-computable.

Definition 12. $x \subseteq \omega$ is $ITRM$-recognizable iff $\{x\}$ is $ITRM$-decidable. $RECOG$ denotes the set of recognizable reals.

Theorem 13. Let $x \in RECOG$ and let y be $ITRM$-c-random such that $x \leq_{ITRM} y$. Then x is $ITRM$-computable.

Proof. Let $x \in RECOG - COMP$ be computable from y, say by program P and let Q be a program that recognizes x. The set $S := \{z \mid P^z \downarrow = x\}$ is meager as in the proof of Theorem 11. But S is decidable: Given a real z, use a halting-problem solver for P (which exists uniformly in the oracle by Theorem 5) to test whether $P^z(i) \downarrow$ for all $i \in \omega$; if not, then $z \notin S$. Otherwise, use Q to check whether the real computed by P^z is equal to x. If not, then $z \notin S$, otherwise $z \in S$. As P^y computes x, it follows that $y \in S$, so that y is an element of an $ITRM$-decidable meager set. Hence y is not $ITRM$-c-random, a contradiction. $\qquad\square$

Remark: Let $(P_i \mid i \in \omega)$ be a natural enumeration of the $ITRM$-programs. Together with the fact that the halting number $h = \{i \in \omega \mid P_i \downarrow\}$ for $ITRM$s is recognizable (see [2]), this implies in particular that the halting problem for $ITRM$s is not $ITRM$-reducible to an $ITRM$-c-random real. In particular, the Kucera-Gacs theorem (see e.g. Theorem 8.3.2 of [5]) does not hold in our setting.

3   An Analogue to van Lambalgen's Theorem

A crucial result of classical algorithmic randomness is van Lambalgen's theorem, which states that for reals a and b, $a \oplus b$ is ML-random iff a is ML-random and b is ML-random relative to a. In this section, we demonstrate an analogous result for $ITRM$-c-randomness.

Lemma 14. Let Q be a deciding $ITRM$-program using n registers and $a \subseteq \omega$. Then $\{y \mid Q^{y \oplus a} \downarrow = 1\}$ is meager iff $Q^{x \oplus a} \downarrow = 0$ for all $x \in L_{\omega_{n+1}^{CK,a}+3}[a]$ that are Cohen-generic over $L_{\omega_{n+1}^{CK,a}+1}[a]$.

Proof. By absoluteness of computations and the bound on $ITRM$-halting times (see Theorem 5), $Q^{x \oplus a} \downarrow = 0$ implies that $Q^{x \oplus a} \downarrow = 0$ also holds in $L_{\omega_{n+1}^{CK,a}}[a]$. As this is expressable by a Σ_1-formula, it must be forced by some condition p by the forcing theorem over KP (see e.g. Theorem 10.10 of [14]).

Hence every y extending p will satisfy $Q^{y \oplus a} \downarrow = 0$. The set C of reals Cohen-generic over $L_{\omega_{n+1}^{CK,a}+1}[a]$ is comeager. Hence, if $Q^{x \oplus a} \downarrow = 0$ for some $x \in C$, then

$Q^{x \oplus a} \downarrow= 0$ for a non-meager (in fact comeager in some interval) set C'. Now, for each condition p, $L_{\omega_{n+1}^{CK,a}+3}[a]$ will contain a generic filter over $L_{\omega_{n+1}^{CK,a}+1}[a]$ extending p (as $L_{\omega_{n+1}^{CK,a}+1}[a]$ is countable in $L_{\omega_{n+1}^{CK,a}+3}[a]$). Hence, if $Q^{x \oplus a} \downarrow= 0$ for all $x \in C \cap L_{\omega_{n+1}^{CK,a}+3}[a]$, then this holds for all elements of C and the complement $\{y | Q^{y \oplus a} \downarrow= 1\}$ is therefore meager.

If, on the other hand, $Q^{x \oplus a} \downarrow= 1$ for some such x, then this already holds for all x in some non-meager (in fact comeager in some interval) set C' by the same reasoning. □

Corollary 15. For a deciding $ITRM$-program Q using n registers, there exists an $ITRM$-program P such that, for all $x, y \in \mathfrak{P}(\omega)$, $P^x \downarrow= 1$ iff $\{y | Q^{x \oplus y} \downarrow= 1\}$ is non-meager.

Proof. From x, compute, using sufficiently many extra registers, a real code for $L_{\omega_{n+1}^{CK,x}+4}$ in the oracle x. This can be done uniformly in x. Then, using the techniques developed in section 6 of [8], identify and check all generics in that structure, according to the last lemma. □

Corollary 16. x is $ITRM$-c-random iff x is Cohen-generic over $L_{\omega_\omega^{CK}}$.

Proof. Let S denote the set of Cohen-generic reals over $L_{\omega_\omega^{CK}}$. Then $x \in S$ iff $x \cap D \neq \emptyset$ for every dense subset $D \in L_{\omega_\omega^{CK}}$ of Cohen-forcing. Clearly, for every such D, $G_D := \{y \mid y \cap D \neq \emptyset\}$ is comeager and $ITRM$-decidable, so every $ITRM$-c-random real must be in every G_D and hence in S.

On the other hand, if $x \in S$ and $P^x \downarrow= 1$ for some deciding $ITRM$-program P, then there is some finite $p \subseteq x$ such that $P^y \downarrow= 1$ for every $p \subset y \in S$, so the set decided by P is not meager as in the proof of Lemma 15. Hence x is not an element of any $ITRM$-decidable meager set, so x is $ITRM$-c-random. □

Our proof of the $ITRM$-analogue for van Lambalgen's theorem now follows a general strategy inspired by that used in [5], Theorem 6.9.1 and 6.9.2:

Theorem 17. Assume that a and b are reals such that $a \oplus b$ is not $ITRM$-c-random. Then a is not $ITRM$-c-random or b is not $ITRM$-c-random relative to a.

Proof. As $a \oplus b$ is not $ITRM$-c-random, let X be an $ITRM$-decidable meager set of reals such that $a \oplus b \in X$. Suppose that P is a program deciding X.

Let $Y := \{x | \{y \mid x \oplus y \in X\}$ non-meager$\}$. By Corollary 15, Y is $ITRM$-decidable.

We claim that Y is meager. First, Y is provably Δ_1^1 and hence has the Baire property (see e.g. Exercise 14.5 of [10]). Hence, by the Kuratowski-Ulam-theorem (see e.g. [11], Theorem 8.41), Y is meager. Consequently, if $a \in Y$, then a is not $ITRM$-c-random.

Now suppose that $a \notin Y$. This means that $\{y \mid a \oplus y \in X\}$ is meager. But $S := \{y \mid a \oplus y \in X\}$ is easily seen to be $ITRM$-decidable in the oracle a and $b \in S$. Hence b is not $ITRM$-c-random relative to a. □

Theorem 18. Assume that a and b are reals such that $a \oplus b$ is $ITRM$-c-random. Then a is $ITRM$-c-random and b is $ITRM$-c-random relative to a.

Proof. Assume first that a is not $ITRM$-c-random, and let X be an $ITRM$-decidable meager set with $a \in X$. Then $X \oplus [0,1]$ is also meager and $X \oplus [0,1]$ is $ITRM$-decidable. As $a \in X$, we have $a \oplus b \in X \oplus [0,1]$, so $a \oplus b$ is not $ITRM$-c-random, a contradiction.

Now suppose that b is not $ITRM$-c-random relative to a, and let X be a meager set of reals such that $b \in X$ and X is $ITRM$-decidable in the oracle a. Let Q be an $ITRM$-program such that Q^a decides X. Our goal is to define a deciding program $\tilde{Q}$ such that $\tilde{Q}^a$ still decides X, but also $\{x | \tilde{Q}^x \downarrow = 1\}$ is meager. This suffices, as then $\tilde{Q}^{a \oplus b} \downarrow = 1$ and $\{x | \tilde{Q}^x \downarrow = 1\}$ is $ITRM$-decidable. $\tilde{Q}$ operates as follows: Given $x = y \oplus z$, check whether $\{w \mid Q^{y \oplus w}\}$ is meager, using Corollary 15. If that is the case, carry out the computation of Q^x and return the result. Otherwise, return 0. This guarantees (since X is meager) that $\{y \mid \tilde{Q}^{x \oplus y} \downarrow = 1\}$ is meager and furthermore that $\tilde{Q}^{a \oplus x} \downarrow = 1$ iff $Q^{a \oplus x} \downarrow = 1$ iff $x \in X$ for all reals x, so that $\{x | \tilde{Q}^{a \oplus x} \downarrow = 1\}$ is just X, as desired. $\square$

Combining Theorem 17 and 18 gives us the desired conclusion:

Theorem 19. Given reals x and y, $x \oplus y$ is $ITRM$-c-random iff x is $ITRM$-c-random and y is $ITRM$-c-random relative to x. In particular, if x and y are $ITRM$-c-random, then x is $ITRM$-c-random relative to y iff y is $ITRM$-c-random relative to x.

We note that a classical Corollary to van Lambalgen's theorem continues to hold in our setting:

Corollary 20. Let x, y be $ITRM$-c-random. Then x is $ITRM$-c-random relative to y iff y is $ITRM$-c-random relative to x.

Proof. Assume that y is $ITRM$-c-random relative to x. By assumption, x is $ITRM$-c-random. By Theorem 19, $x \oplus y$ is $ITRM$-c-random. Trivially, $y \oplus x$ is also $ITRM$-c-random. Again by Theorem 19, x is $ITRM$-c-random relative to y. By symmetry, the corollary holds. $\square$

4 Some Consequences for the Structure of $ITRM$-Degrees

In the new setting, we can also draw some standard consequences of van Lambalgen's theorem.

Definition 21. If $x \leq_{ITRM} y$ but not $y \leq_{ITRM} x$, we write $x <_{ITRM} y$. If $x \leq_{ITRM} and$ $y \leq_{ITRM} x$, then we write $x \equiv_{ITRM} y$. If neither $x \leq_{ITRM} y$ nor $y \leq_{ITRM} x$, we call x and y incomparable and write $x|_{ITRM}y$.

Clearly, $\equiv_{ITRM}$ is an equivalence relation. We may hence form, for each real x, the $\equiv_{ITRM}$-equivalence class $[x]_{ITRM}$ of x, called the $ITRM$-degree of x. It is easy to see that $\leq_{ITRM}$ respects $\equiv_{ITRM}$, so that $[x]_{ITRM} \leq_{ITRM} [y]_{ITRM}$ etc. are well-defined and have the obvious meaning.

Corollary 22. If a is $ITRM$-c-random, $a = a_0 \oplus a_1$, then $a_0|_{ITRM}a_1$.

Proof. By Theorem 19, a_0 and a_1 are mutually $ITRM$-c-random. If a_0 was $ITRM$-computable from a_1, then $\{a_0\}$ would be decidable in the oracle a_1, meager and contain a_0, so a_0 would not be $ITRM$-c-random relative to a_1, a contradiction. By symmetry, the claim follows.   □

Lemma 23. Let h be a real coding the halting problem for $ITRM$s as in the remark following Theorem 13. Then there is an $ITRM$-c-random real $x \leq_{ITRM} h$.

Proof. Given h, we can compute a code for $L_{\omega_\omega^{CK}+2}$, which contains a real x which is Cohen-generic over $L_{\omega_\omega^{CK}+1}$. Hence, x itself is $ITRM$-computable from h. Assume that x is not $ITRM$-c-random, so there exists a decidable meager set $X \ni x$. Let P be a program deciding X. Then $P^x \downarrow= 1$. By the forcing theorem for Cohen-forcing, this must be forced by some condition $p \subseteq x$. The set Y of $y \supseteq p$ which are Cohen-generic over $L_{\omega_\omega^{CK}+1}$ is non-meager (see above) and $p \subseteq y$ implies $p \Vdash P^y \downarrow= 1$. As P decides X, we must have $Y \subseteq X$, a contradiction to the assumption that X is meager.   □

As a corollary, we obtain an analogue to the Kleene-Post-theorem on Turing degrees between 0 and 0' (see e.g. Theorem VI.1.2 of [16]) for $ITRM$s.

Corollary 24. With h as in Lemma 23, there are x_0, x_1 such that $[0]_{ITRM} <_{ITRM} [x_0]_{ITRM}, [x_1]_{ITRM} \leq h$ and $x_0|_{ITRM}x_1$. In particular, there is a real x_0 such that $[0]_{ITRM} <_{ITRM} [x_0]_{ITRM} <_{ITRM} h$.

Proof. Pick x as in Lemma 23, let $x = x_0 \oplus x_1$, and use Corollary 22.   □

5 Conclusion and Further Work

The most pressing issue is certainly to strengthen the parallelism between $ITRM$-randomness and ML-randomness by studying the corresponding notion for sets of Lebesgue measure 0 rather than meager sets.

Still, $ITRM$-c-randomness shows an interesting behaviour, partly analogous to ML-randomness, though by quite different arguments. Similar approaches are likely to work for other machine models of generalized computations, in particular $ITTM$s ([6]) (which were shown in [3] to obey the analogue of the non-meager part of Theorem 7) and ordinal Turing Machines ([12]) (for which the analogues of both parts of Theorem 7 turned out to be independent from ZFC) which we study in ongoing work. This further points towards a more general background theory of computation that allow unified arguments for all these various models as well as classical computability. Furthermore, we want to see whether the remarkable conceptual stability of ML-randomness (for example the equivalence with Chaitin randomness or unpredictabiliy in the sense of r.e. Martingales, see e.g. sections 6.1 and 6.3 of [5]) carries over to the new context.

Acknowledgements. I am indebted to Philipp Schlicht for several helpful discussions of the results and proofs and suggesting various crucial references. I also thank the referees for suggesting several simplifications and clarifications.

References

[1] Carl, M.: The distribution of ITRM-recognizable reals. To appear in: Annals of Pure and Applied Logic. In: Special Issue from CiE (2012)

[2] Carl, M.: Optimal Results on ITRM-recognizability. arXiv:1306.5128v1 (preprint)

[3] Carl, M., Schlicht, P.: Infinite Computations with Random Oracles. arXiv:1307.0160v3 (submitted)

[4] Cutland, N.: Computability - An introduction to recursive function theory. Cambridge University Press (1980)

[5] Downey, R.G., Hirschfeldt, D.: Algorithmic Randomness and Complexity. Theory and Applications of Computability. Springer LLC (2010)

[6] Hamkins, J., Lewis, A.: Infinite Time Turing Machines. Journal of Symbolic Logic 65(2), 567–604 (2000)

[7] Hjorth, G., Nies, A.: Randomness in effective descriptive set theory. Journal of the London Mathematical Society (to appear)

[8] Koepke, P., Miller, R.: An enhanced theory of infinite time register machines

[9] Carl, M., Fischbach, T., Koepke, P., Miller, R., Nasfi, M., Weckbecker, G.: The basic theory of infinite time register machines

[10] Kanamori, A.: The higher infinite. Springer (2005)

[11] Kechris, A.: Measure and category in effective descriptive set theory. Annals of Mathematical Logic 5(4), 337–384 (1973)

[12] Koepke, P.: Turing computations on ordinals. Bulletin of Symbolic Logic 11, 377–397 (2005)

[13] Koepke, P.: Ordinal Computability. In: Ambos-Spies, K., Löwe, B., Merkle, W. (eds.) CiE 2009. LNCS, vol. 5635, pp. 280–289. Springer, Heidelberg (2009)

[14] Mathias, A.R.D.: Provident sets and rudimentary set forcing (preprint), https://www.dpmms.cam.ac.uk/~ardm/fifofields3.pdf

[15] Sacks, G.: Higher recursion theory. Springer (1990)

[16] Soare, R.I.: Recursively Enumerable Sets and Degrees. Springer (1987)

Constraint Logic Programming
for Resolution of Relative Time Expressions

Henning Christiansen

Research group PLIS: Programming, Logic and Intelligent Systems
Department of Communication, Business and Information Technologies
Roskilde University, Denmark
henning@ruc.dk

Abstract. Translating time expression into absolute time points or durations is a challenge for natural languages processing such as text mining and text understanding in general. We present a constraint logic language $CLP(Time)$ tailored to text usages concerned with time and calendar. It provides a simple and flexible formalism to express relationships between different time expressions in a text, thereby giving a recipe for resolving them into absolute time. A constraint solver is developed which, as opposed to some earlier approaches, is independent of the order in which temporal information is introduced, and it can give meaningful output also when no exact reference time is available.

1   Introduction

Humans often prefer relative time expressions in text instead of explicitly time stamping every event. Wordings like "two days later" are preferred when the reference date is known, and it often gives good sense to the human reader even without a reference date. For automated document analysis, correct identification of the actual time or date may be important for text understanding and data mining, where the goal may be to provide a time stamped list of events.

A document may also contain expressions that are relative to an implicit time of writing. Here it may be interesting also to identify the time of writing. References to known events may give a clue as in "Two years ago, when the last Venus transit for the next 100 years took place, ...", where a knowledge base of astronomical events can help concluding that the text was written in 2014.

There are several problems involved in annotating a text with correct time stamps. First of all, the time expressions, both relative and absolute, must be identified; this a task for taggers and parsers for natural language. Secondly, the relationship between the different time expressions must be determined ("two years later than *what?*"), and thirdly, a bit of reasoning is needed to calculate the correct time stamps from these relationships.

For the third task, constraint programming presents several advantages compared to ad-hoc techniques. Constraints have been suggested for this earlier, but has not gained popularity in any major text processing systems. We demonstrate here how constraint *logic* programming (CLP) may give rise to an effective

A. Beckmann, E. Csuhaj-Varjú, and K. Meer (Eds.): CiE 2014, LNCS 8493, pp. 93–102, 2014.

mechanism for the third step, and also provide a flexible language in which to express relationships between time expressions, thus also overlapping sub-task two. CLP introduces a well-defined semantics, meaning that the calculated times are correct solutions to the network of relationships set up by steps one and two. It implies a robust and incremental evaluation scheme that is independent of the order in which time expressions occur in the text: it can still manage if the only absolute time is given at the very end of a text, or even if no absolute anchor point in time is available. It also integrates in a natural way with knowledge bases of known events which may help to situate a document in time.

We present a constraint logic language *CLP(Time)* equipped with a constraint solver tailored to the pragmatics of standard usages related to the Gregorian calendar. It is implemented in the programming language of Constraint Handling Rules [1], which provides a modular, rules-based and easily extendible architecture. While time and duration in principle can be represented by integer and interval arithmetic, we develop specific datatypes relating to calendric notions, so we may, e.g., add two months to a given month without knowing the duration of those months or the year. It is language independent, but may be extended with new sorts of constraints that reflect special usages. *CLP(Time)* is currently being developed to support arbitrary intervals, such as "in the late nineties" or "between May and August", which is not described here.

Related work and state of the art are reviewed in section 2 as a background for the present approach. The facilities of the constraint language are introduced in section 3, and it is briefly shown how it can be tested together with Prolog's Definite Clause Grammars and with pre-tagged text. Section 4 demonstrates how it applies for different sorts of text. The implementation of a constraint solver with Constraint Handling Rules is sketched in section 5. Some conclusions and ideas for future work are given in the final section.

2　Related Work

To resolve time expressions in a text, one needs to 1) identify those expressions, 2) assign a formula to each such expression that sets the relationship to the context of other time expressions in the text, and 3) to evaluate these formulas.[1]

The HeidelTime system [2,3], which is considered state of the art, is based on a tagger which, via a specific rule format, can be adapted to different languages and specific usages. Such rules are manually crafted, and they combine matching of textual patterns with the building of a limited sort of arithmetic expressions; this may involve relative distance to an assumed anchor time not specified explicitly in the rule. Evaluation has two different modes. In "narratives' mode", all expressions are evaluated sequentially from the start of the document and

---

[1] In the literature, e.g., [3], the term "normalization" has been used for the third phase. Constraints provide more flexibility in phase 2 for specifying relationships. Phase 3 is a matter of a correct implementation of a constraint solver. Thus constraints tend to make "normalization" a combination of phases 2 and 3.

in "news' mode" every expression is evaluated relative to a fixed document creation time. Evaluation is problematic for narratives until a first absolute time is met, and for news articles when document creation time is missing, but good recall and precision figures are reported for selected classes of documents [3]. Handling of inconsistency (over-specification) is not described. Machine learning have been used for identifying the time of writing for news articles from mentions of historical facts, e.g., [4,5,6]; see also [7] for an overview and more references.

HeidelTime and other works referenced above do not use constraint technology for specification and evaluation of interdependencies, which could lead to simpler and more transparent formulations; in fact constraint techniques are not mentioned at all. Constraints, and especially constraint logic programming, provide a uniform framework for expressing different dependencies and (under-) specifications, which otherwise may give rise to a complex nomenclature (as demonstrated by, e.g., [8,9]). Standard machine learning techniques have been used to train recognizers of time expressions, but it helps only little for learning how to evaluate them; we shall refrain from giving a literature overview of these directions as the goals are different from ours.

Logically based formalisms for reasoning about time exist such as temporal logics, the event calculus [10] or Allen's theory [11], but they do not relate to calendar conventions and everyday usages concerned with time and date. Constraint solvers related to time, dates and calendric data have been seen suggested, e.g., [12,13]; these approaches involve very complex solving algorithms and do not approach the problem of finding partial solutions in case of inconsistency.

No work has been found on constraint *logic* programming related to temporal information in language usage. The work of [14] uses Constraint Handling Rules (CHR) for relative time expressions in text already tagged with the methods of [2,3] in order to correct for mistakes and unresolved expressions. This approach used CHR as a programming language for flexible search back and forth in the text in order to find reference points, but did not develop a proper constraint solver as suggested in the present paper.

CHR has been used for semantic-pragmatic language analysis in combination with parsers written in Prolog's Definite Clause Grammar notation [15,16,17] or using CHR for parsing [18,19]. CHR based techniques have been used for extracting UML diagram from use case text [20,21]; this work includes an (although simplistic) approach to pronoun resolution that has similarities to relative or indirect time indications. Other applications of CHR for language processing include parsing from Property Grammars [22], analyzing biological sequences [23,24] and Chinese Word Segmentation [25].

3   A Constraint Language for Time Expressions

A constraint language called *CLP(Time)* is defined upon Prolog using its extension of Constraint Handling Rules [1]. We take over the basic nomenclature of Prolog such as its terms, variables and operators, which may greatly enhance readability. In its present form, *CLP(Time)* can be tested immediately in language analyzers written with Prolog's Definite Clause Grammars and with pre-tagged text imported from other systems. An implementation under development is available at http://www.ruc.dk/~henning/clptime.

3.1 Datatypes and Basic Constraints of *CLP(Time)*

In the present version, the finest granule of time corresponds to dates, and distinguished types of terms are used to represent different units of time.

⟨date⟩ ::= date(⟨month⟩, n)   ⟨decade⟩ ::= decade(⟨century⟩, n)
⟨month⟩ ::= month(⟨year⟩, n)   ⟨century⟩ ::= century(n)
⟨year⟩ ::= year(⟨decade⟩, n)

The occurrences of "n" represent integer numbers of relevant size. For example, the term date(month(year(decade(century(20),1),4),1),1) represents the date of New Years day 2014.[2] Terms or subterms may be replaced by variables, so that decade(C,9) may represent "the nineties" in a yet unknown century. Such terms should only be instantiated as to represent legal times according to the Gregorian calendar since 1582. Type constraints are available, e.g., month(M) states that variable M can only be bound to terms of type month (type constraints for variables can be left out when the type is clear from context).

Expressions of the different types can be formed by adding or subtracting units of a number of granules of similar size. Examples:

```
date(month(year(decade(century(20),1),4),1),1) + 3 days
month(Y,1) + 12 months
```

The first one refers to the 4th of January 2014 and the second one to the month of January in the year following whatever year Y may end up representing. Equality terms of the same type can expressed using constraints *=* and order of time by *<* and *=<*. Examples:

```
Y1 *=* Y0 + 1 years
month(Y,3) *=<* M, M *<* month(Y1,5), Y1 *=* Y + 1 years
month(Y,3) *=<* M, M *<* month((Y + 1 years),5)
```

The first one means that Y1 and Y0 are years with Y1 being one greater than T0. Two next ones state in different ways that M is a month between March in the year given by Y and April the following year; the comma understood as conjunction. Restrictions on possible values for numerical variables can be specified by interval notation as follows; notice that this compound expression denotes a single year with some uncertainty and not an interval of several years.

```
Y *=* year(decade(century(20),1),4) + N years, N in [3;6]
```

---

[2] This notation makes it easy to write different conditions involving time, but may seem clumsy for writing specific dates. Utilities are supplied for this so that the date shown can be created and assigned to variable D by mk_date(2014-01-01,D).

Additional constraints are available for stating the day of the week, a leap year, etc.. For example,

```
D1 *=* D0 + N days, N in [1;7], dayOfWeek(D1,2)
```

means that date D1 stands for "next Tuesday" relative to D0.

A constraint `failed(···)` is used for handling inconsistencies caused by problems in the text; it should not be used in specifications of dependencies, but used solely by the constraint solver; described in the end of section 5.

3.2 Using *CLP(Time)* with Definite Clause Grammars and Prolog

CLP(Time) can be used directly from Prolog programs, in particular its grammar notation as demonstrated in the following fragment.

```
event(E,D) --> event(E), [happened, on], date(D).
date(D) --> [the], ordinal(O), [of], month(Mn), year(Yn),
            {mk_year(Yn,Y), D *=* date(month(Y,Mn),O}.
year(_) --> [].
year(N) --> [N], {integer(N)}.
event(venus_transit) --> [the, transit, of, 'Venus'].
```

Parsing the fragment such as "The transit of Venus happened on the sixth of June", produces a syntax tree node `event(e,d)`, where e describes the event and d a date whose value is to be determined by the constraint solver. Calling a constraint within the curly bracket part of a grammar rule means to cast it off into the constraint store, so that the constraint solver can evaluate it.

Pre-tagged text may be converted into a Prolog list and processed by grammars as above, adapted to take the different tags into account. CHR can also be used for traversing a text represented as token constraints indexed by position numbers as done by [14] or using CHR Grammars [19].

4 Semantics and Evaluation of Time Expressions

Here we demonstrate how *CLP(Time)* can model time dependencies in different sorts of texts. Since this paper is not about syntax analysis, we show text fragments (first column below) together with constraints (second column) modelling their content with respect to time and indicate the solutions (third column) produced incrementally by the solver. The function symbols that define the data types for the different time units are abbreviated to save space.

4.1 Narrative with Initial Time Indication

The following shows a text where every time expression can be evaluated immediately from the value of the previous one, similarly to the narrative mode of HeidelTime [2,3].

It all began in 1864 ...	Y0 *=* y(de(c(18),6),4)	Y0 = y(de(c(18),6),4)
...three years later ...	Y1 *=* Y0 + 3 years	Y1 = y(de(c(18),6),7)
...the 17th of May that year ...	D2 *=* d(m(Y1,5),17)	D2 = d(m(y(de(c(18),6),7),5),17)

4.2 Narrative without Anchor Time or with Anchor Time at the End

Also without a given anchor time, $CLP(Time)$ and its constraint solver still give meaningful output as it propagates also partly know information as far as possible.

...some year ...	Y0	Y0 (uninstantiated)
...three years later ...	Y1 *=* Y0 + 3 years	Y1 *=* Y0 + 3 years
...the 17th of May that year ...	D2 *=* d(m(Y1,5),17)	D2 = d(m(Y1,5),17)
...the 18th of June the following year ...	Y3 *=* Y1 + 1 years D3 *=* d(m(Y3,6),18)	Y3 *=* Y0 + 4 years D3 = d(m(Y3,6),18)

...which was 1867 ...	Y3 *=* y(de(c(18),6),8)	Y3 = y(de(c(18),6),8)
		Y1 = y(de(c(18),6),7)
		Y0 = y(de(c(18),6),4)
		D3 =
		  d(m(y(de(c(18),6),8),6),18)
		D2 =
		  d(m(y(de(c(18),6),7),5),17)

If the story had ended immediately before the last phrase, the collected result can still be taken as meaningful output. When a time point that serves as anchor is introduced at the end, everything resolves into definite times.

The following examples show that the solver adds offsets to dates whenever it is safe, and propagates other safe information, but needs to delay in case an end of a month, whose number of days is uncertain, is exceeded.

—	D1 *=*
	  d(m(y(De,7),2),20) + 10 days
		D1 =
		  d(m(y(De,7),3),2)

—	D2 *=*
	  d(m(y(c(19,De),Y),2),20)
	  + 10 days
		D2 *=*
		  d(m(y(c(19,De),Y),2),20)
		  + 10 days
		D2 =
		  d(m(y(c(19,De),Y),3),_)

In the first example, we add 10 days to Feb. 20 in a year ending with -7 so we know that the month has 28 days, and the addition and shift of month is safe. In the second we do the same but concerned with a February month in some unknown year in the 20th century. The addition cannot be made, but the year and new month being 3=March can be propagated into D2 which may trigger yet other evaluations to be made.

4.3 News Style Article with Mixed Anchor Points

Here we illustrate a news style text which also has narrative style relationships.

(day of newspaper
is 2014-06-23)
	DayOfPrint *=*
	  d(m(y(de(c(20),1),4),6),23)
		DayOfPrint =
		  d(m(y(de(c(20),1),4),6),23)

...2 years ago ...	DayOfPrint *=* d(YearOfPrint,_) Y0 *=* YearOfPrint - 2 years	YearOfPrint = y(de(c(20),1),4) Y0 = y(de(c(20),1),2)

...the Venus transit took
place June 6 ...
	D1 *=*
	  d(m(Y0,6),6)
		D1 =
		  d(m(y(de(c(20),1),2),6),6)

...and a week later ...	D2 *=* D1 + 7 days	D2 = d(m(y(de(c(20),1),2),6),13)

This example may be varied, assuming that the issue date for the newspaper is not known, but there is a knowledge base about astronomical events and the dates when they occurred in terms of a predicate event(*event*, *date*). Then we might have as result in the second line that Y0 is still unknown, and then in the third line a call event(venus_transit,d(m(Y0,6),6)) would instantiate Y0=y(de(c(20),1),2) and following that YearOfPrint=y(de(c(20),1),4).

5   A Constraint Solver for *CLP*(*Time*)

As mentioned, the constraint solver is implemented in Constraint Handling Rules [1] (CHR). For reasons of space, we can only give a very brief sketch of the principles. CHR can be understood as rewriting rules over constraint stores; it has different sorts of rules, but in the fragment shown below we use only simplification rules. A rule of form *constraints-before* <=> *guard* | *constraints-after* can apply if a collection of constraints matching the pattern *constraints-before* is found in the store and the test in *guard* succeeds; in that case *constraints-before* are replaced by *constraints-after*. The *after* part may also refer to auxiliary code written in Prolog. We show here some of the rules for processing constraints of the form "··· *=* *date* + *n* days" (they are preceded by rules that brings all applications of plus into this form). The version shown here is slightly simplified as it ignores constraints of the form "in *numeric-interval*".

```
T *=* date(M,Dn) + N days  <=>
     ground(Dn), ground(N), DnN is Dn+N,
     lastSafeDateInMonth(M,Max),
     DnN =< Max
   |
     T *=* date(M,DnN).


T *=* date(M,Dn) + N days  <=>
     ground(Dn), ground(N), DnN is Dn+N,
     lastDateInMonth(M, Max), DnN =< Max
   |
     T *=* date(M,DnN).


T *=* date(M,Dn) + N days  <=>
     ground(Dn), ground(N), DnN is Dn+N,
     lastDateInMonth(M,Max), DnN > Max
   |
     Dn1 is DnN-Max-1,
     T *=* date(M1,1) + Dn1 days,
     M1 *=* M + 1 months.
```

The two first rules apply when an addition can be made giving a new day-of-month guaranteed not to lead into the following month. The first one uses the auxiliary predicate lastSafeDateInMonth; if the month is completely unspecified or is a Feb. in an unknown year, it returns 28, and when more information is present it returns the highest safe number (28 or 29 for Feb., 30 or 31 for others). The second rule takes care of cases not caught by the first rule using a more precise test (due to the first rule, actually only involved when the incremented date will be a 29th, 30 or 31). The last rule applies when the incremented date is in a later month.

Inconsistency Handling

An inconsistent set of constraints may arise due to misconceptions in a text. A basic constraint solver, incapable of handling inconsistency, may include the following rule.

```
T *=* month(Y,Mn) <=> T=month(Y,Mn).
```

It is executed when an absolute date is entered or an increment has been successfully added. An inconsistency manifests itself by the left and righthand sides being non-unifiable, and thus the execution of the equality predicate will result in the whole computation failing – which is logically correct as there is no solution to the total set of constraints. We can avoid this and still get a partial solution using a constraint `failed(···)` in the following way.

```
T *=* month(Y,Mn) <=> (T=month(Y,Mn) -> true ; failed((T=month(Y,Mn))).
```

The arrow-semicolon notation stands for if-then-else, so if the unification succeeds, everything is fine, otherwise we record that failure would have occurred if the unification had been enforced. Evaluations before and after the critical point are still executed. The constraint solver can be extended, so it indicates the position of a possible source of inconsistency in the text.

6   Conclusions

A constraint logic language *CLP(Time)* is introduced which can specify a wide range of dependencies between time expression in a natural language text. A constraint solver is demonstrated that can evaluate these expressions independently of the order in which anchoring times may be introduced, and it can produce meaningful results also when such anchors are absent. These properties, which do not hold for some state-of-the-art systems, are inherent in constraint solving, so the main message of this paper is to advocate constraint technologies for resolving time expressions.

CLP(Time) is intended to be used together with language analyzers capable of setting up relevant constraints, which is not a trivial task, and the results obtained will critically depend on the quality of the language analyzer. However, realistic tests together with a capable analyzer are yet to be made. Fragments of the constraint solver programmed in a rule-based fashion were shown, indicating a highly modular and easily extendible structure. This makes it a reasonable task to add new facilities, for example to match special usages in a particular language or more specific, idiomatic expressions used in a particular corpus.

For the implementation, we avoided finite domain techniques that have some drawbacks for calendric data. These techniques are not fitted for an incremental, detailed propagation of values necessary to figure out, say, that adding two days to a date which is the 28th of some yet unknown month in the 1900 years will preserve the century (and carry over variables referring to decade, year in decade and perhaps to the month), as we have demonstrated. Furthermore, a finite domain constraint solver typically works in two phases, first it collects and simplifies constraints, then at the end there is a grounding phase that attempts to assign concrete values to the variables; this is opposite to the incremental propagation of as much information as possible as we have aimed at. The modular structure of an implementation in Constraint Handling Rules makes it possible to add rules one by one for different special cases taking care of the particularities of calendric data and relationships that distinguish them from plan integer arithmetic. As mentioned, *CLP(Time)* and its constraint solver is currently being extended to handle different sorts of time intervals that we did not describe in the present paper. Our approach to handle inconsistency introduces a new research topic of identifying a best or minimal repair of an inconsistent set of time constraints, where our current version just pics an arbitrary one determined by the constraint solver's internal evaluation order.

References

1. Frühwirth, T.: Constraint Handling Rules. Cambridge University Press (2009)
2. Strötgen, J., Gertz, M.: Heideltime: High quality rule-based extraction and normalization of temporal expressions. In: Proceedings of the 5th International Workshop on Semantic Evaluation, SemEval 2010, pp. 321–324. Association for Computational Linguistics, Stroudsburg (2010)
3. Strötgen, J., Gertz, M.: Multilingual and cross-domain temporal tagging. Language Resources and Evaluation 47(2), 269–298 (2013)
4. de Medeiros Caseli, H., Villavicencio, A., Teixeira, A.J.S., Perdigão, F.: Computational Processing of the Portuguese Language. In: de Medeiros Caseli, H., Villavicencio, A., Teixeira, A.J.S., Perdigão, F. (eds.) PROPOR 2012. LNCS, vol. 7243, pp. 1–11. Springer, Heidelberg (2012)
5. Hovy, D., Fan, J., Gliozzo, A.M., Patwardhan, S., Welty, C.A.: When did that happen? - Linking events and relations to timestamps. In: Daelemans, W., Lapata, M., Màrquez, L. (eds.) EACL, pp. 185–193. The Association for Computer Linguistics (2012)
6. Chambers, N.: Labeling documents with timestamps: Learning from their time expressions. In: ACL (1), pp. 98–106. The Association for Computer Linguistics (2012)
7. Verhagen, M., Gaizauskas, R.J., Schilder, F., Hepple, M., Moszkowicz, J., Pustejovsky, J.: The TempEval challenge: identifying temporal relations in text. Language Resources and Evaluation 43(2), 161–179 (2009)
8. Pustejovsky, J., Castaño, J.M., Ingria, R., Sauri, R., Gaizauskas, R.J., Setzer, A., Katz, G., Radev, D.R.: Timeml: Robust specification of event and temporal expressions in text. In: Maybury, M.T. (ed.) New Directions in Question Answering, pp. 28–34. AAAI Press (2003)
9. Mazur, P.P., Dale, R.: LTIMEX: representing the local semantics of temporal expressions. In: Ganzha, M., Maciaszek, L.A., Paprzycki, M. (eds.) FedCSIS, pp. 201–208 (2011)
10. Eshghi, K.: Abductive planning with event calculus. In: Kowalski, R.A., Bowen, K.A. (eds.) ICLP/SLP, pp. 562–579. MIT Press (1988)
11. Allen, J.F.: Towards a general theory of action and time. Artif. Intell. 23(2), 123–154 (1984)

12. Han, B., Lavie, A.: A framework for resolution of time in natural language. ACM Trans. Asian Lang. Inf. Process. 3(1), 11–32 (2004)
13. Bry, F., Rieß, F.A., Spranger, S.: CaTTS: calendar types and constraints for web applications. In: Ellis, A., Hagino, T. (eds.) WWW, pp. 702–711. ACM (2005)
14. van de Camp, M., Christiansen, H.: Resolving Relative Time Expressions in Dutch Text with Constraint Handling Rules. In: Duchier, D., Parmentier, Y. (eds.) Constraint Solving and Language Processing. LNCS, vol. 8114, pp. 166–177. Springer, Heidelberg (2013)
15. Christiansen, H., Dahl, V.: HYPROLOG: A New Logic Programming Language with Assumptions and Abduction. In: Gabbrielli, M., Gupta, G. (eds.) ICLP 2005. LNCS, vol. 3668, pp. 159–173. Springer, Heidelberg (2005)
16. Christiansen, H.: Executable specifications for hypothesis-based reasoning with Prolog and Constraint Handling Rules. J. Applied Logic 7(3), 341–362 (2009)
17. Christiansen, H.: Constraint programming for context comprehension (to appear, 2014)
18. Christiansen, H.: Abductive language interpretation as bottom-up deduction. In: Wintner, S. (ed.) Natural Language Understanding and Logic Programming, Roskilde, Denmark. Datalogiske Skrifter, vol. 92, pp. 33–47 (2002)
19. Christiansen, H.: CHR Grammars. Int'l Journal on Theory and Practice of Logic Programming 5(4-5), 467–501 (2005)
20. Christiansen, H., Have, C.T., Tveitane, K.: From use cases to UML class diagrams using logic grammars and constraints. In: RANLP 2007: Proc. Intl. Conf. Recent Adv. Nat. Lang. Processing, pp. 128–132 (2007)
21. Christiansen, H., Have, C.T., Tveitane, K.: Reasoning about use cases using logic grammars and constraints. In: CSLP 2007: Proc. 4th Intl. Workshop on Constraints and Language Processing. Roskilde University Computer Science Research Report, vol. 113, pp. 40–52 (2007)
22. Dahl, V., Blache, P.: Implantation de grammaires de propriétés en CHR. In: Mesnard, F. (ed.) Actes des 13èmes Journées Francophones de Programmation en Logique avec Contraintes, Hermes (2004)
23. Bavarian, M., Dahl, V.: Constraint based methods for biological sequence analysis. Journal of Universal Computing Science 12(11), 1500–1520 (2006)
24. Dahl, V., Gu, B.: A CHRG analysis of ambiguity in biological texts. In: CSLP 2007: Proc. 4th Intl. Workshop on Constraints and Language Processing. Roskilde University Computer Science Research Report, vol. 113, pp. 53–64 (2007)
25. Christiansen, H., Li, B.: Approaching the chinese word segmentation problem with CHR grammars. In: CSLP 2011: Proc. 4th Intl. Workshop on Constraints and Language Processing. Roskilde University Computer Science Research Report, vol. 134, pp. 21–31 (2011)

Maximal Parallelism in Membrane Systems with Generated Membrane Boundaries

Zoltán Ernő Csajbók[1] and Tamás Mihálydeák[2]

[1] Department of Health Informatics, Faculty of Health, University of Debrecen,
Sóstói út 2-4, H-4400 Nyíregyháza, Hungary
csajbok.zoltan@foh.unideb.hu
[2] Department of Computer Science, Faculty of Informatics, University of Debrecen,
Kassai út 26, H-4028 Debrecen, Hungary
mihalydeak.tamas@inf.unideb.hu

Abstract. In membrane systems, the maximal parallelism is a useful tool for modeling real biotic/chemical interactions. After all, there are many attempts to relax maximal parallelism at the definition level, e.g. minimal parallelism, bounded parallelism etc., or even at the system level as the metabolic P system. By the help of topological means, membrane computations and maximal parallelism can be controlled. Besides, in natural processes, the events represented by communication rules take place in the vicinity of the membranes. The authors, motivated by natural phenomena, propose a framework in which the abstract notion of boundaries along membranes is modeled. In this paper, behaviors of communication rules restricted to these membrane boundaries are presented, in particular, showing how these restrictions affect the maximal parallelism.

Keywords: Membrane computing, communication P systems, maximal parallelism, multiset theory, partial approximation of multisets.

1 Introduction

Parallelism is a fundamental aspect of real life processes. Moreover, it is of great importance, e.g., in software design and database management systems.

In membrane systems, invented by Păun [14,15,17], the maximal parallelism is a useful tool for modeling real biotic/chemical interactions. However, designing, controlling parallelized computational processes, measuring the degree of parallelism (DoP), etc. are complicated tasks [1,2,5,18]. There are many attempts to relax maximal parallelism in order to find more realistic P systems from a biological point of view and, last but not least, be able to handle it exactly. Just a few examples are: minimal or bounded parallelism [3,6] at the definition level; assigning priorities, preferences to evolution/communication rules [18] at the computation level; or even at the system level, e.g., the metabolic P system [8,7]. Membrane computations and maximal parallelism can be controlled with the help of topological means [4] as well.

Besides, the events in natural phenomena which are represented by communication rules in P systems take place in the vicinity of the membranes.

A. Beckmann, E. Csuhaj-Varjú, and K. Meer (Eds.): CiE 2014, LNCS 8493, pp. 103–112, 2014.
© Springer International Publishing Switzerland 2014

In particular, an object actually has to be close enough to a membrane in order to be able to pass through it. Here, the words "vicinity", "close" do not necessarily refer to a space-like notion. Consequently, an abstract notion of boundaries along the membranes in P systems is required.

In [9,11], such an abstract, not necessarily space-like, notion of "to be close to a membrane" was proposed. Restricting the communication rules to these boundaries, the movements of objects through membranes can locally be controlled to a certain extent during the membrane computations. Although this approach restricts, it preserves the two important principles of the basic version of P systems, namely, the maximal parallelism and the nondeterminism. Thus, this method may be a means to capture maximal parallelism in P systems.

In this paper, Section 2 and 3 outline the fundamentals of multiset approximation framework and communication P systems. In Section 4, the abstract notion of membrane boundaries is presented. Then, in Section 5, the behavior of communication rules in the presence of membrane boundaries is investigated, showing how the restrictions of communication rules affect maximal parallelism.

2   Basics of Multiset Approximation Framework

2.1   Multisets

Let U be a finite nonempty set and $\mathbb{N}$ denote the set of natural numbers. A *multiset* M over U, or mset M for short, is a mapping $M : U \to \mathbb{N} \cup \{\infty\}$.

Let $\mathcal{MS}(U)$ denote the set of all msets over U.

M is *finite* if $M(a) < \infty$ $(a \in U)$. M is the *empty mset*, denoted by $\emptyset$, if $M(a) = 0$ for all $a \in U$. If M is finite, it can be represented by all permutations of the string $w = a_{k_1}^{M(a_{k_1})} a_{k_2}^{M(a_{k_2})} \ldots a_{k_l}^{M(a_{k_l})}$ if $M \neq \emptyset$, and $w = \lambda$ otherwise (λ is the empty string). Any permutation of the string w also can represent M.

A set $\mathcal{M}$ of finite msets is called a *macroset* over U. The following two fundamental macrosets are defined: $\mathcal{MS}^n(U)$ $(n \in \mathbb{N})$, the set of all msets M over U such that $M(a) \leq n$ for all $a \in U$, and $\mathcal{MS}^{<\infty}(U) = \bigcup_{n=0}^{\infty} \mathcal{MS}^n(U)$.

Let $M, M_1, M_2 \in \mathcal{MS}(U)$ and $\mathcal{M} \subseteq \mathcal{MS}(U)$.

Mset *equality* relation is: $M_1 = M_2$ if $M_1(a) = M_2(a)$ for all $a \in U$; mset *inclusion* relation is: $M_1 \sqsubseteq M_2$ if $M_1(a) \leq M_2(a)$ for all $a \in U$.

Mset *intersection* $M_1 \sqcap M_2$ is defined by $(M_1 \sqcap M_2)(a) = \min\{M_1(a), M_2(a)\}$ $(a \in U)$, and $(\sqcap \mathcal{M})(a) = \min\{M(a) \mid M \in \mathcal{M}\}$ $(a \in U)$. *Set-type union* $M_1 \sqcup M_2$ is defined by $(M_1 \sqcup M_2)(a) = \max\{M_1(a), M_2(a)\}$ $(a \in U)$, and $(\sqcup \mathcal{M})(a) = \sup\{M(a) \mid M \in \mathcal{M}\}$ $(a \in U)$. By definition, $\sqcup \emptyset = \emptyset$.

Mset addition and *mset subtraction* is defined by $(M_1 \oplus M_2)(a) = M_1(a) + M_2(a)$ $(a \in U)$ and $(M_1 \ominus M_2)(a) = \max\{M_1(a) - M_2(a), 0\}$ $(a \in U)$.

For $n \in \mathbb{N}$, *n-times addition* of M, denoted by $\oplus_n M$, is given by the following inductive definition: 1) $\oplus_0 M = \emptyset$; 2) $\oplus_1 M = M$; 3) $\oplus_{n+1} M = \oplus_n M \oplus M$.

The *n-times inclusion* relation ($\sqsubseteq^n$) is defined for any $n \in \mathbb{N}$. Let $M_1(\neq \emptyset)$, $M_2 \in \mathcal{MS}(U)$. $M_1 \sqsubseteq^n M_2$ if $\oplus_n M_1 \sqsubseteq M_2$ but $\oplus_{n+1} M_1 \not\sqsubseteq M_2$.

2.2   Multiset Approximation Framework

In order to model the boundary zones along membranes in P systems, the approach of rough set theory (RST) [12,13] should be a plausible opportunity. However, RST works within conventional set theory, therefore, its ideas have to be generalized for multisets. It is called the *multiset approximation framework*.

A multiset approximation framework has five basic components [10,11]:

- a finite set of abstract objects called the *alphabet*;
- a set of msets over the alphabet as the *domain*;
- some distinguished msets of the domain for the *basis* of approximations;
- *definable msets* derived from base msets;
- an *approximation pair* which determines lower and upper approximations.

Informally, the alphabet models different constituents of biotic/chemical entities. Msets are unordered collections of these abstract objects in which the multiplicities are allowed. The domain can be thought of as a set of observable msets. Some distinguished members of the domain are chosen in order to keep and use them together as a unit. It can be viewed as a *stable* joint occurrence of one or more objects and it exactly is what is called the base mset. If a base mset is stable then its n-times additions can also be viewed stable. Base msets and their n-times additions can be taken as the representation of different kinds of coexistence in chemical processes and symbiosis in living nature.

Formally, the ordered 5-tuple $\mathsf{MAS}(U) = \langle \mathcal{MS}^{<\infty}(U), \mathfrak{B}, \mathfrak{D}_{\mathfrak{B}}, \mathsf{l}, \mathsf{u} \rangle$ is an *mset approximation space* over a finite alphabet U with the domain $\mathcal{MS}^{<\infty}(U)$ if

1. $\mathfrak{B} \subseteq \mathcal{MS}^{<\infty}(U)$ and if $B \in \mathfrak{B}$, then $B \neq \emptyset$. $\mathfrak{B}$ is called the *base system*, its members are called the *base msets*;
2. $\mathfrak{D}_{\mathfrak{B}} \subseteq \mathcal{MS}^{<\infty}(U)$ is an extension of $\mathfrak{B}$ with the minimal requirement: if $B \in \mathfrak{B}$, $\oplus_n B \in \mathfrak{D}_{\mathfrak{B}}$ $(n \in \mathbb{N})$; members in $\mathfrak{D}_{\mathfrak{B}}$ are called *definable msets*;
3. the functions $\mathsf{l}, \mathsf{u} : \mathcal{MS}^{<\infty}(U) \to \mathcal{MS}^{<\infty}(U)$ (called *lower* and *upper approximations*) form a *weak approximation pair* $\langle \mathsf{l}, \mathsf{u} \rangle$ if
   (C0) $\mathsf{l}(\mathcal{MS}^{<\infty}(U)), \mathsf{u}(\mathcal{MS}^{<\infty}(U)) \subseteq \mathfrak{D}_{\mathfrak{B}}$ (*definability* of l, u);[1]
   (C1) the functions l and u are monotone (*monotonicity* of l, u);
   (C2) $\mathsf{u}(\emptyset) = \emptyset$ (*normality* of u);
   (C3) if $M \in \mathcal{MS}^{<\infty}(U)$, then $\mathsf{l}(M) \sqsubseteq \mathsf{u}(M)$ (*weak approximation property*).

A weak approximation pair $\langle \mathsf{l}, \mathsf{u} \rangle$ is

(C4) *granular* if $B \in \mathfrak{B}$, then $\mathsf{l}(\oplus_n B) = \oplus_n B$ $(n \in \mathbb{N})$ (l is granular);
(C5) *standard* if $D \in \mathfrak{D}_{\mathfrak{B}}$, then $\mathsf{l}(D) = D$ (l is *standard*).
(C6) *lower semi-strong* if $\mathsf{l}(M) \sqsubseteq M$ $(M \in \mathcal{MS}^{<\infty}(U))$ (l is *contractive*);
(C7) *upper semi-strong* if $M \sqsubseteq \mathsf{u}(M)$ $(M \in \mathcal{MS}^{<\infty}(U))$ (u is *extensive*);
(C8) *strong* if it is lower and upper semi-strong simultaneously.

The general mset approximation space $\mathsf{MAS}(U)$ is a *weak/granular/standard/ lower semi-strong/upper semi-strong/strong* mset approximation space if the approximation pair $\langle \mathsf{l}, \mathsf{u} \rangle$ is weak/granular/standard/lower semi-strong/upper semi-strong/ strong, respectively.

---

[1] $\mathsf{l}(\mathcal{MS}^{<\infty}(U))$, $\mathsf{u}(\mathcal{MS}^{<\infty}(U))$ denote the ranges of the functions l and u, respectively.

3   Communication P Systems

There are some different variants of the basic version of P system. Among them, the posed problem can adequately be discussed in communication P systems. In the proposed model, communication rules are the classical symport/antiport types which are applied in a nondeterministic and maximally parallel way.

A membrane structure μ of degree m $(m \geq 1)$ is a rooted tree with m nodes identified with the integers $1, \ldots, m$. It can be represented by the set $R_\mu \subseteq \{1, \ldots, m\} \times \{1, \ldots, m\}$. $\langle i, j \rangle \in R_\mu$ means that there is an edge from i (parent) to j (child) of the tree μ which is formulated by $\mathsf{parent}(j) = i$.

Let μ be a membrane structure with m nodes and V be a finite alphabet.

Definition 1. *The tuple $\Pi = \langle V, E, \mu, w_1, w_2, \ldots, w_m, R_1, R_2, \ldots, R_m \rangle$ is a communication P system, or P system for short, where*

1. *$E(\subseteq V)$ and objects from E are available in the environment in infinite multiplicities;*
2. *$w_i \in \mathcal{MS}^{<\infty}(V)$ represent regions $(i = 1, 2, \ldots, m)$;*
3. *R_i is a finite set of rules $(i = 1, 2, \ldots, m)$ such that if $r \in R_i$*
   (a) symport rules: $\langle u, in \rangle$, $\langle u, out \rangle$, where $u \neq \lambda$ and there is an mset $M \in \mathcal{MS}^{<\infty}(V)$ such that u represents M; or
   (b) antiport rule: $\langle u, in; v, out \rangle$, where $u \neq \lambda, v \neq \lambda$ and there are msets $M_1, M_2 \in \mathcal{MS}^{<\infty}(V)$ such that u, v represent M_1, M_2, respectively.

Example 1. Let us define a running example P system Π as follows.

$$\Pi = \langle V, E, \mu, w_1, w_2, w_3, w_4, w_5, R_1, R_2, R_3, R_4, R_5 \rangle,$$

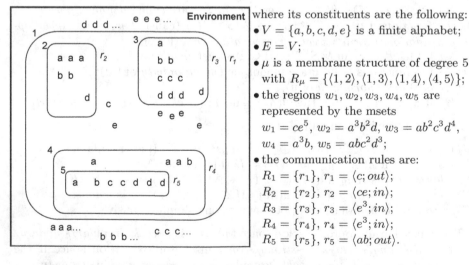

where its constituents are the following:
- $V = \{a, b, c, d, e\}$ is a finite alphabet;
- $E = V$;
- μ is a membrane structure of degree 5 with $R_\mu = \{\langle 1, 2 \rangle, \langle 1, 3 \rangle, \langle 1, 4 \rangle, \langle 4, 5 \rangle\}$;
- the regions w_1, w_2, w_3, w_4, w_5 are represented by the msets
  $w_1 = ce^5$, $w_2 = a^3b^2d$, $w_3 = ab^2c^3d^4$, $w_4 = a^3b$, $w_5 = abc^2d^3$;
- the communication rules are:
  $R_1 = \{r_1\}$, $r_1 = \langle c; out \rangle$;
  $R_2 = \{r_2\}$, $r_2 = \langle ce; in \rangle$;
  $R_3 = \{r_3\}$, $r_3 = \langle e^3; in \rangle$;
  $R_4 = \{r_4\}$, $r_4 = \langle e^3; in \rangle$;
  $R_5 = \{r_5\}$, $r_5 = \langle ab; out \rangle$.

Fig. 1. An example P system represented as a set of nested membranes

4 Membrane Boundaries

The proposed model is a two-component structure consisting of a P system and a multiset approximation space. Let Π be a communication P system as the first component of the model. In order to define the second component, a special multiset approximation space is formed over the finite alphabet V of Π.

Let $\mathsf{MAS}(V) = \langle \mathcal{MS}^{<\infty}(V), \mathfrak{B}, \mathfrak{D}_{\mathfrak{B}}, \mathsf{l}, \mathsf{u} \rangle$ be an mset approximation space with the finite alphabet V of Π. Let $\mathfrak{B}^{\oplus} = \{\oplus_n B \mid B \in \mathfrak{B},\ n = 1, 2, \dots\}$.

Definition 2. $\mathsf{MAS}(V)$ *is a* strictly set-union type *mset approximation space if* $\mathfrak{D}_{\mathfrak{B}}$ *is given by the following inductive definition:*

1. $\emptyset \in \mathfrak{D}_{\mathfrak{B}}$ $\mathfrak{B}^{\oplus} \subseteq \mathfrak{D}_{\mathfrak{B}}$, *and*
2. *if* $\mathfrak{B}' \subseteq \mathfrak{B}^{\oplus}$, *then* $\bigsqcup \mathfrak{B}' \in \mathfrak{D}_{\mathfrak{B}}$.

Definition 3. *Let* $\mathsf{MAS}(V)$ *be a strictly set-union type mset approximation space. The functions* $\mathsf{l}, \mathsf{u}, \mathsf{b} : \mathcal{MS}^{<\infty}(V) \to \mathcal{MS}^{<\infty}(V)$ *form a* Pawlakian mset approximation pair $\langle \mathsf{l}, \mathsf{u} \rangle$ *and the* boundary *if for any mset* $M \in \mathcal{MS}^{<\infty}(V)$

1. $\mathsf{l}(M) = \bigsqcup\{\oplus_n B \mid n \in \mathbb{N}^+, B \in \mathfrak{B} \text{ and } B \sqsubseteq^n M\}$,
2. $\mathsf{b}(M) = \bigsqcup\{\oplus_n B \mid B \in \mathfrak{B}, B \not\sqsubseteq M,\ B \sqcap M \neq \emptyset \text{ and } B \sqcap M \sqsubseteq^n M\}$,
3. $\mathsf{u}(M) = \mathsf{l}(M) \sqcup \mathsf{b}(M)$.

A strictly set-union type approximation space with a Pawlakian mset approximation pair is called a *Pawlakian mset approximation space.*

Proposition 1. *If* $\mathsf{MAS}(V)$ *is a Pawlakian mset approximation space, it is lower semi-strong and granular.*

If the P system $\Pi = \langle V, E, \mu, w_1, w_2, \dots, w_m, R_1, R_2, \dots, R_m \rangle$ is given, let $\mathsf{MAS}(\Pi) = \langle \mathcal{MS}^{<\infty}(V), \mathfrak{B}, \mathfrak{D}_{\mathfrak{B}}, \mathsf{l}, \mathsf{u} \rangle$ be a Pawlakian mset approximation space. $\mathsf{MAS}(\Pi)$ is called a *joint (mset) approximation space* of the P system Π.

Lower/upper approximations and boundaries of the msets $w_1, w_2, \dots, w_m$ can be formed in $\mathsf{MAS}(\Pi)$. They are called *region lower/upper approximations* and *boundaries*. Region upper approximations and boundaries associated with not skin membrane have to be adjusted to the membrane structure.

Definition 4. *Let* Π *be a* P *system and* $\mathsf{MAS}(\Pi)$ *be its joint membrane approximation space. If* $B \in \mathfrak{B}$ *and* $i = 1, 2, \dots, m$, *let*

$$
N(B, i) = \begin{cases} 0, & \text{if } B \sqsubseteq w_i \text{ or } B \sqcap w_i = \emptyset; \\ n, & \text{if } i = 1 \text{ and } B \sqcap w_1 \sqsubseteq^n w_1; \\ \min\{k, n \mid B \sqcap w_i \sqsubseteq^k w_i, \text{ and } B \ominus w_i \sqsubseteq^n w_{\mathsf{parent}(i)}\}, & \text{otherwise.} \end{cases}
$$

Then, the functions membrane boundaries, outside *and* inside membrane boundaries *are defined as follows* $(i = 1, \dots, m)$:

$$\mathsf{bnd}(w_i) = \bigsqcup\{\oplus_{N(B,i)} B \mid B \in \mathfrak{B}\};$$
$$\mathsf{bnd}^{\mathsf{out}}(w_i) = \mathsf{bnd}(w_i) \ominus w_i;$$
$$\mathsf{bnd}^{\mathsf{in}}(w_i) = \mathsf{bnd}(w_i) \ominus \mathsf{bnd}^{\mathsf{out}}(w_i).$$

Corollary 1. $\mathsf{b}(w_1) = \mathsf{bnd}(w_1)$ *and* $\mathsf{bnd}(w_i) \sqsubseteq \mathsf{b}(w_i)$ $(i = 2, \dots, m)$.

Using membrane boundaries, a rule $r \in R_i$ of a membrane i $(i = 1, \ldots, m)$ has to work only in the membrane boundary. More precisely,

- symport rules of the form $\langle u, in \rangle$ and $\langle v, out \rangle$ are executed only in the case when $u \sqsubseteq \mathsf{bnd}^{out}(w_i)$ and $v \sqsubseteq \mathsf{bnd}^{in}(w_i)$, respectively;
- an antiport rule of the form $\langle u, in; v, out \rangle$ is executed only in the case when $u \sqsubseteq \mathsf{bnd}^{out}(w_i)$ and $v \sqsubseteq \mathsf{bnd}^{in}(w_i)$.

Example 2. Let Π be the P system as in Example 1. Let us define its joint approximation space $\mathsf{MAS}(\Pi) = \langle \mathcal{MS}^{<\infty}(V), \mathfrak{B}, \mathfrak{D}_\mathfrak{B}, l, u \rangle$ as follows:

- $V = \{a, b, c, d, e\}$ is the finite alphabet as in Π;
- $\mathcal{MS}^{<\infty}(V)$ is the domain of the approximation space;
- $\mathfrak{B} = \{a^2, ab, b, cde\}$ is the base system;
- $\mathfrak{D}_\mathfrak{B}$ is the set of definable sets in such a way that
  - $\emptyset \in \mathfrak{D}_\mathfrak{B}$;
  - $\mathfrak{B}^\oplus = \{a^2, a^4, a^6, \ldots, ab, a^2b^2, a^3b^3, \ldots, b, b^2, b^3, \ldots,$ $cde, c^2d^2e^2, c^3d^3e^3, \ldots\}$, and for any $\mathfrak{B}' \subseteq \mathfrak{B}^\oplus$, $\bigsqcup \mathfrak{B}' \in \mathfrak{D}_\mathfrak{B}$;
  - $\mathfrak{D}_\mathfrak{B}$ does not have any other member;
- $\langle l, u \rangle$ is a Pawlakian approximation pair.

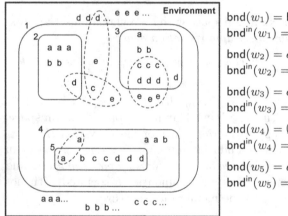

$\mathsf{bnd}(w_1) = \mathsf{b}(w_1) = cde,$
$\mathsf{bnd}^{in}(w_1) = ce, \ \mathsf{bnd}^{out}(w_1) = d;$

$\mathsf{bnd}(w_2) = cde,$
$\mathsf{bnd}^{in}(w_2) = d, \ \mathsf{bnd}^{out}(w_2) = ce;$

$\mathsf{bnd}(w_3) = c^3d^3e^3$
$\mathsf{bnd}^{in}(w_3) = c^3d^3, \ \mathsf{bnd}^{out}(w_3) = e^3;$

$\mathsf{bnd}(w_4) = \emptyset,$
$\mathsf{bnd}^{in}(w_4) = \emptyset, \ \mathsf{bnd}^{out}(w_4) = \emptyset;$

$\mathsf{bnd}(w_5) = a^2,$
$\mathsf{bnd}^{in}(w_5) = a, \ \mathsf{bnd}^{out}(w_5) = a.$

Fig. 2. Membrane boundaries determined by Definition 4

5 Maximal Parallelism in P Systems in the Presence of Membrane Boundaries

Communication rules in P systems compete for objects, so their maximal parallel execution mode has to be defined on the level of the whole system. To this end, sets of communication rules associated with different membranes are considered disjoint. It can be achieved by labeling the rules by distinct labels [16].

First, for the sake of comparison, the maximal parallel executions of communication rules are determined without membrane boundaries.

Let $\Pi = \langle V, E, \mu, w_1, w_2, \ldots, w_m, R_1, R_2, \ldots, R_m \rangle$ be a P system and $\mathsf{MAS}(\Pi) = \langle \mathcal{MS}^{<\infty}(V), \mathfrak{B}, \mathfrak{D}_{\mathfrak{B}}, \mathsf{l}, \mathsf{u} \rangle$ be its joint membrane approximation space.

Let $Appl(\Pi)$ denote the set of all multisets of communication rules which are applicable in Π *regardless of membrane boundaries*. Then, the set of multisets of communication rules from $Appl(\Pi)$ applicable to Π in the maximal parallel mode can be defined as follows [16]:

$$Appl_{max}(\Pi) = \{r \mid r \in Appl(\Pi) \text{ and there is no } r' \in Appl(\Pi) \text{ with } r' \subsetneqq r\}.$$

In this case, communication rules associated with neighboring membranes compete for objects in "shared" regions.

Example 3. Let the P system Π be as in Example 1. The set of multisets of communication rules applicable to Π regardless of membrane boundaries is:

$$Appl(\Pi) = \{r_1, r_2, r_3, r_4, r_5, r_1 r_3, r_1 r_4, r_1 r_5, r_2 r_3, r_2 r_4, r_2 r_5, r_3 r_5, r_4 r_5,$$
$$r_1 r_3 r_5, r_1 r_4 r_5, r_2 r_3 r_5, r_2 r_4 r_5\}.$$

Then, $Appl_{max}(\Pi) = \{r_1 r_3 r_5, r_1 r_4 r_5, r_2 r_3 r_5, r_2 r_4 r_5\}$.

When the communication rules are executed in the maximally parallel mode *in the presence of membrane boundaries*, they also compete for objects in general. However, in this case, the competition for objects takes place not in "shared" regions but "shared" inside/outside membrane boundaries. Consequently, the set of all multisets of communication rules applicable to Π depends on the mutual relation of neighboring inside and/or outside membrane boundaries.

Let $ApplBnd^{C_i}(\Pi)$ $(i \in I)$ denote the set of all multisets of communication rules applicable to Π in the presence of membrane boundaries, where I is a finite index set which may be empty. $ApplBnd^{C_i}(\Pi)$'s pertain to the *different relations*, configurations for short, of neighboring inside and/or outside membrane boundaries.

$ApplBnd^{C_i}_{max}(\Pi)$'s can be defined similar to $Appl_{max}(\Pi)$:

$$ApplBnd^{C_i}_{max}(\Pi) = \{r \mid r \in ApplBnd^{C_i}(\Pi) \text{ and there is no } r' \in ApplBnd^{C_i}(\Pi)$$
$$\text{with } r' \subsetneqq r\} \quad (i \in I).$$

It is straightforward that $ApplBnd^{C_i}_{max}(\Pi) \subseteq ApplBnd^{C_i}(\Pi)$ and any multiset from $ApplBnd^{C_i}(\Pi)$ can be executed regardless of membrane boundaries. For the sake of easy reference, these observations are formulated in a lemma.

Lemma 1. $ApplBnd^{C_i}_{max}(\Pi) \subseteq ApplBnd^{C_i}(\Pi) \subseteq Appl(\Pi)$ $(i \in I)$.

However, $ApplBnd^{C_i}_{max}(\Pi) \nsubseteq Appl_{max}(\Pi)$ in general.

If $r \in ApplBnd^{C_i}_{max}(\Pi)$, $r \in Appl(\Pi)$ also holds by Lemma 1. Then, by the maximal property of $Appl_{max}(\Pi)$, the following statement immediately follows.

Proposition 2. *For any* $r \in ApplBnd^{C_i}_{max}(\Pi)$ $(i \in I)$ *there is an* $r' \in Appl_{max}(\Pi)$ *in such a way that* $r \subseteq r'$.

Any maximal multiset of communication rules $r \in ApplBnd^{C_i}_{max}(\Pi)$ starting from the initial msets in the regions $w_1, \ldots, w_m$ leads to the same new msets as if r should be executed without membrane boundaries. However, r have a significant influence on the inside/outside boundaries of $w_1, \ldots, w_m$. Thus, after every transition step, membrane boundaries and $ApplBnd^{C_i}(\Pi)$, $ApplBnd^{C_i}_{max}(\Pi)$ have to be recalculated.

Example 4. Let the P system Π be as in Example 1 and its joint mset approximation space MAS(Π) be as in Example 2.

In this example, in the presence of membrane boundaries, membrane computation in Π is investigated starting from its initial configuration, see Fig. 1.

The communication rules r_4 and r_5 cannot be executed since $\mathsf{bnd}^{out}(w_4) = \emptyset$ and $ab \not\sqsubseteq \mathsf{bnd}^{in}(w_5) = a$, respectively (Fig. 2).

The communication rules r_1, r_2, r_3 can be executed (see Fig. 2) because

- $c \sqsubseteq \mathsf{bnd}^{in}(w_1) = ec$, i.e., $r_1 = \langle c; out \rangle$ is applicable;
- $ce \sqsubseteq \mathsf{bnd}^{out}(w_2) = ce$, i.e., $r_2 = \langle ce; in \rangle$ is applicable;
- $e^3 \sqsubseteq \mathsf{bnd}^{out}(w_3) = e^3$, i.e., $r_3 = \langle e^3; in \rangle$ is applicable;.

One can observe that r_1, r_2, r_3 compete for objects. More precisely, r_1, r_2 compete for one c, and r_2, r_3 compete for e's.

In order to determine the set of all multisets of communication rules applicable to Π, first, we have to take into account how rules r_1, r_2, r_3 compete for objects. There are six possibilities denoted by C_i ($i \in I = \{1, 2, 3, 4, 5, 6\}$).

In all configurations C_i's, $\mathsf{bnd}^{in}(w_1)$ and $\mathsf{bnd}^{out}(w_2)$ have one joint clone of c. The further features of different configurations are the following.

If $\mathsf{bnd}^{in}(w_1)$, $\mathsf{bnd}^{out}(w_2)$ have two different clones of e (Fig. 3(1)-(4)),

C_1: $\mathsf{bnd}^{out}(w_3)$ have other three clones of e;
C_2: $\mathsf{bnd}^{in}(w_1)$, $\mathsf{bnd}^{out}(w_3)$ have one joint clone of e, $\mathsf{bnd}^{out}(w_3)$ have other two clones of e;
C_3: $\mathsf{bnd}^{out}(w_2)$, $\mathsf{bnd}^{out}(w_3)$ have one joint clone of e, $\mathsf{bnd}^{out}(w_3)$ have other two clones of e;
C_4: $\mathsf{bnd}^{in}(w_1)$, $\mathsf{bnd}^{out}(w_3)$ have one joint clone of e, $\mathsf{bnd}^{out}(w_2)$, $\mathsf{bnd}^{out}(w_3)$ have one joint clone of e, $\mathsf{bnd}^{out}(w_3)$ have another clone of e.

If $\mathsf{bnd}^{in}(w_1)$, $\mathsf{bnd}^{out}(w_2)$ have one joint clone of e (Fig. 3(5)-(6)),

C_5: $\mathsf{bnd}^{out}(w_3)$ have other three clones of e;
C_6: $\mathsf{bnd}^{in}(w_1)$, $\mathsf{bnd}^{out}(w_2)$, $\mathsf{bnd}^{out}(w_3)$ have one joint clone of e, $\mathsf{bnd}^{out}(w_3)$ have other two clones of e.

The possible multisets of communication rules applicable to Π are:

- $ApplBnd^{C_1}(\Pi) = \{r_1, r_2, r_3, r_1r_3, r_2r_3\}$, $ApplBnd^{C_1}_{max}(\Pi) = \{r_1r_3, r_2r_3\}$;
- $ApplBnd^{C_2}(\Pi) = \{r_1, r_2, r_3, r_1r_3, r_2r_3\}$, $ApplBnd^{C_2}_{max}(\Pi) = \{r_1r_3, r_2r_3\}$;
- $ApplBnd^{C_3}(\Pi) = \{r_1, r_2, r_3, r_1r_3\}$, $ApplBnd^{C_3}_{max}(\Pi) = \{r_2, r_1r_3\}$;
- $ApplBnd^{C_4}(\Pi) = \{r_1, r_2, r_3, r_1r_3\}$, $ApplBnd^{C_4}_{max}(\Pi) = \{r_2, r_1r_3\}$;
- $ApplBnd^{C_5}(\Pi) = \{r_1, r_2, r_3, r_1r_3, r_2r_3\}$, $ApplBnd^{C_5}_{max}(\Pi) = \{r_1r_3, r_2r_3\}$;
- $ApplBnd^{C_6}(\Pi) = \{r_1, r_2, r_3, r_1r_3\}$, $ApplBnd^{C_6}_{max}(\Pi) = \{r_2, r_1r_3\}$.

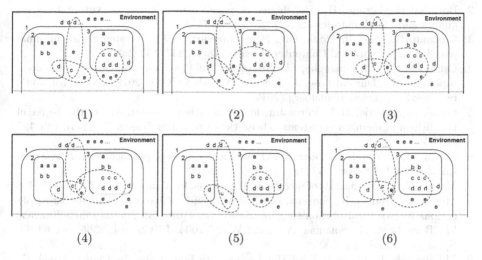

Fig. 3. Mutual relations between inside/outside boundaries

Note that although $ApplBnd_{max}^{C_1}(\Pi) = ApplBnd_{max}^{C_2}(\Pi) = ApplBnd_{max}^{C_5}(\Pi)$ and $ApplBnd_{max}^{C_3}(\Pi) = ApplBnd_{max}^{C_4}(\Pi) = ApplBnd_{max}^{C_6}(\Pi)$, after the executions of the same maximal multiset of rules setting from different configurations, the result inside/outside membrane boundaries may be different fundamentally.

6 Conclusion and Future Work

In this paper, a two-component structure consisting of a communication P system and a multiset approximation space with common finite alphabet has been proposed. Execution of communication rules are restricted to membrane boundaries generated by the multiset approximation space. Although this approach restricts, it preserves the two important principles of the original model of P system, the maximal parallelism and the nondeterminism.

In the future, the mutual relation between communication P systems and approximation spaces will be investigated.

Acknowledgements. The publication was supported by the TÁMOP–4.2.2.C–11/1/KONV–2012–0001 project. The project has been supported by the European Union, co–financed by the European Social Fund.

The authors are thankful to György Vaszil for his valuable comments.

References

1. Adorna, H., Păun, Gh., Pérez-Jiménez, M.J.: On communication complexity in evolution-communication P systems. Romanian Journal of Information Science and Technology 13(2), 113–130 (2010)

2. Bordihn, H., Fernau, H.: The degree of parallelism. Journal of Automata, Languages and Combinatorics 12(1-2), 25–47 (2007)
3. Ciobanu, G., Pan, L., Păun, Gh., Pérez-Jiménez, M.J.: P systems with minimal parallelism. Theoretical Computer Science 378(1), 117–130 (2007)
4. Csuhaj-Varjú, E., Gheorghe, M., Stannett, M.: P systems controlled by general topologies. In: Durand-Lose, J., Jonoska, N. (eds.) UCNC 2012. LNCS, vol. 7445, pp. 70–81. Springer, Heidelberg (2012)
5. Gutiérrez-Naranjo, M.A., Pérez-Jiménez, M.J., Riscos-Núñez, A.: On the degree of parallelism in membrane systems. Theoretical Computer Science 372(2-3), 183–195 (2007)
6. Ibarra, O.H., Yen, H.C., Dang, Z.: On various notions of parallelism in P systems. Int. J. Found. Comput. Sci. 16(4), 683–705 (2005)
7. Manca, V.: Fundamentals of metabolic P systems. In: Păun et al., [17], pp. 475–498
8. Manca, V., Bianco, L., Fontana, F.: Evolution and oscillation in P systems: Applications to biological phenomena. In: Mauri, G., Păun, Gh., Jesús Pérez-Jímenez, M., Rozenberg, G., Salomaa, A. (eds.) WMC 2004. LNCS, vol. 3365, pp. 63–84. Springer, Heidelberg (2005)
9. Mihálydeák, T., Csajbók, Z.E.: Membranes with boundaries. In: Csuhaj-Varjú, E., Gheorghe, M., Rozenberg, G., Salomaa, A., Vaszil, Gy. (eds.) CMC 2012. LNCS, vol. 7762, pp. 277–294. Springer, Heidelberg (2013)
10. Mihálydeák, T., Csajbók, Z.E.: Partial approximation of multisets and its applications in membrane computing. In: Lingras, P., Wolski, M., Cornelis, C., Mitra, S., Wasilewski, P. (eds.) RSKT 2013. LNCS, vol. 8171, pp. 99–108. Springer, Heidelberg (2013)
11. Mihálydeák, T., Csajbók, Z.E., Takács, P.: Communication rules working in generated membrane boundaries. In: Alhazov, A., Cojocaru, S., Gheorghe, M., Rogozhin, Y. (eds.) Proceedings of the 14th International Conference on Membrane Computing CMC 2014, August 20-23, pp. 241–254. Institute of Mathematics and Computer Science, Academy of Sciences of Moldova, Chisinău, Republic of Moldova (2013)
12. Pawlak, Z.: Rough sets. International Journal of Computer and Information Sciences 11(5), 341–356 (1982)
13. Pawlak, Z.: Rough Sets: Theoretical Aspects of Reasoning about Data. Kluwer Academic Publishers, Dordrecht (1991)
14. Păun, Gh.: Computing with membranes. Journal of Computer and System Sciences 61(1), 108–143 (2000)
15. Păun, Gh.: Membrane Computing. An Introduction. Springer, Berlin (2002)
16. Păun, Gh., Rozenberg, G.: An introduction to and an overview of membrane computing. In: Păun et al. [17], pp. 1–27
17. Păun, Gh., Rozenberg, G., Salomaa, A. (eds.): The Oxford Handbook of Membrane Computing. Oxford Handbooks. Oxford University Press, Inc., New York (2010)
18. Raman, A., Kim, H., Oh, T., Lee, J.W., August, D.I.: Parallelism orchestration using DoPE: the degree of parallelism executive. In: Hall, M.W., Padua, D.A. (eds.) Proceedings of the 32nd ACM SIGPLAN Conference on Programming Language Design and Implementation, PLDI 2011, San Jose, CA, USA, June 4-8, pp. 26–37. ACM (2011)

Learnability Thesis Does Not Entail Church's Thesis

Marek Czarnecki, Michał Tomasz Godziszewski, and Dariusz Kalociński

Department of Logic, Institute of Philosophy, University of Warsaw
ul. Krakowskie Przedmieście 3, 00-047 Warsaw, Poland
{marek.czarnecki,mtgodziszewski,dariusz.kalocinski}@gmail.com

Abstract. We consider the notion of intuitive learnability and its rela-
tion to intuitive computability. We briefly discuss the Church's Thesis.
We formulate the Learnability Thesis. Further we analyse the proof of the
Church's Thesis presented by M. Mostowski. We indicate which assumptions of the Mostowski's argument implicitly include that the Church's
Thesis holds. The impossibility of this kind of argument is strengthened
by showing that the Learnability Thesis does not imply the Church's
Thesis. Specifically, we show a *natural* interpretation of intuitive computability under which intuitively learnable sets are exactly algorithmically learnable but intuitively computable sets form a proper superset of
recursive sets.

Keywords: computability, algorithmic learnability, potential infinity,
FM-representability, low sets, Learnability Thesis, Church's Thesis.

1 Intuitive Computability and the Church's Thesis

Before the emergence of computability theory as a branch of modern logic, many
algorithms had been known. Historically, the first non-trivial algorithm, the Euclidean algorithm, dates back to circa 300 BC when the Greek mathematician,
Euclid of Alexandria, formulated his method for calculating the greatest common
divisor. In 1900, shortly before the appearance of the first mathematical models
of computation, Hilbert formulated his tenth problem of finding an algorithm
for deciding whether a given equation is solvable in integers. These and many
other historical examples convince us that even before the era of computability
we had some intuitive notion of algorithm, precise enough to be incorporated
by science. The era of the computability theory started in the 1930s and was
marked with the appearance of the first mathematical models of computation
[1], [4], [9], [13]. Almost immediately the following question arose: are the notions of intuitive computability and, for example, λ-definability or, what comes
to the same thing, Turing-computability, equivalent? In other words, is the class
of intuitively computable sets equal to the class of recursive sets? The affirmative
answer to this question is known as the Church's Thesis and was first formulated

A. Beckmann, E. Csuhaj-Varjú, and K. Meer (Eds.): CiE 2014, LNCS 8493, pp. 113–122, 2014.
© Springer International Publishing Switzerland 2014

in [1], [13].[1] The Church's Thesis, if not treated as definition, and we actually do not treat it as such,[2] is a statement about the equality of two classes of objects. From now on, by $\mathcal{IC}$ we mean a subset of $\mathcal{P}(\omega)$, consisting of intuitively computable sets of natural numbers. The class of recursive sets is known to be Δ_1^0 in arithmetical hierarchy.[3] Having this notation, the Church's Thesis presents shortly as follows:

Thesis 1 (Church's Thesis). $\mathcal{IC} = \Delta_1^0$.

The inclusion $\Delta_1^0 \subseteq \mathcal{IC}$ is generally accepted as a rule. The whole mystery lies in $\mathcal{IC} \subseteq \Delta_1^0$. $\mathcal{IC}$ is not fully understood. We have some intuitions based on practice in devising intuitive algorithms and writing computer programs. Our intuitions are strengthened by deep insights of computability theory. However, it is still possible, though unlikely, that $\Delta_1^0 \neq \mathcal{IC}$. This possibility is essentially used in the proof of the main theorem (Theorem 6) which is based on the notion of intuitive learnability.

2   Intuitive Learnability and the Learnability Thesis

The notion of intuitive learnability is based entirely on the notion of intuitive computability. Given the intuitive notion of an algorithm, one can define the notion of intuitive learnability as follows:

Definition 1 (Intuitive Learnability). *A decision problem is intuitively learnable if there is an (possibly infinite) intuitive algorithm that for each example of the problem produces a finite sequence of yeses and nos such that the last answer in the sequence is correct.*

The origins of the notion of intuitive learnability can be traced back to the same Euclid of Alexandria that is known as the author of the first non-trivial algorithm. His *Elements* contains the first exposition of the axiomatic method. The search for a proof of a sentence in a given axiomatic system may be viewed as an example of an intuitive algorithm that generates a finite sequence of answers as to whether the input sentence is provable. At the beginning the negative answer is produced. Then the space of proofs is systematically explored. If the input sentence is provable, the exploration finishes once the proof is found, the positive answer is produced, the algorithm stops and the generated sequence of answers is "no", "yes", with the last answer being correct. If the input sentence is not provable, the exploration goes on forever, and the generated sequence is always "no". This intuitive algorithm shows that the set of theorems of a recursive set of axioms is intuitively learnable.

---

[1] Another formulation of the Church's Thesis, in terms of functions, states that the class of intuitively computable functions is identical with the class of partial recursive functions. We restrict ourselves to the first formulation.

[2] Observe that treating $\mathcal{IC} = \Delta_1^0$ as a definition of $\mathcal{IC}$ strips away the whole problem, since then, $\mathcal{IC} \subseteq \Delta_1^0$ holds.

[3] For a detailed exposition of arithmetical hierarchy see, for example, [11].

Modern science, dating back to 17th century, provides another example of intuitive learnability. Consider a simplified model of the activity of a modern scientist. The scientist proposes a system of hypotheses. The system is to describe the world correctly. Initially, the positive answer is produced, meaning that hypotheses are considered true. Then the scientist proceeds to testing. If hypotheses are correct, testing goes on forever and the generated sequence of answers is always "yes". If hypotheses are incorrect, some test fails, the negative answer is produced, the activity stops and the generated sequence of answers is "yes", "no", with the last answer being correct. This intuitive algorithm shows that the problem of whether a system of empirical hypotheses describes the world correctly is intuitively learnable.[4]

The axiomatic method and the scientific method had appeared long before 1960s when algorithmic learning theory was established. The emergence and endurance of these sophisticated learning techniques provide a rationale that we had some intuitive understanding of learnability in times preceding its mathematical models.

Mathematical notion of learnability is due to Gold [3] and Putnam [10]. Here is Putnam's definition of algorithmic learnability that accounts for a mathematical counterpart of an intuitive idea of a set "decidable" by a mind-changing procedure:

Definition 2 (Algorithmic Learnability). *Let $A \subseteq \omega$. A is algorithmically learnable if there is a total computable function $g : \omega^2 \to \{0,1\}$ such that for all $x \in \omega$: $\lim_{t \to \infty} g(t, x) = 1 \Leftrightarrow x \in A$ and $\lim_{t \to \infty} g(t, x) = 0 \Leftrightarrow x \notin A$.*

Algorithmic learnability is equivalent with many natural notions. One of them is the notion of FM-representability proposed by Mostowski in [8]. His research was motivated by computational foundations of mathematics and the search for the semantics under which first-order sentences would be interpreted in potentially infinite domains. Potentially infinite domains are understood as growing sequences of finite models. We consider the latter to have purely relational vocabulary and initial segments of natural numbers as universes. Let $R \subseteq \omega^r$. Then by $R^{(n)}$ we denote $R \cap \{0, 1, \ldots, n\}^r$. For any model on natural numbers $\mathcal{A}$ over the signature $\sigma = (R_1, \ldots, R_k)$ we define the FM-domain of $\mathcal{A}$ as follows: $\mathrm{FM}(\mathcal{A}) = \{\mathcal{A}_n : n \in \omega\}$, where $\mathcal{A}_n = (\{0, 1, \ldots, n\}, R_1^{(n)}, \ldots, R_k^{(n)})$. By $\mathbb{N}$ we denote the standard model of arithmetic $(\omega, R_+, R_\times)$ of the vocabulary $\sigma = (R_+, R_\times)$, where instead of function symbols $+$, $\times$, we have corresponding relational symbols R_+, $R_\times$, interpreted in the same way as $+$, $\times$.

Definition 3 (FM-representability). *We say that the relation $R \subseteq \omega^r$ is FM-represented in $\mathrm{FM}(\mathcal{A})$ by a formula $\varphi(x_1, \ldots, x_r)$ if and only if for each $a_1, \ldots, a_r \in \omega$ both of the following conditions hold:*

$$R(a_1, \ldots, a_r) \text{ if and only if } \exists m \; \forall k \geq m \; \mathcal{A}_k \models \varphi(a_1, \ldots, a_r) \qquad (1)$$

---

[4] Our description is simplified. However, it seems, that it captures the main idea, that the system of empirical hypotheses cannot be conclusively justified but can be conclusively rejected (falsified).

$$\neg R(a_1, \ldots, a_r) \text{ if and only if } \exists m \, \forall k \geq m \, \mathcal{A}_k \models \neg\varphi(a_1, \ldots, a_r) \qquad (2)$$

*We say that R is **FM-representable** in* $\mathrm{FM}(\mathcal{A})$ *if there is a formula* φ *such that it FM-represents R in* $\mathrm{FM}(\mathcal{A})$. *If a relation is FM-representable in* $\mathrm{FM}(\mathbb{N})$ *we say that it is **FM-representable**.*

FM-representability is a good model of the semantic meaningfulness of mathematical concepts that we learn. The simplest argument is that objects, concepts and phenomena that are in the scope of cognitive accessibility and computational tractability for a human mind are of a finite character. Even if it is actually infinite, we may experience only its finite parts – hence we assume that the only epistemically reasonable notion of infinity we may adopt is the notion of potential infinity, explicated within the framework of FM-domains.

Subsequent theorem is a collection of notions that turned out to be equivalent to algorithmic learnability.

Theorem 1 (Limit Lemma). *Let* $R \subseteq \omega^r$. *Then the following are equivalent:*[5]

1. *R is recursive with recursively enumerable oracle,*
2. $\deg(R) \leq 0'$,
3. *R is algorithmically learnable,*
4. *R is* Δ_2^0,
5. *R is FM-representable.*

Algorithmic learnability and equivalent notions given in the Limit Lemma are of mathematical nature. However, as we showed in the Definition 1, the notion of learnability has also a very clear intuitive content that may be formulated using the notion of intuitive computability. Therefore we actually have two notions of learnability: the intuitive one, given in the Definition 1, and the mathematical one, given by any of equivalent statements in the Limit Lemma. And just as in the case of the notions of intuitive and, for example, Turing-computability, we face the question of whether the notion of intuitive learnability is equivalent to the notion of algorithmic learnability. In other words, is the class of intuitively learnable sets equal to the class of algorithmically learnable sets? We put forward a claim, under the name of Learnability Thesis, that intuitive learnability is equivalent to algorithmic learnability. From now on, by $\mathcal{IL}$ we mean a subset of $\mathcal{P}(\omega)$, consisting of intuitively learnable sets of natural numbers. The class of algorithmically learnable sets is, by the Limit Lemma, Δ_2^0. Having this notation, the Learnability Thesis presents shortly as follows:

Thesis 2 (Learnability Thesis). $\mathcal{IL} = \Delta_2^0$.

At this point, a natural question to ask is: why should we accept this claim? It is not our main purpose to argue in favour of the Learnability Thesis (we need it in our argumentation for the impossibility of the specific kind of proof

---

[5] The equivalence between 1, 2 and 4 is due to Shoenfield [12]. The equivalence between 3 and 4 is due to Gold [3] and Putnam [10]. The equivalence between 1-4 and 5 is due to Mostowski [5], [8] and is called the FM-representability theorem.

of the Church's Thesis). Nevertheless, as the Limit Lemma indicates, algorithmically learnable sets form a very natural class of objects. So far, the class has been discovered by researchers from three different domains: computability theory (Shoenfield), artificial intelligence (Gold), logic and philosophy (Putnam, Mostowski). Moreover, it is easy to see, that $\Delta_2^0 \subseteq \mathcal{IL}$ – the argument goes analogously to the one that shows $\Delta_1^0 \subseteq \mathcal{IC}$. The tricky part is $\mathcal{IL} \subseteq \Delta_2^0$. However, assuming the Church's Thesis, the argument trivialises (we provide it only for illustrative purposes).

Proposition 1. *The Church's Thesis entails the Learnability Thesis.*

Proof. Assume the Church's Thesis.

($\Delta_2^0 \subseteq \mathcal{IL}$) Let $A \in \Delta_2^0$. Let $y : \omega^2 \rightarrow \{0, 1\}$ be as in the Definition 2. By the Church's Thesis, g is an intuitively computable total function. Devise an intuitive infinite procedure for A, satisfying the Definition 1. Let $x \in \omega$. Set $t = 0$. In the infinite loop do: intuitively compute $g(t, x)$, output the result in case it differs from the result obtained previously, increment t. This shows $A \in \mathcal{IL}$.

($\mathcal{IL} \subseteq \Delta_2^0$) Let $A \in \mathcal{IL}$. Then there is an intuitive algorithm, say G, satisfying the Definition 1. Without loss of generality, G never stops. Devise an intuitive algorithm G' that takes (t, x) as an input and returns the last answer generated by G on the input x up to t steps of the intuitive computation. By the Church's Thesis, the function intuitively computed by G' is recursive. Let g be that function. Clearly, g is total and satisfies the Definition 2. Hence, by the Limit Lemma, A is Δ_2^0.

The main theorem of this paper (Theorem 6) states that the reverse implication does not hold. Before we give a proof, we analyse the proof of the Church's Thesis presented by Mostowski [6]. The proof of Mostowski goes in the direction that the Theorem 6 considers impossible. Of course, the proof of Mostowski uses some additional assumptions. We carefully discuss them and indicate their weak points.

3   Analysis of the Proof of Mostowski

In [6] M. Mostowski gives an argument for the Church's Thesis. The argument is based on three assumptions. **Ontological assumption**: there exist finitely, but potentially infinitely many objects. **Semantical assumption**: satisfaction and truth relations in finite models are recursive. **Epistemological assumption**: there exists a recursive enumeration of the FM-domain. It is namely assumed that cognitively accessible reality is finite, but potentially infinite, that our knowledge is expressible in our language and that it is decidable whether a given (without loss of generality - arithmetical) formula is satisfied in a finite, but sufficiently large (arithmetical) model and that enlarging the domain of the finite model we perform the computations (more generally: cognitive activity) in is recursive. Further, it is argued by the FM-representability theorem that the class of concepts that may be meaningfully described in a potentially infinite domain with recursive truth relation and recursive enumeration of finite approximations of the model is identical to the class of Δ_2^0 sets. Finally, an epistemological criterion separates computable relations from other FM-representable ones. The key notion employed in Mostowski's justification of the Church's Thesis is the notion of a **testing formula**.

Definition 4 (Testing Formula). *Let $R \subseteq \omega^n$ and $\varphi(x_1, \ldots, x_n)$ be a formula. A formula $\psi(x_1, \ldots, x_n)$ is a testing formula for $\varphi(x_1, \ldots, x_n)$ and R if:*

- *for each $a_1, \ldots, a_n \in \omega$ there is $n_0 \in \omega$ such that for each finite model M, $M \models \psi(a_1, \ldots, a_n)$ if and only if $|M| \geq n_0$,*
- *for each $a_1, \ldots, a_n \in \omega$ and each finite model M, if $M \models \psi(a_1, \ldots, a_n)$, then $R(a_1, \ldots, a_n)$ if and only if $M \models \varphi(a_1, \ldots, a_n)$.*

The conditions defining the notion of a testing formula for φ and R may be read as an explication of the concept of *knowing the answer (and achieving the answer effectively)* to the query of the form: *is a tuple $a_1, \ldots, a_n$ in the relation R?* Testing formulae then serve the abovementioned epistemological criterion of separating decidable relations from other FM-representable notions. This is justified by the following theorem.

Theorem 2 (Mostowski [7]). *Let $R \subseteq \omega^n$. R is decidable if and only if there are formulae $\varphi(x_1, \ldots, x_n)$, $\psi(x_1, \ldots, x_n)$ such that $\psi(x_1, \ldots, x_n)$ is a testing formula for $\varphi(x_1, \ldots, x_n)$ and R.*

Proof. Fix $R \subseteq \omega^n$.

($\Rightarrow$) Let $T(e, x_1, \ldots, x_n, c)$ be the Kleene predicate meaning that c is the code of the computation of the algorithm with code e on input $x_1, \ldots, x_n$ (note that every quantifier occurring in T is bounded by c). Let $U(c, y)$ mean that a computation with code c accepts if $y = 1$ or rejects if $y = 0$. Suppose that R is decidable and let e be the code of an algorithm deciding R. We define:

$$\psi(x_1, \ldots, x_n) = \exists c\, T(e, x_1, \ldots, x_n, c),$$

$$\varphi(x_1, \ldots, x_n) = \exists c\, (T(e, x_1, \ldots, x_n, c) \wedge U(c, 1)).$$

Fix $\bar{a} = a_1, \ldots, a_n \in \omega$. We show that ψ is a testing formula for φ and R. We have $\mathbb{N} \models \exists c\, T(e, \bar{a}, c)$ thus for some $n_0 \in \omega$ it holds that $\mathbb{N} \models T(e, \bar{a}, n_0)$. Since the computation of e on $\bar{a}$ is unique, so is n_0. Therefore for $m \in \omega$, $\mathbb{N}_m \models \psi(\bar{a})$ if and only if $m \geq n_0$.

Now fix $m \in \omega$ such that $\mathbb{N}_m \models \psi(\bar{a})$. Let $n_0 \in \omega$ be such that $\mathbb{N} \models T(e, \bar{a}, n_0)$. Then for every $m \geq n_0$ it holds that $\mathbb{N}_m \models T(e, \bar{a}, n_0)$. If $R(\bar{a})$, then $\mathbb{N} \models U(n_0, 1)$ and $\mathbb{N}_m \models \varphi(\bar{a})$. On the other hand if $\neg R(\bar{a})$, then $\mathbb{N} \models U(n_0, 0)$ and $\mathbb{N}_m \models \neg\varphi(\bar{a})$.

Therefore $\psi(x_1, \ldots, x_n)$ is a testing formula for φ and R.

($\Leftarrow$) Let $\psi(x_1, \ldots, x_n)$ be a testing formula for $\varphi(x_1, \ldots, x_n)$ and R. The algorithm deciding R is the following.

Algorithm 1. Algorithm deciding R

Input: $a_1, \ldots, a_n \in \omega$
Output: truth value of $R(a_1, \ldots, a_n)$
1: $i \leftarrow 0$
2: **while** $\mathbb{N}_i \not\models \psi(a_1, \ldots, a_n)$ **do**
3:    $i \leftarrow i + 1$
4: **end while**
5: **return** truth value of $\mathbb{N}_i \models \varphi(a_1, \ldots, a_n)$

The algorithm implicitly uses subroutines to compute $i \mapsto \ulcorner \mathbb{N}_i \urcorner$ and $\mathbb{N}_i \models \alpha$ which are both recursive. It also always halts since $\psi(x_1, \ldots, x_n)$ is a testing formula for $\varphi(x_1, \ldots, x_n)$ and R. This ends the proof.

It is clear now that the Theorem 2 enables to identify recursive relations as the class for which we are able to *know* the model in which the truth of the relation's representing formula fixes. As we see, the proof of the Theorem 2 depends on two following statements:

1. There is a recursive enumeration of finite models,
2. Every finite model $\mathbb{N}_m$ has a recursive satisfaction relation.

While the second assumption is not controversial we take a closer look at 1. This takes us directly to key considerations needed in the proof of the Theorem 6. It is worth noting that the main assumptions of Mostowski's argument (namely the abovementioned ontological one and semantical one) taken together with the FM-representability theorem are actually equivalent to a version of the Learnability Thesis. It is so, since by those assumptions we model relations that can be meaningfully described in potentially infinite by an appropriate growing sequence of finite models with **computable** satisfaction relation. To put it in an even stronger way, one might say that any formal model compatible with ontological and semantical assumptions of Mostowski (which by the way seem to be plausible philosophical statements in general) shall be a class of finite models such that *meaningful* concepts are computed in the limit. In particular, such semantics gives us a class of formulae *decidable in the limit*, i.e. such that their interpretations stabilise after finitely many steps within an (potentially) infinite trial-and-error **computable** procedure. Such formulae express exactly intuitively learnable concepts. By the FM-representability theorem the set of such concepts is identical to the set of Δ_2^0 relations.

4   Learnability Thesis Does Not Entail Church's Thesis

So far, we have worked in relational arithmetical vocabulary $\sigma = (R_+, R_\times)$. Now we extend it to $\sigma' = \sigma \cup \{A\}$, where A is an additional 1-place predicate.

Theorem 3. *Let $(\mathbb{N}, A)$ be any σ'-model, $R \subseteq \omega^n$. R is decidable in A if and only if there are σ'-formulae $\varphi(x_1, \ldots, x_n), \psi(x_1, \ldots, x_n)$ such that $\psi(x_1, \ldots, x_n)$ is a testing formula in $\mathrm{FM}((\mathbb{N}, A))$ for $\varphi(x_1, \ldots, x_n)$ and R.*

Proof. The proof is an easy generalisation of the proof of the Theorem 2.

($\Rightarrow$) It suffices to consider the Kleene predicate $T^A(e, x_1, \ldots, x_n, c)$ for oracle machines, meaning that c is the code of the computation of the oracle algorithm with Gödel number e on input $x_1, \ldots, x_n$ using A as an oracle.

($\Leftarrow$) The algorithm deciding R is essentially the same as the one from the proof of the Theorem 2, but since the map $i \mapsto \ulcorner(\mathbb{N}_i, A^{(i)})\urcorner$ is recursive in A, R is recursive in A (rather than just recursive as in the original proof).

Taking FM($\mathbb{N}$) as our formal model is aimed at distinguishing exactly those properties that are essential for performing intuitive computations. It seems that considering the FM-domain of the finite cuts of an arithmetical model in which all predicate symbols have recursive interpretations, just as in case of FM($\mathbb{N}$), is actually equivalent to assuming that intuitively computable relations are exactly recursive ones, namely the Church's Thesis itself. Observe that if we admit the existence of some non-recursive but intuitively computable relations, we could intuitively compute the function $i \mapsto \ulcorner(\mathbb{N}_i, A^{(i)})\urcorner$ and by the theorem 2 exactly those relations which are recursive in A have testing formulae.

The arithmetical hierarchy can be naturally relativised to capture notions concerning computations relative to oracles. By extending the arithmetical vocabulary by an additional predicate and interpreting it as an oracle we obtain a relativised arithmetical hierarchy of definable notions relative to the oracle. A relation R is Δ_2^A if it is definable both by Σ_2^A and Π_2^A formulae i.e.:

$$R(\bar{a}) \equiv \exists x \forall y\, R(x, y, \bar{a}), \tag{3}$$

$$R(\bar{a}) \equiv \forall x \exists y\, S(x, y, \bar{a}), \tag{4}$$

for some recursive in A predicates R and S. The following theorem is obvious by the relativisation of the Limit Lemma:

Theorem 4. *Let $R \subseteq \omega^n$. Then R is FM-representable in $\mathrm{FM}((\mathbb{N}, A))$ if and only if R is Δ_2^A.*

Definition 5 (Low Sets). *Let $A \subseteq \omega$. A is low if $\deg(A)' = 0'$.*

Of course every recursive set is low, but the converse does not hold. The existence of non-recursive low sets is a folklore (see for example [2]).

Theorem 5. *Let A be a low set. Then $\Delta_2^A = \Delta_2^0$.*

Proof. Fix a low set A. The non-obvious inclusion is $\Delta_2^A \subseteq \Delta_2^0$.

Fix a Δ_2^A relation R. Then for some recursive in A predicates R and S we have:

$$R(\bar{a}) \equiv \exists x \underbrace{\forall y\, R(x, y, \bar{a})}_{\leq \deg(A)'}, \tag{5}$$

$$R(\bar{a}) \equiv \forall x \underbrace{\exists y\, S(x, y, \bar{a})}_{\leq \deg(A)'}. \tag{6}$$

Since A is low, $\deg(A)' = 0'$. Therefore by the generalised Post's theorem R is recursive in $0'$ and thus, by the Limit Lemma, R is Δ_2^0.

Now, by an easy application of Theorems 4 and 5, we obtain:

Corollary 1. *Let A be a low set and $R \subseteq \omega^n$. Then R is FM-representable in* FM$((\mathbb{N}, A))$ *if and only if R is Δ_2^0.*

By the Corollary 1, adding any low set A to the FM-domain does not affect the class of FM-representable relations and therefore the Learnability Thesis itself.

We are ready to prove our main theorem:

Theorem 6. *The Learnability Thesis does not entail the Church's Thesis.*

Proof. Let A be a low, non-recursive set. Let the interpretation of $\mathcal{IC}$ be $\{R : R \leq_T A\}$. Therefore under such an interpretation the Church's Thesis fails. On the other hand consider an FM-domain FM$((\mathbb{N}, A))$. We may consider such an FM-domain since $A \in \mathcal{IC}$. By the Corollary 1 relations FM-representable in FM$((\mathbb{N}, A))$ are exactly those which are Δ_2^0. Therefore the Learnability Thesis holds in such a model. We have shown that there is an interpretation of $\mathcal{IC}$ such that $\mathcal{IC} \neq \Delta_1^0$ and $\mathcal{IL} = \Delta_2^0$. Therefore the Learnability Thesis does not entail Church's Thesis.

5   Concluding Remarks

In this paper we have described the Learnability Thesis and argued that an attempt of justifying the Church's Thesis based only on the Learnability Thesis must fail, by the Theorem 6. The clue of the argument is that there exists an interpretation of intuitive computability consistent with the Learnability Thesis such that certain intuitively computable sets are by no means recursive.

One of the paths of criticism towards our main result could proceed by questioning the *naturality* of our interpretation of $\mathcal{IC}$, namely that it is only theoretically admissible.[6] This is why we have performed the proof of the Theorem 6 in the framework that Mostowski used in his argument. This enabled us to justify the *naturality* of the interpretation of $\mathcal{IC}$ as $\{R : R \leq_T A\}$, for some low set A.[7] Mostowski used a very natural notion of a testing formula to show that recursive relations are exactly those FM-representable relations (equivalently - intuitively learnable) which have testing formulae. We have pointed out a flaw in his argument to show that if we admit some non-recursive but intuitively computable

---

[6] The discussion on the naturality of the interpretation of $\mathcal{IC}$ started with our first attempt to prove that the Learnability Thesis does not entail Church's Thesis in which we considered $\mathcal{IC} = \{R : R \leq_T A$, for any low set $A\}$. Such an interpretation of intuitive computability, however, would have very unnatural properties since for instance there are low sets A, B such that their recursive sum $A \oplus B$ is Turing-equivalent to $0'$. Therefore a very natural operation such as taking a recursive sum of some two intuitively computable sets would lead to intuitively non-computable set (assuming the Learnability Thesis).

[7] Under such an interpretation, $\mathcal{IC}$ is closed under Turing-reducibility and therefore also under recursive sums, hence it is more *natural*.

relations we are able to consider FM-domains expanded with their interpretations. This has led to singling out the relations recursive in A as those which have testing formulae in $FM((\mathbb{N}, A))$. On the other hand, by the Corollary 1, relations FM-representable in such FM-domain are still Δ_2^0.

References

1. Church, A.: An unsolvable problem of elementary number theory. American Journal of Mathematics 58(2), 345–363 (1936)
2. Epstein, R.L.: Degrees of Unsolvability: Structure and Theory. Springer (1979)
3. Gold, E.M.: Limiting recursion. Journal of Symbolic Logic 30, 28–48 (1965)
4. Kleene, S.C.: General recursive functions of natural numbers. Mathematische Annalen 112, 727–742 (1936)
5. Mostowski, M.: On representing concepts in finite models. Mathematical Logic Quarterly 47, 513–523 (2001)
6. Mostowski, M.: Potential infinity and the Church Thesis. Fundamenta Informaticae 81, 241–248 (2007)
7. Mostowski, M.: On representability in finite arithmetics. In: Cordon-Franco, A., Fernandez-Margarit, A., Lara-Martin, F.F. (eds.) JAF26, 26eme Journees sur les Arithmetiques Faibles (26th Weak Arithmetics Days), pp. 53–64. Fenix Editora, Sevilla (2008)
8. Mostowski, M., Zdanowski, K.: FM-representability and beyond. In: Cooper, S.B., Löwe, B., Torenvliet, L. (eds.) CiE 2005. LNCS, vol. 3526, pp. 358–367. Springer, Heidelberg (2005)
9. Post, E.L.: Finite combinatory processes – formulation 1. The Journal of Symbolic Logic 1(3), 103–105 (1936)
10. Putnam, H.: Trial and error predicates and the solution to a problem of Mostowski. J. Symbolic Logic 30, 49–57 (1965)
11. Rogers, H.: Theory of Recursive functions and effective computability. McGraw-Hill, New York (1967)
12. Shoenfield, J.: On degrees of unsolvability. Annals of Mathematics, vol. 69 (1959)
13. Turing, A.M.: On computable numbers, with an application to the Entscheidungsproblem. Proceedings of the London Mathematical Society 42, 230–265 (1936)

Phase Transitions Related to the Pigeonhole Principle

Michiel De Smet* and Andreas Weiermann**

[1] Smidsestraat 192
B 9000 Ghent
Belgium
mds@michieldesmet.eu

[2] Department of Mathematics
Ghent University
Building S22
Krijgslaan 281
B 9000 Ghent
Belgium
Andreas.Weiermann@UGent.be

Abstract. Since Jeff Paris introduced them in the late seventies [Par78], densities turned out to be useful for studying independence results. Motivated by their simplicity and surprising strength we investigate the combinatorial complexity of two such densities which are strongly related to the pigeonhole principle. The aim is to miniaturise Ramsey's Theorem for 1-tuples. The first principle uses an unlimited amount of colours, whereas the second has a fixed number of two colours. We show that these principles give rise to Ackermannian growth. After parameterising these statements with respect to a function $f \colon \mathbb{N} \to \mathbb{N}$, we investigate for which functions f Ackermannian growth is still preserved.

Keywords: Ackermann function, pigeonhole principle, Ramsey theory, phase transitions.

1 Introduction

The pigeonhole principle is one of the most well-know combinatorial principles, due to both its simplicity and usefulness. The principle is also known as the chest-of-drawers principle or Schubfachprinzip and is attributed to Dirichlet in 1834. The pigeonhole principle can also be considered as a finite instance of Ramsey's theorem for 1-tuples. So, if RT^n_k stands for Ramsey's Theorem for n dimensions and k colours, i.e.

$$\mathrm{RT}^n_k \leftrightarrow \text{For every } G \colon [\mathbb{N}]^n \to k \text{ there exists an infinite set } H$$
$$\text{such that } G \restriction [H]^n \text{ is constant,}$$

---

* Research supported in part by Fonds Wetenschappelijk Onderzoek (FWO) - Flanders.
** Research supported in part by(FWO) and the John Templeton Foundation.

A. Beckmann, E. Csuhaj-Varjú, and K. Meer (Eds.): CiE 2014, LNCS 8493, pp. 123–132, 2014.

then the pigeonhole principle is a finite instance of $RT^1_{<\infty} = \forall k \, RT^1_k$. In this paper we will investigate miniaturisations of the statements $RT^1_{<\infty}$ and RT^1_2. Let us recall some results from reverse mathematics: for any fixed natural number k, $RCA_0 \vdash RT^1_k$, whereas $WKL_0 \nvdash RT^1_{<\infty}$. Both results are due to Hirst (see [Hir87], Theorem 6.3 and Theorem 6.5). In addition, there it is also proved that $RT^1_{<\infty}$ does not imply ACA_0 over RCA_0. As it does not fit nicely into the programme of reverse mathematics, one might be tempted to think that $RT^1_{<\infty}$ is of little importance. However, it pops up every now and then in the literature. It is, for instance, equivalent to Rado's Lemma over RCA_0 (see [Hir87], Theorem 6.6).

For miniaturising $RT^1_{<\infty}$ and RT^1_2 we define two notions of density, n-density and $(\alpha, 2)$-density, which are parametrised by a function $f \colon \mathbb{N} \to \mathbb{N}$. Using these notions we define two first order assertions and study their provability with respect to $I\Sigma_1$, the first-order part of RCA_0.

We show which f give rise to Ackermannian growth and determine the exact phase transition. In case of n-density Ackermannian growth is obtained for $f(i) = i^{\frac{1}{A_\omega^{-1}(i)}}$, whereas for $f(i) = i^{\frac{1}{A_d^{-1}(i)}}$ it is not. Here A_d denotes the d-th branch of the Ackermann function A_ω. Our proof will show that in these results A_ω (A_d) could be replaced by any non decreasing unbounded non primitive recursive function (resp. by any non decreasing unbounded primitive recursive function). In the case of $(\alpha, 2)$-density we restrict ourselves to only two colours and strength disappears, as expected. Surprisingly, iterating up to ω^2 suffices to gain proof theoretic strength again. It turns out that $f(i) = \frac{1}{A_d^{-1}(i)} \log(i)$ gives rise to no more than primitive recursive growth, but $f(i) = \frac{1}{A_\omega^{-1}(i)} \log(i)$ does. Our proof will show that also in these results A_ω (A_d) could be replaced by any non decreasing unbounded non primitive recursive function (resp. by any non decreasing unbounded primitive recursive function).

We would like to mention that the n-density threshold functions are exactly the same as those for the parameterised Kanamori-McAloon principle, whereas the $(\omega^2, 2)$-density functions equal those for the parameterised Paris-Harrington principle [KLOW08, WVH2012]. It is our hope that by investigating miniaturisations of $RT^1_{<\infty}$ and RT^2_2 one could obtain insights into the seemingly difficult question whether RT^2_2 does or does not prove the totality of the Ackermann function (the so called Ramsey for pairs problem).

For related work we also we also refer to [DSW08] and the unpublished PhD thesis of the first author [DS11].

2 n-Density

Henceforth, let $f \colon \mathbb{N} \to \mathbb{N}$ be any (elementary recursive function), such that $1 \le f(x) \le x$, for x large enough. We define the functions $F_{f,k}$ and F_f, depending on f, by

$$F_{f,0}(n) := n + 1$$
$$F_{f,k+1}(n) := \underbrace{F_{f,k}(\ldots(F_{f,k}(n))\ldots)}_{f(n) \text{ times}} := F_{f,k}^{f(n)}(n)$$
$$F_f(n) := F_{f,n}(n),$$

for every $k, n \in \mathbb{N}$.

If it is clear which f we are working with, we leave out the subscript f and simply write F_k and F, instead of $F_{f,k}$ and F_f, respectively. We also consider functions f with non integer values. It is then understood that we round a value $f(i)$ down to $\lfloor f(i) \rfloor$, the biggest natural number below $f(i)$. Moreover we assume that f has always values at least as big as 1. It is easy to verify that the functions $F_{f,k}$ and F_f are strictly monotonic increasing if the parameter function f is non decreasing.

In case of $f(i) = i$ we write A_ω for F_f and A_d for $F_{f,d}$. A_ω is a standard choice for the well known Ackermann function, which is a recursive but not a primitive recursive function. The function A_d is called the d-th branch of the Ackermann function. Every function A_d is primitive recursive. In [OW09] a classification is given of those functions f for which F_f is primitive or non primitive recursive.

Let us define n-density, the first density notion related to the pigeonhole principle. In this case the number of colours depends on the minimum of X and the function f.

Definition 1. X is called 0-dense(f) if $|X| \geq \max\{f(\min X), 3\}$. X is called $(n+1)$-dense(f) if for all $G : X \to f(\min X)$, there exists $Y \subseteq X$, such that Y is homogeneous for G and Y is n-dense(f).

Lemma 1. Assume that $k \leq l$ and that $X \subseteq [k, l]$ and that $f, g : [k, l] \to \mathbb{N}$ are two functions such that $f(i) \leq g(i)$ for all $i \in [k, l]$. If X is n-dense(g) then X is n-dense(f).

Proof. One verifies the claim easily by induction on n.

2.1 Upper Bound

Lemma 2. Let f be non decreasing. Let $n \in \mathbb{N}$ and $X \subseteq \mathbb{N}$ be a finite set. If X is n-dense(f), then $\max X \geq F_{f,n}(\min X)$.

Proof. Being of no importance for the proof itself, we leave out the subscript f. Henceforth, let $x_0 = \min X$ and $c = f(x_0)$. The proof goes by induction on n.

If X is 0-dense(f), then $|X| \geq \max\{f(x_0), 3\}$. Thus, $\max X \geq x_0 + 2 \geq F_0(x_0)$.

Secondly, assume the statement is proven for n and X is $(n+1)$-dense(f). Consider the following partition of $X = \cup_{0 \leq i < c} Y_i$, where Y_i is defined by

$$Y_i = \{x \in X | F_n^i(x_0) \leq x < F_n^{i+1}(x_0)\}$$

for $0 \leq i < c - 1$ and $Y_{c-1} = \{x \in X | F_n^{c-1}(x_0) \leq x\}$. Now, define $G : X \to c$, as follows

$$G(x) := i,$$

for $x \in Y_i$. Since X is $(n+1)$-dense(f), there exists a subset Y of X, such that Y is n-dense(f) and homogeneous for G. By contradiction assume $Y \subseteq Y_{i_0}$ for some i_0 with $0 \leq i_0 < c - 1$. The n-density of Y and the monotonicity of F_n yield

$$F_n^{i_0+1}(x_0) - 1 \geq \max Y_{i_0} \geq \max Y \geq F_n(\min Y) \geq$$
$$F_n(\min Y_{i_0}) \geq F_n(F_n^{i_0}(x_0)) = F_n^{i_0+1}(x_0),$$

a contradiction. So $Y \subseteq Y_{c-1}$, which implies

$$\max X = \max Y_{c-1} \geq \max Y \geq F_n(\min Y) \geq$$
$$F_n(\min Y_{c-1}) \geq F_n(F_n^{c-1}(x_0)) = F_n^c(x_0) = F_n^{f(x_0)}(x_0) = F_{n+1}(x_0),$$

by the n-density of Y. This concludes the induction argument.

Definition 2. *Define* $\mathrm{PHP}_f : \mathbb{N} \to \mathbb{N}$ *by*

$$\mathrm{PHP}_f(n) := \min\{n' \in \mathbb{N} | [n, n'] \text{ is } n\text{-dense}(f)\}.$$

Let $f(i) = i^{\frac{1}{A_\omega^{-1}(i)}}$, where A_ω denotes the Ackermann fuction. Then F_f is Ackermannian, due to Theorem 1 in [OW09]. If f would be non decreasing then Lemma 2 would yield

$$\mathrm{PHP}_f(n) \geq F_{f,n}(n) = F_f(n),$$

for all $n \in \mathbb{N}$, hence also PHP_f would Ackermannian. Since the provably total functions of $I\Sigma_1$ are exactly the primitive recursive functions, we would immediately obtain that PHP_f would not be provably recursive in $I\Sigma_1$. We now show how to overcome this problem.

Theorem 1. *If* $f(i) = i^{\frac{1}{A_\omega^{-1}(i)}}$, *then*

$$I\Sigma_1 \nvdash (\forall n)(\forall a)(\exists b)([a, b] \text{ is } n\text{-dense}(f)).$$

Proof. Let $p(n) := 4 + 3^{n+1} + (n+1)^{n+1}$ and let $f_k(i) := i^{\frac{1}{k}}$. It suffices to show that $\mathrm{PHP}_f(p(n)) > A_\omega(n)$ Assume that $\mathrm{PHP}_f(p(n)) \leq A_\omega(n)$. Then for $i \leq A_\omega(n)$ one has that $A_\omega^{-1}(i) \leq n$ which yields $f(i) \geq f_n(i)$ for all $i \leq \mathrm{PHP}_f(p(n))$. The proof of Claim 2.12 from [KLOW08] yields $F_{f_{n+1},n+n^2+4n+5}(p(n)) > A_\omega(n)$. Together with Lemma 1 this yields

$$\mathrm{PHP}_f(p(n)) \geq \mathrm{PHP}_{f_n}(p(n)) \geq F_{f_{n+1},n+n^2+4n+5}(p(n)) > A_\omega(n)$$

which is a contradiction.

2.2 Lower Bound

Let $f(i) = i^{\overline{A_d^{-1}(i)}}$, where A_d denotes the d-th branch of the Ackermann function A_ω. This function is not weakly increasing on its domain. For $i \in [A_d(k), A_d(k+1) - 1]$ one has that $A_d^{-1}(i) = k$ and on such intervals f will be non decreasing. In intervals of the form $i \in [A_d(k) - 1, A_d(k)]$ the function A_d^{-1} jumps from $k-1$ to k. But since the intervals of the form $[A_d(k), A_d(k+1) - 1]$ are rather long it is very easy to find enough points b such that $f(b) \geq f(i)$ for all $i \leq b$. One simply has to choose b so large that f majorizes $f(c)$ where c is the initial point of a last jump interval which comes before b. With this caveat we can consider f basically as non decreasing function although it in fact is not.

Theorem 2. *If* $f(i) = i^{\overline{A_d^{-1}(i)}}$, *then*

$$I\Sigma_1 \vdash (\forall n)(\forall a)(\exists b)([a, b] \text{ is } n\text{-dense}(f)).$$

Proof. Assume that n and a are given. Put $b := 2^{A_d(a2^{n+1})2^{n+1}}$. Then $f(i) \leq f(b)$ for all $i \leq b$. We claim that any $Y \subseteq [a, b]$ with $|Y| > 2^{A_d(a2^{n+1})2^k}$ is k-dense(f). To prove the claim we proceed by induction on k.

Assume the claim holds for $k-1$ and consider $Y \subseteq [a, b]$ with $|Y| > 2^{A_d(a2^{n+1})2^k}$. Since $2^{A_d(a2^{n+1})2^{n+1}} > A_d(2^{n+1})$, we have

$$f(\min Y) < f(b) = \left(2^{A_d(a2^{n+1})2^{n+1}}\right)^{\frac{1}{A_d^{-1}(2^{A_d(a2^{n+1})2^{n+1}})}}$$

$$\leq \left(2^{A_d(a2^{n+1})2^{n+1}}\right)^{\frac{1}{A_d^{-1}(A_d(2^{n+1}))}}$$

$$= \left(2^{A_d(a2^{n+1})2^{n+1}}\right)^{\frac{1}{2^{n+1}}} = 2^{A_d(a2^{n+1})} \leq 2^{A_d(a2^{n+1})2^{k-1}}.$$

Let $c = f(\min Y)$ and $G : Y \to c$ be any function. Consider the partition of Y induced by G, i.e.

$$Y = \cup_{0 \leq i < c} Y_i,$$

with $Y_i = \{y \in Y | G(y) = i\}$. By contradiction, assume that $|Y_i| \leq 2^{A_d(a2^{n+1})2^{k-1}}$ for every $0 \leq i < c$. Then

$$2^{A_d(a2^{n+1})2^k} < |Y| \leq c \cdot 2^{A_d(a2^{n+1})2^{k-1}} = f(\min Y) \cdot 2^{A_d(a2^{n+1})2^{k-1}}$$

$$< 2^{A_d(a2^{n+1})2^{k-1}} \cdot 2^{A_d(a2^{n+1})2^{k-1}} = 2^{A_d(a2^{n+1})2^{k-1} + A_d(a2^{n+1})2^{k-1}} = 2^{A_d(a2^{n+1})2^k},$$

a contradiction. Thus, there exists an index $i_0 \in \{0, \ldots, c-1\}$, such that $|Y_{i_0}| > 2^{A_d(a2^{n+1})2^{k-1}}$. The induction hypothesis yields that Y_{i_0} is $(k-1)$-dense(f) and by definition Y_{i_0} is homogeneous for G, so Y is k-dense(f).

If $k = 0$ then $|Y| > 2^{A_d(a2^{n+1})} > f(\min Y)$, which completes the induction argument and proves the claim.

Now return to $[a, b]$. $[a, b]$ is n-dense(f) since $|[a, b]| \geq 2^{A_d(a2^{n+1})2^{n+1}} - a \geq 2^{A_d(a2^{n+1})2^n}$. Remarking that the function $E : \mathbb{N} \times \mathbb{N} \to \mathbb{N}$, defined by $E(a, n) = 2^{A_d(a2^{n+1})2^{n+1}}$ is primitive recursive, completes the proof.

Let PHP_f stand for "$(\forall n)(\forall a)(\exists b)([a, b] \text{ is } n\text{-dense}(f))$". Then we obtain the following picture.

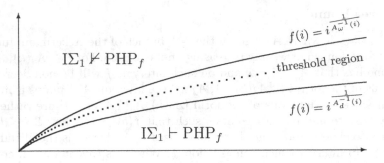

Fig. 1. Phase transition for PHP$_f$

3 $(\alpha, 2)$-Density

In this section we work with a fixed number of colours, namely two. For a limit ordinal not exceeding ω^2 we define the fundamental sequence as follows. We put $\omega \cdot (k+1)[n] = \omega \cdot k + n$ and we set $\omega^2[n] = \omega \cdot n$.

Definition 3. X *is called* $(0, 2)$-*dense(f) if* $|X| \geq \max\{f(\min X), 3\}$. X *is called* $(\alpha + 1, 2)$-*dense(f) if for all* $G : X \to 2$ *there exists* $Y \subseteq X$, *such that* Y *is* $(\alpha, 2)$-*dense(f) and* Y *is homogeneous for* G. *If* λ *is a limit ordinal, then* X *is called* $(\lambda, 2)$-*dense(f) if for all* $G : X \to 2$ *there exists* $Y \subseteq X$, *such that* Y *is* $(\lambda[\min X], 2)$-*dense(f) and* Y *is homogeneous for* G.

Lemma 3. *Let* $k \leq l$ *and* $f, g : [k, l] \to \mathbb{N}$ *be non decreasing such that* $f(i) \leq g(i)$ *for all* $i \in [k, l]$. *If* $X \subseteq [k, l]$ *is* $(n, 2)$-*dense(g) then* X *is is* $(n, 2)$-*dense(g)*.

3.1 Upper Bound

In this section we will use another hierarchy which we call $B_{f,\alpha}$ and which turns out to be related to $F_{f,k}$. We define B_α, depending on f, by

$$B_{f,0}(n) := n + 1$$
$$B_{f,\alpha+1}(n) := B_{f,\alpha}(B_{f,\alpha}(n)) := B_{f,\alpha}^2(n)$$
$$B_{f,\lambda}(n) := B_{f,\lambda[f(n)]}(n),$$

for all $n \in \mathbb{N}$ and ordinals α and λ, with the latter a limit ordinal. As for F_f, we leave out the subscript f and write B_α if it is clear which f we are working with. We first show a simple lemma concerning the relation between the two hierarchies defined in this paper. This lemma might be considered as folklore.

Lemma 4. *Let* k, l *and* m *be natural numbers. Then* $B_{f, \omega \cdot k + l}(m) = F_{2^f, k}^{2^l}(m)$.

Proof. We proceed by main induction on k and subsidiary induction on l.
     If k equals l equals zero, we have $B_{f,0}(m) = m + 1 = F_{2^f, 0}(m)$.

Assume the statement is proven for $k-1$, we will prove it for k by subsidiary induction on l.

If $l = 0$, then the main induction hypothesis yields

$$B_{f,\omega \cdot (k-1)}(m) = F_{2^f, k-1}(m).$$

Assume the claim is proven for $l-1$. We have

$$B_{f,\omega \cdot (k-1)+l}(m) = B_{f,\omega \cdot (k-1)+l-1}(B_{f,\omega \cdot (k-1)+l-1}(m))$$
$$= F_{2^f, k-1}^{2^{l-1}}(F_{2^f, k-1}^{2^{l-1}}(m)) = F_{2^f, k-1}^{2^l}(m),$$

which proves the statement for $k-1$ and every l. Using this fact, we obtain

$$B_{f,\omega \cdot k}(m) = B_{f,\omega \cdot (k-1)+\omega[f(m)]}(m)) = B_{f,\omega \cdot (k-1)+f(m)}(m))$$
$$= F_{2^f, k-1}^{2^{f(m)}}(m) = F_{2^f, k}(m),$$

which concludes the main induction and proves the statement.

Lemma 5. *Let f be non decreasing. Let α be any ordinal not extenting ω^2. If $X \subseteq \mathbb{N}$ is $(\alpha, 2)$-dense(f), then $\max X \geq B_{f,\alpha}(\min X)$.*

Proof. Being of no importance for the proof itself, we leave out the subscript f. Henceforth, let $x_0 = \min X$. The proof goes by transfinite induction on α.

If X is $(0, 2)$-dense(f), then $|X| \geq \max\{f(x_0), 3\}$. Thus, $\max X \geq x_0 + 2 > x_0 + 1 = B_0(x_0)$.

Assume the statement is proven for α and X is $(\alpha + 1, 2)$-dense(f). Define $G : X \to 2$ as follows

$$G(x) := \begin{cases} 0 & \text{if } x_0 < x < B_\alpha(x_0) \\ 1 & \text{if } B_\alpha(x_0) \leq x \end{cases},$$

for all $x \in X$. Since X is $(\alpha + 1, 2)$-dense(f), there exists a subset Y of X, such that Y is $(\alpha, 2)$-dense(f) and Y is homogeneous with respect to G. By contradiction, assume G takes colour 0 on Y. Then, by the induction hypothesis,

$$B_\alpha(x_0) - 1 \geq \max Y \geq B_\alpha(\min Y) = B_\alpha(x_0),$$

a contradiction. So, the colour needs to be 1, which implies

$$Y \subseteq \{x \in X | B_\alpha(x_0) \leq x\}.$$

The induction hypothesis yields

$$\max X \geq \max Y \geq B_\alpha(\min Y) = B_\alpha(B_\alpha(x_0)) = B_{\alpha+1}(x_0).$$

Finally, assume the statement is proven for all $\alpha < \lambda$, with λ a limit ordinal, and X is $(\lambda, 2)$-dense(f). There exists a subset Y which is $(\lambda[f(x_0)], 2)$-dense(f). We obtain by the induction hypothesis

$$\max X \geq \max Y \geq B_{\lambda[f(x_0)]}(\min Y) \geq B_{\lambda[f(x_0)]}(x_0) = B_\lambda(x_0).$$

This completes the proof.

Definition 4. *Define* $PHP2_f : \mathbb{N} \to \mathbb{N}$ *by*

$$PHP2_f(n) := \min\{n' \in \mathbb{N}|[n, n'] \text{ is } (\omega^2, 2)\text{-}dense(f)\}.$$

Fix $f(i) = \frac{1}{A_\omega^{-1}(i)} \log(i)$ for the rest of this subsection.

Lemma 6. *Let* $f_n(i) := \frac{1}{n} \cdot \log_2(i)$. *Then* $PHP2_{f_n}(2^{n^2}) \geq F_{2^{f_n}}(n)$ *holds for every* $n \in \mathbb{N}$.

Proof. Let $X = [2^{n^2}, PHP2_f(2^{n^2})]$. Define $G : X \to 2$ by $G(x) = 0$ for every $x \in X$. Since X is $(\omega^2, 2)$-dense(f_n) there exists $Y \subseteq X$, such that Y is $(\omega^2[f_n(\min X)], 2)$-PHP2-dense$(f_n)$, i.e. $(\omega \cdot f_n(2^{n^2}), 2)$-dense$(f_n)$. Lemma 4 and Lemma 5 yield

$$PHP2(2^{n^2}) \geq \max Y \geq B_{f_n, \omega \cdot f(\min Y)}(\min Y) \geq B_{f_n, \omega \cdot f(2^{n^2})}(2^{n^2})$$

$$= F_{2^{f_n}, f_n(2^{n^2})}(2^{n^2}) \geq F_{2^{f_n}, n}(n) = F_{2^{f_n}}(n),$$

since $f_n(2^{n^2}) = n$.

Corollary 1. *If* $f(i) = \frac{1}{A_\omega^{-1}(i)} \log(i)$, *then*

$$I\Sigma_1 \nvdash (\forall a)(\exists b)([a, b] \text{ is } (\omega^2, 2)\text{-}dense(f)).$$

Proof. Let $p(n) = 4 + 3^{n+1} + (n+1)^{n+1}$ and $f_k(i) := \frac{1}{k} \log_2(i)$. It suffices to show that $PHP2_f(2^{p(n)^2}) > A_\omega(n)$. Assume for a contradiction that $PHP2_f(2^{p(n)^2}) \leq A_\omega(n)$. For $i \leq A_\omega(n)$ one has $A_\omega^{-1}(i) \leq n$ hence $f(i) \geq f_n(i)$ for all $i \leq PHP2_f(2^{p(n)^2})$. This yields $PHP2_f(2^{p(n)^2}) \geq PHP2_{f_n}(2^{p(n)^2}) \geq F_{2^{f_n}}(p(n)) > A_\omega(n)$. Contradiction!

3.2 Lower Bound

As in Section 2.2 let $f(i) = \frac{1}{A_d^{-1}(i)} \log(i)$, where A_d denotes the dth branch of the Ackermann function A_ω. Recall from Section 2.2 that f is almost non decreasing and that it is easy to identify the jumps for f.

Theorem 3. *If* $f(i) = \frac{1}{A_d^{-1}(i)} \log(i)$, *then*

$$I\Sigma_1 \vdash (\forall a)(\exists b)([a, b] \text{ is } (\omega^2, 2)\text{-}dense(f)).$$

Proof. Assume that a is given. Put $b := 2^{A_d(2^{a+2})2^{a+1}}$. Then $f(i) \leq f(b)$ for all $i \leq b$. We claim that any $Y \subseteq [a, b]$, with $|Y| > 2^{A_d(2^{a+2})2^k}$ is $(\omega \cdot k, 2)$-dense(f). The proof goes by induction on k.

Let $k = 0$. Since $2^{A_d(2^{a+2})2^{a+1}} > A_d(2^{a+2})$, we have

$$f(\min Y) < f(b) = \frac{1}{A_d^{-1}(2^{A_d(2^{a+2})2^{a+1}})} \log(2^{A_d(2^{a+2})2^{a+1}})$$

$$< \frac{1}{2^{a+2}} A_d(2^{a+2})2^{a+1} < A_d(2^{a+2}),$$

so, $|Y| > 2^{A_d(2^{a+2})} > \max\{f(\min Y), 3\}$, i.e. Y is $(0, 2)$-dense(f).

Assume the assertion holds for $k - 1$ and consider $Y \subseteq [a, b]$ with $|Y| > 2^{A_d(2^{a+2})2^k}$. We claim that if $Z \subseteq Y$ and $|Z| > 2^{A_d(2^{a+2})2^{k-1}+l}$, then Z is $(\omega \cdot (k - 1) + l, 2)$-dense$(f)$. The proof goes by subsidiary induction on l.

If $l = 0$, then the claim follows by the main induction hypothesis. Assume the claim holds for $l - 1$ and $|Z| > 2^{A_d(2^{a+2})2^{k-1}+l}$. Let $G : Z \to 2$ be any function. Consider the partition of Z induced by G, i.e.

$$Z = Z_0 \cup Z_1,$$

with $Z_i = \{z \in Z | G(z) = i\}$. By contradiction, assume that

$$|Z_i| \le 2^{A_d(2^{a+2})2^{k-1}+l-1},$$

for $i = 0, 1$. Then

$$2^{A_d(2^{a+2})2^{k-1}+l} < |Z| \le 2 \cdot 2^{A_d(2^{a+2})2^{k-1}+l-1} = 2^{A_d(2^{a+2})2^{k-1}+l},$$

a contradiction. Thus, there exists an index $i_0 \in \{0, 1\}$, such that $|Z_{i_0}| > 2^{A_d(2^{a+2})2^{k-1}+l-1}$. The induction hypothesis yields Z_{i_0} is $(\omega \cdot (k - 1) + l - 1, 2)$-dense$(f)$, and so Z is $(\omega \cdot (k - 1) + l, 2)$-dense$(f)$, since Z_{i_0} is homogeneous for G. This proves the latter claim.

Now return to Y. Let $G : Y \to 2$ be any function. Consider the partition of Y induced by G, i.e.

$$Y = Y_0 \cup Y_1,$$

with $Y_i = \{y \in Y | G(y) = i\}$. In the same way as above, one can prove there exists an index $i_0 \in \{0, 1\}$, such that

$$|Y_{i_0}| > 2^{A_d(2^{a+2})2^k - 1} = 2^{A_d(2^{a+2})2^{k-1}+A_d(2^{a+2})2^k - 1}.$$

Since

$$A_d(2^{a+2})2^{k-1} \ge A_d(2^{a+2}) \ge f(\min Y) + 1,$$

we have $|Y_{i_0}| > 2^{A_d(2^{a+2})2^{k-1}+f(\min Y)}$. The latter claim yields Y_{i_0} is $(\omega \cdot (k - 1) + f(\min Y), 2)$-dense$(f)$, i.e. $(\omega \cdot k[f(\min Y)], 2)$-dense$(f)$. Thus Y is $(\omega \cdot k, 2)$-dense(f), since Y_{i_0} is homogeneous for G.

We finally prove that $[a, b]$ is $(\omega^2, 2)$-dense(f). Let $G : [a, b] \to 2$ be any function and consider the partition of $[a, b]$ induced by G, i.e.

$$[a, b] = Y_0 \cup Y_1,$$

with $Y_i = \{y \in [a, b] | G(y) = i\}$. Remark that $|[a, b]| > 2^{A_d(2^{a+2})2^{a+1}} - a \ge 2^{A_d(2^{a+2})2^a+1}$. Similarly as before, there exists an index $i_0 \in \{0, 1\}$, such that

$$|Y_{i_0}| > 2^{A_d(2^{a+2})2^a} \ge 2^{A_d(2^{a+2})2^{f(a)}}.$$

The main claim yields Y is $(\omega \cdot f(a), 2)$-dense(f), i.e. $(\omega^2[f(a)], 2)$-dense(f). In combination with Y_{i_0} being homogeneous for G, this implies $[a, b]$ is $(\omega^2, 2)$-dense(f).

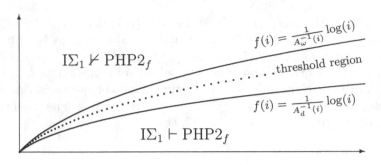

Fig. 2. Phase transition for PHP2$_f$

Let PHP2$_f$ stand for "$(\forall n)(\forall a)(\exists b)([a,b]$ is $(\omega^2, 2)$-dense$(f))$". Once again, we obtain the following picture.

In accordance with the referees (for which we are grateful for valuable comments) we expect that it will be not too hard to show that for natural choices of f the principles PHP$_{\log \circ f}$ and PHP2$_f$ are equivalent over IΣ_1.

References

[DSW08]   De Smet, M., Weiermann, A.: Phase Transitions for Weakly Increasing Sequences. In: Beckmann, A., Dimitracopoulos, C., Löwe, B. (eds.) CiE 2008. LNCS, vol. 5028, pp. 168–174. Springer, Heidelberg (2008)

[DS11]   De Smet, M.: Unprovability and phase transitions in Ramsey theory. PhD thesis, Ghent (2011)

[Hir87]   Hirst, J.L.: Combinatorics in subsystems of second-order arithmetic. PhD thesis, Pennsylvania State University (1987)

[KLOW08]   Kojman, M., Lee, G., Omri, E., Weiermann, A.: Weiermann: Sharp thresholds for the phase transition between primitive recursive and Ackermannian Ramsey numbers. JCT A. 115(6), 1036–1055 (2008)

[OW09]   Omri, E., Weiermann, A.: Classifying the phase transition threshold for Ackermannian functions. Ann. Pure Appl. Logic 158(3), 156–162 (2009)

[WVH2012]   Weiermann, A., Hoof, W.V.: Sharp phase transition thresholds for the Paris Harrington Ramsey numbers for a fixed dimension. Proc. Amer. Math. Soc. 140(8), 2913–2927 (2012)

[Par78]   Paris, J.B.: Some independence results for Peano arithmetic. J. Symbolic Logic 43(4), 725–731 (1978)

Generic Parallel Algorithms

Nachum Dershowitz and Evgenia Falkovich

School of Computer Science, Tel Aviv University, Tel Aviv, Israel
nachum.dershowitz@cs.tau.ac.il, jenny.falkovich@gmail.com

Abstract. We develop a nature-inspired generic programming language for parallel algorithms, one that works for all data structures and control structures. Any parallel algorithm satisfying intuitively-appealing postulates can be modeled by a collection of cells, each of which is an abstract state machine, augmented with the ability to spawn new cells. All cells run the same algorithm and communicate via a shared global memory.

1 Introduction

Evolving systems – physical, biological, or computational – are typically viewable on many distinct levels of abstraction. Let us imagine some closed ecosystem as an example. An *ecologist* views species, populations, and their interactions; population growth and shrinkage may be modeled, say, by predator-prey and other resource equations. A *biologist* takes a different viewpoint, based on the individual organisms; she may develop a kinetic model for swarming behavior, for instance. On a lower level still, a *biochemist* sees interacting cell systems; he might use a diffusion-reaction equation to describe the development of the colorings on an animal's coat. The *chemist* looks at reactions on the molecular level; the *physicist* sees atoms and their constituents. The common denominator of all these views is one of a complex of objects that evolve over time and that interact with each other and with their environment according to a set of rules. It is this generic notion of a *system of interacting objects* that we seek to capture.

It has been convincingly argued by Gurevich [15] (presaged by Post [16]) that logical structures are the right way to view evolving algorithmic states, just as they are ideal for capturing the salient features of static entities. The structure stores the values (taken from the structure's domain) of components of the state that are updated during the computation (variables and program counters) as well as the state's functional capabilities (like arithmetic).

That there are multiple levels at which to understand the same overall system necessitates an abstraction mechanism. Atomic physics is of no relevance to the ecologist; the ecologist's view of the system is the same regardless of quantum physics. This means that the behavior of the entities at the ecological level should be modeled independently of the underlying physical model, which translates into the requirement that states qua structures are isomorphism-closed (making them oblivious as to how the domain values they deal with are in fact implemented) and that their evolution respects those isomorphisms. The importance of isomorphism-invariance for purposes of abstraction has been repeatedly emphasized [7,13,15].

A. Beckmann, E. Csuhaj-Varjú, and K. Meer (Eds.): CiE 2014, LNCS 8493, pp. 133–142, 2014.

On each level of the ecological system there are interacting entities: populations interact on the ecology level, organisms on the biological level, cells on the biochemical level, etc. The interacting entities need not all have "algorithmic" behavior. Aspects of the external environment (such as weather conditions) can also be treated as entities with which algorithmic components trade information. Accordingly, we need a model of communication between entities, which we shall refer to as "cells", in addition to a model of their individual evolution. To that end, we can allow the control of one cell to access values in another cell – a shared-memory viewpoint, or request values from another cell – a message-based framework. Similarly, we can allow one cell to set values in another cell or to request those changes from the other cell (depending again on one's viewpoint). Interaction and coöperation have been considered within Gurevich's framework [2,4]. We take the shared memory viewpoint here and assume that cells work in discrete time with a shared clock.

Many systems, be they natural or artificial, create new entities as they evolve in time. We will, therefore, need to model the "birth" of new component cells. But we will not, in this paper, consider changes in channels of communication (the "topology") other than at birth (cf. [12]). Were it not for possible interaction with external agents and for the birth of new components, one might have been tempted to view a software system as one large evolving global "organism", rather than as a conglomerate of many interacting individual cells.

In the next two sections, we characterize parallel algorithms and their cells. Then, in Sect. 4, we give a description of a parallel programming language based on abstract state machines (ASMs) [14]. Section 5 proves that all parallel algorithms, as characterized here, can be programmed with the constructs of the proposed language. We conclude with a brief discussion.

2   System Evolution

Informally, a *parallel algorithm* consists of a (finite or infinite) set of cells, whose individual *states* all evolve according to the same algorithm. The state of each *cell*, at any moment, is a (logical) structure with a tripartite vocabulary $F \uplus F' \uplus G$ consisting of *private (internal)* operations F, *public (global)* G, and *embryonic* F', the latter having the same similarity type as F. (There could be any fixed number of embryonic copies $F', F'', \ldots, F^{(k)}$, but let us leave it simple for now, one child at a time.) The individual cells all run a "classical" (sequential) algorithm in the sense of [15,8].

Initially, all cells agree on G and their F' are pristine (completely undefined). A single *global step* of the algorithm comprises of the following stages.

1. First, each cell C takes one classical step, producing a set of *updates* U.
2. Cells' private operations F and embryonic operations F' are updated per U.
3. Then the union of all the cell's public updates together are applied to every cell's public G. If there is any disagreement between cells regarding updates to G (the same location getting contradictory new values), the whole system

aborts. (Abortion could be replaced with nondeterministic behavior, should one prefer.)

4. Assuming there are no conflicts, *mitosis* takes place as follows: Each cell C in which the values of the operations F' were modified splits into two, a mother C and daughter C'. The daughter C' inherits G, as updated, from her mother; her F is a copy of her mother's F'. For both mother and daughter, F' is reinitialized to *undefined*.

5. If one wishes, an individual cell can be allowed to *die* and be dropped from the global organism whenever it has no next state, as when it suffers an internal clash.

3   Parallel Algorithms

An algorithm $\mathcal{A}$, in general, is normally viewed as a state-transition system composed of a collection (set, class) $\mathcal{S}$ of states, a (partial) *transition function* $\tau : \mathcal{S} \rightharpoonup \mathcal{S}$ and a (nonempty) subset $\mathcal{S}_0 \subseteq \mathcal{S}$ of *initial states*. We first explain what states of a parallel algorithm look like and then discuss algorithmic transitions.

As explained above, states should be formalized as (first-order) logical structures over some (fixed by the algorithm) vocabulary. On the other hand, we need for systems to comprise multiple local processes, what we called "cells". Each cell has its own unique identity (id), taken from some index set (or class) $\mathcal{I}$. Since we are dealing with parallel algorithms, with both private and shared memory, each cell has a *local* state, which is a structure over a (finite) vocabulary $G \uplus F$, where the (current) values of operations in G are stored (conceptually, at least) in *global* locations, accessible to all cells i, while private data is stored as values of operations in its personal copy of F. The *global* state of the algorithm will be an algebra over the combined (possibly infinite) vocabulary $V = G \cup F^*$, were $F^* = \cup_{i \in \mathcal{I}} F_i$, where each $F_i = \{f_i^1, \ldots, f_i^k\}$ consists of the k local operations of cell i, with f_i^j $(j = 1, \ldots, k)$ of the same arity for all cells. It will be convenient in what follows to denote $F^j = \cup_{i \in \mathcal{I}} \{f_i^j\}$.

With this intuition in mind, we define a global state of *federacy* $\mathcal{I}$, for a given (countable or uncountable) set of *identities* $\mathcal{I}$, to be an algebra X over vocabulary $V = G \cup F^*$. A *global (transition) system* $\mathcal{A}$ *(of federacy $\mathcal{I}$)* is composed of a collection $\mathcal{S}$ of global states of federacy $\mathcal{I}$, all over the same vocabulary, a (partial) *transition function* $\tau : \mathcal{S} \rightharpoonup \mathcal{S}$ and a subset $\mathcal{S}_0 \subseteq \mathcal{S}$ of states assumed to be the full collection of *initial states* of $\mathcal{A}$. If τ is undefined for some $X \in \mathcal{S}$ we will say that X is a *terminal state* of $\mathcal{A}$.

Let X be a state of $\mathcal{A}$ with domain $\mathcal{U}$. Let g be some function in V (either in G or in $F^* = F^1 \cup \ldots \cup F^k$) of arity n and $\bar{u} = (u_1, \ldots, u_n)$ be an n-tuple over that domain. If $g(\bar{u}) = w$ in X, we denote this by $g_X(\bar{u}) = w$ or, alternatively, will say that $\langle g, \bar{u}, w \rangle$ is a *location-value* of X. For any ground term t, we write $t_X = w$ to mean that the value of t (as interpreted) in X is w.

We intend that cell i operate over vocabulary $G \cup F_i$ only. So we define the ith *localization* X_i of global state X of federacy $\mathcal{I}$ to be the restriction of X to $G \cup F_i$. The ith cell is expected to manipulate this ith localization only, identifying its private F with the global F_i. We say that cell (local state) X_i is *empty* if f_i^j is undefined ($\perp$) for all j; on the other hand, we say that X is an *i-cell* if $X = X_i$ and is nonempty. When state X with transition τ is not terminal, we say that $\delta = \langle g, \bar{u}, w \rangle$ is an *update* of X if τ *changes* the value of $g(\bar{u})$ to be w. We define by $\Delta_\tau(X)$ the set of all updates of X.

To compare different cells we should ignore their individual identities. So we define a *depersonalization* operator $\sharp$; its application wipes out the id, dropping the id-index from function symbols. Thus, the depersonalized $X_i^\sharp$ is obtained from a cell X_i by replacing its f_i^j symbols by f^j, for all j. So we write $X_i = Y_k$ if $X_i^\sharp = Y_k^\sharp$ for states X and Y and localizations X_i and Y_k. Similarly, we say that transition τ generates the same updates for X_i and Y_k if $\Delta_\tau(X_i)^\sharp = \Delta_\tau(Y_k)^\sharp$. In this case, we will use the notation $\Delta_\tau(X_i) = \Delta_\tau(Y_k)$. We denote by $\Delta_\tau^i(X)$ the set of all updates of locations of $f_i^1, \ldots, f_i^k$ in X. And again, we say that transition τ generates the same $\Delta_\tau^i(X) = \Delta_\tau^l(Y)$ if $\Delta_\tau^i(X)^\sharp = \Delta_\tau^l(Y)^\sharp$.

To capture the uniform behavior of cells, we introduce *templates*, which are terms over an unadorned vocabulary $G \cup \{f^1, \ldots, f^k\}$, where f^i is a symbol of the same arity as the $f_i^j \in F$. For each $i \in \mathcal{I}$, the template t induces a term t_i, obtained by replacing each occurrence of f^j by f_i^j. Given states X and Y from the same transition system and given a template t, we say that $X =_T Y$ if $t_{iX} = t_{iY}$ for any $i \in \mathcal{I}$ (i.e. every term defined by t has the same value in both X and Y). To compare different cells we should again ignore their identities. So let X_i and X_m be distinct localizations of global state X. We say that $X_i =_T X_m$ if $t_{iX} = t_{mX}$ for each $t \in T$. Similarly, we may compare localizations of two distinct global states. Letting X_i be a localization of X and Y_m a localization of Y, we write $X_i =_T Y_m$ if $t_{iX} = t_{mY}$ for each $t \in T$.

Let $\mathcal{A} = (\mathcal{S}, \mathcal{S}_0, \tau)$ be a transition system of federacy $\mathcal{I}$ over vocabulary $V = G \cup F^1 \cup \ldots \cup F^k$. We deem a parallel process $\mathcal{A}$ to be *algorithmic* if it satisfies several postulates, which we now proceed to explicate.

Postulate 1 (Genericity). *The set of states (and also the sets of initial states and of terminal states) is closed under isomorphism (of first-order structures). The set of states is also closed under localizations: if X is a state of $\mathcal{A}$ then X_i is also a state of $\mathcal{A}$, for each $i \in \mathcal{I}$. Transitions preserve the domain (universe) of states, and, furthermore, isomorphic states are either both terminal (have no transition) or else their next states are isomorphic (via the same isomorphism).*

States as structures make it possible to consider any data structure sans encodings. In this sense, algorithms are generic. The structures are "first-order" in syntax, though domains may include sequences, or sets, or other higher-order objects, in which case the state would provide operations for dealing with those objects. (States with infinitary operations, like the supremum of infinitely many objects, are precluded.) Closure under isomorphism ensures that the algorithm can operate on the chosen level of abstraction and that states' internal representation of data is invisible to the algorithm. This means that the behavior of an *algorithm*, in contradistinction with its "implementation" as a program in some particular programming language, cannot depend on the memory address of some variable.

It must be possible to describe the effect of transitions in terms of the information in the current state.

Postulate 2 (Describability). *There exists a finite set T of critical templates such that $\Delta_\tau(X) = \Delta_\tau(Y)$ if $X =_T Y$ for any states X and Y of $\mathcal{A}$.*

The critical templates are those locations in the state named by the algorithm (or program). If every referenced location has the same value in two states, then the behavior of the algorithm must be the same for both those states. This, the essence of what makes a process algorithmic, is a crucial insight of [15].

The updates created by an individual cell may not depend on its id, but only on global and local locations that are available to it. Furthermore, each cell is fully responsible for its dates, and no other cell may change them. Also, all updates of a global state are generated by local cells only.

Postulate 3 (Locality). *If $X_i =_T Y_j$ for two states X and Y and localizations $i, j \in \mathcal{I}$, then $\Delta_\tau(X_i) = \Delta_\tau(Y_j)$ and $\Delta_\tau^i(X) = \Delta_\tau^j(Y)$.*

Postulate 4 (Globality). *$\Delta_\tau(X) = \cup_{i \in \mathcal{I}} \Delta_\tau(X_i)$ for all states X.*

If some localization of X is empty but is not empty for $\tau(X)$, this indicates that a child has been born.

Postulate 5 (Fertility). *There exists a (input-independent) bound $n \in \mathbb{N}$ such that $\tau(X)$ has at most n non-empty localizations for any local i-cell X, $i \in \mathcal{I}$.*

The idea is that in one step a cell may participate in the creation of only a bounded number of new cells. And each newborn cell has exactly one mother:

Postulate 6 (Motherhood). *For every state X, if a localization X_i is empty, but is non-empty for $\tau(X)$, then there is a $j \in \mathcal{I}$ such that $\Delta_\tau^i(X) \subseteq \Delta_\tau(X_j)$.*

With the above requirements in place, we state what a parallel algorithm is.

Definition 1. *A global system is* algorithmic *if it satisfies Postulates 1–6.*

We say that two parallel algorithmic systems are "congruent" if they are identical, up to permutation of identities $\mathcal{I}$.

Definition 2. *A parallel algorithm* **A** *is a family of all parallel algorithmic systems, congruent with some algorithmic system $\mathcal{A}$.*

Proposition 1. *Let $\mathcal{A}$ be an algorithmic system over a* finite *vocabulary. Then $\mathcal{A}$ may be described as an ordinary algorithm.*

Proof. Imagine $\mathcal{A}$ is an algorithmic system over a *finite* vocabulary. Then instead of $V = G \cup F^*$, we may assume that we only have $V = G$ (and we required that G be finite). So for this case, Postulates 3–6 are redundant, and $\mathcal{A}$ is only required to satisfy the **geniricity** and **describability** postulates. Also, our final set of critical templates T is just a finite set of terms over $V = G$. Then $\mathcal{A}$ is a *classical (sequential) algorithm* with *critical* terms T, as defined in [15]. $\qquad\square$

4   Parallel Programs

The two basic program statements are assignment and creation. These may be composed in parallel and guarded by conditions.

Assignment. An *atomic assignment* is a rule of the form $h(t^1, \ldots, t^n) := t^0$, where $t^0, \ldots, t^n$ are templates and $h \in G \cup F$.

Let X_i be a localization of X, and suppose that $t^j_{iX} = u^j_i$ for $j = 0, \ldots, n$. If $h \in G$, then application of the assignment on X for i generates a global update $\Delta_a(X_i) = \{\langle h, (u^1_i, \ldots, u^n_i), u^0_i \rangle\}$. If $h \in F$, then the application generates dates $\Delta_a(X_i) = \{\langle h_i, (u^1_i, \ldots, u^n_i), u^0_i \rangle\}$. If any of the t^j is undefined in X_i ($t^j_{X_i} = \bot$) , then $\Delta_a(X_i) = \varnothing$. The application of assignment a to global state X generates the update set $\Delta_a(X) = \cup_{i \in \mathcal{I}} \Delta_a(X_i)$.

Parallel assignment. More generally, a *parallel assignment* rule is a set $\{a_1, a_2, \ldots, a_n\}$ of atomic assignments, written with $\parallel$ between the atomic a_j.

The update set generated by such parallel assignment a is $\Delta_a(X) = \cup_{j=1}^n \Delta_{a_j}(X)$. If $\Delta_a(X)$ has conflicting updates (different values assigned to the same location), then the rule fails.

Creation. The creation rule $\nu.a$ takes the form **new** a, where a is a parallel assignment, atomic assignments in a are of the form $f^j(t^1, \ldots, t^n) := t^0$, and n is the arity of symbol $f^j \in F$.

Let X_i be a localization, and suppose $t^j_{iX} = u^j_i$ for $j = 0, \ldots, n$ for an atomic assignment. The transition initializes some empty localization X_{k_i} with location-value $\langle f^j_{k_i}, (u^1_i, \ldots, u^n_i), u^0_i \rangle$. So $\Delta_{\nu.a}(X_i) = \{\langle f^j_{k_i}, (u^1_i, \ldots, u^n_i), u^0_i \rangle\}$. If any one of the t^j is undefined in X_i, then $\Delta_{\nu.a}(X_i) = \varnothing$. For each cell i, the transition chooses a unique k_i and the cell's updates are appended to the total set of updates $\Delta_{\nu.a}(X) = \cup_{i \in \mathcal{I}} \Delta_{\nu.a}(X_i)$. If a is a parallel assignment $a_1 \parallel a_2 \parallel \cdots \parallel a_n$, then application of $\nu.a$ chooses a unique empty X_{k_i} for each X_i in which all arguments t^j are defined, and $\Delta_{\nu.a}(X_i) = \cup_{\ell=1}^n \Delta_{\nu.a_\ell}(X)$. If there is no way to choose k_i for all i so that the rule applies, then it is not applied at all.

Guard. An *atomic guard* is a condition of the form $s = t$ or $s \neq t$. Guard $t = s$ evaluates to T (true) for localization X_i if $t_{iX} = s_{iX}$. Similarly, a guard $t \neq s$ is T if $t_{iX} \neq s_{iX}$. More generally, a *guard* g may be a conjunction of atomic guards $g_1 \& g_2 \& \cdots \& g_n$, which is T for X_i if each g_j is.

Guarded assignment. This is a rule $g : a$ of form **if** g **then** a, where g is a guard and a is a parallel assignment. Application of $g : a$ to X generates the set of updates $\Delta_{g:a}(X) = \cup \{\Delta_a(X_i) : i \in \mathcal{I} \text{ s.t. } g_{X_i} = \text{T}\}$.

Guarded creation. This is a rule $g : \nu.a$ of form **if** g **then new** a. The rule $\nu.a$ is executed on each X_i for which g evaluates to T.

Definition 3 (Program). *A (parallel) program is a finite set P of rules r_i as above, written $r_1 \parallel \cdots \parallel r_n$. To execute P on state X, all rules are executed in parallel (simultaneously), that is, $\Delta_P(X) = \cup_{r_i \in P} \Delta_{r_i}(X)$. If $\Delta_P(X)$ has conflicting updates, then no updates are applied at all.*

Note that for each application of creation, the program chooses in some fashion new unused indices from $\mathcal{I}$. So for each given initial state, the program may have multiple runs, depending on the choices made. Each choice is possible and none is preferred. And it does not affect the computation's final result or running time (number of steps). So for each state X, we denote by $P(X)$ any one of the possible (congruent) states obtained by application of P to X.

5   Representation Theorem

A parallel program P is a *characteristic program* of algorithmic system $\mathcal{A}$ if $P(X) = \tau(X)$ for each state X of $\mathcal{A}$. A parallel program P is a *characteristic program* of parallel algorithm $\mathbb{A}$ if it is a characteristic program for each algorithmic system $\mathcal{A}$ in $\mathbb{A}$. We shall presume for simplicity that $\mathcal{A}$ is over a vocabulary $G \cup F^1$ only and denote it by $G \cup F$. We will also assume that in Postulate 5 we have at most one child born per step ($n = 2$). All proofs can be easily extended to the general case.

By **globality**, $\Delta_\tau(X) = \cup_i \Delta_\tau(X_i)$. So we start with i-cells. We first prove that the transitions of any i-cell can be described by a rule composed of assignment and creation rules.

Let X be an i-cell of system $\mathcal{A}$. According to the above simplifying assumption, $\mathcal{A}$ has only one local function. Since X is an i-cell, its non-default locations are over $G \cup \{f_i\}$. Furthermore, any cell may have at most one child in one transition. Hence, non-default locations of $\tau(X)$ are over $G \cup \{f_i, f_j\}$ for some $j \in \mathcal{I}$. So we may consider X and $\tau(X)$ as ordinary states of an ordinary algorithm over finite vocabulary $G \cup \{f_i, f_j\}$ with critical terms $T_i \cup T_j$.

Let $\delta = \langle h, (u_1, \ldots, u_n), w \rangle$ be an update in $\Delta_\tau(X)$. According to [17, Lemma 5] for each $k = 0, \ldots, n$ there exists $t^k \in T_i \cup T_j$ such that $t^k_X = u_k$. Let X_δ be an ordinary assignment rule $h(t^1, \ldots, t^n) := t^0$. Then $\Delta_{X_\delta}(X) = \delta$. We will call X_δ an *ordinary characteristic assignment* of δ. Denote by $\mathbf{X}$ the ordinary assignment obtained by parallel composition of X_δ for all $\delta \in \Delta_\tau(X)$. Obviously, $\Delta_{\mathbf{X}}(X) = \Delta_\tau(X)$. Take a look at $X_\delta = h(t^1, \ldots, t^n) := t^0$. As we said before, $t^k_X = u_k$ for all $k = 0, \ldots, n$. In particular, t^k is defined over X. Since all non-trivial locations of X are over $G \cup \{f_i\}$, we may conclude that t^k are over T_i for all $k = 0, \ldots, n$. And since all defined locations of $\tau(X)$ are over $G \cup \{f_i, f_j\}$, we may conclude that $h \in G \cup \{f_i, f_j\}$. We partition $\mathbf{X}$ into two parallel assignment rules: a_X for all those rules with $h \in G \cup \{f_i\}$ and n_X for all the rest, rules of the form $f_j(t^1, \ldots, t^n) := t^0$. Obviously, $\mathbf{X} = a_X \| n_X$.

Let $a^\sharp_X$ be obtained from a_X by replacing f_i with f. Then $a^\sharp_X$ is an assignment rule over templates T. From the definition of parallel-assignment application we obtain that $\Delta_{a^\sharp_X}(X) = \Delta_{a_X}(X)$. Let $n^\sharp_X$ be obtained from n_X by replacing f_i and f_j with f. From the definitions of parallel ν-rule application and of comparing updates for different cells, we obtain that $\Delta_{\nu.n^\sharp_X}(X) = \Delta_{n_x}(X)$. Define $\mathbb{X} = a^\sharp_X \| \nu.n^\sharp_X$. Then $\Delta_{\mathbb{X}}(X) = \Delta_{a^\sharp_X}(X) \cup \Delta_{\nu.n^\sharp_X}(X) = \Delta_{\mathbf{X}}(X)$.

Proposition 2. *Let X be an i-cell of $\mathcal{A}$. Then $\mathbb{X}(X) = \mathbf{X}(X) = \tau(X)$.*

Proof. That $\mathbb{X}(X)$ is $\tau(X)$ follows from the above. That $\mathbf{X}(X)$ is $\tau(X)$ follows from [17, Lemma 11]. $\qquad\square$

Updates of i-cells depend on the values of critical terms only.

Proposition 3. *Let X and Y be i-cells of $\mathcal{A}$, such that $X =_T Y$. Then $\mathbf{Y}(X) = \tau(X)$. And $\mathbb{Y}(X) = \tau(X)$.*

Proof. Since $\mathbb{Y}$ is a rule over T it generates updates based on the values of T only. Hence $\Delta_{\mathbb{Y}}(X) = \Delta_{\mathbb{Y}}(Y)$, since $X =_T Y$. It follows from the earlier discussion that $\mathbb{Y}(Y) = \tau(Y)$. According to **locality**, we have that $\Delta_{\tau}(Y) = \Delta_{\tau}(X)$, again since $X =_T Y$. Combining all, we conclude $\mathbb{Y}(X) = \tau(X)$. Hence, we get the following implication for i-cell X and j-cell Y: $X =_T Y \Rightarrow \mathbb{Y}(X) = \tau(X)$. $\qquad\square$

Let X be an i-cell. We define an equivalence relation $\sim_X$ on T by $t \sim_X s$ iff $t_X = s_X$. We next show that updates of i-cell X depend on $\sim_X$ only. Let X be an i-cell and Y be a j-cell of $\mathcal{A}$. We write $X \approx_T Y$ if $\sim_X = \sim_Y$.

Proposition 4. *Let X be an i-cell of $\mathcal{A}$ and let Y be a j-cell of $\mathcal{A}$ such that $X \approx_T Y$ and $\mathbf{Y}(X) = \tau(X)$. Then $\mathbb{Y}(X) = \tau(X)$.*

Proof. First assume that $i = j$, i.e. that both X and Y are i-cells. Consider X and Y to be ordinary states over finite vocabulary $G \cup \{f_i\}$ (as we did at the start of this section). By assumption, X and Y each have one child cell in a single transition. Define $\sim_X^i$ on T_i by $t \sim_X s$ iff $t_X = s_X$ for any $t, s \in T_i$. Then $\sim_X^i = \sim_Y^i$. It follows from [17, Lemma 13] that $\mathbf{Y}(X) = \tau(X)$. And by Proposition 2 we conclude that $\mathbb{Y}(X) = \tau(X)$. Recall that we defined $X = Y$ for parallel states X and Y of the same system if they are equal up to a permutation of identities of their cells. The general case follows immediately. $\qquad\square$

Lemma 1. *For each parallel algorithmic transition system there exists a characteristic parallel program.*

Proof. Let $\sim$ be some binary relation on T. Then for any pair of distinct term-templates $s, t \in T$ we have that either $s \sim t$ or $s \not\sim t$. For each $s, t \in T$ we define $\beta_\sim(s, t)$ to be an atomic guard $s = t$ if $s \sim t$ and $s \neq t$ otherwise. Define a guard $\beta_\sim$ to be a conjunction of all atomic guards $\beta_\sim(s, t)$ for all $s, t \in T$. Choose an i-cell X of $\mathcal{A}$ for some $i \in \mathcal{I}$ such that $\sim_X$ is $\sim$. Denote it by $X_\sim$. Define a rule $R_\sim = $ **if** $\beta_\sim$ **then** $\mathbb{X}_\sim$. Obviously $\beta_\sim$ evaluates to $\top$ on X. and hence $R_\sim(X_\sim) = \mathbb{X}_\sim$. According to Proposition 4, we have that $\mathbb{X}_\sim = \tau(X_\sim)$ and so $R_\sim(X_\sim) = \tau(X_\sim)$.

Define P to be a parallel program consisting of rules $R_\sim$ for all binary relations $\sim$ of T. Note that since T is finite, it has only finitely many distinct binary relations and so program P is finite. We claim that P is a characteristic program of $\mathcal{A}$, that is, $P(X) = \tau(X)$ for any state X of $\mathcal{A}$. Assume first that $X = X_\sim$ for some binary relation $\sim$ on T. Let $\sim'$ be another binary relation on T, distinct from $\sim$. Then for some $s, t \in T$ we have that $\beta_\sim(s, t) \neq \beta_{\sim'}(s, t)$. So $\beta_{\sim'}(s, t)$ is false for X and so is $\beta_{\sim'}$. Then $\Delta_{R_{\sim'}}(X) = \varnothing$ and that is for any binary relation on T other than $\sim$. Hence $P(X) = R_\sim(X)$ and according to the previous discussion $R_\sim(X) = \tau(X)$, as desired. Assume next that X is an i-cell for some i. Denote $\sim_X$ by $\sim$. As in the previous item, $\Delta_{R_{\sim'}}(X) = \varnothing$ for any binary relation $\sim'$ on T, other then $\sim$. And so $P(X) = R_\sim(X)$. Let $X_\sim$ be as above. Then $R_\sim(X) = \mathbb{X}_\sim(X)$ (by the definition of $R_\sim$). According to Proposition 4 we have that $\mathbb{X}_\sim(X) = \tau(X)$. Combining everything together we conclude that in this case again we have that $P(X) = \tau(X)$. Assume finally that X is a general state of $\mathcal{A}$. According to **globality**, the update of X is a union of updates of all its localizations X_i, i.e. $\Delta_\tau(X) = \cup_{i \in \mathcal{I}} \Delta_\tau(X_i)$. By the **genericity** axiom, X_i is a state in $\mathcal{A}$. According to **locality**, updates for X_i do not depend on whether X_i is considered as a standalone state or a localization of a general state. So it is enough to show that $\Delta_P(X_i) = \Delta_\tau(X_i)$ for all $i \in \mathcal{I}$. And that follows from the previous paragraph. □

Theorem 1 (Main). *For each parallel algorithm, there exists a characteristic parallel program.*

Proof. Let $\mathcal{A}$ be an algorithmic system in $\mathbb{A}$. By Lemma 1, there exists a characteristic parallel program $P_\mathcal{A}$ of $\mathcal{A}$. If $\mathcal{B}$ is another algorithmic system in $\mathbb{A}$, then B is identical to $\mathbb{A}$, up to permutation of indices in $\mathcal{I}$. Then, obviously, $P_\mathcal{A}$ is a characteristic program of B as well. □

6 Discussion

The starting point for this research was the desire to characterize parallel computation in as generic a form as possible, with an eye especially towards the effective special case. Blass and Gurevich [1,3] successfully characterized parallel algorithms within the abstract-state-machine framework, but their approach is not easily restricted to the effective case. In their setup, an unbounded number of children may be created by a single cell in a single step.

Our model is simpler than Blass and Gurevich for the cases we consider. As we do not have message passing, algorithms need not deal with process ids. Though we bound the number of new cells created by a cell in a step, an infinite number of initial cells for a non-effective parallel algorithm poses no problem. For example, one can imagine a cell for each of uncountably many points on a line segment in 3D space and an algorithm that applies, in parallel, an affine transformation to the coordinates of each point, resulting in a translated segment.

We've considered discrete-time systems, where all cells progress in lockstep with each other, as in [1,3]. We plan to expand this work in several directions:

– Characterize what makes a parallel algorithm effective. Analogous to prior work on classical effectiveness [11,5], we need to demand that the initial global state be finitely describable. This decomposes into two main requirements: (i) each cell itself be an effective classical algorithm; (ii) there be only finitely many cells initially, though their number may depend on the input.
– Prove the extended Church-Turing thesis for parallel algorithms: all effective parallel models of computation can be polynomially simulated by a standard model (like PRAM), as has been done for classical algorithms [10].

– Distributed systems, where cells each progress at their own rate, require separate treatment. This will require a sense of identity for cells and a means of communication between them. Cf. [4].
– Systems that evolve in continuous time are a subject of ongoing research [6,9].

References

1. Blass, A., Gurevich, Y.: Abstract state machines capture parallel algorithms. ACM Trans. on Computation Logic 4, 578–651 (2003)
2. Blass, A., Gurevich, Y.: Ordinary interactive small-step algorithms (Parts I–III). ACM Trans. on Computational Logic 7, 363–419; 8: art. 15–16 (2006-2007)
3. Blass, A., Gurevich, Y.: Abstract state machines capture parallel algorithms: Correction and extension. ACM Trans. on Computation Logic 9, Art. 19 (2008)
4. Blass, A., Gurevich, Y., Rosenzweig, D., Rossman, B.: Interactive small-step algorithms (Parts I–II). LMCS 3: ppr. 3; 4: ppr. 43 (2007)
5. Boker, U., Dershowitz, N.: Three paths to effectiveness. In: Blass, A., Dershowitz, N., Reisig, W. (eds.) Fields of Logic and Computation. LNCS, vol. 6300, pp. 135–146. Springer, Heidelberg (2010)
6. Bournez, O., Dershowitz, N., Falkovich, E.: Towards an axiomatization of simple analog algorithms. In: Agrawal, M., Cooper, S.B., Li, A. (eds.) TAMC 2012. LNCS, vol. 7287, pp. 525–536. Springer, Heidelberg (2012)
7. Chandra, A.K., Harel, D.: Computable queries for relational data bases. Journal of Computer and System Sciences 21, 156–178 (1980)
8. Dershowitz, N.: The generic model of computation. In: Proc. DCM, pp. 59–71 (2012)
9. Dershowitz, N.: Res Publica: The universal model of computation. In: Computer Science Logic 2013, Turin, Italy. Leibniz International Proceedings in Informatics, vol. 23, pp. 5–10 (2013)
10. Dershowitz, N., Falkovich, E.: A formalization and proof of the Extended Church-Turing Thesis. In: Proc. 7th International Workshop on Developments in Computational Models. EPTCS, vol. 88, pp. 72–78 (2011)
11. Dershowitz, N., Gurevich, Y.: A natural axiomatization of computability and proof of Church's Thesis. Bulletin of Symbolic Logic 14, 299–350 (2008)
12. Dowek, G.: Around the physical Church-Turing thesis: Cellular automata, formal languages, and the principles of quantum theory. In: Dediu, A.-H., Martín-Vide, C. (eds.) LATA 2012. LNCS, vol. 7183, pp. 21–37. Springer, Heidelberg (2012)
13. Gandy, R.: Church's thesis and principles for mechanisms. In: The Kleene Symposium, vol. 101, pp. 123–148. North-Holland (1980)
14. Gurevich, Y.: Evolving algebras 1993: Lipari guide. In: Börger, E. (ed.) Specification and Validation Methods, pp. 9–36. Oxford University Press, Oxford (1995)
15. Gurevich, Y.: Sequential abstract state machines capture sequential algorithms. ACM Trans. on Computational Logic 1, 77–111 (2000)
16. Post, E.L.: Absolutely unsolvable problems and relatively undecidable propositions. In: Davis, M. (ed.) Solvability, Provability, Definability: The Collected Works of Emil L. Post, pp. 375–441. Birkhaüser, Boston (1994)
17. Reisig, W.: On Gurevich's theorem on sequential algorithms. Acta Informatica 39, 273–305 (2003)

Isomorphisms of Non-Standard Fields
and Ash's Conjecture

Rumen Dimitrov[1], Valentina Harizanov[2], Russell Miller[3], and K.J. Mourad[4]

[1] Department of Mathematics, Western Illinois University, Macomb, IL 61455, USA
rd-dimitrov@wiu.edu
[2] Department of Mathematics, George Washington University, Washington,
DC 20052, USA
harizanv@gwu.edu
[3] Department of Mathematics, Queens College and Graduate Center,
City University of New York, New York, NY 10016, USA
Russell.Miller@qc.cuny.edu
[4] Department of Mathematics, Georgetown University, Washington, DC 20005, USA
kjm57@georgetown.edu

Abstract. Cohesive sets play an important role in computability theory. Here we use cohesive sets to build nonstandard versions of the rationals. We use Koenigsmann's work on Hilbert's Tenth Problem to establish that these nonstandard fields are rigid. As a consequence we obtain results about automorphisms of the lattices of computably enumerable vector spaces arising in the context of Ash's conjecture.

1   Introduction

This paper is motivated by the 30-year open problem of finding automorphisms of the lattice $\mathcal{L}^*(V_\infty)$. As in Metakides and Nerode [16], the space V_∞ is the canonical computable $\aleph_0$-dimensional vector space over a computable field F. The lattice of computably enumerable (c.e.) subspaces of V_∞ is denoted by $\mathcal{L}(V_\infty)$. The lattice $\mathcal{L}(V_\infty)$ modulo finite dimension is denoted by $\mathcal{L}^*(V_\infty)$. For all undefined notions, as well as more background on computability theory and c.e. vector spaces, the reader can consult [18] and [16], correspondingly. Guichard [7] established that there are countably many automorphisms of $\mathcal{L}(V_\infty)$ because they are generated by computable semilinear transformations. Ash conjectured that the automorphisms of $\mathcal{L}^*(V_\infty)$ are generated by special computable semilinear transformations.

Definition 1. *An automorphism of $\mathcal{L}^*(V_\infty)$ is called an Ash automorphism if it is generated by a semilinear transformation with finite dimensional kernel and co-finite dimensional image in V_∞.*

Conjecture 1. (Ash) Every automorphism of $\mathcal{L}^*(V_\infty)$ is an Ash automorphism.

Definition 2. *(1) An infinite set $C \subset \omega$ is cohesive if for every c.e. set W either $W \cap C$ or $\overline{W} \cap C$ is finite.*

A. Beckmann, E. Csuhaj-Varjú, and K. Meer (Eds.): CiE 2014, LNCS 8493, pp. 143–152, 2014.
© Springer International Publishing Switzerland 2014

(2) A set M is maximal *if M is c.e. and $\overline{M}$ is cohesive.*

(3) A set B is quasimaximal *if it is the intersection of finitely many maximal sets.*

For sets A and B we use $A =^* B$ to denote that A and B differ on at most finitely many elements, and $A \subset_* B$ to denote that all but finitely many elements of A are also elements of B. For vector spaces we use the same notation where "finitely many elements" is replaced by "finite dimension." Let A be a computable basis of V_∞ and let B be a quasimaximal subset of A. Let $\mathcal{E}^*(B, \uparrow)$ denote the principal filter of B in the lattice $\mathcal{E}^*$ of c.e. sets modulo $=^*$. It is known that $\mathcal{E}^*(B, \uparrow)$ is isomorphic to a finite Boolean algebra $\mathbf{B_n}$. Let $V = cl(B)$ be the closure of B in V_∞. In contrast to $\mathcal{E}^*(B, \uparrow)$, the principal filter of V in $\mathcal{L}^*(V_\infty)$, $\mathcal{L}^*(V, \uparrow)$, is not always isomorphic to $\mathbf{B_n}$. Rather, as shown in [2] and [3], these filters are isomorphic to either:

(1) a finite Boolean algebra,

(2) a lattice of subspaces of an n-dimensional vector space W over a certain extension of F (denoted by $L(n, \widetilde{F})$), or

(3) a finite product of structures from the previous two cases.

The extension $\widetilde{F}$ of F mentioned in (2), which is denoted by $\prod_C F$, is called the cohesive power of F, and is defined below. In the context of computable vector spaces the main interesting cases occur when F is finite or $F = \mathbb{Q}$. For finite F we have $\prod_C F \cong F$. The first key result in this paper is that $\prod_{M_1} \mathbb{Q} \cong \prod_{M_2} \mathbb{Q}$ iff the maximal sets M_1 and M_2 are of the same 1-degree up to finitely many elements (see Definition 7). This result implies the following theorem when $F = \mathbb{Q}$.

Theorem 1. *(i) The principal filters $\mathcal{L}^*(V, \uparrow)$ of type (2) fall into infinitely many non-isomorphic classes even when these filters are isomorphic to lattices of subspaces of finite dimensional vector spaces of the same dimension (≥ 3).*

(ii) Every automorphism of $\mathcal{L}^*(V_\infty)$ preserves m-degrees of the spaces in (2).

A bijective semilinear map Φ on a vector space W over a field F is defined by

$$\Phi(av + bw) = f(a)\Phi(v) + f(b)\Phi(w), \tag{4}$$

where f is an automorphism of F. Such a bijective semilinear map Φ on the space W in (2) above generates an automorphism τ_Φ of $L(n, \widetilde{F})$. Moreover, by the fundamental theorem of projective geometry, all automorphisms of $L(n, \widetilde{F})$ for $n \geq 3$ are generated by such semilinear maps. By (2) above we can regard τ_Φ as an automorphism of $\mathcal{L}^*(V, \uparrow)$. When Φ is merely semilinear, τ_Φ is not the restriction of any Ash automorphism to $\mathcal{L}^*(V, \uparrow)$. When Φ is linear, τ_Φ has a natural extension $\overline{\tau_\Phi}$ to an automorphism of $\mathcal{L}^*(V_\infty)$ as described in the construction in the proof of Theorem 2.1 in [5]. This $\overline{\tau_\Phi}$ is an Ash automorphism of $\mathcal{L}^*(V_\infty)$. In certain cases we may hope to generalize this construction in the case when Φ is merely semilinear and thereby generating a non-Ash automorphism. However, our second key result is that $\prod_C F$ has only the trivial automorphism when $F = \mathbb{Q}$ and we can establish the following results.

Theorem 2. *(i) Any automorphism of $\mathcal{L}^*(V_\infty)$ of the form $\overline{\tau_\Phi}$ for every bijective semilinear map Φ is an Ash automorphism.*

(ii) Any automorphism of $\mathcal{L}^*(V, \uparrow)$, where $\mathcal{L}^*(V, \uparrow)$ is of type (2) and $n \geq 3$, can be extended to an automorphism of $\mathcal{L}^*(V_\infty)$.

The result that the cohesive power $\prod_M \mathbb{Q}$ is rigid, and its proof are of independent interest. Our proof uses a recent number-theoretic result about definability of $\mathbb{Z}$ in $\mathbb{Q}$ by Koenigsmann [11] (see [1] for additional background). This result allows us to apply work in nonstandard models of arithmetic to our problem.

In Section 2, we define a cohesive power of a computable structure $\mathcal{A}$ over a cohesive set C of natural numbers, $\prod_C \mathcal{A}$. We give a natural way of embedding $\prod_C \mathbb{N}$ into $\prod_C \mathbb{Q}$. In Section 3, we prove that $\prod_C \mathbb{N}$ is definable in both $\prod_C \mathbb{Z}$ and $\prod_C \mathbb{Q}$. The main result in this section implies that if M_1 and M_2 are maximal sets of natural numbers, then $\prod_{M_1} \mathbb{Q} \cong \prod_{M_2} \mathbb{Q}$ iff $M_1 \equiv_1^* M_2$. Finally, in Section 4, we prove that if C is a co-maximal (hence co-c.e.) set, then $\prod_C \mathbb{Q}$ is rigid.

2   Effective Ultraproducts and Isomorphisms

Homomorphic images of the semiring of computable functions have been studied as models of fragments of arithmetic in [6], [8], and [12]. Let C be an r-cohesive set. Fefferman, Scott, and Tennenbaum considered the quotient structure $\mathcal{R}/\sim_C$, where $\mathcal{R}$ is the set of all unary (total) computable functions and $\sim_C$ is the equivalence relation on $\mathcal{R}$ defined by:

$$f \sim_C g \Leftrightarrow C \subseteq^* \{n \in \omega \mid f(n) = g(n)\}. \tag{5}$$

They proved that there is a specific Π_3^0 sentence σ such that $\mathbb{N} \models \sigma$ but $\mathcal{R}/\sim_C \nvDash \sigma$ (see Theorem 2.1 in [12]). Lerman [12] further proved that if $R_1 \equiv_m R_2$ are r-maximal sets, then $\mathcal{R}/\sim_{\overline{R_1}} \cong \mathcal{R}/\sim_{\overline{R_2}}$. Moreover, Corollary 2.4 in [12] states that if M_1 and M_2 are maximal sets of different m-degrees, then $\mathcal{R}/\sim_{\overline{M_1}}$ and $\mathcal{R}/\sim_{\overline{M_2}}$ are not even elementary equivalent. These models of fragments of arithmetic have been further studied by Hirschfeld, Wheeler, and McLaughlin in [8], [9], [13], [14], and [15], and are special cases of what we call cohesive powers. The cohesive powers of fields, which were used in [3] to characterize the principal filters of quasimaximal spaces, motivated the following general definition in [4]. As usual, we will denote the equality of partial functions by $\simeq$.

Definition 3. *Let $\mathcal{A}$ be a computable structure with domain A in a computable language L, and and let $C \subset \omega$ be a cohesive set. The cohesive power of $\mathcal{A}$ over C, denoted by $\prod_C \mathcal{A}$, is a structure $\mathcal{B}$ for L with domain B such that the following holds.*

1. The set $B = (D/ =_C)$, where $D = \{\varphi \mid \varphi : \omega \to A$ is a partial computable function, and $C \subseteq^* dom(\varphi)\}$.
   For $\varphi_1, \varphi_2 \in D$, we have $\varphi_1 =_C \varphi_2$ iff $C \subseteq^* \{x : \varphi_1(x) \downarrow= \varphi_2(x) \downarrow\}$.
   The equivalence class of φ with respect to $=_C$ will be denoted by $[\varphi]_C$, or simply by $[\varphi]$ (when the reference to C is clear from the context).

2. If $f \in L$ is an n-ary function symbol, then f^B is an n-ary function on B such that for every $[\varphi_1], \ldots, [\varphi_n] \in B$, we have $f^B([\varphi_1], \ldots, [\varphi_n]) = [\varphi]$, where for every $x \in A$,

$$\varphi(x) \simeq f^A(\varphi_1(x), \ldots, \varphi_n(x)). \tag{6}$$

If $P \in L$ is an m-ary predicate symbol, then P^B is an m-ary relation on B such that for every $[\varphi_1], \ldots, [\varphi_m] \in B$,

$$P^B([\varphi_1], \ldots, [\varphi_m]) \quad iff \quad C \subseteq^* \{x \in A \mid P^A(\varphi_1(x), \ldots, \varphi_m(x))\}. \tag{7}$$

If $c \in L$ is a constant symbol, then c^B is the equivalence class of the (total) computable function on A with constant value c^A.

Remark 1. Let C and B be as in Definition 3.
   (i) The requirement that C is cohesive can be weakened to C being r-cohesive.
   (ii) If C is co-c.e., then for every $[\varphi] \in B$ there is a computable function f such that $f =_C \varphi$. In this case the structures $\prod_C \mathbb{N}$ and $\mathcal{R}/ \sim_C$ are isomorphic.

Versions of restricted Łoś's theorem were given in [13], [14] for models of fragments of arithmetic. The version of Łoś's theorem for cohesive powers of computable structures was given in [4] and was called the fundamental theorem of cohesive powers. Here is a part of the theorem that we will use in the proof of Proposition 1.

Theorem 3. [4] If $\Phi(y_1, \ldots, y_n)$ is a formula in L that is a Boolean combination of Σ_1^0 and Π_1^0 formulas and $[\varphi_1], \ldots, [\varphi_n] \in B$, then

$$\prod_C A \models \Phi([\varphi_1], \ldots, [\varphi_n]) \quad iff \quad C \subseteq^* \{x : A \models \Phi(\varphi_1(x), \ldots, \varphi_n(x))\}. \tag{8}$$

The structure $\prod_C \mathbb{N}$ can be embedded naturally into $\prod_C \mathbb{Q}$ by mapping the equivalence class of $[\varphi] \in \prod_C \mathbb{N}$ to the larger equivalence class of the same $[\varphi]$ in $\prod_C \mathbb{Q}$. With the following general approach we can also obtain $\prod_C \mathbb{Q}$ from $\prod_C \mathbb{N}$.

Definition 4. Let $\mathcal{M}_1$ be a structure for the language $L = \{+, \cdot, 0, 1\}$, which satisfies the commutative semiring axioms.
   Let $\mathcal{M}_2$ be a ring with domain $(M_1 \times M_1)_{=^+}$ where $(a_1, b_1) \equiv^+ (a_2, b_2)$ iff $a_1 + b_2 = b_1 + a_2$. Suppose that the natural definition of the ring operations of $\mathcal{M}_2$ is such that $\mathcal{M}_2$ is an integral domain.
   Let $\mathcal{M}_3$ be a field with domain $(M_2 \times M_2)_{\equiv^{\cdot}}$ where $(a_1, b_1) \equiv^{\cdot} (a_2, b_2)$ iff $a_1 \cdot b_2 = b_1 \cdot a_2$ and the field (of quotients) operations of $\mathcal{M}_3$ are naturally defined.

Remark 2. Let $\mathcal{M}_1$ be as in Definition 4. If $\mathcal{M}_1 = \prod_C \mathbb{N}$, then $\mathcal{M}_2 \cong \prod_C \mathbb{Z}$ and $\mathcal{M}_3 \cong \prod_C \mathbb{Q}$. The natural embedding of $\mathcal{M}_1$ into $\mathcal{M}_3$ yields the natural embedding of $\prod_C \mathbb{N}$ into $\prod_C \mathbb{Q}$ mentioned earlier.

Lemma 1. *(i) Any automorphism of $\mathcal{M}_i$ induces an automorphism of $\mathcal{M}_j$ for $i < j \leq 3$.*

(ii) If $\mathcal{M}_i$ is definable in $\mathcal{M}_j$ for $i < j \leq 3$, then any automorphism of $\mathcal{M}_j$ induces an automorphism of $\mathcal{M}_i$.

(iii) If $\mathcal{M}_j$ is rigid, then so is $\mathcal{M}_i$ for $i < j \leq 3$.

(iv) If $\mathcal{M}_i$ is definable in $\mathcal{M}_j$ for $i < j \leq 3$ and $\mathcal{M}_i$ is rigid, then so is $\mathcal{M}_j$.

Proof. We will prove only (ii) and (iv) and leave the rest of the theorem to the reader.

(ii) Let Γ be an automorphism of $\mathcal{M}_3$. To define an automorphism

$$\Gamma_1 : \mathcal{M}_2 \to \mathcal{M}_2, \tag{9}$$

let $a \in M_2$. The natural embedding of $\mathcal{M}_2$ into $\mathcal{M}_3$ maps a to $[(a, 1)]_{\equiv}$. Let ϕ be a first-order formula in L that defines the set $\{\, [(x, 1)]_{\equiv} \mid x \in M_2 \}$ in $\mathcal{M}_3$. Then $\mathcal{M}_3 \models \phi([(a, 1)]_{\equiv})$, so $\mathcal{M}_3 \models \phi(\Gamma([(a, 1)]_{\equiv}))$. Then $\Gamma([(a, 1)]_{\equiv}) = [(c, 1)]_{\equiv}$ for a unique $c \in M_2$. Let $\Gamma_1(a) = c$. The proofs for the other cases are similar.

(iv) We will only prove that $\mathcal{M}_3$ is rigid provided that $\mathcal{M}_2$ is rigid. Suppose that Γ is an automorphism of $\mathcal{M}_3$. Since M_2 is first-order definable in $\mathcal{M}_3$, Γ_1 defined in (ii) is an automorphism of $\mathcal{M}_2$. Let $a \in M_3$, and let $b_1, b_2 \in M_2$ be such that $a = [(b_1, b_2)]_{\equiv}$. Then

$$\Gamma(a) = \Gamma((b_1, 1)) \cdot \Gamma((1, b_2)) = [(\Gamma_1(b_1), \Gamma_1(b_2))]_{\equiv} = [(b_1, b_2)]_{\equiv} = a \tag{10}$$

because $\mathcal{M}_2$ is rigid.

Remark 3. Note that if $\mathcal{M}_1$ is rigid in a language L and a relation R is definable in $\mathcal{M}_1$, then $\mathcal{M}_1$ is rigid in $L \cup \{R\}$. We will later use this fact when the relation R is $<$.

3 Definability and Isomorphisms

We will now prove that $\prod_C \mathbb{N}$ is definable both in $\prod_C \mathbb{Z}$ and $\prod_C \mathbb{Q}$. The definability of $\mathbb{Z}$ in $\mathbb{Q}$ (by a Π_3^0 formula) has been established by J. Robinson in [10]. More recently, Koenigsmann [11] gave a Π_1^0 definition of $\mathbb{Z}$ in $\mathbb{Q}$. He proved that there is a positive integer n and a polynomial $p \in \mathbb{Z}[y, z_1, \ldots, z_n]$ such that

$$y \in \mathbb{Z} \Leftrightarrow \forall z_1 \cdots \forall z_n [p(y, z_1, \ldots, z_n) \neq 0]. \tag{11}$$

We note that the intended range of all quantified variables in formulas (11) through (14) is $\mathbb{Q}$. The proof of Proposition 1 below essentially uses the Koenigsmann's definition and cannot work with a definition of higher complexity.

The definability of $\mathbb{N}$ in $\mathbb{Z}$ (by various Σ_1^0 formulas) has been established by R. Robinson in [17]. We will use the formula that defines the natural numbers as sums of squares of four integers. Using these results we obtain that $\mathbb{N}$ can be defined in $\mathbb{Q}$ as follows:

$$x \in \mathbb{N} \Leftrightarrow \exists y_1 \cdots \exists y_4 [\bigwedge_{i \leq 4} y_i \in \mathbb{Z} \wedge x = y_1^2 + y_2^2 + y_3^2 + y_4^2] \tag{12}$$

$$x \in \mathbb{N} \Leftrightarrow \exists y_1 \cdots \exists y_4 \forall z_1 \cdots \forall z_n [\bigwedge_{i \leq 4} p(y_i, z_1, \ldots, z_n) \neq 0 \wedge x = y_1^2 + y_2^2 + y_3^2 + y_4^2], \tag{13}$$

which we will abbreviate as

$$x \in \mathbb{N} \Leftrightarrow \exists \overline{y} \forall \overline{z} \theta(x, \overline{y}, \overline{z}), \tag{14}$$

where $\phi(x, \overline{y}, \overline{z})$ is a quantifier-free formula in the language of rings $L = \{+, \cdot, 0, 1\}$. Note that there is a natural embedding of $\prod_C \mathbb{N}$ into $\prod_C \mathbb{Z}$, and of $\prod_C \mathbb{Z}$ into $\prod_C \mathbb{Q}$.

Proposition 1. *The natural embedding of $\prod_C \mathbb{N}$ is definable in $\prod_C \mathbb{Q}$ by the same formula $\exists \overline{y} \forall \overline{z} \theta(x, \overline{y}, \overline{z})$ that defines $\mathbb{N}$ in $\mathbb{Q}$.*

Proof. First, assume that for some $[\varphi] \in \prod_C \mathbb{Q}$ we have

$$\prod_C \mathbb{Q} \models \exists \overline{y} \forall \overline{z} \theta([\varphi], \overline{y}, \overline{z}), \tag{15}$$

and that $y_i = [\psi_i]$ are such that

$$\prod_C \mathbb{Q} \models \forall \overline{z} \theta([\varphi], \overline{[\psi_i]}, \overline{z}). \tag{16}$$

By Theorem 3, we have that:

$$C \subseteq^* \{n : \mathbb{Q} \models \forall \overline{z} \theta(\varphi(n), \overline{\psi_i(n)}, \overline{z})\}. \tag{17}$$

Using the definition of $\theta(x, \overline{y}, \overline{z})$ we obtain that $C \subseteq^* \{n : \varphi(n) \in \omega\}$, which means that $[\varphi] \in \prod_C \mathbb{N}$.

Now, assume that $[\varphi] \in \prod_C \mathbb{N}$. We will prove that

$$\prod_C \mathbb{Q} \models \exists \overline{y} \forall \overline{z} \theta([\varphi], \overline{y}, \overline{z}). \tag{18}$$

Define the partial computable functions $\xi_i : \omega \to \mathbb{Q}$ $(i \leq 4)$ as follows. If at stage s we have $\varphi^s(n) = m$ and $m \in \omega$, then find the least $(b_1, \ldots, b_4) \in \omega^4$ such that $m = \sum_{i=1}^4 b_i^2$ and let $\xi_i(n) = b_i$.

By the definition of the functions y_i, we have that

$$C \subseteq^* \{n : \mathbb{Q} \models [\bigwedge_{i \leq 4}(\xi_i(n) \in \mathbb{Z}) \wedge \varphi(n) = \sum_{i \leq 4} \xi_i(n)^2]\}. \tag{19}$$

Again, by Theorem 3, we obtain that

$$\prod_C \mathbb{Q} \models \forall \bar{z}\theta([\varphi], [y_1], \dots, [y_4], \bar{z}), \tag{20}$$

which implies that

$$\prod_C \mathbb{Q} \models \exists \bar{y} \forall \bar{z}\theta([\varphi], \bar{y}, \bar{z}). \tag{21}$$

For convenience we will introduce additional notation. Let
(1) $\varphi_1(x) =_{def} \exists y_1 \cdots \exists y_4 [x = y_1^2 + y_2^2 + y_3^2 + y_4^2]$, and
(2) $\varphi_2(x) =_{def} \forall z_1 \cdots \forall z_n [p(x, z_1, \dots, z_n) \neq 0]$.

Definition 5. *Let $\varphi(\bar{x})$ be a formula in a prenex normal form. Define $\varphi^*(\bar{x})$ inductively as follows:*

(1) $\varphi^*(\bar{x}) =_{def} \varphi(\bar{x})$ if φ is a quantifier-free formula,

(2) $\varphi^*(\bar{x}) =_{def} \exists y[\varphi_2(y) \wedge \psi^*(\bar{x}, y)]$ if $\varphi(\bar{x}) = \exists y\psi(\bar{x}, y)$,

(3) $\varphi^*(\bar{x}) =_{def} \forall y[\varphi_2(y) \Rightarrow \psi^*(\bar{x}, y)]$ if $\varphi(\bar{x}) = \forall y\psi(\bar{x}, y)$.

Note that in this case: $x \in N \Leftrightarrow \varphi_1^*(x) \Leftrightarrow \exists \bar{y} \forall \bar{z}\theta(x, \bar{y}, \bar{z})$.

Definition 6. *Let $\varphi(\bar{x})$ be a formula in a prenex normal form. Define $\varphi^\dagger(\bar{x})$ inductively as follows:*

(1) $\varphi^\dagger(\bar{x}) =_{def} \varphi(\bar{x})$ if φ is a quantifier-free formula

(2) $\varphi^\dagger(\bar{x}) =_{def} \exists y[\varphi_1^*(y) \wedge \psi^\dagger(\bar{x}, y)]$ if $\varphi(\bar{x}) = \exists y\psi(\bar{x}, y)$,

(3) $\varphi^\dagger(\bar{x}) =_{def} \forall y[\varphi_1^*(y) \Rightarrow \psi^\dagger(\bar{x}, y)]$ if $\varphi(\bar{x}) = \forall y\psi(\bar{x}, y)$.

The idea for this definition is that $\phi^\dagger(\bar{x})$ essentially expresses the formula $\phi(\bar{x})$ with the scope of its quantifiers limited from $\mathbb{Q}$ to N (and from $\prod_C \mathbb{Q}$ to $\prod_C N$ because of Proposition 1).

Proposition 2. $\prod_C \mathbb{Q}$ *and* $\mathbb{Q}$ *are not elementary equivalent.*

Proof. Let T be Kleene's predicate. By a result by Fefferman, Scott, and Tennenbaum (see Theorem 2.1 in [12]), we know that for

$$\phi = \forall x \exists t \forall e \forall z[(e < x \wedge T(e, x, z)) \Rightarrow z < t], \tag{22}$$

$$\mathbb{N} \models \phi \quad \text{but} \quad \prod_C \mathbb{N} \nvDash \phi. \tag{23}$$

We can assume that ϕ is a sentence in $L = (+, \cdot, 0, 1)$ since both Kleene's T predicate and $<$ are definable in $\mathbb{N}$. Then $\mathbb{Q} \models \phi^\dagger$ iff $\mathbb{N} \models \phi$, and $\prod_C \mathbb{Q} \nvDash \phi^\dagger$ iff $\prod_C \mathbb{N} \nvDash \phi$. This finally gives us that

$$\mathbb{Q} \not\equiv \prod_C \mathbb{Q}. \tag{24}$$

Definition 7. ([4]) *The sets $A \subseteq \omega$ and $B \subseteq \omega$ have the same 1-degree up to $=^*$ (denoted by $A \equiv_1^* B$) if there are $C =^* A$ and $D =^* B$ such that $C \equiv_1 D$.*

Remark 4. Using Myhill's Isomorphism Theorem (see [18], p. 24), we conclude that $A \equiv_1^* B$ iff there is a computable permutation σ of ω such that $\sigma(A) =^* B$.

Proposition 3. *Let $M_1 \subseteq \omega$ and $M_2 \subseteq \omega$ be maximal sets.*

1. *If $M_1 \equiv_1^* M_2$, then $\prod_{\overline{M_1}} \mathbb{Q} \cong \prod_{\overline{M_2}} \mathbb{Q}$.*
2. *If $M_1 \not\equiv_1^* M_2$, then $\prod_{\overline{M_1}} \mathbb{Q} \not\equiv \prod_{\overline{M_2}} \mathbb{Q}$.*

Proof. (1) This fact has been proven in [4] for an arbitrary computable structure $\mathcal{A}$. If σ is a computable permutation of ω such that $\sigma(M_1) =^* M_2$, then the map $\Phi : \prod_{\overline{M_1}} \mathcal{A} \to \prod_{\overline{M_2}} \mathcal{A}$ such that $\Phi([\psi]) = [\psi \circ \sigma]$ is an isomorphism.

(2) Note that for maximal sets we have $M_1 \equiv_1^* M_2$ iff $M_1 =_m M_2$. For the proof in the nontrivial direction, assume that $M_1 \leq_m M_2$ via f and $M_2 \leq_m M_1$ via g. Since $\overline{M_1}$ is cohesive, $g \circ f(\overline{M_1}) \cap \overline{M_1}$ is infinite and, by Proposition 2.1 in [12], $g \circ f|_{\overline{M_1}}$ and $I|_{\overline{M_1}}$ differ only on finitely many elements. Then, to define the computable permutation σ, we enumerate M_1 and let

$$\sigma(n) = \begin{cases} n, & \text{if n is enumerated into } M_1 \text{ first;} \\ f(n), & \text{if } g(f(n)) = n. \end{cases}$$

Note that $\sigma(n)$ will be defined for almost every $n \in \omega$, and let $\sigma(n) = n$ in the finitely many remaining cases.

If $M_1 \neq_m M_2$, then we apply Theorem 2.3 from [12]. In fact, Lerman provided a specific sentence θ (originally in the language $L_< = \{+, \cdot, 0, 1, <\}$) for which $\prod_{\overline{M_1}} \mathbb{N} \models \theta$ while $\prod_{\overline{M_2}} \mathbb{N} \models \neg\theta$. As before, we can assume that the sentence θ is equivalent to a sentence in the language L. Thus, for the relativisation $\theta^\dagger$, we have $\prod_{\overline{M_1}} \mathbb{Q} \models \theta^\dagger$, while $\prod_{\overline{M_2}} \mathbb{Q} \models \neg\theta^\dagger$.

4   Automorphisms

We now assume that C is a co-maximal (co-c.e. and cohesive) set and will prove that the field $\prod_C \mathbb{Q}$ is rigid (i.e., it has only the trivial automorphism). To do this we will show that $\prod_C \mathbb{N}$ is a special case of arithmetic (exactly Δ_1^0) ultrapowers studied by Hirschfeld, Wheeler, and McLaughlin. Specifically, they studied the structures $\mathcal{F}_n/U$, where U is a non-principal ultrafilter in the Boolean algebra of Δ_n^0 sets, and $\mathcal{F}_n$ is the set of all total functions with Σ_n^0 graphs. In Theorem 2.11 of [14], McLaughlin proved that $\mathcal{F}_n/U$ is rigid for the language $L_< = \{+, \cdot, 0, 1, <\}$. To apply McLaughlin's result we need to make a few observations. First, clearly, the theorem also holds for the language $L = \{+, \cdot, 0, 1\}$ because of the definability of the relation "$<$". Second, we will see how the equivalence relation induced by the co-maximal set C is equivalent to the one induced by a Δ_1^0 ultrafilter. Finally, the domain of $\prod_C \mathbb{N}$ consists of partial computable functions, while the functions in $\mathcal{F}_1$ are total. The last two points are addressed in the following proposition.

Proposition 4. *(1)* $U_C = \{R \mid R \in \Delta_1^0 \text{ and } C \subseteq^* R\}$ *is an ultrafilter in the Boolean algebra* Δ_1^0.
(2) $\prod_C \mathbb{N} \cong \mathcal{F}_1/U_C$

Proof. (1) It is straightforward to show that U_C is a filter. Since C is cohesive, we have $(\forall R \in \Delta_1^0)[C \subseteq^* R \lor C \subseteq^* \overline{R}]$ and therefore, U_C is maximal.

To prove (2) we will show that for every partial computable function φ for which $C \subseteq^* dom(\varphi)$, there is a computable function f_φ such that $[\varphi]_C = [f_\varphi]_C$. We simply define
$$f_\varphi(n) = \begin{cases} \varphi(n), & \text{if } \varphi(n) \downarrow \text{ first;} \\ 0, & \text{if n is enumerated into } \overline{C} \text{ first.} \end{cases}$$
Obviously, $f_\varphi(n)$ is defined for all but finitely many n. For the finitely many n for which $f_\varphi(n)$ is not defined, we let $f_\varphi(n) = 0$. It is immediate that f_φ is computable and $[\varphi]_C = [f_\varphi]_C$. It is also immediate that if $[\varphi]_C = [\psi]_C$, then $A = \{n : f_\varphi(n) = f_\psi(n)\}$ is a computable set such that $C \subseteq^* A$, and so $A \in \mathcal{U}$ and $[f_\varphi] =_\mathcal{U} [f_\psi]$.

Then the map $\Phi : \prod_C \mathbb{N} \to \mathcal{F}_1/U_C$ given by $\Phi([\varphi]_C) = [f_\varphi]_{U_C}$ is an isomorphism.

Corollary 1. *The structure* $\prod_C \mathbb{N}$ *is rigid.*

Theorem 4. *The structure* $\prod_C \mathbb{Q}$ *is rigid.*

Proof. If $\mathcal{M}_1 = \prod_C \mathbb{N}$ and $\mathcal{M}_3 \cong \prod_C \mathbb{Q}$, then $\mathcal{M}_1$ is definable in $\mathcal{M}_3$ by Proposition 1, and $\mathcal{M}_1$ is rigid by Proposition 4. Then the rigidity of $\mathcal{M}_3$ follows from Lemma 1, part (4).

Corollary 2. *If co-maximal powers* $\prod_{M_1} \mathbb{Q}$ *and* $\prod_{M_2} \mathbb{Q}$ *are isomorphic, then there is a unique isomorphism between them.*

Proof. If f_1 and f_2 are isomorphisms that map $\prod_{\overline{M_1}} \mathbb{Q}$ to $\prod_{\overline{M_2}} \mathbb{Q}$, then $f_2^{-1} \circ f_1$ must be the identity automorphism of $\prod_{\overline{M_1}} \mathbb{Q}$.

References

1. Davis, M., Matiyasevich, Y., Robinson, J.: Hilbert's tenth problem: Diophantine equations: positive aspects of a negative solution, Mathematical developments arising from Hilbert problems. In: Proc. Sympos. Pure Math., vol. XXVIII, Northern Illinois Univ, De Kalb, Ill (1974); Amer. Math. Soc., Providence, RI, pp. 323–378 (1976)
2. Dimitrov, R.D.: Quasimaximality and principal filters isomorphism between $\mathcal{E}^*$ and $\mathcal{L}^*(V_\infty)$. Arch. Math. Logic 43, 415–424 (2004)
3. Dimitrov, R.D.: A class of Σ_3^0 modular lattices embeddable as principal filters in $\mathcal{L}^*(V_\infty)$. Arch. Math. Logic 47(2), 111–132 (2008)
4. Dimitrov, R.D.: Cohesive powers of computable structures, Annuare De L'Universite De Sofia "St. Kliment Ohridski". Fac. Math. and Inf., Tome 99, 193–201 (2009)
5. Dimitrov, R.D.: Extensions of certain partial automorphisms of $\mathcal{L}^*(V_\infty)$, Annuare De L'Universite De Sofia "St. Kliment Ohridski". Fac. Math. and Inf., Tome 99, 183–191 (2009)
6. Feferman, S., Scott, D.S., Tennenbaum, S.: Models of arithmetic through function rings. Notices Amer. Math. Soc. 6, 173. Abstract #556-31 (1959)
7. Guichard, D.R.: Automorphisms of substructure lattices in recursive algebra. Ann. Pure Appl. Logic 25(1), 47–58 (1983)
8. Hirschfeld, J.: Models of arithmetic and recursive functions. Israel Journal of Mathematics 20(2), 111–126 (1975)
9. Hirschfeld, J., Wheeler, W.: Forcing, arithmetic, division rings. Lecture Notes in Mathematics, vol. 454. Springer, Berlin (1975)
10. Robinson, J.: Definability and decision problems in arithmetic. Journal of Symbolic Logic 14(2), 98–114 (1949)
11. Koenigsmann, J.: Defining $\mathbb{Z}$ in $\mathbb{Q}$, forthcoming in the Annals of Mathematics, http://arxiv.org/abs/1011.3424
12. Lerman, M.: Recursive functions modulo co-r-maximal sets. Transactions of the American Mathematical Society 148(2), 429–444 (1970)
13. McLaughlin, T.: Some extension and rearrangement theorems for Nerode semirings, Zeitschr. f. math. Logic und Grundlagen d. Math. 35, 197–209 (1989)
14. McLaughlin, T.: Sub-arithmetical ultrapowers: a survey. Annals of Pure and Applied Logic 49(2), 143–191 (1990)
15. McLaughlin, T.: Δ_1 ultrapowers are totally rigid. Archive for Mathematical Logic 46, 379–384 (2007)
16. Metakides, G., Nerode, A.: Recursively enumerable vector spaces. Annals of Mathematical Logic 11, 147–171 (1977)
17. Robinson, R.: Arithmetical definitions in the ring of integers. Proceedings of the American Mathematical Society 2(2), 279–284 (1951)
18. Soare, R.I.: Recursively Enumerable Sets and Degrees. A Study of Computable Functions and Computably Generated Sets. Springer, Berlin (1987)

Modeling Life as Cognitive Info-computation

Gordana Dodig-Crnkovic

School of Innovation, Design and Engineering, Mälardalen University, Västerås, Sweden
gordana.dodig-crnkovic@mdh.se

Abstract. This article presents a naturalist approach to cognition understood as a network of info-computational, autopoietic processes in living systems. It provides a conceptual framework for the unified view of cognition as evolved from the simplest to the most complex organisms, based on new empirical and theoretical results. It addresses three fundamental questions: *what cognition is, how cognition works and what cognition does at different levels of complexity of living organisms*. By explicating the info-computational character of cognition, its evolution, agent-dependency and generative mechanisms we can better understand its life-sustaining and life-propagating role. The info-computational approach contributes to rethinking cognition as a process of natural computation in living beings that can be applied for cognitive computation in artificial systems.

1 Introduction

It is a remarkable fact that even after half a century of research in cognitive science, cognition still lacks a commonly accepted definition [1]. E.g. Neissers description of cognition as "all the processes by which sensory input is transformed, reduced, elaborated, stored, recovered and used" [2] is so broad that it includes present day robots. On the other hand, the Oxford dictionary definition: "the mental action or process of acquiring knowledge and understanding through thought, experience, and the senses" applies only to humans. Currently the field of cognitive robotics is being developed where we can learn by construction *what cognition might be* and then, returning to cognitive systems in nature find out what solutions nature has evolved. The process of two-way learning [3] starts from nature by reverse engineering existing cognitive agents, while simultaneously trying to design cognitive computational artifacts. We have a lot to learn from natural systems about how to engineer cognitive computers. [4]

Until recently only humans were commonly accepted as cognitive agents (*anthropogenic* approach in Lyon). Some were ready to ascribe certain cognitive capacities to all apes, and some perhaps to all mammals. The lowest level cognition for those with the broadest view of cognition included all organisms with nervous system. Only a few were prepared to go below that level. Among those very few, the first who were ready to acknowledge a cognitive agency of organisms without nervous system were Maturana and Varela [5][6], who argued that *cognition and life are identical processes*. Lyons classification, besides describing the anthropogenic approach, includes a *biogenic* approach based on self-organizing complex systems and autopoiesis. The adoption in the present paper of the biogenic approach through the definition of Maturana and Varela is motivated by the wish to provide a theory that includes all living organisms and artificial cognitive agents within the same framework.

A. Beckmann, E. Csuhaj-Varjú, and K. Meer (Eds.): CiE 2014, LNCS 8493, pp. 153–162, 2014.

2   The Computing Nature, Computational Naturalism and Minimal Cognition

Naturalism is the view that *nature is the only reality*. It describes nature through its structures, processes and relationships using a scientific approach. Naturalism studies the evolution of the entire natural world, including the life and development of humanity as a part of nature. Computational naturalism (pancomputationalism, naturalist computationalism) is the view that the nature is a huge network of computational processes which, according to physical laws, computes (dynamically develops) its own next state from the current one. Representatives of this approach are Zuse, Fredkin, Wolfram, Chaitin and Lloyd, who proposed different varieties of computational naturalism. According to the idea of computing nature, one can view the time development (dynamics) of physical states in nature as information processing (natural computation). Such processes include self-assembly, self-organization, developmental processes, gene regulation networks, gene assembly, protein-protein interaction networks, biological transport networks, social computing, evolution and similar processes of morphogenesis (creation of form). The idea of computing nature and the relationships between two basic concepts of information and computation are explored in [7] and [8].

In computing nature, cognition should be studied as a natural process. If we adopt the biogenetic approach to cognition, the important question is *what is the minimal cognition*? Recently, a number of empirical studies have revealed an unexpected richness of cognitive behaviors (perception, information processing, memory, decision making) in organisms as simple as bacteria. Single bacteria are too small to be able to sense anything but their immediate environment, and they live too briefly to be able to memorize a significant amount of data. On the other hand bacterial colonies, swarms and films exhibit an unanticipated complexity of behaviors that can undoubtedly be characterized as biogenic cognition, [9][10][11][12][13][14].

Apart from bacteria and similar organisms without nervous system (such as e.g. slime mold, multinucleate or multicellular Amoebozoa, which recently has been used to compute shortest paths), even plants are typically thought of as living systems without cognitive capacities. However, plants too have been found to possess memory (in their bodily structures that change as a result of past events), the ability to learn (plasticity, ability to adapt through morphodynamics), and the capacity to anticipate and direct their behavior accordingly. Plants are argued to possess rudimentary forms of knowledge, according to [15] p. 121, [16] p. 7 and [17] p. 61.

In this article we focus on primitive cognition as the totality of processes of self-generation, self-regulation and self-maintenance that enables organisms to survive using information from the environment. The understanding of cognition as it appears in degrees of complexity can help us better understand the step between inanimate and animate matter from the first autocatalytic chemical reactions to the first autopoietic proto-cells.

3   Informational Structure of Reality for a Cognitive Agent

When we talk about computing nature, we can ask: what is the hardware for this computation? We, as cognizing agents interacting with nature through information exchange, experience nature cognitively as information. Informational structural realism [18][19][20] is a framework that takes information as the fabric of the universe (for an agent). The physicists Zeilinger [21] and Vedral [22] suggest that information and reality are one. For a cognizing agent in the informational universe, the dynamical changes of its informational structures make it a huge computational network where computation is understood as information dynamics (information processing). Thus the substrate, the "hardware", is information that defines data-structures on which computation proceeds.

Info-computationalism is a synthesis of informational structural realism and natural computationalism (pancomputationalism) - the view that the universe computes its own next state from the previous one [23]. It builds on two basic complementary concepts: information (structure) and computation (the dynamics of informational structure) as described in [24] [25] and [26].

The world for a cognizing agent exists as potential information, corresponding to Kants *das Ding an sich*. Through interactions, this potential information becomes actual information, "a difference that makes a difference" [27]. Shannon describes the process as the conversion of latent information into manifest information [28]. Even though Batesons definition of information as a difference that makes a difference (for an agent) is a widely cited one, there is a more general definition that includes the fact that information is relational and subsumes Batesons definition:

> "Information expresses the fact that a system is in a certain configuration
> that is correlated to the configuration of another system. Any physical system
> may contain information about another physical system." Hewitt [29] p. 293

Combining the Bateson and Hewitt insights, at the basic level, information is a difference in one physical system that makes a difference in another physical system.

When discussing cognition as a bioinformatic process of special interest, there is the notion of agent, i.e. a system able to act on its own behalf [26]. Agency has been explored in biological systems by Kauffman and Deacon [30] [31] [32]. The world as it appears to an agent depends on the type of interaction through which the agent acquires information, [33]. Agents communicate by exchanging messages (information) that help them coordinate their actions based on the (partial) information they possess (a form of social cognition).

4   Information Self-Structuring through Morphological/Physical/Intrinsic Computation and PAC Algorithms

Regarding computational models of biological phenomena, we must emphasize that within the info-computational framework computation is defined as information processing. This differs from the traditional Turing machine model of computation that is an algorithm/effective procedure/recursive function/formal language. The Turing machine is a logical construct, not a physical device. Cooper [34] points towards definability as a form of higher order computation, and its relationship to embodiment. Modeling computing nature adequately, including biological information processing with its self-generating and learning real-time properties, requires new models of computation such as interactive and networked concurrent computation models, as argued in [7] and [35] with reference to [36] and [37].

Computation in nature can be described as a self-generating system consisting of networks of programs [38], a model inspired by the self-modifying systems of [39]. In the course of the development of the general theory of networked physical information processing, the idea of computation becomes generalized. Examples of new computing paradigms include natural computing [40] [41] [42] [43]; superrecursive algorithms [44]; interactive computing [45]; actor model [36] and similar "second generation" models of computing [37].

Among novel models of computation of special interest are Valiants ecorythms or algorithms satisfying "Probably Approximately Correct" criteria (PAC) as they explicitly model natural systems "learning and prospering in a complex world". [46] The difference between PAC learning algorithms and the Turing machine model is that the latter does not interact with the environment, and thus does not learn. It has unlimited resources, both space (memory) and time, and even though it is sequential, it does not operate in real time. In order to computationally model living nature, we need suitable resource-aware learning algorithms, such as ecorithms, described by Valiant:

> "The model of learning they follow, known as the probably approximately correct model, provides a quantitative framework in which designers can evaluate the expertise achieved and the cost of achieving it. These ecorithms are not merely a feature of computers. I argue in this book that such learning mechanisms impose and determine the character of life on Earth. The course of evolution is shaped entirely by organisms interacting with and adapting to their environments." [46] p. 8

A different approach to evolution is taken by Chaitin, who argues for Darwins theory from the perspective of gene-centric metabiology [47]. The interesting basic idea that life is software (executable algorithms) run by physics is applied in the search for biological creativity (in the form of increased fitness). Darwins idea of common descent and the evolution of organisms on earth is strongly supported by computational models of self-organization through information processing i.e. morphological computing. [48]

The cognitive capacity of living systems depends on *the specific morphology of organisms* that enables perception, memory, information processing and agency. As argued in [48], morphology is the central idea connecting computation and information. The process of mutual evolutionary shaping between an organism and its environment is a result of information self-organization. Here, both the physical environment and the physical body of an agent can be described by their informational structure that consists of data as atoms of information. Intrinsic computational processes, which drive changes in informational structures, result from the operation of physical laws. The environment provides an organism with a variety of inputs in the form of both information and matter-energy, where the difference between information and matter-energy is not in the kind, but in the use the organism makes of it. As *there is no information without representation* [49], *all information is carried by some physical carrier* (light, sound, radio-waves, chemical molecules, etc.). The same physical object can be used by an organism as a source of information and as a source of nourishment/matter/energy. In general, the simpler the organism, the simpler the information structures of its body, the simpler the information carriers it relies on, and the simpler its interactions with the environment.

5 Cellular Computation

The environment is a *resource*, but at the same time it also imposes *constraints* that limit an agents space of possibilities. In an agent that can be described as a complex informational structure, constraints imposed by the environment drive the time development (computation) of its structures to specific trajectories. This relationship between an agent and its environment is called *structural coupling* by Maturana and Varela[5]. Experiments with bacteria performed by Ben-Jacob and Bassler show that bacteria interact with the environment, sense it, and extract its latent/potential information. This information triggers cognitive processes ("according to internally stored information") that result in changes of their structure, function and behavior. Moreover, Ben-Jacob explains how information can be seen as inducing "an internal condensed description (model of usable information)" of the environment, which directs its behavior and function. This is a process of intracellular computation, which proceeds via "gene computation circuits or gene logical elements", that is gene circuits or regulatory pathways. As bacteria multiply by cell division, complex colony forms.

Every single bacterium is an autonomous system with internal information management capabilities: *interpretation*, *processing* and *storage* of information. Ben-Jacob has found that complex forms emerge as a result of the communication between bacteria as interplay of the micro-level vs. macro-level (single organism vs. colony). Chemical sign-processes used by bacteria for signaling present a rudimentary form of language. Waters and Bassler [14] describe the process of "quorum sensing" and communication between bacteria that use two kinds of languages – intra-species and inter-species chemical signalling. That is how they are capable of building films consisting of a variety of species.

Experiments show that the colony as a whole "behaves much like a multi-cellular organism" governed by the distributed information processing with message broadcasting that stimulates changes in individual bacteria (plasticity). *Communication, cooperation and self-organization* within a swarm/colony enable decision-making at the group level as a form of *social cognition*.

> "The cells thus co-generate new information that is used to collectively assume newly engineered cell traits and abilities that are not explicitly stored in the genetic information of the individuals. Thus, the bacteria need only have genetically stored the guidelines for producing these capabilities." [12] p. 88

A bacteria colony changes its morphology and organization through natural distributed information processing and thus learns from experience (such as encounters with antibiotics). Ben-Jacob concludes that they " possibly alter the genome organization or even create new genes to better cope with novel challenges." All those processes can be modelled as distributed concurrent computation in networks of networks of programs, where individual bacteria form networks and bacteria themselves can be modelled as networks of programs (processes or executing algorithms).

Empirical studies of the cognitive abilities of bacteria swarms, colonies and films confirm the result of Harms [50], proving a theorem that natural selection will always lead a population to accumulate information, and so to 'learn' about its environment. Okasha points out that

> "any evolving population 'learns' about its environment, in Harms' sense, even if the population is composed of organisms that lack minds entirely, hence lack the ability to have representations of the external world at all." [51]

Experimental results by [10][11][12][13][14] have shown that bacteria indeed learn from the environment even though the mechanisms of bacterial cognition are limited to relatively simple chemical information processes.

6   Self-organization, Cognitive Info-computation and Evolution of Life

In computational (information processing) models of bacterial cognition, the biological structure (hardware) is at the same time a program (software) that controls the behavior of that hardware both internally and in the interactions with the environment. Already in 1991 Kampis proposed a unified model of computation as the mechanism underlying biological processes through self-generation of information by non-trivial change (self-modification) of systems [39]. This process of self-organization and self-generation of information is what is elsewhere described as morphological computation on different levels of organization of natural systems. Current research in adaptive networks goes in the same direction, [7].

However, understanding of the basic evolutionary mechanisms of information accumulation, with resulting increase in information-processing capacities of organisms (memory, anticipation, computational efficiency), is only the first step towards a fully-fledged evolutionary understanding of cognition, though it is probably the most difficult one, as it requires a radical redefinition of fundamental concepts of information, computation and cognition in naturalist terms. According to Maturana:

> "A cognitive system is a system whose organization defines a domain of interactions in which it can act with relevance to the maintenance of itself, and the process of cognition is the actual (inductive) acting or behaving in this domain. Living systems are cognitive systems, and living as a process is a process of cognition. This statement is valid for all organisms, with and without a nervous system." [6] p. 13

The role of cognition for a living agent, from bacteria to humans is *to efficiently deal with the complexity of the world*, helping an agent to survive and thrive. The world is inexhaustible and largely complex and exceeds by all accounts what a cognizing agent can take in. Cognition is then the mechanism that enables cognizing agents to control their own behavior in order to deal with the complexity of the environment, make sense of the world and use it as a resource for survival, [52] p. 234. In this view, " cognition ' shades off' into basic biological processes such as metabolism."

Through autopoietic processes with structural coupling (interactions with the environment) a biological system changes its structures and thereby the information processing patterns in a self-reflective, recursive manner [5]. But self-organisation with natural selection of organisms, responsible for nearly all information that living systems have built up in their genotypes and phenotypes, is a simple albeit costly method to develop. Higher organisms (which are more expensive to evolve in terms of resources) have developed language and reasoning as a more efficient way of learning. The step from genetic learning (typical of more primitive forms of life) to the acquisition of cognitive skills on higher levels of organisation of the nervous system (such as found in vertebrata) will be the next step to explore in the project of cognitive info-computation, following Jablonka and Lamb [53] who distinguish *genetic, epigenetic, behavioral, and symbolic evolution.* The studies of bacterial cognition suggest that there are some important processes that operate during evolution such as self-organization and auto-poiesis, which guarantee growth of order, and the propagation of structures in spite of the randomness of environmental influences. Also, colonies, swarms and films seem to play a prominent role in bacterial evolution (as swarm intelligence, i.e. distributed cognition).

Interesting question arises in connection to AI and AL which are not based on chemical processes: *is molecular computation necessary for cognition?* For example [9] proposed that minimal cognition can be identified with sensorimotor coordination. However, even though fundamental, sensorimotor coordination is not enough to explain cognition in biological systems. Chemical processes of autopoiesis based on molecular computation (information processing) are essential, not only for simple organisms like bacteria, but also for the functioning of the human nervous system. In the words of Shapiro:

> "molecular biology has identified specific components of cell sensing, information transfer, and decision-making processes. We have numerous precise molecular descriptions of cell cognition, which range all the way from bacterial nutrition to mammalian cell biology and development." [54] p. 24

Info-computational approach provides an appropriate framework for studying the above question of minimal cognition. The advantages of info-computational approaches to the modeling of cognition are that they bridge the Cartesian gap between matter and mind, providing a unified naturalist framework for a vast range of phenomena, and they are testable. Dennett declared in a talk at the International Computers and Philosophy Conference, Laval, France in 2006: "AI makes philosophy honest." Paraphrasing Dennett we can say that info-computational models make cognition honest - transparent and open for critical investigation and experimentation. In that sense parallel research in biology and cognitive robotics present a "reality check" where our understanding of cognition, information processing and morphological computation can be tested in a rigorous manner.

7 Conclusions

Studied as a natural phenomenon, cognition can be seen as info-computational processes in living systems. The aim of this article is to present methodological and practical grounds for a naturalist computational approach to cognition supported by new experimental results on cognition of simplest living organisms such as bacteria. The hope is to contribute to the elucidation of the following fundamental questions according to [55] [1] and [9]:

What cognition is. The nature of cognition, the question about how the concept of cognition should be defined. In the info-computational framework it becomes transformed into the question: *what in the computing nature is cognition?* Cognition for an adaptive, developing and evolving living agent is the process of learning that operates according to the PAC (Probably Approximately Correct) strategy [46]. Results from the studies of natural cognitive systems will help resolve the question concerning artifactual computational cognition.

How cognition works. Cognition as information processing happening in an informational network of cognizing agents with distributed computational dynamics connects the agents intrinsic structures with the outside world of potential information, through interactions. Those interactions include all four levels on which evolution operates: *genetic, epigenetic, behavioral, and symbolic* [53]. We have shown in the example of bacterial cognition how all four levels contribute.

What cognition does. By elucidating the info-computational and evolutionary character of cognition we can understand its agent-dependency, its generative mechanisms and its life-sustaining and life-propagating role. Cognition is the mechanism that enables cognizing agents to deal with the complexity of the environment, through control of their own behavior, [52] p. 234.

The info-computational approach can contribute to rethinking cognition as information self-organising processes of morphological/chemical/molecular/natural computation *in all living beings*. Thus, we can start to learn how to adequately computationally model living systems, which has up to now been impossible, [33]. "Second generation computational models" [37] under current development promise to enble us to frame theoretically, simulate and study living organisms in their full complexity. Based on current work in the related fields such as information science, computability and theory of computing, logic, molecular biology, and evolution, a new more coherent picture of cognition can be expected to emerge. As a complement to Woframs idea of *mapping and mining the computational universe* [56] this article suggests *mapping and mining the biological universe* by computational tools with the goal to reverse engineer cognition and find smart cognitive computational strategies.

References

1. Lyon, P.: The biogenic approach to cognition. Cognitive Processing 7, 11–29 (2005)
2. Neisser, U.: Cognitive psychology. Appleton-Century Crofts (1967)
3. Rozenberg, G., Kari, L.: The many facets of natural computing. Communications of the ACM 51, 72–83 (2008)

4. Modha, D.S., Ananthanarayanan, R., Esser, S.K., Ndirango, A., Sherbondy, A.J., Singh, R.: Cognitive computing. Communications of the ACM 54(8), 62–71 (2011)
5. Maturana, H., Varela, F.: Autopoiesis and cognition: the realization of the living. D. Reidel Pub. Co. (1980)
6. Maturana, H.: Biology of Cognition. Defense Technical Information Center (1970)
7. Dodig-Crnkovic, G., Giovagnoli, R.: Computing Nature. Springer (2013)
8. Dodig-Crnkovic, G., Burgin, M.: Information and Computation. World Scientific Pub. Co. Inc. (2011)
9. van Duijn, M., Keijzer, F., Franken, D.: Principles of minimal cognition: Casting cognition as sensorimotor coordination. Adaptive Behavior 14, 157–170 (2006)
10. Ben-Jacob, E., Shapira, Y., Tauber, A.: Seeking the foundations of cognition in bacteria. Physica A 359, 495–524 (2006)
11. Ben-Jacob, E.: Social behavior of bacteria: from physics to complex organization. The European Physical Journal B 65(3), 315–322 (2008)
12. Ben-Jacob, E.: Learning from bacteria about natural information processing. Annals of the New York Academy of Sciences 1178, 78–90 (2009)
13. Ng, W.L., Bassler, B.L.: Bacterial quorum-sensing network architectures. Annual Review of Genetics 43, 197–222 (2009)
14. Waters, C.M., Bassler, B.L.: Quorum sensing: Cell-to-cell communication in bacteria. Annual Review of Cell and Developmental Biology 21, 319–346 (2005)
15. Pombo, O., Torres, J., Symons, J. (eds.): Special Sciences and the Unity of Science. Springer (2012)
16. Rosen, R.: Anticipatory Systems. Pergamon Press (1985)
17. Popper, K.: All Life is Problem Solving. Routledge (1999)
18. Floridi, L.: Informational realism. In: Weckert, J., Al-Saggaf, Y. (eds.) Selected Papers from Conference on Computers and Philosophy, vol. 37, pp. 7–12. Australian Computer Society, Inc. (2003)
19. Sayre, K.M.: Cybernetics and the Philosophy of Mind. Routledge and Kegan Paul (1976)
20. Stonier, T.: Information and meaning: an evolutionary perspective. Springer (1997)
21. Zeilinger, A.: The message of the quantum. Nature 438, 743–743 (2005)
22. Vedral, V.: Decoding reality: the universe as quantum information. Oxford University Press (2010)
23. Chaitin, G.: Epistemology as information theory: From leibniz to omega. In: Dodig Crnkovic, G. (ed.) Computation, Information, Cognition, The Nexus and The Liminal, pp. 2–17. Cambridge Scholars Pub. (2007)
24. Dodig-Crnkovic, G.: Dynamics of information as natural computation. Information 2(3), 460–477 (2011)
25. Dodig-Crnkovic, G.: Investigations into Information Semantics and Ethics of Computing. Mälardalen University Press (2006)
26. Dodig-Crnkovic, G.: Information, computation, cognition. Agency-based hierarchies of levels. In: Müller, V.C. (ed.) Fundamental Issues of Artificial Intelligence (Synthese Library). Springer (forthcoming, 2014)
27. Bateson, G.: Steps to an Ecology of Mind: Collected Essays in Anthropology, Psychiatry, Evolution, and Epistemology. University of Chicago Press (1972)
28. McGonigle, D., Mastrian, K.: Introduction to information, information science, and information systems. In: Nursing Informatics and the Foundation of Knowledge. Jones & Bartlett (2012)
29. Hewitt, C.: What is Commitment? Physical, Organizational, and Social. In: Noriega, P., Vázquez-Salceda, J., Boella, G., Boissier, O., Dignum, V., Fornara, N., Matson, E. (eds.) COIN 2006. LNCS (LNAI), vol. 4386, pp. 293–307. Springer, Heidelberg (2007)

30. Kauffman, S.: At Home in the Universe: The Search for Laws of Self-Organization and Complexity. Oxford University Press (1995)
31. Kauffman, S.: Origins of Order: Self-Organization and Selection in Evolution. Oxford University Press (1993)
32. Deacon, T.: Incomplete Nature. How Mind Emerged from Matter. W. W. Norton and Company (2011)
33. Dodig-Crnkovic, G., Müller, V.: A dialogue concerning two world systems: Info-computational vs. mechanistic. In: Dodig Crnkovic, G., Burgin, M. (eds.) Information and Computation, pp. 149–184. World Scientific (2011)
34. Cooper, S.B.: Turing's Titanic Machine? Communications of the ACM 55(3), 74–83 (2012)
35. Dodig-Crnkovic, G.: Significance of Models of Computation from Turing Model to Natural Computation. Minds and Machines 21(2), 301–322 (2011)
36. Hewitt, C.: What is computation? Actor model versus Turing's model. In: Zenil, H. (ed.) A Computable Universe, Understanding Computation and Exploring Nature As Computation. World Scientific Publishing Company/Imperial College Press (2012)
37. Abramsky, S.: Information, processes and games. In: Benthem van, J., Adriaans, P. (eds.) Philosophy of Information, pp. 483–549. North-Holland (2008)
38. Goertzel, B.: Chaotic Logic. Language, Thought, and Reality from the Perspective of Complex Systems Science. Plenum Press (1994)
39. Kampis, G.: Self-Modifying Systems in Biology and Cognitive Science: A New Framework for Dynamics, Information, and Complexity. Pergamon Press (1991)
40. Rozenberg, G., Bäck, T., Kok, J. (eds.): Handbook of Natural Computing. Springer (2012)
41. MacLennan, B.: Natural Computation and Non-Turing Models of Computation. Theoretical Computer Science 317(1), 115–145 (2004)
42. Nunes de Castro, L.: Fundamentals of natural computing: An overview. Physics of Life Reviews 4, 1–36 (2007)
43. Cardelli, L.: Artificial biochemistry. In: Condon, A., Harel, D., Kok, J., Salomaa, A., Winfree, E. (eds.) Algorithmic Bioprocesses, pp. 429–462. Springer (2009)
44. Burgin, M.: Super-Recursive Algorithms. Springer-Verlag New York Inc. (2005)
45. Wegner, P.: Interactive foundations of computing. Theoretical Computer Science 192(2), 315–351 (1998)
46. Valiant, L.: Probably Approximately Correct: Nature's Algorithms for Learning and Prospering in a Complex World. Basic Books (2013)
47. Chaitin, G.: Life as evolving software. In: Zenil, H. (ed.) A Computable Universe, Understanding Computation and Exploring Nature As Computation. World Scientific (2012)
48. Dodig-Crnkovic, G.: The info-computational nature of morphological computing. In: Müller, V.C. (ed.) Theory and Philosophy of Artificial Intelligence, pp. 59–68. Springer (2012)
49. Landauer, R.: Information is physical. Physics Today 44, 23–29 (1991)
50. Harms, W.F.: Naturalizing epistemology: Prospectus 2006. Biological Theory 1, 23–24 (2006)
51. Okasha, S.: Review of William F. Harms, Information and Meaning in Evolutionary Processes. Notre Dame Philosophical Reviews 12 (2005)
52. Godfrey-Smith, P.: Environmental complexity and the evolution of cognition. In: Sternberg, R., Kaufman, J. (eds.) The Evolution of Intelligence, pp. 233–249. Lawrence Erlbaum Associates (2001)
53. Jablonka, E., Lamb, M.: Evolution in Four Dimensions: Genetic, Epigenetic, Behavioral, and Symbolic Variation in the History of Life. MIT Press (2005)
54. Shapiro, J.A.: Evolution: A View from the 21st Century. FT Press Science (2011)
55. Bechtel, W.: Representations and cognitive explanations: Assessing the dynamicist's challenge in cognitive science. Cognitive Science 22(3), 295–318 (1998)
56. Wolfram, S.: A New Kind of Science. Wolfram Media (2002)

Deciding the Borel Complexity
of Regular Tree Languages

Alessandro Facchini* and Henryk Michalewski

University of Warsaw
{A.Facchini,H.Michalewski}@mimuw.edu.pl

Abstract. We show that it is decidable whether a given a regular tree language belongs to the class Δ_2^0 of the Borel hierarchy, or equivalently whether the Wadge degree of a regular tree language is countable.

1 Introduction

In [14] there was given an algorithm which for a *deterministic* parity tree automaton $\mathcal{A}$ decides whether the language $L(\mathcal{A})$ is Borel. This was further extended to a finer classification in [12] and finally to a full Wadge classification in [13]. The algorithms look for a pattern in the graph of the automaton and decide the Borel and Wadge classes upon finding of these special patterns.

Similar problems for *non-deterministic* parity tree automata seem to be much harder. Recently in [2] was provided an algorithm which decides for a given non-deterministic parity tree automaton $\mathcal{A}$, whether $L(\mathcal{A})$ is a Boolean combination of open sets. For other Borel classes there was no known algorithm. This paper provides a relatively simple extension of the result in [2] to the class of $\Delta_2^0 = \Sigma_2^0 \cap \Pi_2^0$ sets, that is the sets which are simultaneously presentable as countable unions of closed sets and countable intersections of open sets. This result is presented in Section 4 in Theorem 1. The proofs in [2] are based on an analysis of an algebraic structure computable from $\mathcal{A}$ and the main result states that the language $L(\mathcal{A})$ is a Boolean combination of open sets if and only if a certain finite number of algebraic requirements hold. Since the class Δ_2^0 is bigger, in order to characterize this class, the set of algebraic requirements must be relaxed. In this paper we show that indeed this is the case. Our proofs closely follow the proofs from [2] with some necessary adjustments. In particular the crucial concept of the topological cutting game introduced in [2] is considered in this paper not only in the finite, but also in the infinite case.

The approach presented in [2] and in the present paper in a certain sense is a reminiscent of the approach applied to deterministic automata in [12,13,14]. Namely, the algebraic structure computed from a given automaton $\mathcal{A}$ induces a graph with edges reflecting the algebraic properties. In the deterministic case it is possible to decide Borel

* The author is supported by the *Expressiveness of Modal Fixpoint Logics* project realized within the 5/2012 Homing Plus programme of the Foundation for Polish Science, co-financed by the European Union from the Regional Development Fund within the Operational Programme Innovative Economy ("Grants for Innovation").

A. Beckmann, E. Csuhaj-Varjú, and K. Meer (Eds.): CiE 2014, LNCS 8493, pp. 163–172, 2014.
© Springer International Publishing Switzerland 2014

and Wadge classes analyzing patterns in the graph of the automaton, in the present paper we are looking for patterns in the algebraic graph.

Finally let us mention results which provide information about the set-theoretical complexity of a language accepted by a non-deterministic automaton $\mathcal{A}$ assuming some additional properties of $\mathcal{A}$:

- Rabin in [16] proved, that if L and its complement are accepted by a non-deterministic Büchi tree automata, then L is weakly definable, in particular it is Borel.
- Recently in [5] it was shown using decidability results about the cost functions, that for a given non-deterministic Büchi tree automaton it is decidable whether the language is weakly definable.
- In [9] the decidability results regarding deterministic automata were lifted to a more general context of game automata.

This paper consists of four Sections: the Introduction, a preliminary Section 2 introducing automata, set-theoretical and algebraic notations, a Section 3 introducing topological games and linking these games to the Wadge hierarchy and Section 4 containing the main result. All missing proofs can be found in the long version of the present paper [8].

2   Preliminaries

Trees and Contexts. Given a finite alphabet A, a *tree over* A is a partial function $t : \{0, 1\}^* \to A$ such that its domain $\mathrm{dom}(t)$ is prefix closed. A node of a tree t is an element $v \in \mathrm{dom}(t)$. A left child of a node v of t is the node $v0$, while its right child is $v1$. A leaf of a tree is a node without children. We denote by T_A the family of all trees over A. A set of trees over A is called a tree language, or simply a language. A *multi-context* over A is a tree c over $A \cup \{\star\}$, where

- $\star \notin A$, and
- $\star$ only labels some leaves of c.

A leaf of c labelled by $\star$ is called a *port*. Notice that a multi-context may have infinitely many ports. For a multi-context c and a function η mapping each port of c to a tree t over A, by $c[\eta]$ we denote the tree given by inserting into every port x a tree $\eta(x)$. When $\eta(x) = t$ for each port x, we just write $c[t]$. We say that a tree t extends a multi-context c if there is a mapping η such that $c[\eta] = t$. Given a multi-context c and a language L, by $[c]^{-1}L$ we denote the language of trees $t \in L$ extending c. The class generated by c and all possible mappings η is denoted by $c[T_A]$. A finite multi-context is called a *prefix*. A multi-context with only one port is called a *context*.

Topology. For a finite alphabet A, we equip the class T_A of all trees over A with the prefix topology. That is the basic open sets are sets of the form $p[T_A]$, for a prefix p over A, and thus the open sets are of the form $\bigcup_{p \in P} p[T_A]$ for some set P of prefixes.

The class of Borel tree languages of T_A is the closure of the class of open sets of T_A with respect to countable unions and complementations. Given T_A, the initial finite levels of the Borel hierarchy are defined as follows:

- $\Sigma_1^0(T_A)$ is the class of open subsets of T_A,
- $\Pi_n^0(T_A)$ consists of complements of sets from $\Sigma_n^0(T_A)$,
- $\Sigma_{n+1}^0(T_A)$ consists of countable unions of sets from $\Pi_n^0(T_A)$.

A much finer measure of the topological complexity is the *Wadge degree* (see [10, Chapter 21.E]). If $L \subseteq T_A$ and $M \subseteq T_B$, we say that L is *continuously (or Wadge) reducible* to M, if there exists a continuous function $f : T_A \to T_B$ such that $L = f^{-1}(M)$. We write $L \leq_W M$ iff L is continuously reducible to M. This pre-ordering is called the *Wadge ordering*. If $L \leq_W M$ and $M \leq_W L$, then we write $L \equiv_W M$. If $L \leq_W M$ but not $M \leq_W L$, then we write $L <_W M$. The Wadge hierarchy is the partial order induced by $<_W$ on the equivalence classes given by $\equiv_W$. A language L is called *self dual* if it is equivalent to its complement, otherwise it is called *non self dual*.

Given a certain family of sets $\mathcal{C}$, we say that M is $\mathcal{C}$-hard if $L \leq_W M$ for every $L \in \mathcal{C}$. A $\mathcal{C}$-hard set L is said to be $\mathcal{C}$-complete if moreover $L \in \mathcal{C}$.

Algebra. The Wadge hierarchy of the regular languages of infinite words is well understood thanks to a classification result by K. Wagner ([18]). In particular from Wagner's result one can derive an algorithm which decides whether a given regular language of infinite words is a Boolean combination of open sets. Being a Boolean combination of open sets is equivalent to being in Δ_2^0 class in the context of regular languages of infinite words. This is not true for the regular languages of infinite trees (see [2, Section 4.1] and Proposition 7 in [8] - the long version of the present paper - for an analysis of one special case). A natural algebraic interpretation of Wagner's result can be found in [15, Theorem V.6.2]. In the case of languages of infinite trees, the algebraic theory is not yet fully developed. As a general reference may serve papers [1,2,3,4]. For details of the approach applied in the present paper refer to [2, Section 3].

Following the approach presented in [2], the family of all trees T_A is divided into finitely many Myhill-Nerode equivalence classes H_L. Similarly, there are finitely many equivalence classes V_L of contexts. The same holds for multi-contexts with a fixed number of holes. Starting from an automaton accepting language L, one can compute families H_L and V_L. The equivalence class of a tree t or a context v is denoted $\alpha_L(t)$, $\alpha_L(v)$, respectively. For a given tree t and contexts v_1, v_2 multiplication of contexts and trees $v_1 t$, $v_1 v_2$ naturally induces multiplication between elements of H_L and V_L. Similarly, for a given context v the operation of infinite power $v^\infty = vv \ldots$ induces a mapping from V_L to H_L.

Given a regular language L, its *strategy graph* G_L is the pair $(V_L \times H_L, E)$ such that $((v, h), (v', h')) \in E$ iff there exists a tree t of type h such that t can be decomposed as the concatenation of a context of type v and another tree, and each prefix of t can be completed into a context of type v'. We thus say that the strategy graph is *recursive* if there exists a strongly connected component that contains two nodes (v, h) and (v', h') with $h \neq h'$. For a more formal approach to the strategy graph refer to [2, Section G]. We will need the following

Proposition 1 ([2]). *If there exists a path from (v, h) to (v', h') in G_L, then there exists an edge from (v, h) to (v', h').*

3   Topological Complexity and Games

Topological Games. Let L and M be two languages. The *Wadge game* $\mathcal{W}(L, M)$ is an infinite two-player game between Player I and Player II. It is defined as follows. During a play Player I constructs a tree t and Player II a tree t'. At the first round Player I plays a root of t and Player II plays a root of t', and at each consecutive round both players add a level to their corresponding tree (thus either Player adds some child to a leaf or Player signalizes that the node will be also a leaf of the final resulting tree of the play by not adding any children to it). Player I plays first and Player II is allowed to skip her turn but not forever. Player II wins the game iff $t \in L \Leftrightarrow t' \in M$. The game was designed precisely in order to obtain a characterisation of continuous reducibility.

Lemma 1 ([17]). *Let L, M be two languages. Then $L \leq_W M$ iff Player II has a winning strategy in the game $\mathcal{W}(L, M)$.*

From Borel determinacy ([11]), if both L and M are Borel, then $\mathcal{W}(L, M)$ is determined. The ordering $<_W$ restricted to the Borel sets is well-founded (see [10, Theorem 21.15]). The *Wadge degree* for sets of finite Borel rank can be defined inductively. First, we remark that since every self dual set A is Wadge equivalent to the disjoint union of a certain non self dual set B and its complement $B^{\complement}$, it is enough to start associating a Wadge degree only to non self dual sets and say that the Wadge degree of A equals the Wadge degree of B. For each degree there are exactly three equivalence classes with the same degree, represented by L, $L^{\complement}$ and $L^{\pm}$ — the disjoint union of L and $L^{\complement}$. Clearly $L, L^{\complement} <_W L^{\pm}$ and $L^{\pm}$ is self dual.

In [6], J. Duparc showed that for non self dual sets, it is possible to determine its sign, $+$ or $-$, which specifies precisely the $\equiv_W$-class. For instance, $\emptyset$ and complete open sets have sign $-$, while the whole space and complete closed sets have sign $+$. All self dual sets by definition have sign $\pm$. Let κ be the length of Wadge hierarchy of Borel sets of finite rank. Thus an ordinal $\alpha < \kappa$ determines a $\equiv_W$-class, denoted $[\alpha]^{\epsilon}$ for $\epsilon \in \{+, -, \pm\}$. In the same paper, in the context of Wadge degrees, Duparc defined set-theoretical counterparts of ordinal multiplication by a countable ordinal, and (quasi) exponentiation of base ω_1. From now on $[\alpha]^{\epsilon}$ will also denote the canonical sets of Wadge degree generated with Duparc's operations.

Cutting Games. Below we define a family of two-player games of perfect information, called *cutting games*. These games were introduced in [2]. For the argument in [2] the most important was the finite version of the game. In the present paper we will consider both infinite and finite versions of the cutting game.

Let L_i ($i = 1, 2, \dots$) be languages over the alphabet A, and let p be a prefix over the alphabet A. The *simple cutting game* of length k, denoted $\mathcal{H}_k^p(L_1, \dots, L_k)$ is played by two players, Constrainer and Alternator. For each $i \in \{1, \dots, k\}$ the i-th round of the game is played as follows:

- Alternator chooses a tree $t_i \in L_i$ extending the prefix chosen in the previous round by the Constrainer; in the first round of the game Alternator must choose an extension of the given prefix p,
- Constrainer chooses a prefix of the tree t_i.
- If Alternator cannot move, she loses, but if she survives k rounds then she wins.

The *infinite cutting game*, denoted by $\mathcal{H}_\infty^p(L_1, \dots)$, is played just like a simple game but without the restriction to a fixed given number of rounds. Alternator wins iff she can make infinitely many moves.

Let X be a language over the alphabet A. The *X-delayed cutting game*, denoted by $\mathcal{H}_\omega^X(L_1, \dots)$ is similar to a simple cutting game, except that a mini game is played to determine the prefix p and the length k of the match. The mini game goes as follows. Firstly, Alternator chooses a tree $t \in X$. Then Constrainer chooses a prefix p of t and a finite ordinal k. Finally the two players start to play the simple cutting game $\mathcal{H}_k^p(L_1, \dots, L_k)$.

When $L_{2i} = L$ and $L_{2i+1} = L^{\complement}$, then we simply write $\mathcal{H}_k^p(L, L^{\complement})$, $\mathcal{H}_\infty^p(L, L^{\complement})$ and $\mathcal{H}_\omega^X(L, L^{\complement})$. It was verified in [2] that a given language M has a Wadge degree less than ω iff Constrainer has a winning strategy in $\mathcal{H}_k^\varepsilon(M, M^{\complement})$, for all but finitely many $k < \omega$. In [2] it was also remarked that the language L described in [2, Section 4.1] and in Proposition 7 in [8], even if it is such that Alternator has a winning strategy in every corresponding finite cutting game, she looses the infinite one. In the next two propositions we establish a link between delayed cutting games and infinite Wadge degrees on the one hand, and infinite simple cutting games and uncountable Wadge degrees on the other hand.

Proposition 2. *Let L be a tree language, $[\omega]^+ \leq_W L$ iff Alternator has a winning strategy in $\mathcal{H}_\omega^L(L^{\complement}, L)$.*

Proposition 3. *Let L be a tree language. For every prefix p, $d_W([p]^{-1}L) \geq \omega_1$ iff Alternator has a winning strategy in $\mathcal{H}_\infty^p(L, L^{\complement})$.*

4   A Characterization of Languages of Uncountable Degree

Games on Types and Strategy Trees. Following [2], for a given regular language of trees L, a prefix p and types $h_i \in H_L$ ($i = 1, 2, \dots$) we define games on types $\mathcal{H}_k^p(h_1, \dots, h_k)$ and $\mathcal{H}_\infty^p(h_1, h_2, \dots)$. The Constrainer plays as in the simple and infinite cutting games and the task of the Alternator is to play in the i-th round a tree of type h_i, that is an element of $\alpha_L^{-1}(h_i)$.

A type tree for L is a tree over the finite alphabet H_L. For a given tree t, there is a type tree σ_t induced by t such that for every node $w \in \text{dom}(\sigma_t)$,

$$\sigma_t(w) \text{ is the type of the tree } t.w. \tag{1}$$

Let σ be a type tree, and t a tree. We say that a type tree σ is *locally consistent* with a tree t if $\text{dom}(\sigma) = \text{dom}(t)$ and for every node $w \in \text{dom}(t)$ such that $t(w) = a$,

- if w is a leaf, then $\sigma(w)$ is the type of a,
- if w has two children m_ℓ and m_r, then $\sigma(w)$ is the type obtained by applying a to the pair $(\sigma(m_\ell), \sigma(m_r))$.

Definition 1. *A finite strategy tree is a tuple $\mathfrak{s} = (t, \sigma_1, \dots, \sigma_k)$ where*

- *t is a tree, the support of the strategy and $\sigma_1 = \sigma_t$,*

- σ_ℓ is locally consistent with t, for each $\ell \leq k$,
- for each $w \in \mathrm{dom}(t)$, Alternator has a winning strategy in $\mathcal{H}_k^\varepsilon(\sigma_1(w), \ldots, \sigma_k(w))$.

An infinite strategy tree $\mathfrak{s} = (t, \sigma_1, \sigma_2, \ldots)$ is defined analogously.

The root sequence of a strategy tree $\mathfrak{s} = (t, \sigma_1, \sigma_2, \ldots)$ is the sequence of types $(\sigma_1(\varepsilon), \sigma_2(\varepsilon), \ldots)$. We define the *alternation* of a sequence $(h_1, \ldots, h_\ell)$ of types as the cardinality of the set $\{i : h_i \neq h_{i+1}\}$. The same definition applies to infinite sequences of types. Let $\mathfrak{s}$ be a finite strategy tree. The *root alternation* of $\mathfrak{s}$ is the alternation of the root sequence, while the *limit alternation* of $\mathfrak{s}$ is the maximal number k such that infinitely many subtrees of $\mathfrak{s}$ have root alternation at least k. We say that a set $\mathfrak{S}$ of finite strategy trees has *bounded root alternation* if there is a k such that the root alternation of each $\mathfrak{s} \in \mathfrak{S}$ is at most k, unbounded otherwise. Analogously for limit alternation.

A finite or infinite strategy tree $\mathfrak{s} = (t, \sigma_1, \ldots)$ is *locally optimal* if for every strategy tree $\mathfrak{s}' = (t, \sigma_1', \ldots)$ with same root sequence, and every $i > 1$, the depth at which σ_i and σ_{i+1} first differ is greater than or equal to the depth at which σ_i' and σ_{i+1}' first differ. The next Proposition is a very important technical point of [2].

Proposition 4 (Lemma G.2 in Appendix of [2]). *For a regular tree language L, if $\mathfrak{S}$ is a set of locally optimal finite strategy trees with both root and unbounded limit alternation, then the strategy graph G_L is recursive.*

The next Proposition establishes an important link between infinite cutting games and strategy trees.

Proposition 5. *Assume Alternator has a winning strategy in $\mathcal{H}_\infty^\varepsilon(L, L^{\complement})$. Then there is an infinite strategy tree $\mathfrak{s}^\infty$ with infinite root alternation.*

Proof. Assume Alternator has a winning strategy f in $\mathcal{H}_\infty^\varepsilon(L, L^{\complement})$. The infinite strategy tree $\mathfrak{s}^\infty$ is constructed as follows. First of all, we can represent f as a tree satisfying the following properties:

- the root is labelled by ε, and its unique child is labelled by Alternator's move obtained by applying the winning strategy f at the first round of the game,
- if a node v is labelled with a tree t, then for every prefix p of t there is a unique child of v labelled by p,
- if a node v is labelled with a prefix p, then v has a unique child, and such a child is labelled by the answer obtained by applying the winning strategy f to the position in the cutting game given by the labels of the path from the root to v.

Notice that nodes at odd depth represent Alternator's moves (according to f) and are therefore labelled by trees, while nodes at even depth represent Constrainer's move and are thus labelled by prefixes. From now on, we always identify f and the aforementioned tree.

Claim. For every node v of f labelled by a prefix p, there is an infinite sequence of strategy trees $(\mathfrak{s}_\ell^v : \ell < \omega)$ such that for each ℓ

1. $\mathfrak{s}_\ell^v = (t, \sigma_1, \ldots, \sigma_\ell)$, with the type $\sigma_{2k+1}(\varepsilon)$ included in L and the type $\sigma_{2k}(\varepsilon)$ included $L^{\complement}$ if v is at depth $2i$ with i even, else dually. In particular this means that $\sigma_{2k+1}(\varepsilon) \neq \sigma_{2k}(\varepsilon)$;

2. $\mathfrak{s}_{\ell+1}^n$ extends $\mathfrak{s}_\ell^v$, that is $\mathfrak{s}_{\ell+1}^v = (t, \sigma_1, \ldots, \sigma_\ell, \sigma_{\ell+1})$ and $\mathfrak{s}_\ell^n = (t, \sigma_1, \ldots, \sigma_\ell)$.

Given the Claim, from Property 1 we have that for each node v labelled by a prefix p, and each $\ell = 1, 2, \ldots$, $\mathfrak{s}_\ell^v$ has root alternation ℓ and defines a winning strategy for Alternator in $\mathcal{H}_\ell^p(L, L^C)$ if v is at depth $2i$ with i even, in $\mathcal{H}_\ell^p(L^C, L)$ otherwise. Let

$$\mathfrak{s}_\ell^\varepsilon = (t, \sigma_1, \ldots, \sigma_\ell) \text{ for } \ell = 1, 2, \ldots.$$

The required infinite stategy tree is defined as $\mathfrak{s}^\infty = (t, \sigma_1, \ldots)$. It remains to prove the Claim. Firstly, by induction with respect to $\ell = 1, 2, \ldots$ we will assign a strategy tree $\mathfrak{s}_\ell^v$ to each node v of f labelled by a prefix. In the process of inductive construction we will also verify that Property 1 of the Claim is satisfied. Verification of Property 2 will be done later. Let us start from a remark that given an infinite sequence of type trees $(\sigma_1, \ldots)$, by compactness there is a converging subsequence $(\sigma'_1, \ldots)$. We assume that every time we have to choose a converging subsequence $(\sigma'_1, \ldots)$ of a given sequence $(\sigma_1, \ldots)$, we always choose the same subsequence and denote it's limit as $\mathrm{limit}(\sigma_1, \ldots)$. We also assume that given a tree t, we have fixed an enumeration $(p_1, \ldots)$ of all its prefixes such that sequence $(p_k)_{k=1,2,\ldots}$ converge to the tree t. For $\ell = 1$, it is enough to take for each node v

$$\mathfrak{s}_1^v = (t, \sigma_t),$$

where t is given by applying f to the considered position and σ_t is defined by formula (1) at the beginning of this section. By choice of σ_t, Property 1 is satisfied. For $\ell > 1$ we proceed as follows. We assume the construction performed for $\ell - 1$. Fix any node v labelled by a prefix p. Assume that a tree t is the answer given by f at the position in the game given by the path from the root to the node v. To every prefix p of t corresponds a child w of v to which we already associated a strategy tree $\mathfrak{s}_{\ell-1}^w = (t^p, \sigma_2^p, \ldots, \sigma_\ell^p)$. Let us thence consider the sequence $(p_1, \ldots)$, with limit t and the sequences $(t^{p_1}, \ldots)$, $(\sigma_2^{p_1} \ldots), \ldots, (\sigma_\ell^{p_1} \ldots)$. The limits $\mathrm{limit}(\sigma_2^{p_k}), \ldots, \mathrm{limit}(\sigma_\ell^{p_k})$ were chosen in advance and are equal $\sigma_2^*, \ldots, \sigma_\ell^*$. Since each t^{p_k} extends p_k, the limit t^* of $(t^{p_1}, t^{p_2}, \ldots)$ is t. Now, for each p, the type trees $(\sigma_2^p, \ldots, \sigma_\ell^p)$ are locally consistent with t^p. Furthermore, given a sequence of trees $(t_1, \ldots)$ that converges to t^* and a sequence of type trees $(\sigma_1, \ldots)$ that converges to σ^*, if σ_k is locally consistent with t_k for every k, then σ^* is locally consistent with t^*. From this fact follows that the limits $\sigma_2^*, \ldots, \sigma_\ell^*$ are locally consistent with t. Finally, define σ_1^* to be σ_t as in formula (1). We have just proved that $\mathfrak{s}_\ell^v = (t, \sigma_1^*, \ldots, \sigma_\ell^*)$ is a strategy tree. From induction hypothesis together with definition of σ_t and preservation of Property 1 under limits follows that $\mathfrak{s}_\ell^v$ also satisfies Property 1.

We now verify that the described procedure preserves Property 2. For $\ell = 1$ there is nothing to check. For the induction step, we reason as follows. Assume the Property holds for each node and for each $k < \ell$. Now, let us consider an arbitrary node v. We have to prove that $\mathfrak{s}_{\ell+1}^v$ extends $\mathfrak{s}_\ell^v$. By induction hypothesis, $\mathfrak{s}_{\ell-1}^w = (t^p, \sigma_2^p, \ldots, \sigma_\ell^p)$ and $\mathfrak{s}_\ell^w = (t^p, \sigma_2^p, \ldots, \sigma_\ell^p, \sigma_{\ell+1}^p)$, for every node w in the described procedure. Since the limits have been fixed in advance, we have that $\mathfrak{s}_\ell^w = (t, \sigma_t, \sigma_2^*, \ldots, \sigma_\ell^*)$ and $\mathfrak{s}_{\ell+1}^v = (t, \sigma_t, \sigma_2^*, \ldots, \sigma_\ell^*, \sigma_{\ell+1}^*)$, meaning that the latter extends the former. This concludes the proof of the Claim.

Using the above Proposition, we can generalize to infinite games Proposition 5.2 from [2]:

Proposition 6. *For a regular language L the following conditions are equivalent.*

1. *Alternator wins the game $\mathcal{H}_\infty^\varepsilon(L, L^\complement)$,*
2. *There are tree types $h, g \in H_L$, such that $h \neq g$ and Alternator wins $\mathcal{H}_\infty^\varepsilon(h, g)$.*

We will use the following Lemma, presented in [2] for finite strategy trees, with proof extending straightforwardly to infinite strategy trees.

Lemma 2. *For every finite or infinite strategy tree, there is a locally optimal strategy tree with same root sequence.*

The next Lemma follows immediately from the definition of a strategy tree.

Lemma 3. *Let $\mathfrak{s} = (t, \sigma_1, \ldots, \sigma_\ell)$ be a strategy tree. For the game $\mathcal{H}_\ell^\varepsilon(\sigma_1(\varepsilon), \ldots, \sigma_\ell(\varepsilon))$ and a strategy of Constrainer given by always cutting at level i, Alternator wins by playing as follows:*

– *at first, Alternator plays t, then*
– *for each port w at level i of the multi context given by Constrainer's move, Alternator plugs in the tree given by her winning strategy $\mathcal{H}_\ell^\varepsilon(\sigma_1(w), \ldots, \sigma_\ell(w))$.*

In particular, if from a certain $j < \ell$ on $\sigma_k(w) = \sigma_{k+1}(w)$, $j \leq k < \ell$, then for each round k such that $j < k < \ell$ Alternator always plugs in the same tree of type $\sigma_j(w)$ chosen at round j.

An Effective Characterization. Everything now is ready to prove the main result of this paper.

Theorem 1. *Let L be a regular tree language given by a non-deterministic tree automaton $\mathcal{A}$. The following conditions are equivalent:*

1. *The strategy graph G_L is recursive.*
2. *$d_W(L) \geq \omega_1$*

In particular, since the graph G_L is computable from the automaton $\mathcal{A}$, it is decidable whether the language accepted by $\mathcal{A}$ is of Wadge degree greater than or equal to ω_1.

Proof. $(1) \Rightarrow (2)$. Assume the strategy graph is recursive. This means that there exists a strongly connected component that contains two nodes (v, h) and (v', h') with $h \neq h'$. Thanks to Proposition 1, if there exists a path between (v, h) and (v', h'), there is also an edge between (v, h) and (v', h'). Moreover, for vertices $(v_1, h_1), (v_2, h_2), \ldots,$ if for every $i = 1, 2, \ldots$ there is an edge from (v_i, h_i) to (v_{i+1}, h_{i+1}), this means that Alternator has a winning strategy in $\mathcal{H}_\infty^\varepsilon(h_1, h_2, \ldots)$. So, take $(v_i, h_i) = (v, h)$ for i even, and $(v_i, h_i) = (v', h')$ for i odd. This shows that Alternator has a winning strategy in $\mathcal{H}_\infty^\varepsilon(h, h')$. By Proposition 6 Alternator has a winning strategy in $\mathcal{H}_\infty^\varepsilon(L, L^\complement)$.

$(2) \Rightarrow (1)$. By Propositions 3 and 4, it is enough to verify that if Alternator has a winning strategy in $\mathcal{H}_\infty^\varepsilon(L, L^\complement)$ then there is a set $\mathfrak{S}$ of locally optimal finite strategy trees with both root and limit unbounded alternation. Assume Alternator has a winning strategy f in $\mathcal{H}_{\infty}^{\varepsilon}(L, L^{\complement})$. From Proposition 5 there is a strategy tree $\mathfrak{s}^{\infty} = (t, \sigma_1, \dots)$ with infinite root alternation. By Lemma 2 we can assume that $\mathfrak{s}^{\infty}$ is locally optimal. Let us define

$$\mathfrak{S} = \{(t, \sigma_1, \dots, \sigma_k) : k = 1, 2, \dots\}.$$

Note that each element of $\mathfrak{S}$ is locally optimal. Now, assume limit alternation of $\mathfrak{S}$ is bounded. From this fact and since every element of $\mathfrak{S}$ is a prefix of $\mathfrak{s}^{\infty}$, it holds that with respect to $\mathfrak{s}^{\infty}$, the set of subtrees of t with infinite root alternation has to be finite. This means that $\mathfrak{s}^{\infty}$ satisfies the following property:

(*) there is a finite set X of nodes of t satisfying the following properties:
 – the root is included in X, and each node of X is at most at depth i in t,
 – $\sigma_k(v) \neq \sigma_{k'}(v)$, for every node v in the set X, and $\sigma_k(w) = \sigma_{k'}(w)$, for every node w of t of depth $i + 1$, for some k, k', with $k < k' \leq j$.

The strategy tree $\mathfrak{s} = (t, \sigma_1, \dots, \sigma_j)$ from $\mathfrak{S}$, where j is given by the previous property, also satisfies the property (*) above (for the same X and the same k, k').

Let us consider the game $\mathcal{H}_j^{\varepsilon}(\sigma_1(\varepsilon), \dots, \sigma_j(\varepsilon))$ where at first Alternator plays t and then Constrainer uses the strategy given by cutting always at level $i+1$. We can therefore apply Lemma 3 and assume that Alternator plays the winning strategy described there. This implies that the trees played at round k and k' are the same, say t' (from the root to level i they are the same, because the Constrainer insists on this and below they are the same, because the Alternator plays the same answers in rounds k and k'). But by local consistency, since $\sigma_k(\varepsilon) \neq \sigma_{k'}(\varepsilon)$, the two trees should have two different types, a contradiction. We therefore conclude that limit alternation of $\mathfrak{S}$ is unbounded.

5 Conclusion

The algorithm provided in [2] decides whether a given non-deterministic automaton accepts a language which is a Boolean combination of open sets or equivalently is of a Wadge degree smaller than ω. By the same approach we showed an algorithm which decides whether a given non-deterministic automaton accepts a language in $\mathbf{\Delta}_2^0$ or equivalently, a language of a Wadge degree smaller than ω_1. We propose for further research the following three generalizations of the result presented in this paper:

1. For a given $n = 1, 2, \dots$ there are natural topological games which characterize languages of Wadge degrees smaller than ω^n. Moreover, there are known examples of regular languages of degree ω^n. It would be a desirable and perhaps more involved extension of results in [2] if for a given n one can provide an algorithm deciding whether a given non-deterministic automaton accepts a language of degree smaller than ω^n.

2. In the absence of examples of regular languages between Wadge degree ω^ω and Wadge degree ω_1, one could reasonably expect, that the decidability result in the present paper should show that indeed any regular language of countable Wadge degree is of Wadge degree smaller than ω^ω. However, this question still remains open.

3. Regarding higher Borel classes, in particular regular languages which are Boolean combinations of $\mathbf{\Sigma}_2^0$ sets, the following extension of the method in [2] seems to be

plausible. The cutting game is based around restrictions of moves by prefixes, that is by languages in Δ_1^0. Its topological counterpart on the next Borel level is a game, where the Constrainer is allowed to play constraints which are regular languages in Δ_2^0. This leads to a natural topological characterization similar to the results in Section 3, but the algebraic counterpart of this generalized cutting game is not yet fully understood.

References

1. Blumensath, A.: An algebraic proof of Rabin's theorem. Theoretical Computer Science 478, 1–21 (2013)
2. Bojańczyk, M., Place, T.: Regular Languages of Infinite Trees That Are Boolean Combinations of Open Sets. In: Czumaj, A., Mehlhorn, K., Pitts, A., Wattenhofer, R. (eds.) ICALP 2012, Part II. LNCS, vol. 7392, pp. 104–115. Springer, Heidelberg (2012)
3. Bojańczyk, M.: Algebra for trees. In: Handbook of Automata Theory, European Mathematical Society Publishing House (to appear)
4. Bojańczyk, M., Idziaszek, T.: Algebra for infinite forests with an application to the temporal logic EF. In: Bravetti, M., Zavattaro, G. (eds.) CONCUR 2009. LNCS, vol. 5710, pp. 131–145. Springer, Heidelberg (2009)
5. Colcombet, T., Kuperberg, D., Löding, C., Vanden Boom, M.: Deciding the weak definability of Büchi definable tree languages. In: CSL 2013, pp. 215–230 (2013)
6. Duparc, J.: Wadge Hierarchy and Veblen Hierarchy Part 1: Borel Sets of Finite Rank. Journal of Symbolic Logic 66(1), 56–86 (2001)
7. Duparc, J., Murlak, F.: On the Topological Complexity of Weakly Recognizable Tree Languages. In: Csuhaj-Varjú, E., Ésik, Z. (eds.) FCT 2007. LNCS, vol. 4639, pp. 261–273. Springer, Heidelberg (2007)
8. Facchini, A., Michalewski, H.: Deciding the Borel complexity of regular tree languages (2014), http://arxiv.org/abs/1403.3502
9. Facchini, A., Murlak, F., Skrzypczak, M.: Rabin-Mostowski index problem: a step beyond deterministic automata. In: LICS 2013 (2013)
10. Kechris, A.: Classical Descriptive Set Theory. Springer (1995)
11. Martin, D.A.: Borel determinacy. The Annals of Mathematics 102, 363–371 (1975)
12. Murlak, F.: On deciding topological classes of deterministic tree languages. In: Ong, L. (ed.) CSL 2005. LNCS, vol. 3634, pp. 428–441. Springer, Heidelberg (2005)
13. Murlak, F.: The Wadge Hierarchy of Deterministic Tree Languages. In: Bugliesi, M., Preneel, B., Sassone, V., Wegener, I. (eds.) ICALP 2006. LNCS, vol. 4052, pp. 408–419. Springer, Heidelberg (2006)
14. Niwiński, D., Walukiewicz, I.: A gap property of deterministic tree languages. Theoretical Comput. Sci. 303, 215–231 (2003)
15. Perrin, D., Pin, J.E.: Infinite Words: Automata, Semigroups, Logic and Games. Academic Press (2004)
16. Rabin, M.O.: Weakly definable relations and special automata. In: Bar-Hillel, Y. (ed.) Foundations of Set Theory, pp. 1–23 (1970)
17. Wadge, W.W.: Reducibility and Determinateness on the Baire Space, Ph.D. Thesis, Berkeley (1984)
18. Wagner, K.: On ω-regular sets. Inform and Control 43, 123–177 (1979)

Chemical Production and Molecular Computing in Addressable Reaction Compartments

Harold Fellermann[1,2] and Natalio Krasnogor[1,3]

[1] School of Computing Science, Newcastle University, Newcastle upon Tyne
NE1 7RU, United Kingdom
[2] Complex Systems Lab, Barcelona Biomedical Research Park, Dr. Aiguadé 88,
08003 Barcelona, Spain
[3] European Centre for Living Technology, Ca' Minich,
S. Marco 2940, 30124 Venezia, Italy

Abstract Biological systems employ compartmentalisation in order to orchestrate a multitude of biochemical processes by simultaneously enabling "data hiding" and modularisation. In this paper, we present recent research projects that embrace compartmentalisation as an organisational programmatic principle in synthetic biological and biomimetic systems. In these systems, artificial vesicles and synthetic minimal cells are envisioned as nanoscale reactors for programmable biochemical synthesis and as chassis for molecular information processing. We present P systems, brane calculi, and the recently developed *chemtainer* calculus as formal frameworks providing data hiding and modularisation and thus enabling the representation of highly complicated hierarchically organised compartmentalised reaction systems. We demonstrate how compartmentalisation can greatly reduce the complexity required to implement computational functionality, and how addressable compartments permit the scaling-up of programmable chemical synthesis.

1 Introduction

Biological systems employ compartmentalisation in order to orchestrate a multitude of biochemical processes. Such organisation is prominently featured in the cytoplasm, where a multitude of biochemical compounds is highly organised in vesicular compartments that co-locate reactants of desired reactions while separating those of undesired reactions. Surface markers on these compartments are used for vesicular trafficking, as well as vesicle budding and fusion, thereby allowing for the fine-tuned control of biochemical reaction cascades [1,2].

The desire to harvest compartmentalisation as a way of providing data hiding and modular organisation for next generation chemical synthesis and molecular computation has led to various approaches. For example Chaplin et al. [3] have demonstrated that photochromic molecules such as NitroBIPS, a kind of spiropyran, can be localised inside a collection of static polydimethylsiloxane silicone polymer (PDMS) microwells and used to implement registers, logic gates and circuits. Other studies have focused on utilising supermolecular compartments as nanoscale "nano-bioreactors" [4,5,6,7]. In this regard, several pathways

A. Beckmann, E. Csuhaj-Varjú, and K. Meer (Eds.): CiE 2014, LNCS 8493, pp. 173–182, 2014.
© Springer International Publishing Switzerland 2014

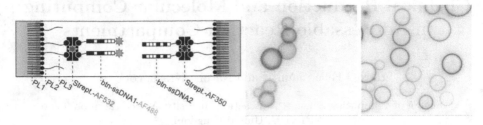

Fig. 1. DNA mediated compartment association – here in the case of emulsion droplets. Compartment surfaces are decorated with two different DNA single strands via biotin-streptavidin linkers (left) and stained according to the respective DNA sequence. If the DNA sequences are complementary, droplets bind with each other (center). If they are non-complementary, they do not associate (right). The different arrangements impact the rate of compartment fusion such that associated compartments are more likely to fuse than dissociated ones. Figures taken from [10].

for vesicle aggregation [8,9,10], budding [11], and fusion [12,13,14] have been suggested.

In particular, Hadorn et al. employ DNA tags for the specific interaction of reaction compartments. In their approach, the surface of either phospholipid vesicles [10] or oil-in-water emulsion droplets [9] are decorated non-covalently with short single stranded DNA sequences of typically 15 nucleotides via biotin-streptavidin linkers (see Figure 1). If the DNA markers of two compartments are complementary, they force the compartments to aggregate as a consequence of DNA hybridization, whereas aggregation of compartments with non-complementary DNA markers is prohibited.

In this paper, we present recent research in computer science that embraces compartmentalization in synthetic biological and biomimetic systems. In Section 2, we review state-of-the-art frameworks that are capable of expressing reactions in and of hierarchically compartmentalized reaction systems, and present a novel theoretical framework that closely follows the work of Hadorn et al. in Section 3. The novel framework is put to the test by applying it to approaches in programmable chemical synthesis in Section 4 as well as molecular information processing in Section 5.

2   Formal Frameworks for Compartmentalized Reactions

While chemistry in well-stirred reaction vessels has been described mathematically already in the late 19th century [15], formal frameworks for compartmentalised reaction systems have been suggested only within the last 15 years. The two main branches that concern themselves with compartmentalized reaction systems are *P systems*, proposed by Păun in 1998 [16], and *brane calculi*, proposed by Cardelli in 2005 [17].

Both frameworks employ a language of balanced parentheses to capture the topological organisation of nested compartments. Each compartment can hold a multiset of arbitrary chemicals as well as other compartments. In P *systems*, compartments are traditionally labeled and the molecular content is given by associating a multiset of molecules to each compartment label. In *brane calculi*, the grammar for parentheses is extended to include molecules directly. Thus, the formal language is defined by the recursive production rules

$$P := \emptyset \mid P + P \mid \langle\!\langle\, P \,\rangle\!\rangle \mid m_j \tag{1}$$

for the non-terminal and start symbol P, the terminals $\{\emptyset, +, \langle\!\langle\,, \,\rangle\!\rangle\} \cup \{m_j \mid j \in J\}$ and the broken vertical bar indicating choice. Here, $\mathcal{M} = \{m_j \mid j \in J\}$ is a set of chemicals for some index set J and the halfmoon parentheses denote compartments. Structural equivalence relations are introduced to give $(\mathcal{M}, +, \emptyset)$ associative and commutative semantics with the neutral element $\emptyset$.

Over the state space defined by the grammar in Eq. (1), chemical reactions can be introduced as bulk reactions

$$P \longrightarrow P' \tag{2}$$

for multisets $P = \sum_i \nu_i m_i$ and $P' = \sum_j \mu_j m_j$ of simple educt and product molecules with stoichiometric coefficients ν_i and μ_j. In addition, reactions can describe trans-membrane processes

$$P + \langle\!\langle\, Q \,\rangle\!\rangle \longrightarrow P' + \langle\!\langle\, Q' \,\rangle\!\rangle, \tag{3}$$

where P, P', Q, Q' are again simple multisets of molecules. For example, the action of a sodium potassium pump could be codified by the reaction

$$2\,\mathrm{K}^+ + \langle\!\langle\, 3\,\mathrm{Na}^+ + \mathrm{ATP} \,\rangle\!\rangle \longrightarrow 3\,\mathrm{Na}^+ + \langle\!\langle\, 2\,\mathrm{K}^+ + \mathrm{ADP} \,\rangle\!\rangle. \tag{4}$$

In traditional P systems, all compartments of a membrane structure are labeled and each compartment has its specific set of reaction rules. Brane calculi also associate reaction rules with individual membranes but employ a dedicated syntax that directly attaches the rule to the respective membrane. For example, reaction (3) would be associated to an empty compartment by writing

$$!P(Q) \triangleright P'(Q') \langle\!\langle\, \,\rangle\!\rangle. \tag{5}$$

Importantly, the brane calculus defines a process algebra for sequential and parallel composition of rules. For example, the exclamation mark in the above notation signifies that the rule is not consumed upon application.

Avoiding explicit labeling of compartments has the advantage, that one can introduce reactions that fundamentally alter the arrangement of the compartments themselves, such as compartment fusion

$$\mathrm{mate}_i \langle\!\langle\, P \,\rangle\!\rangle + \mathrm{mate}_i^\mathsf{T} \langle\!\langle\, Q \,\rangle\!\rangle \longrightarrow \langle\!\langle\, P + Q \,\rangle\!\rangle \tag{6}$$

where two compartments with contents P and Q fuse to form a single compartment of mixed content. Fusion is triggered by a *mate* action associated with one compartment and a respective co-action associated with the other compartment. Compartments fuse if and only if the two actions match. Similarly, a *drip* action can initiate compartment fission:

$$^{\mathrm{drip}}(\!(\, P + Q \,)\!) \longrightarrow (\!(\, P \,)\!) + (\!(\, Q \,)\!) \tag{7}$$

Note that this brief summary barely touches the surfaces of both frameworks. For a complete introduction to P systems and brane calculi the reader is advised to consult Refs. [16] and [17]. While both frameworks where originally developed with non-deterministic semantics, the subsequent development of stochastic semantics [18] now allows for their use e.g. in stochastic simulation.

3   Compartments as Programmable Nano-bioreactors

Recently, Fellermann and Cardelli [19] have developed the *chemtainer calculus* in an attempt to bring the theoretical framework of brane calculi closer to a potential experimental implementation in a biomimetic setup: following the experimental accomplishments of Hadorn et al. [9,10] the chemtainer calculus employs membrane associated DNA markers to govern the fate of supramolecular compartments, where two different types of DNA markers are used. On the one hand, single stranded, or simple, DNA tags are used for compartment recognition and fusion similar to the mate action in the original brane calculus:

$$^{\sigma}(\!(\, P \,)\!) + {}^{\sigma^{\mathsf{T}}}(\!(\, Q \,)\!) \longrightarrow {}^{\sigma^{\|}}(\!(\, P + Q \,)\!) \tag{8}$$

Here, σ signifies a DNA single strand with a specific sequence, σ^{T} its complement, and $\sigma^{\|}$ the conjugated double strand. Hybridization of the double strand is controlled by temperature, and the two single strands are recovered when surpassing the melting temperature of the double strand:

$$^{\sigma^{\|}}(\!(\, P \,)\!) \longleftrightarrow {}^{\sigma + \sigma^{\mathsf{T}}}(\!(\, P \,)\!) \tag{9}$$

Join gates, on the other hand, are multi-strand DNA constructs that allow for the specification of DNA-based computational processes [20]. In this approach to DNA computation, one DNA strand that is composed of several logical domains, with all but one domain being hybridised to one or more complementary strands. The only initially exposed single strand domain of the gate is a short toehold region. This toehold can reversibly bind a complementary signal strand which is designed to be longer than the toehold domain and complementary to the next domain(s) of the template. The newly binding signal is then able to hybridise to all matching domains of the template, thereby displacing strands that where previously bound and possibly exposing new toeholds. The displaced strands can either be output signals, or signals that bind to toehold regions of downstream gates. By choosing domains of appropriate length, it can be guaranteed that toehold binding is reversible, whereas the total strand displacement process is effectively irreversible, thus computation is energetically downhill and kinetically irreversible, if and only if all correct input strands are present.

Disregarding the details of strand displacement, the chemtainer calculus encodes join gates with the syntax

$$s^* \rhd s',$$
(10)

where s^* denotes the set of input strands that the gate recognizes, and s' denotes the gates' output strand. The action of the gate is then given by the transition

$$s^* \rhd s' + s^* \longrightarrow s'.$$
(11)

That is, the gate exposes its output strand s' if and only if all its inputs are present, and inputs are consumed upon firing of the gate. Just like simple DNA tags, join gates are membrane bound.

Note that the original chemtainer calculus uses slightly different fusion semantics, has the additional notion of physical locations, and defines a small programming language for external nano-bioreactor manipulation. Here, we restrict ourselves to the elements of the calculus that are used in the applications that follow. For a concise exposition, the reader is referred to Ref. [19].

4   Programmable Chemical Synthesis

In Refs. [21] and [19], we have employed the chemtainer calculus for the programmed synthesis of oligosaccharides. Oligosaccharides are branched heteropolymers composed of typically five to ten individual sugar monomers such as mannose, galactose, and glucose. This diverse class of biochemicals is involved in various physiological processes pertaining e.g to cell-cell recognition, intra- and intercellular trafficking, and metabolic modulation [22]. However, their combinatorial richness poses a challenge for chemical oligosaccharide synthesis based on conventional chemical manufacturing techniques [23].

Chemical one-pot synthesis of a given target structure is challenging, because repetition of bindings sites in the oligomer structure can lead to undesired side products. The number of potential side products can be controlled, however, by forcing some reaction steps to occur sequentially while others are allowed to proceed in parallel.Weyland et al. [21] present an algorithm that identifies such optimal reaction cascades. For example, assume that the structure P_3 shown in Figure 2 can be produced with the reaction cascade

$$\text{Gal} + \text{E}^*_{\text{Gal-4}}\text{Gal} + \text{E}^*_{\text{Gal-4}}\text{Man} \longrightarrow P_0 + 2\,\text{E}_{\text{Gal-4}}$$
(12)

$$\text{E}^*_{\text{Man-6}}\text{Man} + \text{E}^*_{\text{Man-6}}\text{Glc} + \text{E}^*_{\text{Man-3}}\text{Glc} \longrightarrow \text{E}^*_{\text{Man-6}}P_1 + \text{E}_{\text{Man-6}} + \text{E}_{\text{Man-3}}$$
(13)

$$\text{E}^*_{\text{Man-3}}\text{Man} + \text{E}^*_{\text{Man-2}}\text{Gal} + \text{E}^*_{\text{Gal-4}}\text{Glc} \longrightarrow \text{E}^*_{\text{Man-3}}P_2 + \text{E}_{\text{Man-2}} + \text{E}_{\text{Gal-4}}$$
(14)

$$P_0 + \text{E}^*_{\text{Man-6}}P_1 + \text{E}^*_{\text{Man-3}}P_2 \longrightarrow P_3 + \text{E}_{\text{Man-6}} + \text{E}_{\text{Man-3}},$$
(15)

where each reaction combines three reagents to create either an intermediate or the final product. It has to be ensured that reactions (12) through (14) occur in isolation and prior to reaction (15) in order to avoid undesired side products.

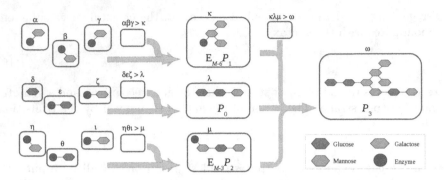

Fig. 2. Programmed chemical synthesis of the target oligosaccharide P_3 by subsequent co-location of intermediate reactants. Thick outlines indicate compartments and gray arrows show chemical reactions triggered by programmed compartment fusion. Figure modified from [19].

Assuming that all initial reactants are provided in uniquely DNA tagged reaction compartments, we can introduce a single empty compartment that is decorated with complementary DNA strands as well as join gates. While the simple DNA tags orchestrate co-location of reactants via fusion, join gates reflect the outcome of the respective chemical reaction by altering the DNA markers on the surface. This is shown here for reaction (12):

$$\alpha^T+\beta^T+\gamma^T+\alpha\beta\gamma\triangleright\kappa \llbracket \ \rrbracket + {}^{\alpha}\llbracket \text{Gal} \rrbracket + {}^{\beta}\llbracket \text{E}^*_{\text{Gal-4}}\text{Gal} \rrbracket + {}^{\gamma}\llbracket \text{E}^*_{\text{Gal-4}}\text{Man} \rrbracket$$

$$\alpha^{\|}+\beta^T+\gamma^T+\alpha\beta\gamma\triangleright\kappa \llbracket \text{Gal} \rrbracket + {}^{\beta}\llbracket \text{E}^*_{\text{Gal-4}}\text{Gal} \rrbracket + {}^{\gamma}\llbracket \text{E}^*_{\text{Gal-4}}\text{Man} \rrbracket$$

$$\alpha^{\|}+\beta^{\|}+\gamma^T+\alpha\beta\gamma\triangleright\kappa \llbracket \text{Gal} + \text{E}^*_{\text{Gal-4}}\text{Gal} \rrbracket + {}^{\gamma}\llbracket \text{E}^*_{\text{Gal-4}}\text{Man} \rrbracket$$

$$\alpha^{\|}+\beta^{\|}+\gamma^{\|}+\alpha\beta\gamma\triangleright\kappa \llbracket \text{Gal} + \text{E}^*_{\text{Gal-4}}\text{Gal} + \text{E}^*_{\text{Gal-4}}\text{Man} \rrbracket \qquad (16)$$

$$\alpha+\alpha^T+\beta+\beta^T+\gamma+\gamma^T+\alpha\beta\gamma\triangleright\kappa \llbracket \text{Gal} + \text{E}^*_{\text{Gal-4}}\text{Gal} + \text{E}^*_{\text{Gal-4}}\text{Man} \rrbracket$$

$$\alpha^T+\beta^T+\gamma^T+\kappa \llbracket P_0 + 2\,\text{E}_{\text{Gal-4}} \rrbracket.$$

The newly exposed κ tag signifies that the compartment is ready for downstream processing. Here, the complementary tags α^T a.s.o. are effectively garbage, and could easily be avoided by feeding the join gate $\alpha\alpha^T\beta\beta^T\gamma\gamma^T \triangleright \kappa$ instead.

Noteworthy, from a standard library of monomers, we can control a reaction cascade to obtain any desired target compound simply by providing compartments that are decorated with appropriate DNA tags and gates. Paired with automatised equipment such as liquid handling robots and microfluidic technology, this opens up for truly programmable chemical synthesis.

5   Modular Molecular Computing

In Ref. [24], Smaldon et al. construct molecular logical gates from gene regulatory networks. Each gene features a promoter site, an operator site, and a protein

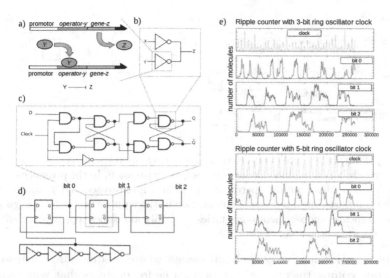

Fig. 3. Gene regulatory network implementation of a 3-bit ripple counter by Smaldon et al. **a)** Modular circuits are built from *operons* comprising a promoter, an operator site, and a gene encoding region. In the absence of the repressor protein Y, protein Z is expressed. In the presence of Y, protein Z is repressed. Thus, the operation implements a NOT gate. **b)** Two NOT gates in parallel build a NAND gate. **c)** Eight NAND gates and one NOT gate can be wired to form a D-flipflop. **d)** Three D-flipflops and a 5-bit ring oscillator built from serial NOT gates form the 3-bit ripple counter. **e)** Trajectories of the output bits obtained from stochastic simulation show that the ripple counter operates reliably with a 5-bit ring oscillator clock (lower data set) whereas the flipflops sometimes fail to switch when driven by a faster, 3-bit ring oscillator clock (upper data set). Figure modified from Ref. [24].

encoding region, with proteins that serve as repressors by binding to operator sites of other genes. The authors demonstrate that such operons can serve as NOT gates and proceed to construct modular gates of successively higher complexity (c.f. Figure 3). The most complex example is a 3-bit ripple counter that is implemented by 56 operons that express 32 mRNAs and corresponding proteins. The overall circuit is thus composed of 120 molecular species that is specified using P-systems and compiled into, and simulated using, a Dissipative Particle Dynamics engine.

Although the authors study the effect of encapsulating logical gates into single vesicles, the full advantage of compartmentalisation becomes apparent when distributing individual logical gates over several compartments. Inspecting the structure of the ripple counter in Figure 3.d shows that the circuit is built from three D-flipflops and one 5-bit ring oscillator – each of them being built from more elementary modules. While the D-flipflop circuit, for example, is built from 17 operons that express nine distinct proteins, it is wired to other modules only by two repressor proteins – one serving as input and the other one as output. The remaining seven regulator proteins serve only for internal wiring of the module.

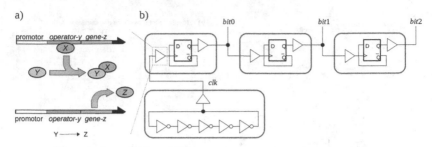

Fig. 4. a) Transducers are built from un-inhibiting operons: in the absence of the activator protein Y, an inhibitor X suppresses gene expression. In the presence of Y, the inhibitor X dissociates from the operon and protein Z is expressed. **b)** Transducers are used in the compartmentalized reimplementation of the ripple counter from Figure 3.d. Thick lines indicate compartment boundaries, and only molecules representing *clk*, *bit0*, *bit1*, and *bit2* are permeable.

This observation motivates a circuit design where modules are encapsulated into separate compartments that are permeable for proteins that wire modules among each other, while being impermeable for proteins that wire gates within each module. Doing so, allows to reuse the same molecular species for multiple instances of a module type, e.g. the D-flipflops in the ripple counter. This can drastically reduce the number of required molecular species.

For the 3-bit ripple counter, we choose permeable proteins for the clock signal and the three counter bits. However, a naïve reimplementation of the circuit shown in Figure 3.d would still require us to use distinct DNA sequences for five of the 17 operons that make up each flipflop, because the module input and output is wired to five operons per flipflop. We can do better by adding transducer gates at the input and output of the original flipflops. One transducer senses an input signal that corresponds to a permeable protein, and expresses the flipflop's original input, which is now also an impermeable protein. The second transducer senses the original output signal and expresses a permeable output signal. Although the number of operons per gate now increases from 17 to 19, the total number of operons decreases from 56 to 28 because operons forming a flipflop module can be reused without modification. The resulting circuit and its encapsulation is shown in Figure 4.

We now employ the chemtainer calculus to assemble the different modules of the ripple counter and their wiring from a library of general parts. Our strategy is to provide a set of compartments that contain flipflops, and counters without any external wiring, and fuse them with compartments that contain transducers to implement the specific wiring among modules.

For the case of a flipflop $F(x, y)$ with original (impermeable) input x and output y, and two transducers $T(clk, x)$ and $T(y, bit0)$ for the permeable species clk and $bit0$, the corresponding operation in the chemtainer calculus reads:

$$\alpha^{\top} + \beta^{\top} + \gamma^{\top} \, ⦇ \, ⦈ + {}^{\alpha}⦇ \, F(x,y) \, ⦈ + {}^{\beta}⦇ \, T(clk,x) \, ⦈ + {}^{\gamma}⦇ \, T(y,bit0) \, ⦈$$

$$\longrightarrow \; {}^{\alpha^{\|} + \beta^{\|} + \gamma^{\|}}⦇ \, F(clk, bit0) \, ⦈. \quad (17)$$

Following this procedure, we can generate the entire ripple counter

$$⦅ C(clk) ⦆ + ⦅ F(clk, bit0) ⦆ + ⦅ F(bit0, bit1) ⦆ + ⦅ F(bit1, bit2) ⦆ \tag{18}$$

where we have omitted the DNA tags.

Importantly, compartments with intricate internal logic, such as the flipflop $^\alpha⦅ F(x, y) ⦆$ or clock $^\delta⦅ C(x) ⦆$ can be provided as a general library, whereas containers that determine the specific wiring, such as $^\beta⦅ T(clk, x) ⦆$ comprise only one internal and one surface associated DNA strand.

6 Conclusion

Recent advances in our understanding of chemical and biological complexity as well as in the availability of powerful computational formalisms for the rapid model prototyping of programmable biological and chemical cells are opening a new frontier in computing science: algorithmic living matter. The availability of new theoretical tools as well as practical implementations for algorithmic living matter, we believe, will lead to a new revolution in computing science based on "programmable everywhere", namely, the ability to program all kinds of materials in all kinds of environments at all scales.

Acknowledgments. The research leading to these results has received partial funding from the European Community's Seventh Framework Programme (FP7/2007-2013) under grant agreement n°249032, the UK's EPSRC under grants EP/J004111/1, EP/L001489/1 and EP/I031642/1.

References

1. Rothman, J.E.: The golgi apparatus: two organelles in tandem. Science 213(4513), 1212–1219 (1981) PMID: 7268428
2. Rothman, J.E.: Mechanisms of intracellular protein transport. Nature 372(6501), 55–63 (1994)
3. Chaplin, J.C., Russell, N.A., Krasnogor, N.: Implementing conventional logic unconventionally: Photochromic molecular populations as registers and logic gates. Biosystems 109(1), 35–51 (2012)
4. Amos, M., Dittrich, P., McCaskill, J., Rasmussen, S.: Biological and chemical information technologies. In: Proceedings from the 2nd European Future Technologies Conference and Exhibition 2011 (FET 2011), pp. 56–60. Procedia Computer Science (2011)
5. Monnard, P.A.: Liposome-entrapped polymerases as models for microscale/nanoscale bioreactors. J. Membr. Biol. 191(2), 87–97 (2003)
6. Noireaux, V., Libchaber, A.: A vesicle bioreactor as a step toward an artificial cell assembly. Proc. Nat. Acad. Sci. USA 101(51), 17669–17674 (2004)
7. Roodbeen, R., van Hest, J.C.M.: Synthetic cells and organelles: compartmentalization strategies. BioEssays 31(12), 1299–1308 (2009)

8. Beales, P.A., Vanderlick, T.K.: Specific binding of different vesicle populations by the hybridization of membrane-anchored DNA. J. Phys. Chem. A 111(49), 12372–12380 (2007)

9. Hadorn, M., Hotz, P.E.: DNA-mediated self-assembly of artificial vesicles. PLoS One 5(3), e9886 (2010)

10. Hadorn, M., Bonzli, E., Fellermann, H., Eggenberger Hotz, P., Hanczyc, M.: Specific and reversible DNA-directed self-assembly of emulsion droplets. Proc. Nat. Acad. Sci. USA 109(47) (2012)

11. Bonifacino, J.S., Glick, B.S.: The mechanisms of vesicle budding and fusion. Cell 116(2), 153–166 (2004)

12. Richard, A., Marchi-Artzner, V., Lalloz, M.N., Brienne, M.J., Artzner, F., Gulik-Krzywicki, T., Guedeau-Boudeville, M.A., Lehn, J.M.: Fusogenic supramolecular vesicle systems induced by metal ion binding to amphiphilic ligands. Proc. Nat. Acad. Sci. USA 101(43), 15279–15284 (2004) PMID: 15492229

13. Caschera, F., Sunami, T., Matsuura, T., Suzuki, H., Hanczyc, M.: Programmed vesicle fusion triggers gene expression. Langmuir 27(21), 13082–13090 (2011)

14. Terasawa, H., Nishimura, K., Suzuki, H., Matsuura, T., Yomo, T.: Coupling of the fusion and budding of giant phospholipid vesicles containing macromolecules. Proc. Nat. Acad. Sci. USA 109(16), 5942–5947 (2012) PMID: 22474340

15. Waage, P., Gulberg, C.M.: Studies concerning affinity. Journal of Chemical Education 63(12), 1044 (1986)

16. Paun, G.: Computing with membranes. Journal of Computer and System Sciences 61(1), 108–143 (2000)

17. Cardelli, L.: Brane calculi – interactions of biological membranes. In: Danos, V., Schachter, V. (eds.) CMSB 2004. LNCS (LNBI), vol. 3082, pp. 257–278. Springer, Heidelberg (2005)

18. Bacci, G., Miculan, M.: Measurable stochastics for brane calculus. Theor. Comp. 431, 117–136 (2012)

19. Fellermann, H., Cardelli, L.: Programmable chemistry in DNA addressable bioreactors. R. Soc. Interface (2014)

20. Cardelli, L.: Strand algebras for DNA computing. Nat. Comput. 10, 407–428 (2011)

21. Weyland, M.S., Fellermann, H., Hadorn, M., Sorek, D., Lancet, D., Rasmussen, S., Fuchslin, R.M.: The MATCHIT automaton: Exploiting compartmentalization for the synthesis of branched polymers. Computational and Mathematical Methods in Medicine, 467428 (December 2013)

22. Varki, A.: Biological roles of oligosaccharides: all of the theories are correct. Glycobiology 3(2), 97–130 (1993) PMID: 8490246

23. Koeller, K.M., Wong, C.: Complex carbohydrate synthesis tools for glycobiologists: enzyme-based approach and programmable one-pot strategies. Glycobiology 10(11), 1157–1169 (2000)

24. Smaldon, J., Romero-Campero, F.J., Trillo, F.F., Gheorghe, M., Alexander, C., Krasnogor, N.: A computational study of liposome logic: towards cellular computing from the bottom up. Systems and Synthetic Biology 4(3), 157–179 (2010)

Visual Modelling of Complex Systems: Towards an Abstract Machine for PORGY

Maribel Fernández[1], Hélène Kirchner[2], Ian Mackie[3], and Bruno Pinaud[4]

[1] King's College London, UK
[2] Inria, Paris-Rocquencourt, France
[3] École Polytechnique, France
[4] University of Bordeaux, France
Maribel.Fernandez@kcl.ac.uk

Abstract. PORGY is a visual modelling tool, where a system is defined by a strategic graph program. In this paper, we provide an operational semantics for strategic graph programs by means of an abstract machine. The semantics specifies the valid transformation steps, providing a link between the model and its implementation in PORGY.

Keywords: port graph, graph rewriting, strategies, operational semantics, abstract machine.

1 Introduction

PORGY [2,9,13] is a visual, interactive environment for the specification, simulation and analysis of complex systems, based on port graph rewriting [1]. *Port graph rewriting systems* are a general class of graph rewriting systems [6], which have been used to model systems in a wide variety of domains (e.g., biochemical systems, interaction nets, games, fractals). PORGY provides a graphical interface (see Fig. 1), where users can define a system and specify its dynamics by means of port graph rewrite rules. To control the rewriting engine, PORGY provides a *strategy language*.

Reduction strategies (see [4] for general definitions) have been extensively studied for the λ-calculus and term rewriting systems. They are present in functional programming languages and can be explicitly defined in, e.g., Stratego [17] and Maude [11]. They are also present in graph transformation tools such as PROGRES [16], AGG [7], Fujaba [12] and GP [15]. A distinctive feature of PORGY's language is that it allows users to define strategies using not only operators to combine graph rewriting rules but also operators to deal with graph traversal and management of rewriting positions in a graph. Port graphs have node, port and edge *attributes*, whose values are taken into account in port graph morphisms (used to define rewriting steps) and in strategy expressions (to control the application of rules).

Strategic graph programs, consisting of an initial port graph (usually called a *model*) and a set of rewrite rules controlled by a strategy, are the essence of PORGY. Our main contribution is a formal operational semantics for strategic

A. Beckmann, E. Csuhaj-Varjú, and K. Meer (Eds.): CiE 2014, LNCS 8493, pp. 183–193, 2014.
© Springer International Publishing Switzerland 2014

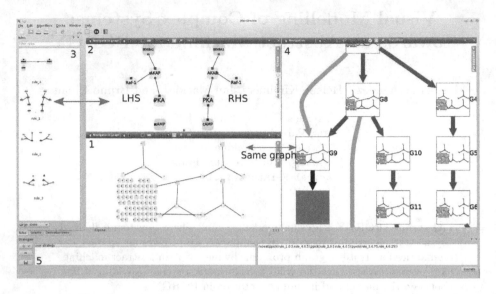

Fig. 1. Overview of PORGY: (1) editing one state of the model (2) editing a rule (3) rewriting rules (4) portion of the derivation tree (5) the strategy editor

graph programs in the form of an abstract machine, which defines, for each program, a set of rewrite derivations – a *derivation tree*.

PORGY provides a visual representation of the derivation tree, and its graphical interface permits to interact with the system and extract strategies that ensure specific behaviours. It includes features such as cycle detection, to facilitate debugging. Strategies are used to select subgraphs of the model as focusing positions for rewriting and to select the rewrite rules to be applied. PORGY's strategy language offers separate primitives for rule selection and position selection, following the separation of concerns principle, which makes programs easier to maintain and adapt.

This paper is organised as follows. In Section 2, we recall port graph rewriting. In Section 3, we present strategic graph programs. Section 4 presents the abstract machine. Section 5 concludes.

2   Port Graph Rewriting

A port graph is a labelled graph where edges are attached to ports. Let $\mathcal{N}, \mathcal{E}, \mathcal{P}$ be sets of nodes, edges and ports, respectively. A *signature* ∇ consists of pairwise disjoint sets $\nabla_{\mathcal{N}}$ (node labels), $\mathcal{X}_{\mathcal{N}}$ (node label variables), $\nabla_{\mathcal{P}}$ (port labels), $\mathcal{X}_{\mathcal{P}}$ (port label variables), $\nabla_{\mathcal{E}}$ (edge labels), $\mathcal{X}_{\mathcal{E}}$ (edge label variables), $\nabla_{\mathcal{A}}$ (attribute labels), $\mathcal{X}_{\mathcal{A}}$ (attribute variables), $\nabla_{\mathcal{V}}$ (values) and $\mathcal{X}_{\mathcal{V}}$ (value variables). Each node label has a finite *interface*: $Interface(N) \subseteq \nabla_{\mathcal{P}}$ for all $N \in \nabla_{\mathcal{N}}$, and $Interface(X) \subseteq \nabla_{\mathcal{P}} \cup \mathcal{X}_{\mathcal{P}}$ for all $X \in \mathcal{X}_{\mathcal{N}}$. Also, each $N \in \nabla_{\mathcal{N}} \cup \mathcal{X}_{\mathcal{N}}$ and each $P \in Interface(N)$ has a finite set of attributes. Similarly, for each $E \in \nabla_{\mathcal{E}} \cup \mathcal{X}_{\mathcal{E}}$, $Attribute(E) \subseteq \nabla_{\mathcal{A}} \cup \mathcal{X}_{\mathcal{A}}$. An attribute may have an associated value in $\nabla_{\mathcal{V}} \cup \mathcal{X}_{\mathcal{V}}$.

Definition 1 (Port graph with attributes). *A port graph with attributes, over a signature ∇, is a tuple $G = (V_G, lv_G, E_G, le_G)$ where:*

- *$V_G \subseteq \mathcal{N}$ is a finite set of nodes.*
- *lv_G is a function that returns, for each $v \in V_G$, a label $N \in \nabla_{\mathcal{N}} \cup \mathcal{X}_{\mathcal{N}}$ (the node's name), a set $\{p_1, \ldots, p_n\} = Interface(N)$ of port labels (each with its own sets of attribute labels and values), and a set of attribute labels (each with a value).*

  We see lv_G as a family of functions: lv_Gname, lv_Gports, lv_Gattributes, lv_GattributeValue, lv_GportAttributes, lv_GportAttributeValue.
- *$E_G \subseteq \mathcal{E}$ is a finite set of undirected edges, each with two attachment ports $(v_1, p_1), (v_2, p_2)$, where $v_i \in V_G, p_i \in Interface(lv_G name(v_i))$. Two nodes may be connected by more than one edge on the same ports.*
- *le_G is a labelling function for edges, which returns for each $e \in E_G$ a label in $\nabla_{\mathcal{E}} \cup \mathcal{X}_{\mathcal{E}}$ (the edge's name), its attachment ports $(v_1, p_1), (v_2, p_2)$ and its set of attribute labels, each with an associated value.*

Note that in a port graph, two nodes with the same node label must have the same set of port and attribute labels, however, the values of the attributes may be different. Similarly, ports with the same port label belonging to nodes with the same node label must have the same set of attribute labels but their values may be different. Label variables are used in rewriting rules (see below). Figure 1 (panel 1) shows a port graph representing four complex molecules and a set of simpler molecules.

Let G and H be two port graphs over ∇, where G may contain variable labels but H does not. A *port graph morphism* $f : G \to H$ instantiates variables in G and maps nodes, ports, edges, node attributes and their values, port attributes and their values, and edge attributes and their values, from G to H, such that all labels are preserved, the attachment of edges is preserved and the set of pairs of attributes and values for nodes, ports and edges are also preserved. Intuitively, the morphism identifies a subgraph of H that is equal to G except for variable occurrences.

To define *port graph rewrite rules* we assume that $\nabla_{\mathcal{N}}$ includes a family of node labels $\Rightarrow_n$, where $Interface(\Rightarrow_n) = \{p_1, \ldots, p_n\}$ and each p_i has an attribute *type*, which can have three different values (*bridge*, *wire* and *blackhole*) to indicate how to connect the ports during rewriting; we give details below.

Definition 2 (Port graph rewrite rule). *A port graph rewrite rule is a port graph consisting of two port graphs L and R over the signature ∇, called the left-hand side and right-hand side, respectively; an arrow node labelled by $\Rightarrow_n$; and a set of arrow-edges that each connect a port of the arrow node to ports in L or R. This set must satisfy the following conditions:*

1. *A port of type bridge must have edges connecting it to L and to R (one edge to L and one or more to R).*
2. *A port of type blackhole must have edges connecting it only to L (at least one edge).*

3. *A port of type wire must have exactly two incident edges from L and no edges connecting it to R.*

The arrow node and arrow-edges are omitted if they are obvious from L and R.

The set of *arrow-edges* is used to control the rewiring, ensuring that no edges are left dangling (see [5]). This definition generalises the one in [1], by including case (3.) above, inspired by interaction nets [10].

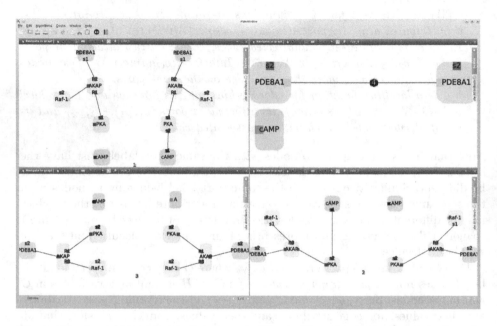

Fig. 2. The four rules of the model shown in Fig. 1. The rule at the top left is the rule shown in the panel 2 of Fig. 1.

Figure 2 shows four examples of rules. The edges in the top right rule are arrow edges connected through an arrow node (represented in black) in the middle of the drawing.

Definition 3 (Rewrite step). *A rewriting step on G using a rule L ⇒ R and morphism g : L → G, written G →$^g_{L⇒R}$ G', transforms G into a new graph G' obtained from G by performing the following operations in three phases: First, a copy R_c = g(R) (i.e., an instantiated copy of the port graph R) is added to G. The rewiring phase then redirects edges to the copy of R as follows:*
  For each port p in the arrow node:

  - *If p is a bridge port and $p_L \in L$ is connected to p*
    for each port $p^i_R \in R$ connected to p,
    find all the ports p^k_G in G that are connected to $g(p_L)$ and are not in $g(L)$, and for each p^k_G create an edge connecting p^k_G and $p^i_{R_c}$, then destroy the edge connecting p^k_G and $g(p_L)$ in G.

- *If p is a wire port connected to two ports p_1 and p_2 in L, then take all the ports outside g(L) that are connected to $g(p_1)$ in G and connect each of them to each port outside g(L) connected by an edge to $g(p_2)$.*
- *If p is a blackhole: for each port $p_L \in L$ connected to p, destroy all the edges connected to $g(p_L)$ in G.*

Finally, g(L) is deleted.

3   Strategic Graph Programs

Definition 4. *A* located graph G_P^Q *is a port graph G with two distinguished subgraphs P and Q, called respectively the* position *and* banned *subgraphs.*

A strategic graph program $[S_{\mathcal{R}}, G_P^Q]$ *(or simply $[S, G_P^Q]$ when $\mathcal{R}$ is obvious) consists of a strategy expression S (see Fig. 3) and a located graph G_P^Q.*

In a located graph G_P^Q, P is the focus of the rewriting and Q represents the part of G where rewriting is forbidden; the intuition is that subgraphs of G that overlap with P may be rewritten, if they are outside Q.

A *located rewrite rule*, defined below, specifies two disjoint subgraphs M and N of the right-hand side R that are used to update the position and banned subgraphs, respectively. Below, we use the operators $\cup, \cap, \setminus$ to denote union, intersection and complement. These operators are defined in the natural way on port graphs considered as sets of nodes, ports and edges.

Definition 5 (Located rewrite rule). *A* located rewrite rule *is given by a port graph rewrite rule $L \Rightarrow R$, and optionally a subgraph W of L and two disjoint subgraphs M and N of R. It is denoted $L_W \Rightarrow R_M^N$. We write $G_P^Q \to_{L_W \Rightarrow R_M^N}^g G'^{Q'}_{P'}$ and say that the located graph G_P^Q rewrites to $G'^{Q'}_{P'}$ using $L_W \Rightarrow R_M^N$ at position P avoiding Q, if $G \to_{L \Rightarrow R} G'$ with a morphism g such that $g(L) \cap P = g(W)$ or simply $g(L) \cap P \neq \emptyset$ if W is not provided, and $g(L) \cap Q = \emptyset$. The new position subgraph P' and banned subgraph Q' are defined as $P' = (P \setminus g(L)) \cup g(M)$, $Q' = Q \cup g(N)$; if M (resp. N) are not provided then we assume $M = R$ (resp. $N = \emptyset$).*

Strategy expressions are generated by the grammar in Figure 3 from S, and combine applications of located rewrite rules generated by A, and position updates generated by U, using *focusing expressions* generated by F.

Focusing. The grammar for F generates expressions that are used to change the rewriting position P and banned subgraph Q (e.g., to specify graph traversals). CrtGraph, CrtPos and CrtBan, applied to a located graph G_P^Q, return respectively G, P and Q. AllNgb, OneNgb and NextNgb denote functions that apply to pairs consisting of a located graph G_P^Q and a subgraph G' of G: If *Pos* denotes a subgraph G' of the current graph G, then AllNgb(*Pos*) denotes the subgraph of G consisting of all immediate successors of the nodes in G', where an immediate successor of a node v is a node that has a port connected to a port of v.

Let $L_W \Rightarrow R_M^N$ be a located rule, m an integer, $p_{i=1...m} \in [0,1]$, $\Sigma_{i=1}^m p_i = 1$
$n \in \nabla_{\mathcal{N}} \cup \nabla_{\mathcal{E}} \cup \nabla_{\mathcal{P}}$, $att \in \nabla_{\mathcal{A}}$ and $v \in \nabla_{\mathcal{V}} \cup \mathcal{X}_{\mathcal{V}}$.

(Strategy)	$S ::= A \mid U \mid S; S \mid \texttt{repeat}(S) \mid \texttt{while}(S)\texttt{do}(S) \mid (S)\texttt{orelse}(S)$
	$\mid \texttt{if}(S)\texttt{then}(S)\texttt{else}(S) \mid \texttt{ppick}(S_1, p_1, \ldots, S_m, p_m)$
(Application)	$A ::= \texttt{Id} \mid \texttt{Fail} \mid \texttt{all}(T) \mid \texttt{one}(T)$
(Transform)	$T ::= L_W \Rightarrow R_M^N$
(Position)	$U ::= \texttt{setPos}(F) \mid \texttt{setBan}(F) \mid \texttt{isEmpty}(F)$
(Focusing)	$F ::= \texttt{CrtGraph} \mid \texttt{CrtPos} \mid \texttt{CrtBan} \mid \texttt{AllNgb}(F) \mid \texttt{OneNgb}(F)$
	$\mid \texttt{NextNgb}(F) \mid \texttt{Property}(\rho, F) \mid F \cup F \mid F \cap F \mid F \setminus F \mid \emptyset$
(Property)	$\rho := (Elem, Expr) \mid (\textsf{Function}, funct\text{-}name)$
	$Elem := \textsf{Node} \mid \textsf{Edge} \mid \textsf{Port}$
	$Expr := \textsf{Label} = n \mid \textsf{Label} \neq n \mid att\ op\ v \mid att\ op\ att$
	$op := = \mid \neq \mid > \mid < \mid \leq \mid \geq$

Fig. 3. Syntax of the Strategy Language

$\texttt{OneNgb}(Pos)$ returns a subgraph of G consisting of one immediate successor of a node in G', chosen non-deterministically. $\texttt{NextNgb}(Pos)$ computes successors of nodes in G' using for each node only the subset of its ports that have the attribute "next"; we call the ports in this distinguished subset the *next* ports. $\texttt{Property}(\rho, F)$ selects a subgraph of a given graph that satisfies a certain property, specified by ρ. Function refers to a user-defined or built-in function to be used to compute the subgraph. In the grammar for $Expr$, Label refers to the label of a node, edge or port.[1] For example, $\texttt{Property}((\textsf{Port}, \textsf{Label} = Principal), F)$ returns all the nodes of the subgraph defined by the expression F that have a port labelled Principal.

Transformations. The most basic transformation is a located rule, which can only be applied to a located graph G_P^Q if at least a part of the redex is in P, and it does not involve Q.

Strategies. Id and Fail are two basic strategies that respectively denote success and failure. $\texttt{all}(T)$ denotes all possible applications of the transformation T on the located graph at the current position, creating a new located graph for each application. $\texttt{one}(T)$ non-deterministically computes only one of the possible applications of the transformation. By default, T stands for $\texttt{one}(T)$. $\texttt{setPos}(F)$ (resp. $\texttt{setBan}(F)$) sets the position subgraph P (resp. Q) to be the graph resulting from the expression F. They always succeed. $\texttt{isEmpty}(F)$ behaves like

---

[1] In the implementation, every node, port and edge has a predefined attribute called *Label*, to store its label, so we assume $Label \notin \nabla_{\mathcal{A}}$.

Id if F returns an empty graph and Fail otherwise; this can be used for instance inside the condition of an if or while. The expression $S;S'$ represents sequential application of S followed by S'. The strategy $\text{ppick}(S_1, p_1, \ldots, S_n, p_n)$ picks one of the strategies for application, according to the given probabilities. $\text{if}(S)\text{then}(S')\text{else}(S'')$ checks if the application of S on (a copy of) G_P^Q returns Id, in which case S' is applied to (the original) G_P^Q, otherwise S'' is applied to the original G_P^Q. $(S)\text{orelse}(S')$ applies S if possible, otherwise applies S'. It fails if both S and S' fail. $\text{while}(S)\text{do}(S')$ keeps on sequentially applying S' while the expression S succeeds on a copy of the graph. If S fails, then Id is returned. $\text{repeat}(S)$ simply iterates the application of S until it fails, then Id is returned.

Examples. The well-known outermost and innermost strategies used in term rewriting languages can be easily defined, and we can also specify an "outside-in" strategy for interaction nets, to compute interface normal forms [8]. We define $start \triangleq \text{Property}((\text{Function}, \textit{Interface}), \text{CrtGraph})$, which selects the subgraph containing nodes with free ports (i.e., the interface), define the *next* ports to be the principal ports, and use the following strategy:

> setPos(*start*);
> while(not(isEmpty(CrtPos)))do(
> if(R)then(R ; setPos(*start*))else(
>   if(isEmpty((CrtPos∪ NextNgb(CrtPos)) \ CrtPos))then(
>   setPos(∅))else(setPos(CrtPos∪ NextNgb(CrtPos)))))

where $\text{not}(S)$ is the strategy $\text{if}(S)\text{then}(\text{Fail})\text{else}(\text{Id})$.

The biochemical network represented in Figure 1 (panel 1) contains different types of molecules; the strategy that explains its behaviour is:

> repeat(
> ppick(r1,0.5,r4,0.5); ppick(r2,0.5,r4,0.5); ppick(r3,0.75,r4,0.25)
>)

4   An Abstract Machine for Strategic Graph Programs

The operational semantics of PORGY is given by an abstract machine: a transition system defined by a set of *configurations* and a *transition relation* [14].

Definition 6. *A configuration is a tuple $\langle T, L, Res, A \rangle$ of a derivation tree T, a list L of current models, a list Res of results, and an auxiliary stack A. L may be empty, written $[]$, or have the form $(c, p) : L'$ where c is a control stack and p a pointer to a leaf in T; Res may be empty or contain pointers to leaves in T. A derivation tree T has the form $Tree(G_P^Q, LT)$ where G_P^Q is a located graph, and LT is a (possibly empty) list of trees. We denote by $T|_p$ the located graph G_P^Q in the node pointed by p in T.*

Stacks are inductively defined: an empty stack is denoted by *nil*, and a non-empty stack $i \cdot c$ is obtained by pushing an element i on a stack c. A list $e_1 : (e_2 : \ldots : (e_n : []) \ldots)$ is abbreviated $[e_1, \ldots, e_n]$. In our implementation, which is built as a Tulip plugin, each node in the derivation tree $Tree(G_P^Q, [T_1, \ldots, T_n])$ contains a pointer to Tulip's heap, where G_P^Q is stored, and a list of pointers to the trees $T_1, \ldots, T_n$. A leaf is a particular case of tree of the form $Tree(G_P^Q, [])$. There is a special leaf, called a Failure node, where the graph G_P^Q has just one node labelled by $Fail$.

Definition 7 (Initial and Final Configurations). *To execute a strategic graph program* $[S, G_P^Q]$*, we start the abstract machine from an* initial *configuration of the form* $\langle Tree(G_P^Q, []), [(S \cdot nil, Root)], [], nil \rangle$ *where Root points to the root of the derivation tree. A configuration of the form* $\langle T, [], Res, nil \rangle$ *is* final.

The transition relation, denoted by $\rightarrow$, is a binary relation between configurations, specified by a set of transition rules. Each transition corresponds to a step of computation, determined by the strategy expression at the top of the control stack in the leftmost current model. We show the transition rules for the deterministic sublanguage.

If a control stack is empty, there are no steps of computation to perform in the corresponding current model, so we have reached a result.

$$\langle T, (nil, p) : L, Res, A \rangle \rightarrow \langle T, L, p : Res, A \rangle$$

For basic strategies Id, Fail, the rules are:

$$\langle T, (\text{Id} \cdot c, p) : L, Res, A \rangle \rightarrow \langle T, (c, p) : L, Res, A \rangle$$
$$\langle T, (\text{Fail} \cdot c, p) : L, Res, A \rangle \rightarrow \langle NewFail(T, p), L, Res, A \rangle$$

where the function $NewFail$ extends the derivation tree T by creating a failure leaf as a child of the node pointed by p (in our implementation, failure leaves are shown in red, see Fig. 1).

$$\langle T, (\text{all}(L_W \Rightarrow R_M^N) \cdot c, p) : L, Res, A \rangle \rightarrow \langle T', (c, p_1) : \ldots : (c, p_n) : L, Res, A \rangle$$

where T' is an extension of T, where a new child of the node pointed by p is created for each *legal reduct* of $T|_p$. Legal reducts are computed by the auxiliary function $LS_{L_W \Rightarrow R_M^N}$ (as explained below). The newly created leaves pointed by p_i $(1 \leq i \leq n)$ become current models and are placed at the front of the current model list, reflecting the fact that we are building the derivation tree in a depth-first fashion. If breadth-first was preferred, then the pairs (c, p_i) should be placed at the back of L. If the list of legal reducts of a current model p is empty, then T' is simply T with an extra failure leaf (as a child of the current model p), which will not be included in the new list of current models.

The function $LS_{L_W \Rightarrow R_M^N}$, when applied to a located graph G_P^Q, computes the list of located graphs $G_{iP_i}^{Q_i}$ $(1 \leq i \leq k)$ such that $G_P^Q \rightarrow_{L_W \Rightarrow R_M^N}^{g_i} G_{iP_i}^{Q_i}$ and $g_1, \ldots, g_k$ are pairwise different. Note that all possible applications of the rule are considered and there is a failure if the rule is not applicable.

Next we give the transition rules for the commands used to specify and update positions via focusing constructs.

$$\langle T, (\text{setPos}(F) \cdot c, p) : L, Res, A \rangle \; \to \; \langle NewPos(F, T, p), (c, p) : L, Res, A \rangle$$
$$\langle T, (\text{setBan}(F) \cdot c, p) : L, Res, A \rangle \; \to \; \langle NewBan(F, T, p), (c, p) : L, Res, A \rangle$$
$$\langle T, (\text{isEmpty}(F) \cdot c, p) : L, Res, A \rangle \to \langle T, (\text{Id} \cdot c, p) : L, Res, A \rangle \quad if \; F(T|_p) = \emptyset$$
$$\langle T, (\text{isEmpty}(F) \cdot c, p) : L, Res, A \rangle \to \langle T, (\text{Fail} \cdot c, p) : L, Res, A \rangle \quad if \; F(T|_p) \neq \emptyset$$

where $NewPos(F, T, p)$ (resp. $NewBan(F, T, p)$) updates the located graph $T|_p$ by setting the position graph (resp. banned graph) to be $F(T|_p)$, where the focusing expression F (see Fig. 3) has a functional semantics: it denotes a function that applies to the current located graph, and computes a subgraph as follows.

$$\text{CrtGraph}(G_P^Q) \quad = G \qquad \text{CrtPos}(G_P^Q) = P \qquad \text{CrtBan}(G_P^Q) = Q$$
$$\text{AllNgb}(F)(G_P^Q) \quad = G' \qquad \text{where } G' \text{ consists of all immediate successors of}$$
$$\text{nodes in } F(G_P^Q)$$
$$\text{NextNgb}(F)(G_P^Q) \quad = G' \qquad \text{where } G' \text{ consists of the immediate successors,}$$
$$\text{via ports labelled ``next'', of nodes in } F(G_P^Q)$$
$$\text{Property}(\rho, F)(G_P^Q) = G' \qquad \text{where } G' \text{ is } \rho(F(G_P^Q))$$
$$(F_1 \; op \; F_2)(G_P^Q) \quad = F_1(G_P^Q) \; op \; F_2(G_P^Q) \text{ where } op \text{ is } \cup, \cap, \backslash$$

The rule for sequences, given below, ensures that S_1 is applied first. The behaviour of the strategy $\text{if}(S_1)\text{then}(S_2)\text{else}(S_3)$ depends on the result of the strategy S_1, and the semantics of the iterative construct is defined using a conditional.

$$\langle T, (S_1; S_2 \cdot c, p) : L, Res, A \rangle \to \langle T, (S_1 \cdot S_2 \cdot c, p) : L, Res, A \rangle$$
$$\langle T, (\text{if}(S_1)\text{then}(S_2)\text{else}(S_3) \cdot c, p) : L, Res, A \rangle \to$$
$$\langle T, [(S_1 \cdot nil, p)], [], \langle T, (S_2 \cdot S_3 \cdot c, p) : L, Res, [] \rangle \cdot A \rangle$$
$$\langle T, [], Res, \langle T', (S_2 \cdot S_3 \cdot c, p) : L, Res', [] \rangle \cdot A \rangle \to \langle T', (S_2 \cdot c, p) : L, Res', A \rangle$$
$$if \; Res \neq []$$
$$\langle T, [], [], \langle T', (S_2 \cdot S_3 \cdot c, p) : L, Res', [] \rangle \cdot A \rangle \to \langle T', (S_3 \cdot c, p) : L, Res', A \rangle$$
$$\langle T, (\text{while}(S_1)\text{do}(S_2) \cdot c, p) : L, Res, A \rangle \to$$
$$\langle T, \text{if}(S_1)\text{then}(S_2; \text{while}(S_1)\text{do}(S_2))\text{else}(\text{Id}) \cdot c, p) : L, Res, A \rangle$$

Definition 8 (Semantics of Strategic Graph Programs). *The program* $[S, G_P^Q]$ *produces the results Res if* $\langle Tree(G_P^Q, []), [(S \cdot nil, Root)], [], nil \rangle \to^*$ $\langle T, [], Res, nil \rangle$.

We can prove that there are no blocked programs: either the abstract machine can perform a transition, or we have reached a final configuration.

Theorem 1 (Progress). *If the abstract machine starting with the initial configuration* $\langle Tree(G_P^Q, []), [(S \cdot nil, Root)], [], nil \rangle$ *stops with a configuration* $\langle T, L, Res, A \rangle$ *then L is empty.*

5   Conclusions

Graph rewriting is implemented in a variety of tools (see, e.g., [7,16,15]). A distinctive feature of PORGY is the fact that the derivation tree is a first-class component of the system, which helps analysing and debugging the model. The implementation follows the operational semantics given in this paper. Some of the transition rules require a copy of the graph, which is done efficiently in PORGY thanks to the cloning functionalities of TULIP [3].

Acknowledgements. This work was partially supported by Inria's "Associated team" programme (PORGY project) and by the French Research Agency ANR via the EVIDEN grant (ANR 2010-JCJC-0201-01).

References

1. Andrei, O.: A Rewriting Calculus for Graphs: Applications to Biology and Autonomous Systems. PhD thesis, Institut National Polytechnique de Lorraine (2008)
2. Andrei, O., Fernández, M., Kirchner, H., Melançon, G., Namet, O., Pinaud, B.: PORGY: Strategy-Driven Interactive Transformation of Graphs. In: 6th Int. Workshop on Computing with Terms and Graphs, vol. 48, pp. 54–68. EPTCS (2011)
3. Auber, D., Archambault, D., Bourqui, R., Lambert, A., Mathiaut, M., Mary, P., Delest, M., Dubois, J., Mélançon, G.: The Tulip 3 Framework: A Scalable Software Library for Information Visualization Applications Based on Relational Data. Technical Report RR-7860. Inria (January 2012)
4. Bourdier, T., Cirstea, H., Dougherty, D.J., Kirchner, H.: Extensional and intensional strategies. In: Proceedings Ninth International Workshop on Reduction Strategies in Rewriting and Programming, vol. 15, pp. 1–19. EPTCS (2009)
5. Corradini, A., Montanari, U., Rossi, F., Ehrig, H., Heckel, R., Löwe, M.: Algebraic approaches to graph transformation - part i: Basic concepts and double pushout approach. In: Handbook of Graph Grammars and Computing by Graph Transformations. Foundations, vol. 1, pp. 163–246. World Scientific (1997)
6. Courcelle, B.: Graph Rewriting: An Algebraic and Logic Approach. In: van Leeuwen, J. (ed.) Handbook of Theoretical Computer Science. Formal Models and Semantics, vol. B, pp. 193–242. Elsevier and MIT Press (1990)
7. Ermel, C., Rudolf, M., Taentzer, G.: The AGG approach: Language and environment. In: Ehrig, H., Engels, G., Kreowski, H.-J., Rozenberg, G. (eds.) Handbook of Graph Grammars and Computing by Graph Transformations. Applications, Languages, and Tools, vol. 2, pp. 551–603. World Scientific (1997)
8. Fernández, M., Mackie, I.: A calculus for interaction nets. In: Nadathur, G. (ed.) PPDP 1999. LNCS, vol. 1702, pp. 170–187. Springer, Heidelberg (1999)
9. Fernández, M., Kirchner, H., Namet, O.: A Strategy Language for Graph Rewriting. In: Vidal, G. (ed.) LOPSTR 2011. LNCS, vol. 7225, pp. 173–188. Springer, Heidelberg (2012)
10. Lafont, Y.: Interaction nets. In: Proceedings of the 17th ACM Symposium on Principles of Programming Languages (POPL 1990), pp. 95–108. ACM Press (1990)
11. Martí-Oliet, N., Meseguer, J., Verdejo, A.: Towards a strategy language for Maude. Electr. Notes Theor. Comput. Sci. 117, 417–441 (2005)

12. Nickel, U., Niere, J., Zündorf, A.: The FUJABA environment. In: ICSE, pp. 742–745 (2000)
13. Pinaud, B., Melançon, G., Dubois, J.: PORGY: A Visual Graph Rewriting Environment for Complex Systems. Computer Graphics Forum 31(3), 1265–1274 (2012)
14. Plotkin, G.D.: A structural approach to operational semantics. J. Log. Algebr. Program. 60-61, 17–139 (2004)
15. Plump, D.: The Graph Programming Language GP. In: Bozapalidis, S., Rahonis, G. (eds.) CAI 2009. LNCS, vol. 5725, pp. 99–122. Springer, Heidelberg (2009)
16. Schürr, A., Winter, A.J., Zündorf, A.: The PROGRES Approach: Language and Environment. In: Handbook of Graph Grammars and Computing by Graph Transformations. Applications, Languages, and Tools, vol. 2, pp. 479–546. World Scientific (1997)
17. Visser, E.: Stratego: A Language for Program Transformation Based on Rewriting Strategies System Description of Stratego 0.5. In: Middeldorp, A. (ed.) RTA 2001. LNCS, vol. 2051, pp. 357–361. Springer, Heidelberg (2001)

Fixed Points and Attractors
of Reaction Systems*

Enrico Formenti[1], Luca Manzoni[1], and Antonio E. Porreca[2]

[1] Univ. Nice Sophia Antipolis, CNRS, I3S, UMR 7271
06900 Sophia Antipolis, France
enrico.formenti@unice.fr, luca.manzoni@i3s.unice.fr
[2] Dipartimento di Informatica, Sistemistica e Comunicazione
Università degli Studi di Milano-Bicocca
Viale Sarca 336/14, 20126 Milano, Italy
porreca@disco.unimib.it

Abstract. We investigate the computational complexity of deciding the occurrence of many different dynamical behaviours in reaction systems, with an emphasis on biologically relevant problems (*i.e.*, existence of fixed points and fixed point attractors). We show that the decision problems of recognising these dynamical behaviours span a number of complexity classes ranging from FO-uniform AC^0 to Π_2^P-completeness with several intermediate problems being either NP or coNP-complete.

1 Introduction

Reaction systems (RS) are a computational model recently introduced by Ehrenfeucht and Rozenberg [5] which was inspired by chemical reactions. Interest in this model has grown due to its ability to be used to investigate practical problems while retaining a formulation clean enough to allow a theoretical investigation of its properties. One of the main research trends in RS is the study of their dynamics, like checking the complexity of the behaviours obtainable with limited resources [4] or the probability of a system to reach a halting state [3]. Other studies focused on understanding the complexity of deciding if a certain dynamical behaviour is present in a given RS or not [14,13].

The present paper follows this trend by extending the first results on complexity proved in [5,14,13], where the idea that RS can be used to evaluate Boolean formulae was introduced. In particular, we investigate the complexity of establishing if a RS admits a fixed point (NP-complete) or a fixed point attractor (NP-complete). We also study the complexity of finding if two RS share all fixed points (coNP-complete), or all fixed point attractors (Π_2^P-complete).

Since RS can be used to model and study biological processes [2], determining if a particular biological system exhibits a certain behaviour is an important task with potential real-life impact. The dynamics of qualitative models (*i.e.*, where

* This work has been partially supported by the French National Research Agency project EMC (ANR-09-BLAN-0164).

A. Beckmann, E. Csuhaj-Varjú, and K. Meer (Eds.): CiE 2014, LNCS 8493, pp. 194–203, 2014.

only the presence or absence of a substance is measured), like Boolean networks, has always been important in the modelling of biological systems. For example, attractors can represent cellular types or cellular states (*cf.*, proliferation or differentiation) [16] and determining the presence of fixed points and cycles is essential when modelling gene regulatory networks [8,1]. Furthermore, in [9] the importance of studying robustness in complex biological systems is highlighted. The identification of attractors is a necessary first step in this direction.

The paper is structured as follows. Section 2 provides the basic notions on RS and a short comparison with related models. Section 3 gives a description in logical terms of the problems we investigate. The decision problems regarding fixed points are collected in Section 4 and the ones regarding fixed point attractors in Section 5. A summary of the results and of possible future developments is given in Section 6.

2   Basic Notions

We recall the definitions of reaction, reaction system, and the associated notation from [5].

Definition 1. *Consider a finite set S, whose elements are called* entities. *A reaction a over S is a triple (R_a, I_a, P_a) of subsets of S. The set R_a is called the set of* reactants, *I_a the set of* inhibitors, *and P_a is the set of* products. *Denote by rac(S) the set of all reactions over S.*

Definition 2. *A reaction system $\mathcal{A}$ is a pair (S, A) where S is a finite set, called the* background set, *and $A \subseteq$ rac(S).*

Given a *state* $T \subseteq S$, a reaction a is said to be *enabled* in T when $R_a \subseteq T$ and $I_a \cap T = \varnothing$. The *result function* $\text{res}_a \colon 2^S \to 2^S$ of a, where 2^S denotes the power set of S, is defined as

$$\text{res}_a(T) = \begin{cases} P_a & \text{if } a \text{ is enabled in } T \\ \varnothing & \text{otherwise.} \end{cases}$$

The definition of res_a naturally extends to sets of reactions. Indeed, given $T \subseteq S$ and $A \subseteq$ rac(S), define $\text{res}_A(T) = \bigcup_{a \in A} \text{res}_a(T)$. The result function res_A of a RS $\mathcal{A} = (S, A)$ is res_A, *i.e.*, it is the result function of the whole set of reactions.

Example 1 (XOR gate). Consider the RS $\mathcal{A} = (\{1_0, 1_1, 1_{out}\}, A)$, where the entities represent the first two inputs and the output when they assume value 1, respectively. The set A contains $(\{1_0\}, \{1_1\}, \{1_{out}\})$ and $(\{1_1\}, \{1_0\}, \{1_{out}\})$. The system, starting from a state that is a subset of $\{1_0, 1_1\}$ encoding the bits set to 1 in the input, produces 1_{out} in one step iff the XOR gate on the same input produces 1.

In the sequel, we are interested in the dynamics of RS, *i.e.*, the study of the successive states of the system under the action of the result function res_A

starting from some initial set of entities. Given a set $T \subseteq S$, the sequence of states visited by the system is $(T, \text{res}_\mathcal{A}(T), \text{res}_\mathcal{A}^2(T), \ldots)$ (*i.e.*, for every $t \in \mathbb{N}$, the t-th element of the sequence is $\text{res}_\mathcal{A}^t(T)$). Since S is finite any sequence of visited states is ultimately periodic, *i.e.*, for any $T \subseteq S$, there exist $h, p \in \mathbb{N}$ such that for all $t \in \mathbb{N}$ we have $\text{res}_\mathcal{A}^{h+pt}(T) = \text{res}_\mathcal{A}^{h+t}(T)$; here h is the *length of the transient*. A state $T \subseteq S$ is part of a *cycle* if the sequence of states starting from T is ultimately periodic with a transient of length 0; in this case, the least p satisfying the previous equation is called the *period* of the cycle. A *fixed point* T is a cycle with period 1 (*i.e.*, $\text{res}_\mathcal{A}(T) = T$). An *attractor* of an RS $\mathcal{A}$ is a cycle $T_1, \ldots, T_p$ for which there exists a state U not belonging to the cycle such that $\text{res}_\mathcal{A}(U) = T_i$ for some $1 \leq i \leq p$. A *fixed point attractor* is a fixed point that is also an attractor. Given a RS $\mathcal{A}$, we say that a state T is a fixed point (resp., attractor) for $\mathcal{A}$ if it is a fixed point (resp., attractor) for $\text{res}_\mathcal{A}$.

2.1   Related Models

Other bio-inspired models having features in common with RS are membrane systems, Boolean networks, and chemical reaction networks.

Membrane systems [11] also provide an idealisation of chemical reactions in the context of a cell. The main difference between RS and membrane systems is the presence of multiplicity, that is, the state of the membrane system is a multiset and not a set, and the rewriting rules consume the substances that they use. Furthermore, the main characteristic of membrane systems is the presence of membranes that partition the system into multiple regions with limited communications. The idea of linking membrane systems and RS is not new and has already been explored [12].

Synchronous Boolean networks [7,15] can be viewed as a generalisation of RS. Indeed, they can be used to simulate RS by associating an entity to each node of the network; the value of a node denotes the absence or presence of an entity in the current state of the simulated RS, that is, the state of the Boolean network is the characteristic vector of the state of the RS. The update function of a node can be written as a Boolean formula in disjunctive normal form that holds iff the entity denoted by the node is generated by some reaction of the RS. The resulting Boolean network has a description of polynomial length with respect to the description of the RS. The converse simulation, while possible, might require an exponential number of reactions, depending on the encoding of the Boolean network.

Chemical reaction networks (CRN) are a model in which a set of entities (called *signals* in CRN) is modified by means of chemical reactions described by reactants, products, and catalysts [18]. Reactants are consumed to generate the products when both they and the catalysts are present in the current state of the system. The operations of CRN can be implemented in multiple ways, for example by means of logical circuits or DNA strand displacement systems. The main differences between CRN and RS are that the state used by the former is a multiset (*i.e.*, the multiplicity is considered) and there are no inhibitors in the reactions.

3   Logical Description

This section provides a tool that will be used in many proofs of the paper. It consists of a logical description of RS and formulae related to their dynamics. This description (or a slight adaptation) will be sufficient for proving membership in many complexity classes. For the background notions of logic and descriptive complexity we refer the reader to Neil Immerman's classical book [6].

In the sequel, we will study several classes of problems over RS, and each of them can be characterised by a logical formula. A RS $\mathcal{A} = (S, A)$ with background set $S \subseteq \{0, \ldots, n - 1\}$ and $|A| \leq n$ can be described by the vocabulary $(S, R_{\mathcal{A}}, I_{\mathcal{A}}, P_{\mathcal{A}})$, where S is a unary relation symbol and $R_{\mathcal{A}}$, $I_{\mathcal{A}}$, and $P_{\mathcal{A}}$ are binary relation symbols. The intended meaning of the symbols is the following: the set of entities is $S = \{i : S(i)\}$ and each reaction $a_j = (R_j, I_j, P_j) \in A$ is described by the sets $R_j = \{i \in S : R_{\mathcal{A}}(i, j)\}$, $I_j = \{i \in S : I_{\mathcal{A}}(i, j)\}$, and $P_j = \{i \in S : P_{\mathcal{A}}(i, j)\}$.

We will also need some additional vocabularies: $(S, R_{\mathcal{A}}, I_{\mathcal{A}}, P_{\mathcal{A}}, T)$, where T is a unary relation representing a subset of S, $(S, R_{\mathcal{A}}, I_{\mathcal{A}}, P_{\mathcal{A}}, T_1, T_2)$ with two additional unary relations representing sets, and $(S, R_{\mathcal{A}}, I_{\mathcal{A}}, P_{\mathcal{A}}, R_{\mathcal{B}}, I_{\mathcal{B}}, P_{\mathcal{B}})$ denoting two RS over the same background set.

The following formulae describe basic properties of $\mathcal{A}$. The first is true if a reaction a_j is enabled in T:

$$\text{EN}_{\mathcal{A}}(j, T) \equiv \forall i (S(i) \Rightarrow (R_{\mathcal{A}}(i, j) \Rightarrow T(j)) \wedge (I_{\mathcal{A}}(i, j) \Rightarrow \neg T(j)))$$

the latter is verified if $\text{res}_{\mathcal{A}}(T_1) = T_2$ for $T_1, T_2 \subseteq S$:

$$\text{RES}_{\mathcal{A}}(T_1, T_2) \equiv \forall i (S(i) \Rightarrow (T_2(i) \Leftrightarrow \exists j (\text{EN}_{\mathcal{A}}(j, T_1) \wedge P_{\mathcal{A}}(i, j)))).$$

Since $\text{EN}_{\mathcal{A}}$ and $\text{RES}_{\mathcal{A}}$ are both first-order (FO) formulae, the following is immediately proved.

Theorem 1. *Given a RS $\mathcal{A} = (S, A)$ and two sets $T_1, T_2 \subseteq S$, deciding whether $\text{res}_{\mathcal{A}}(T_1) = T_2$ is in FO (which is equivalent to FO-uniform AC^0 [6]).*   □

FO logic will quickly prove insufficient for our purposes; therefore we will formulate some problems using stronger logics: existential second order logic SO∃ characterising NP (Fagin's theorem); universally quantified second order logic SO∀ giving coNP; second order logic with one alternation of universal and existential quantifiers (SO∀∃, giving Π_2^P). As an abbreviation, we define the bounded second order quantifiers $(\forall X \subseteq Y) \varphi$ and $(\exists X \subseteq Y) \varphi$ as a shorthand for $\forall X (\forall i (X(i) \Rightarrow Y(i)) \Rightarrow \varphi)$ and $\exists X (\forall i (X(i) \Rightarrow Y(i)) \wedge \varphi)$. We say that a formula is SO∃, SO∀, or SO∀∃ if it is logically equivalent to a formula in the required prenex normal form.

4   Fixed Points

We investigate the complexity of determining if a given state is a fixed point for an RS, if an RS admits fixed points, and if two RS share at least one or all fixed points. First, we are interested in determining if the first-order formula $\text{FIX}_{\mathcal{A}}(T) \equiv \text{RES}_{\mathcal{A}}(T, T)$ holds for a given state T. Substituting $T_2 = T_1$ in Theorem 1, we get the following corollary:

Corollary 1. *Given a RS $\mathcal{A} = (S, A)$ and a state $T \subseteq S$, deciding whether T is a fixed point of $\text{res}_{\mathcal{A}}$ is in* FO. □

As usual, CNF (resp., DNF) means conjunctive (resp., disjunctive) normal form. Given a formula φ in CNF, we denote by $\text{neg}(\varphi)$ (resp., $\text{pos}(\varphi)$) the set of variables that occur negated (resp., non-negated) in φ. The notation $t \vDash \varphi$ means that φ is satisfied by the assignment t.

While it is easy to decide if a point is fixed, determining if a RS admits a fixed point is a vastly more difficult task as proved by the following theorem.

Theorem 2. *Given a RS $\mathcal{A} = (S, A)$, it is* NP-*complete to decide if $\mathcal{A}$ has a fixed point.*

Proof. The problem is in NP, since $(\exists T \subseteq S)\,\text{FIX}_{\mathcal{A}}(T)$ is a $\text{SO}\exists$ formula. In order to show NP-hardness, we reduce SAT [10] to this problem. Given a Boolean formula $\varphi \equiv \varphi_1 \wedge \cdots \wedge \varphi_m$ in CNF over the variables $V = \{x_1, \ldots, x_n\}$, construct a RS $\mathcal{A} = (S, A)$ with $S = V \cup \{\spadesuit, \clubsuit\}$ and the following reactions:

$$(\text{neg}(\varphi_j), \text{pos}(\varphi_j) \cup \{\clubsuit, \spadesuit\}, \{\spadesuit\}) \qquad \text{for } 1 \leq j \leq m \qquad (1)$$
$$(\{x_i\}, \varnothing, \{x_i\}) \qquad \text{for } 1 \leq i \leq n \qquad (2)$$
$$(\{\spadesuit\}, \varnothing, \{\clubsuit\}) \qquad (3)$$
$$(\{\clubsuit\}, \{\spadesuit\}, \{\spadesuit\}). \qquad (4)$$

Given a state $T \subseteq S$, let $X = T \cap V$. The set X encodes an assignment of φ in which the variables having true value are those in X. Reactions of type (1) generate $\spadesuit$ when there exists a clause φ_j not satisfied by X (hence φ itself is not satisfied). Reactions of type (2) preserve the current assignment in the next state. Finally, reactions (3) and (4) rewrite $\spadesuit$ into $\clubsuit$ and $\clubsuit$ into $\spadesuit$ (if $\spadesuit$ is missing). Hence, the RS behaves as follows:

$$\text{res}_{\mathcal{A}}(T) = \begin{cases} (T \cap V) = T & \text{if } T \subseteq V \wedge T \vDash \varphi \\ (T \cap V) \cup \{\spadesuit\} & \text{if } (T \subseteq V \wedge T \nvDash \varphi) \vee (\clubsuit \in T \wedge \spadesuit \notin T) \\ (T \cap V) \cup \{\clubsuit\} & \text{if } \spadesuit \in T \end{cases}$$

i.e., there exists a fixed point if and only if φ is satisfiable. The mapping $\varphi \mapsto \mathcal{A}$ is computable in polynomial time, hence deciding the existence of fixed points is NP-hard. □

A direct consequence of the theorem above is that determining if there exists a state that is a fixed point in common between two RS remains NP-complete.

Corollary 2. *Given two RS $\mathcal{A}$ and $\mathcal{B}$ over the same background set S, deciding if $\mathcal{A}$ and $\mathcal{B}$ have a common fixed point is* NP-*complete.*

Proof. The problem lies in NP, since $(\exists T \subseteq S)(\text{FIX}_\mathcal{A}(T) \wedge \text{FIX}_\mathcal{B}(T))$ is a SO$\exists$ formula. By letting $\mathcal{A} = \mathcal{B}$, NP-hardness follows from Theorem 2. $\square$

Differently from above, determining if two reaction systems have all fixed points in common is in coNP, instead of NP. This is expected since the description of the problem involves universal instead of existential quantification.

Theorem 3. *Given two RS $\mathcal{A} = (S, A)$ and $\mathcal{B} = (S, B)$, it is coNP-complete to decide whether $\mathcal{A}$ and $\mathcal{B}$ share all their fixed points.*

Proof. The problem lies in coNP, since $(\forall T \subseteq S)(\text{FIX}_\mathcal{A}(T) \Leftrightarrow \text{FIX}_\mathcal{B}(T))$ is a SO$\forall$ formula. In order to show coNP-hardness, we reduce TAUTOLOGY (also known as VALIDITY [10]) to this problem. Given a Boolean formula $\varphi = \varphi_1 \vee \cdots \vee \varphi_m$ in DNF over the variables $V = \{x_1, \ldots, x_n\}$, build the RS $\mathcal{A}$ consisting of the background set $S = V \cup \{\heartsuit\}$ and the following reactions:

$$(\text{pos}(\varphi_j) \cup \{\heartsuit\}, \text{neg}(\varphi_j), \{\heartsuit\}) \qquad \text{for } 1 \leq j \leq m \qquad (5)$$
$$(\{x_i, \heartsuit\}, \varnothing, \{x_i\}) \qquad \text{for } 1 \leq i \leq n. \qquad (6)$$

Let T be a state of $\mathcal{A}$ and $X = T \cap V$. When $\heartsuit \in T$, each reaction of type (5) evaluates a term φ_j under the assignment encoded by X, producing $\heartsuit$ when $X \vDash \varphi_j$ (hence $X \vDash \varphi$). Reactions of type (6) preserve the state when $\heartsuit \in T$. Thus the RS behaves as follows:

$$\text{res}_\mathcal{A}(T) = \begin{cases} T & \text{if } T \cap V \vDash \varphi \text{ and } \heartsuit \in T \\ T - \{\heartsuit\} & \text{if } T \cap V \nvDash \varphi \text{ and } \heartsuit \in T \\ \varnothing & \text{if } \heartsuit \notin T. \end{cases}$$

The fixed points of $\mathcal{A}$ are $\varnothing$ and all states of the form $X \cup \{\heartsuit\}$ with $X \subseteq V$ and $X \vDash \varphi$. Now let $\mathcal{B}$ be defined by the following reactions:

$$(\{x_i, \heartsuit\}, \varnothing, \{x_i\}) \qquad \text{for } 1 \leq i \leq n$$
$$(\{\heartsuit\}, \varnothing, \{\heartsuit\}).$$

They preserve the current state T if $\heartsuit \in T$ and yield $\varnothing$ otherwise. Hence, the fixed points of $\mathcal{B}$ are $\varnothing$ and all states of the form $X \cup \{\heartsuit\}$ with $X \subseteq V$.

By construction, the two RS $\mathcal{A}$ and $\mathcal{B}$ share all fixed points exactly when all assignments satisfy φ. Since the mapping $\varphi \mapsto (\mathcal{A}, \mathcal{B})$ is computable in polynomial time, deciding the former property is coNP-hard. $\square$

5   Fixed Point Attractors

In this section we investigate the same problems of Section 4 reformulated for fixed point attractors.

The fact that a set T is a fixed point attractor can be expressed by the following formula: $\text{ATT}_\mathcal{A}(T) \equiv (\exists U \subseteq S)(\text{FIX}_\mathcal{A}(T) \wedge \text{RES}_\mathcal{A}(U, T) \wedge \neg\text{RES}_\mathcal{A}(T, U))$.

Theorem 4. *Given a RS $\mathcal{A} = (S, A)$ and a state $T \subseteq S$, it is* NP-*complete to decide whether T is a fixed point attractor.*

Proof. Since $\text{ATT}_{\mathcal{A}}(T)$ is a $\text{SO}\exists$ formula, the problem lies in NP. We reduce SAT to this problem. Given a formula $\varphi = \varphi_1 \wedge \cdots \wedge \varphi_m$ in CNF over the set of variables $V = \{x_1, \ldots, x_m\}$, let $C = \{\varphi_1, \ldots, \varphi_n\}$ and let $\mathcal{A}$ be the RS having the background set $S = V \cup C \cup \{\spadesuit, \clubsuit\}$ and the following reactions:

$$(\{x\}, C \cup \{\spadesuit, \clubsuit\}, \{\varphi_j\}) \qquad\qquad \text{for } 1 \leq j \leq m \text{ and } x \in \text{pos}(\varphi_j) \quad (7)$$
$$(\varnothing, C \cup \{x, \spadesuit, \clubsuit\}, \{\varphi_j\}) \qquad\qquad \text{for } 1 \leq j \leq m \text{ and } x \in \text{neg}(\varphi_j) \quad (8)$$
$$(C, S - C, C) \qquad\qquad\qquad\qquad\qquad\qquad\qquad\qquad\qquad\qquad (9)$$
$$(\text{neg}(\varphi_j), \text{pos}(\varphi_j) \cup C \cup \{\spadesuit, \clubsuit\}, \{\spadesuit\}) \quad \text{for } 1 \leq j \leq m \qquad\qquad (10)$$
$$(\{\spadesuit\}, \varnothing, \{\clubsuit\}) \qquad\qquad\qquad\qquad\qquad\qquad\qquad\qquad\qquad (11)$$
$$(\{\clubsuit\}, \{\spadesuit\}, \{\spadesuit\}). \qquad\qquad\qquad\qquad\qquad\qquad\qquad\qquad\qquad (12)$$

If the state of $\mathcal{A}$ is $X \subseteq V$, the reactions of kinds (7) and (8) produce the subset of C corresponding to the clauses of φ satisfied by X. If all clauses are generated (*i.e.*, $X \vDash \varphi$), they are preserved by reaction (9). On the other hand, if at least one clause is not satisfied by X, one or more reactions of type (10) are enabled and produce $\spadesuit$. As in the proof of Theorem 2, reactions (11) and (12) generate a cycle between the states $\{\spadesuit\}$ and $\{\clubsuit\}$. The result function of $\mathcal{A}$ is then

$$\text{res}_{\mathcal{A}}(T) = \begin{cases} C & \text{if } T = C \text{ or if } T \subseteq V \text{ and } T \vDash \varphi \\ D \cup \{\spadesuit\} & \text{if } T \subseteq V, D \subsetneq C \text{ and } T \text{ satisfies the clauses in } D \\ & \quad \text{but not the clauses in } C - D \\ \{\clubsuit\} & \text{if } \spadesuit \in T \\ \{\spadesuit\} & \text{if } \clubsuit \in T \text{ and } \spadesuit \notin T \\ \varnothing & \text{otherwise.} \end{cases}$$

Notice that $\mathcal{A}$ has exactly one fixed point, the state C, which is reachable from another state T (*i.e.*, C is an attractor) iff $T \subseteq V$ and $T \vDash \varphi$, *i.e.*, iff φ is satisfiable. Since the mapping $\varphi \mapsto \mathcal{A}$ can be computed in polynomial time, the NP-hardness of the problem follows. $\qquad\square$

As immediate corollaries, finding if a fixed point attractor exists or if it exists as a shared fixed point between two RS, remain NP-complete.

Corollary 3. *Given a RS $\mathcal{A} = (S, A)$, deciding if $\mathcal{A}$ has a fixed point attractor is an* NP-*complete problem.*

Proof. The problem is in NP, since $(\exists T \subseteq S) \text{ ATT}_{\mathcal{A}}(T)$ is a $\text{SO}\exists$ formula. Its NP-hardness follows from the construction in the proof of Theorem 4, where for any Boolean formula φ in CNF, the RS $\mathcal{A}$ has exactly one fixed point, which is an attractor iff φ is satisfiable. $\qquad\square$

Corollary 4. *Given two RS $\mathcal{A}$ and $\mathcal{B}$ with the same background set S, it is* NP-*complete to decide whether $\mathcal{A}$ and $\mathcal{B}$ have a common fixed point attractor.*

Proof. The problem lies in NP since $(\exists T \subseteq S)(\text{ATT}_{\mathcal{A}}(T) \wedge \text{ATT}_{\mathcal{B}}(T))$ is a SO∃ formula. Given a Boolean formula φ in CNF having clauses $C = \{\varphi_1, \ldots, \varphi_m\}$, let $\mathcal{A}$ be the RS in the proof of Theorem 4, and let $\mathcal{B}$ be the RS having $(\varnothing, \varnothing, C)$ as its only reaction. Clearly, $\mathcal{B}$ has C as its only fixed point attractor. Hence, $\mathcal{A}$ and $\mathcal{B}$ share a fixed point attractor iff φ is satisfiable. This proves that the problem is NP-hard. □

Perhaps surprisingly, verifying if two systems share all their fixed point attractors goes one level up in the polynomial hierarchy *w.r.t.* the other problems pertaining fixed point attractors, thus providing a further example of a natural Π_2^P-complete problem.

Theorem 5. *Given two RS $\mathcal{A}$ and $\mathcal{B}$ with a common background set S, it is Π_2^P-complete to decide whether $\mathcal{A}$ and $\mathcal{B}$ share all their fixed point attractors.*

Proof. The problem lies in Π_2^P, since $(\forall T \subseteq S)(\text{ATT}_{\mathcal{A}}(T) \Leftrightarrow \text{ATT}_{\mathcal{B}}(T))$ is a SO∀∃ formula. We prove the Π_2^P-hardness by reduction from the ∀∃SAT problem [17]. Let $V = \{x_1, \ldots, x_n\}$, $V_1 \subseteq V$, and $V_2 = V - V_1$; let $(\forall V_1)(\exists V_2)\varphi$ be a quantified Boolean formula over V with $\varphi = \varphi_1 \wedge \cdots \wedge \varphi_m$ quantifier-free and in CNF. Finally, let $V_1' = \{x' : x \in V_1\}$ and $C = \{\varphi_1, \ldots, \varphi_m\}$. Define a RS $\mathcal{A}$ with background set $S = V \cup V_1' \cup C \cup \{\spadesuit, \clubsuit\}$ and the reactions

$$(\{x\}, C \cup V_1' \cup \{\spadesuit, \clubsuit\}, \{\varphi_j\}) \qquad\qquad \text{for } 1 \le j \le m, \, x \in \text{pos}(\varphi_j) \quad (13)$$

$$(\varnothing, \{x\} \cup C \cup V_1' \cup \{\spadesuit, \clubsuit\}, \{\varphi_j\}) \qquad \text{for } 1 \le j \le m, \, x \in \text{neg}(\varphi_j) \quad (14)$$

$$(\{x\}, C \cup V_1' \cup \{\spadesuit, \clubsuit\}, \{x'\}) \qquad\qquad \text{for } x \in V_1 \quad (15)$$

$$(\text{neg}(\varphi_j), \text{pos}(\varphi_j) \cup C \cup V_1' \cup \{\spadesuit, \clubsuit\}, \{\spadesuit\}) \quad \text{for } 1 \le j \le m \quad (16)$$

$$(\{\spadesuit\}, \varnothing, \{\clubsuit\}) \quad (17)$$

$$(\{\clubsuit\}, \{\spadesuit\}, \{\spadesuit\}) \quad (18)$$

$$(C, V \cup \{\spadesuit, \clubsuit\}, C) \quad (19)$$

$$(\{x'\} \cup C, V \cup \{\spadesuit, \clubsuit\}, \{x'\}) \qquad\qquad \text{for } x' \in V_1'. \quad (20)$$

When the current state of $\mathcal{A}$ is $X \subseteq V$, the reactions of types (13) and (14) produce the set of clauses satisfied by the assignment encoded by X; simultaneously, reactions of type (15) produce "primed" copies of the elements encoding the partial assignment to the universally quantified variables of φ (while the elements encoding the partial assignment to the existentially quantified variables are implicitly discarded). If one of the clauses is not satisfied by X, the corresponding reaction of type (16) is enabled and produces $\spadesuit$. Any state containing $\spadesuit$ or $\clubsuit$ end up in a cycle between $\{\spadesuit\}$ and $\{\clubsuit\}$ by means of reactions (17) and (18). If all clauses appear in the current state, they are preserved by reaction (19), together with any element of V_1' (reactions of type (20)). The inhibitors of reactions (13)–(20) ensure that "bad" states, *i.e.*, those not of the form $X \subseteq V$ or $C \cup U$ with $U \subseteq V_1'$, are mapped to $\{\spadesuit\}$, $\{\clubsuit\}$, or $\varnothing$ (which is a subset of V).

Summarising, the RS $\mathcal{A}$ defines the result function

$$
\mathrm{res}_{\mathcal{A}}(T) = \begin{cases} C \cup U & \text{where } U = \{x' \in V_1' : x \in V_1 \cap T\} \text{ if } T \subseteq V \text{ and } T \vDash \varphi \\ T & \text{if } T = C \cup U \text{ with } U \subseteq V_1' \\ D \cup \{\spadesuit\} & \text{if } T \subseteq V, \, D \subsetneq C \text{ and } T \text{ satisfies the clauses in } D \\ & \text{but not the clauses in } C - D \\ \{\spadesuit\} & \text{if } \clubsuit \in T \text{ and } \spadesuit \notin T \\ \{\clubsuit\} & \text{if } \spadesuit \in T \\ \varnothing & \text{otherwise.} \end{cases}
$$

The RS $\mathcal{A}$ admits $2^{|V_1|}$ fixed points, all of them of the form $C \cup U$ with $U \subseteq V_1'$; a state of this form is an attractor iff there exists a state $X = \{x : x' \in U\} \cup Y$, with $Y \subseteq V_2$, such that $X \vDash \varphi$. Hence, $\mathcal{A}$ has $2^{|V_1|}$ fixed point attractors iff $(\forall V_1)(\exists V_2)\varphi$ is valid. Let $\mathcal{B}$ be a RS having the reactions $(\{x'\}, \varnothing, \{x'\})$ for $x' \in V_1'$, and $(\varnothing, \varnothing, C)$. The result function of $\mathcal{B}$ is $\mathrm{res}_{\mathcal{B}}(T) = C \cup (T \cap V_1')$, having the same fixed points as $\mathcal{A}$; each fixed point of $\mathcal{B}$ is an attractor, since we have $\mathrm{res}_{\mathcal{B}}(C \cup U \cup \{\spadesuit\}) = C \cup U$ for each $U \subseteq V_1'$. Hence, the RS $\mathcal{A}$ and $\mathcal{B}$ have the same fixed point attractors iff $(\forall V_1)(\exists V_2)\varphi$ is valid. Since the mapping $((\forall V_1)(\exists V_2)\varphi) \mapsto (\mathcal{A}, \mathcal{B})$ is computable in polynomial time, the problem is Π_2^P-hard.   □

6   Conclusions

In this paper we have studied the complexity of checking the presence of many different dynamical behaviours of a RS. Deciding if a point is fixed is easy (FO, *i.e.*, FO-uniform AC^0), however it gets increasingly hard to determine the existence of a fixed point (NP-complete), or if two RS have the same fixed points (coNP-complete). When considering fixed point attractors, the majority of the problems are NP-complete, but determining if two RS share all fixed point attractors is one of the few "natural" examples of a Π_2^P-complete problem.

The paper discloses many possible research directions. First of all, it would be very interesting to understand why the comparison of local attractors is a Π_2^P-complete problem and if there are other relevant dynamical properties that populate (supposedly) different levels of the polynomial hierarchy. We are also investigating the complexity of determining the existence of *global* attractors, cycles, and attractor cycles.

The RS studied in the paper are deterministic. However many significant modelling questions involve RS where extra entities are provided externally (*i.e.*, RS with *context*). These RS are, in some sense, non-deterministic, since starting from the same initial state, we can obtain different dynamics depending on the context. It is interesting to understand how the complexity of decision problems about dynamics changes in this case.

Another promising research direction is the study of minimality, *i.e.*, understanding what is the complexity of the problem of deciding if a given RS is the minimal one (*e.g.*, with respect to the number of reactions) having a given dynamical behaviour.

Acknowledgements. We want to thank Daniela Besozzi for pointing out the relevance of the problems on the dynamics of RS to the study of biological systems. We also want to thank Karthik Srikanta for a careful reading of a draft of this manuscript.

References

1. Bornholdt, S.: Boolean network models of cellular regulation: prospects and limitations. J. R. Soc. Interface 5, S84–S94 (2008)
2. Corolli, L., Maj, C., Marini, F., Besozzi, D., Mauri, G.: An excursion in reaction systems: From computer science to biology. Theor. Comp. Sci. 454, 95–108 (2012)
3. Ehrenfeucht, A., Main, M., Rozenberg, G.: Combinatorics of life and death for reaction systems. Int. J. Found. Comput. Sci. 21(3), 345–356 (2010)
4. Ehrenfeucht, A., Main, M., Rozenberg, G.: Functions defined by reaction systems. Int. J. Found. Comput. Sci. 22(1), 167–168 (2011)
5. Ehrenfeucht, A., Rozenberg, G.: Reaction systems. Fundam. Inform. 75, 263–280 (2007)
6. Immerman, N.: Descriptive Complexity. Graduate Texts in Computer Science. Springer (1999)
7. Kauffman, S.A.: Metabolic stability and epigenesis in randomly constructed genetic nets. J. Theor. Biol. 22(3), 437–467 (1969)
8. Kauffman, S.A.: The ensemble approach to understand genetic regulatory networks. Physica A: Statistical Mechanics and its Applications 340(4), 733–740 (2004)
9. Kitano, H.: Biological robustness. Nature Reviews Genetics 5, 826–837 (2004)
10. Papadimitriou, C.H.: Computational Complexity. Addison-Wesley (1993)
11. Păun, G.: Computing with membranes. Journal of Computer and System Sciences 61(1), 108–143 (2000)
12. Păun, Gh., Pérez-Jiménez, M.J., Rozenberg, G.: Bridging membrane and reaction systems – Further results and research topics. Fundam. Inform. 127, 99–114 (2013)
13. Salomaa, A.: Functional constructions between reaction systems and propositional logic. Int. J. Found. Comput. Sci. 24(1), 147–159 (2013)
14. Salomaa, A.: Minimal and almost minimal reaction systems. Natural Computing 12(3), 369–376 (2013)
15. Shmulevich, I., Dougherty, E.R.: Probabilistic boolean networks: the modeling and control of gene regulatory networks. SIAM (2010)
16. Shmulevich, I., Dougherty, E.R., Zhang, W.: From Boolean to probabilistic Boolean networks as models of genetic regulatory networks. Proceedings of the IEEE 90(11), 1778–1792 (2002)
17. Stockmeyer, L.J.: The polynomial-time hierarchy. Theor. Comp. Sci. 3(1), 1–22 (1976)
18. Thachuk, C., Condon, A.: Space and energy efficient computation with DNA strand displacement systems. In: Stefanovic, D., Turberfield, A. (eds.) DNA 2012. LNCS, vol. 7433, pp. 135–149. Springer, Heidelberg (2012)

Fit-Preserving Data Refinement
of Mass-Action Reaction Networks

Cristian Gratie and Ion Petre

Computational Biomodeling Laboratory, Turku Centre for Computer Science and
Department of Information Technologies, Åbo Akademi University
FI-20520 Turku, Finland
{cgratie,ipetre}@abo.fi

Abstract. The systematic development of large biological models can
benefit from an iterative approach based on a refinement process that
gradually adds more details regarding the reactants and/or reactions of
the model. We focus here on data refinement, where the species of the
initial model are substituted with several subspecies in the refined one,
each with its own individual behavior in the model. In this context, we
distinguish between *structural refinement*, where the aim is to generate meaningful refined reactions, and *quantitative refinement*, where one
looks for a data fit at least as good as that of the original model. The
latter generally requires refitting the model and additional experimental
data, a computationally expensive process. A *fit-preserving refinement*,
i.e. one that captures the same species dynamics as the original model,
can serve as a suitable alternative or as initialization for parameter estimation routines. We focus in this paper on the problem of finding all
numerical setups that yield fit-preserving refinements of a given model
and formulate a sufficient condition for it. Our result suggests a straightforward, computationally efficient automation of the quantitative model
refinement process. We illustrate the use of our approach through a discussion of the Lotka-Volterra model for prey-predator dynamics.

Keywords: biomodeling, model fit, parameter estimation, quantitative
model refinement.

1   Introduction

The development of models for large biological systems often starts top-down
with an abstraction of the biological phenomena via a relatively small number of
chemical reactions that illustrate the main mechanisms of the considered process.
A mathematical model is then attached to this abstraction in order to describe
the dynamic behavior of the system. The numerical setup for the mathematical
model comes from various computational procedures that fit the model with
existing experimental data.

The model is then iteratively refined by adding details to it. Refinement can
involve the replacement of one (or more) species with subspecies, in which case
it is called *data refinement*, or the replacement of a generic reaction with a

A. Beckmann, E. Csuhaj-Varjú, and K. Meer (Eds.): CiE 2014, LNCS 8493, pp. 204–213, 2014.
© Springer International Publishing Switzerland 2014

set of reactions that gives more details about the same process by providing intermediary steps, in which case it is called *process refinement* [3].

Instead of refitting the model after every refinement, a computationally expensive process, it would be useful to take advantage of the well-fitted model from the previous step. The iterative step-by-step refinement of a formal specification towards an executable implementation is well established in Computer Science, in connection to software engineering and formal methods, see, e.g., [1]. In Systems Biology, refinement has been considered in the context of rule-based [5] and reaction-based models [9]. The implementation of model refinement using various standard modeling frameworks is discussed in [7].

The method of [9] for the numerical setup of a refined reaction-based model aims to preserve the numerical fit of the original model. The approach applies to the data refinement of models that rely on mass-action kinetics [10] and consists of inspecting the ordinary differential equations (ODEs) of the refined model and assigning parameter values in such a way that the ODEs describing the original model can be recovered as a sum of ODEs from the refined model.

To see why this is worthwhile, let us consider refinement from a machine learning perspective. The original model provides us with an approximate characterization of the dynamic behavior of some system. The refined model, by having more independent parameters, should allow us to obtain a better characterization of the same system. Instead, what we are looking for in fit-preserving data refinement is a model that leads to the same dynamics as the original model, but also accounts for the possible subspecies of species from the original model. Such a refinement is for example appropriate if we are already satisfied with the approximation given by the original model (i.e. it falls within measurement error), if existing experimental data is not enough to support the number of parameters of the refined model, or if fitting the refined model is unfeasible resource-wise. But even when we are looking for a better approximation, fit-preserving refinements can serve as *initialization for iterative algorithms* that estimate the parameters of the refined model. Furthermore, fit-preserving refinement enables us to construct a *hierarchy of models* to describe the same process using various levels of detail.

In this paper we provide a sufficient condition that links the refined parameters to those of the original model and guarantees that the refinement preserves the numerical fit of the original model. The constraints can easily be turned into an automatic procedure for setting the values of the refined parameters without the need for inspecting the ODEs.

Our result also addresses an open problem presented in [4]. Given a refined model for which some of the parameters are known or can be measured experimentally, the question is whether there exists a numerical setup for the unknown parameters such that the refined model preserves the numerical fit of the original model. Our result provides a partial answer to this question: as long as the values that are known do not already violate the fit-preservation constraints that we propose in this paper, there exists at least one solution and, moreover, the corresponding parameter values can be computed automatically.

Furthermore, we provide a more precise mathematical framework for data refinement by separating structural refinement (aimed at generating the refined reactions) from quantitative refinement, where the goal is to obtain at least as good a fit as that of the original model. In this framework, fit-preserving refinement becomes a special kind of quantitative refinement, where we look for a computationally efficient numerical setup at the expense of not improving the fit of the model (but not making it worse either).

The paper is structured as follows. Section 2 provides an introduction to chemical reaction networks and the formal notation that we use throughout the paper, roughly based on [2]. In Section 3 we formally discuss data refinement and state the main result of this paper. We apply our approach to a simple example in Section 4. We discuss the implications of our result in Section 5.

2 Reaction Networks

We first fix some notations used throughout the paper. We denote by $\mathbb{N}$ the set of non-negative integers and by $\mathbb{N}_+$ the set of positive integers. We denote by $\mathbb{R}_{\geq 0}$ the set of non-negative real numbers and by $\mathbb{R}_+$ the set of positive real numbers. For two sets X, Y we denote by X^Y the set of mappings $f : Y \to X$; for a finite set Y, X^Y is also the set of vectors of dimension $|Y|$ with elements from X. Throughout this paper we will always denote vectors with a lower-case bold-faced letter.

We consider in this paper only irreversible reactions; any reversible reaction will be replaced by its 'left-to-right' and 'right-to-left' directions. Reactions are typically denoted as rewriting rules, such as $2A_1 + A_2 \xrightarrow{k_1} A_3$, where k_1 denotes the kinetic rate constant. We formalize a reaction in the style of [2] by denoting the species on its left and right hand side, together with their stoichiometric coefficients, using a vectorial notation. Throughout the paper we denote by $\mathscr{S} = \{S_1, S_2, \ldots, S_m\}$ a finite alphabet whose elements we refer to as *species*. A vector in $\mathbb{N}^{\mathscr{S}}$ is called a *complex* over $\mathscr{S}$. Note that this notion of complex refers to a linear combination of species that may occur on either side of the reaction. It should not be confused with the concept of a chemical complex, which would be represented in our model through a single species.

With this notation, a reaction is defined as a pair of complexes $(\boldsymbol{c}, \boldsymbol{d}) \in \mathbb{N}^m \times \mathbb{N}^m$, standing for its left- and right-hand side complexes, resp.; the reaction will also be written as $\boldsymbol{c} \to \boldsymbol{d}$. For our example above, we define $\mathscr{S} = \{A_1, A_2, A_3\}$, $\boldsymbol{c} = [2, 1, 0]^T$ and $\boldsymbol{d} = [0, 0, 1]^T$. For a reaction $\boldsymbol{c} \to \boldsymbol{d}$, we use $k_{\boldsymbol{c} \to \boldsymbol{d}}$ to denote its *kinetic rate constant*.

We are now ready to give the formal definition of reaction networks.

Definition 1. *A reaction network is a tuple* $N = (\mathscr{S}, \mathscr{C}, \mathscr{R}, k)$, *where* $\mathscr{S}$ *is a finite set of species,* $\mathscr{C} \subseteq \mathbb{N}^{\mathscr{S}}$ *is a finite set of complexes,* $\mathscr{R} \subseteq \mathscr{C} \times \mathscr{C}$ *is a finite set of reactions and* $k : \mathscr{R} \to \mathbb{R}_{\geq 0}$ *gives the kinetic rate constant of each reaction from* $\mathscr{R}$.

Using the notation from [3], let us consider the following reaction network, consisting of n reactions and m species:

$$r_j : \sum_{i=1}^{m} c_{ij} A_i \xrightarrow{k_j} \sum_{i=1}^{m} d_{ij} A_i, \ for \ 1 \leq j \leq n . \tag{1}$$

The dynamic behavior of this network, under mass-action kinetics, see, e.g., [10], can be described by the following system of ODEs:

$$\dot{a}_i = \sum_{j=1}^{n} (d_{ij} - c_{ij}) k_j \prod_{q=1}^{m} a_q^{c_{qj}}, \ for \ 1 \leq i \leq m , \tag{2}$$

where $a_i : \mathbb{R}_{\geq 0} \to \mathbb{R}_{\geq 0}$ stands for the concentration of species A_i at time $t \geq 0$ and $\dot{a}_i$ is used for the derivative of a_i with respect to time. Our goal in what follows is to provide a more concise representation of (2) based on the vectorial notation introduced above. Given two vectors $x, y \in \mathbb{R}^n$, we will denote $x^y = \prod_{i=1}^{n} x_i^{y_i}$. With this, (2) can be rewritten as

$$\dot{a} = \sum_{c \to d \in \mathcal{R}} k_{c \to d} a^c (d - c) . \tag{3}$$

Consider now the initial conditions of our system of ODEs. The differential equation $\dot{x} = F(t, x)$ with initial conditions given by $x(t_0) = x_0$, is known to have a unique solution in a neighborhood of t_0 as long as the function F is continuously differentiable in a neighborhood of (t_0, x_0), see [8]. In the case of reaction networks that follow mass-action dynamics, e.g. equation (3) with initial condition $a(t_0) = a_0$, the function F is a polynomial, and so it is continuously differentiable on its entire domain. Furthermore, since there is no explicit dependence on time in (3), i.e. the system is *autonomous*, it follows that solutions are time-invariant in the following sense: given a solution a' for the initial condition $a'(0) = a_0$, the solution for the initial condition $a(t_0) = a_0$ can be written as $a(t) = a'(t - t_0)$. Thus, without any loss of generality, it suffices to only consider the problem with initial conditions specified at $t = 0$, say $a(0) = \alpha$. Moreover, since the solution depends (even continuously, see [8]) on the actual initial values α, we will make this explicit by writing the solution a as $a[\alpha]$ whenever its dependence on α is relevant.

3   Fit-Preserving Data Refinement

The data refinement of a reaction network is about adding some details into a network, e.g. through replacing one or more species of the network with a set of subspecies carrying more detailed and potentially differentiated behavior. In the general setting, we assume to have two sets of species $\mathcal{S}$ and $\mathcal{S}'$ and a relation $\rho \subseteq \mathcal{S} \times \mathcal{S}'$ that links each species from $\mathcal{S}$ to its corresponding subspecies in $\mathcal{S}'$. The intuition of species refinement is formally captured in Definition 2.

Definition 2. *Let $\mathscr{S}$ and $\mathscr{S}'$ be two sets of species. A relation $\rho \subseteq \mathscr{S} \times \mathscr{S}'$ is a species refinement relation iff it satisfies the following conditions:*

i) for each $A \in \mathscr{S}$ there exists $A' \in \mathscr{S}'$ such that $(A, A') \in \rho$;
ii) for each $A' \in \mathscr{S}'$ there exists exactly one $A \in \mathscr{S}$ such that $(A, A') \in \rho$.

Intuitively, when $(A, A'_1), \ldots, (A, A'_r)$ are all the elements of ρ with A on their left position, we mean that species A is refined and replaced in the refined model by its subspecies $A'_1, \ldots, A'_r$. Each species from the original model should be refined to at least one species in the refined model (more than one in the case of non-trivial refinements) and each species of the refined model should correspond to exactly one "parent" species from the original model.

The species refinement ρ can also be written as a matrix $\boldsymbol{M}_\rho \in \mathbb{R}^{\mathscr{S} \times \mathscr{S}'}$:

$$\boldsymbol{M}_\rho = (m_{A,A'})_{A \in \mathscr{S}, A' \in \mathscr{S}'}, \qquad with\ m_{A,A'} = \begin{cases} 1, & if\ (A, A') \in \rho\ ; \\ 0, & otherwise\ . \end{cases} \tag{4}$$

As a convention, we will denote matrices throughout this paper with upper-case, bold-faced letters.

Definition 3. *Let $\mathscr{S}$ and $\mathscr{S}'$ be two sets of species, and $\rho \subseteq \mathscr{S} \times \mathscr{S}'$ a species refinement relation. Let $\boldsymbol{c}$ and $\boldsymbol{c}'$ be two complexes over $\mathscr{S}$ and $\mathscr{S}'$, respectively. We say that $\boldsymbol{c}'$ is a ρ-refinement of $\boldsymbol{c}$ (or simply a refinement if ρ is clearly understood in that context) if, for every species $S \in \mathscr{S}$, the stoichiometric coefficients of its subspecies in $\boldsymbol{c}'$ add up to its stoichiometric coefficient in $\boldsymbol{c}$. This can be written using matrix notation as $\boldsymbol{M}_\rho \boldsymbol{c}' = \boldsymbol{c}$.*

We say that complex $\boldsymbol{c}$ is ρ-refined to the set of complexes $\mathscr{C}'$ over $\mathscr{S}'$ if all the elements of $\mathscr{C}'$ are ρ-refinements of $\boldsymbol{c}$. The set of all possible ρ-refinements of $\boldsymbol{c}$ is denoted by $\Delta_\rho(\boldsymbol{c})$.

Finally, we say that reaction $(\boldsymbol{c}, \boldsymbol{d})$ over $\mathscr{S}$ is ρ-refined to the set of reactions $\mathscr{R}'$ over $\mathscr{S}'$ if, for every $(\boldsymbol{c}', \boldsymbol{d}') \in \mathscr{R}'$, $\boldsymbol{c}'$ is a ρ-refinement of $\boldsymbol{c}$ and $\boldsymbol{d}'$ is a ρ-refinement of $\boldsymbol{d}$. The set of all possible ρ-refinements of $(\boldsymbol{c}, \boldsymbol{d})$ can be written as $\Delta_\rho(\boldsymbol{c}) \times \Delta_\rho(\boldsymbol{d})$.

Let us consider again the reaction from the previous section $2A_1 + A_2 \to A_3$. Assume we refine A_1 to B_{11} and B_{12}, A_2 to B_2 and A_3 to B_{31}, B_{32} and B_{33}, i.e. we consider the refinement relation $\rho = \{(A_1, B_{11}), (A_1, B_{12}), (A_2, B_2), (A_3, B_{31}), (A_3, B_{32}), (A_3, B_{33})\}$. Then the possible refinements of the original reaction are: $2B_{11} + B_2 \to B_{31}$, $2B_{11} + B_2 \to B_{32}$, $2B_{11} + B_2 \to B_{33}$, $B_{11} + B_{12} + B_2 \to B_{31}$, $B_{11} + B_{12} + B_2 \to B_{32}$, $B_{11} + B_{12} + B_2 \to B_{33}$, $2B_{12} + B_2 \to B_{31}$, $2B_{12} + B_2 \to B_{32}$ and $2B_{12} + B_2 \to B_{33}$.

With Definition 3 we are ready now to introduce the notion of reaction network refinement. Since our notion of a network has two components, the species-reaction structure and the ODE-based quantitative dynamics, our definition of network refinement is given in two parts: a structural refinement and a fit-preserving refinement. The former one is immediate following Definition 3.

Definition 4. *Let* $N = (\mathscr{S}, \mathscr{C}, \mathscr{R}, k)$ *and* $N' = (\mathscr{S}', \mathscr{C}', \mathscr{R}', k')$ *be two reaction networks and* $\rho \subseteq \mathscr{S} \times \mathscr{S}'$ *a species refinement relation. We say that* N' *is a structural refinement of* N *if*

$$\mathscr{C}' = \bigcup_{c \in \mathscr{C}} \Delta_\rho(c) \qquad and \qquad \mathscr{R}' \subseteq \bigcup_{c \to d \in \mathscr{R}} \Delta_\rho(c) \times \Delta_\rho(d) . \qquad (5)$$

We say that N' *is the* full structural refinement *of* N *if we have equality in the definition of* $\mathscr{R}'$ *in* (5).

The quantitative part of our notion of network refinement focuses on preserving the experimental data fit of the original model; in other words, the kinetic rate constants of the model obtained through structural refinement should be set so that the dynamics of a species in the original model is identical to the dynamics of the sum of its subspecies in the refined model. We formalize this condition in what follows.

Definition 5. *Let* $N = (\mathscr{S}, \mathscr{C}, \mathscr{R}, k)$ *and* $N' = (\mathscr{S}', \mathscr{C}', \mathscr{R}', k')$ *be two reaction networks and* $\rho \subseteq \mathscr{S} \times \mathscr{S}'$ *a species refinement relation; we denote by* M_ρ *the matrix representation of* ρ. *Let* $\alpha \in \mathbb{R}_{\geq 0}^{\mathscr{S}}$, $\beta \in \mathbb{R}_{\geq 0}^{\mathscr{S}'}$ *be the initial conditions for* N *and* N', *resp. We say that* β *is a* ρ-refinement *of* α *if* $\alpha = M_\rho\beta$.

Let the ODEs describing the two reaction networks be:

$$\dot{a} = \sum_{c \to d \in \mathscr{R}} k_{c \to d} a^c (d - c) \qquad and \qquad \dot{b} = \sum_{c' \to d' \in \mathscr{R}'} k'_{c' \to d'} b^{c'} (d' - c') , \qquad (6)$$

where $a : \mathbb{R}_{\geq 0} \to \mathbb{R}_{\geq 0}^{\mathscr{S}}$, $b : \mathbb{R}_{\geq 0} \to \mathbb{R}_{\geq 0}^{\mathscr{S}'}$

We say that N' *is a* fit-preserving refinement *of* N *if it is a structural refinement of* N *and, for any initial conditions* $\alpha \in \mathbb{R}_{\geq 0}^{\mathscr{S}}$ *and* $\beta \in \mathbb{R}_{\geq 0}^{\mathscr{S}'}$ *such that* β *is a* ρ-refinement of α, *the solutions* a *and* b *of equations* (6) *satisfy* $a(t) = M_\rho b(t)$ *in a neighborhood of* 0.

Note that in Definition 5 we can assume without loss of generality that N' is the full structural refinement of N. Indeed, any other structural refinement can be extended to the full one by adding to it the missing refined reactions and then setting their kinetic rate constants to 0.

Note also that we required that (6) holds for all values of t in a neighborhood of 0 and not for all $t \geq 0$, as in [3]. This formulation is enough for the formal result we are going to prove and, moreover, it is also applicable when the solution of the ODEs is not defined for all non-negative values of t. More precisely, the solution is unique for its full domain of existence.

The problem that we focus on is how to *effectively construct* a fit-preserving refinement of a given reaction network N when given the species refinement relation ρ. The first part is to build the full structural ρ-refinement of N; this can be done by constructing $\Delta_\rho(c) \times \Delta_\rho(d)$ for all reactions (c, d) of N. The second part is to set the kinetic rate constants of the refined model so that it yields a fit-preserving refinement.

To check whether a given numerical setup of the full structural refinement yields a fit-preserving refinement seems to require in general to solve the systems of ODEs in Equation (6). A difficulty is that those systems of ODEs are non-linear and cannot be solved analytically in general. We recall the following problem of [4].

Problem 1. [4] Let N be a reaction network, ρ a species refinement relation, and N' the full structural ρ-refinement of N. Assuming that numerical values of some of the kinetic rate constants of N' are fixed, find a numerical setup for all its other kinetic rate constants so that N' is a fit-preserving refinement of N.

The following result gives a partial answer, i.e. a sufficient condition, to Problem 1.

Theorem 1. *Let $N = (\mathscr{S}, \mathscr{C}, \mathscr{R}, k)$ be a reaction network, $\mathscr{S}'$ a set of subspecies and $\rho \subseteq \mathscr{S} \times \mathscr{S}'$ a species refinement relation. Let N' be the full structural ρ-refinement of N, $N' = (\mathscr{S}', \mathscr{C}', \mathscr{R}', k')$. If, for every $\boldsymbol{c} \to \boldsymbol{d} \in \mathscr{R}$ and for every $\boldsymbol{c}' \in \Delta_\rho(\boldsymbol{c})$, we have that*

$$\sum_{\boldsymbol{d}' \in \Delta_\rho(\boldsymbol{d})} k'_{\boldsymbol{c}' \to \boldsymbol{d}'} = \binom{\boldsymbol{c}}{\boldsymbol{c}'} k_{\boldsymbol{c} \to \boldsymbol{d}} \ , \ with \ \binom{\boldsymbol{c}}{\boldsymbol{c}'} = \frac{\prod_{i=1}^{|\mathscr{S}|} c_i!}{\prod_{j=1}^{|\mathscr{S}'|} c'_j!} \ , \tag{7}$$

then N' is a fit-preserving data refinement of N.

Due to space constraints we do no provide the proof here, the reader may find it in our technical report [6].

Note that the constraint (7) is not far from what one would expect. Indeed, the rate constants of all refined reactions that share the same left-hand side $\boldsymbol{c}'$ depend on the rate constant of the parent reaction, $k_{\boldsymbol{c} \to \boldsymbol{d}}$, and on its left hand side $\boldsymbol{c}$. The interesting aspect is the linear character of the dependency.

We can apply Theorem 1 for assigning rate constants to a given structural network refinement so that we obtain a fit-preserving refinement. Even if we are given a partially specified structural data refinement (i.e. one where several rate constants are fixed to predefined values, such as zero for reactions that are assumed impossible), we can turn it into a fit-preserving refinement as long as the fixed values do not already lead to a violation of Equation (7). This might happen because rate constants are non-negative so there is no way to reduce a sum that already exceeds the value prescribed by (7). In the absence of any information about the rate constants of the refined model, we can choose a symmetric assignment, i.e. all rate constants in any sum described by (7) are set to be equal, see also the discussion in the next section.

We can also apply Theorem 1 in order to check that a given structural data refinement is fit-preserving; this only gives a partial answer because the condition of Theorem 1 is only sufficient, but not always necessary. Indeed, it is shown in [2] that it is possible to have two reaction networks that differ only in the assignment of rate constants, but describe the same dynamics (translate to the same ODEs). Let $N_1 = (\mathscr{S}, \mathscr{C}, \mathscr{R}, k_1)$ and $N_2 = (\mathscr{S}, \mathscr{C}, \mathscr{R}, k_2)$ be two such networks. A fit-preserving data refinement constructed for N_1 based on Theorem 1 is also a fit-preserving data refinement of N_2, but it may fail to satisfy (7) for N_2.

4   A Simple Example

In this section we are going to use the constraints from Theorem 1 for the data refinement of a simple reaction network, corresponding to the Lotka-Volterra system for modeling prey-predator dynamics [11].

$$A \xrightarrow{k_1} 2A \qquad A + B \xrightarrow{k_2} 2B \qquad B \xrightarrow{k_3} \varnothing \tag{8}$$

In the equations, A stands for the prey and B for the predator. The set of species in this case is $\mathscr{S} = \{A, B\}$ and the set of complexes $\mathscr{C} = \{[1, 0]^T, [2, 0]^T, [1, 1]^T, [0, 2]^T, [0, 1]^T, [0, 0]^T\}$. Now, assume that we can distinguish two different kinds of prey. This corresponds to a species refinement $\rho = \{(A, A_1), (A, A_2), (B, B')\}$. The set of subspecies is $\mathscr{S}' = \{A_1, A_2, B'\}$.

The full structural ρ-refinement of the model (with the same notation as in [3]) can be written as

$$
\begin{array}{lll}
A_1 \xrightarrow{r_1} 2A_1 \;, & A_1 \xrightarrow{r_2} A_1 + A_2 \;, & A_1 \xrightarrow{r_3} 2A_2 \;, \\
A_2 \xrightarrow{r_4} 2A_1 \;, & A_2 \xrightarrow{r_5} A_1 + A_2 \;, & A_2 \xrightarrow{r_6} 2A_2 \;, \\
A_1 + B' \xrightarrow{r_7} 2B' \;, & A_2 + B' \xrightarrow{r_8} 2B' \;, & B' \xrightarrow{r_9} \varnothing \;.
\end{array}
\tag{9}
$$

If we identify each reaction by the name of its rate constant, we can write the corresponding refinement of each reaction from the original model: $k_1 \to \{r_1, r_2, r_3, r_4, r_5, r_6\}, k_2 \to \{r_7, r_8\}, k_3 \to \{r_9\}$.

The sums given by Equation (7) in this case are the following:

$$
r_1 + r_2 + r_3 = \begin{pmatrix} [1, 0]^T \\ [1, 0, 0]^T \end{pmatrix} k_1 = k_1 \;, \quad r_4 + r_5 + r_6 = \begin{pmatrix} [1, 0]^T \\ [0, 1, 0]^T \end{pmatrix} k_1 = k_1 \;,
$$

$$
r_7 = \begin{pmatrix} [1, 1]^T \\ [1, 0, 1]^T \end{pmatrix} k_2 = k_2 \;, \quad r_8 = \begin{pmatrix} [1, 1]^T \\ [0, 1, 1]^T \end{pmatrix} k_2 = k_2 \;, \tag{10}
$$

$$
r_9 = \begin{pmatrix} [0, 1]^T \\ [0, 0, 1]^T \end{pmatrix} k_3 = k_3 \;.
$$

Equation (10) can now be used in a versatile way to construct very different refined models. For example, assume that all reactions involving both subspecies A_1 and A_2 should be eliminated from the refined model based on biological arguments. We can easily remove them from the model and still obtain a fit-preserving refinement by taking $r_2 = r_3 = r_4 = r_5 = 0$ and then consider the problem of finding values for the remaining rate constants so that (10) is satisfied. In this particular case, the solution obtained is $r_1 = r_6 = k_1$, $r_7 = r_8 = k_2$, and $r_9 = k_3$.

As another example, assume that we may allow reactions involving both A_1 and A_2, but with no quantitative distinction between them; in this case the constants in each of the sums of Equation (10) could be set equal, as follows, and as set also in [3]: $r_1 = r_2 = r_3 = k_1/3, r_4 = r_5 = r_6 = k_1/3, r_7 = k_2, r_8 = k_2, r_9 = k_3$.

A similar approach can be taken when some of the rate constants of the refined model are known or can be measured/estimated experimentally. In such cases, for each sum in (7), all known rate constants are subtracted, and the remainder, if non-negative, can be split (perhaps evenly) between the unknown rate constants. If, e.g., we know from the literature that $r_1 = r_4 = k_1/2$, then we can obtain a fit-preserving refinement by choosing any value for the remaining constants so that $r_2 + r_3 = k_1/2$ and $r_5 + r_6 = k_1/2$.

5   Conclusions

In this paper we introduced a new mathematical framework for the notion of reaction network model and model refinement. We focused on the problem of finding fit-preserving refinements and we proposed a sufficient condition for the rate constants of a structural refinement of a reaction network so that the resulting model describes the same dynamics as the original model (with respect to the species of the original model). Our result is versatile and can be combined with partial information about some of the rate constants of the refined model. In particular, this leads to an algorithmic assignment of rate constant values that can be performed automatically and be used as initialization for parameter estimation software.

Since the constraint (7) of Theorem 1 only partitions the rate constants of the refined model into sets that should add up to particular values that depend on the original model, this means that in general there is still a lot of freedom left in the choice of the actual values. On one hand, this is useful for the initialization of parameter estimation algorithms, since it provides room for randomization in order to avoid local optima. On the other hand, this also raises an interesting question for further research, namely whether among all possible assignments there are some that offer other desirable properties in addition to fit-preservation. Such properties may, for example, be based on the internal structure of species and some desired conservation laws and, in such cases, the symmetric assignment of the remaining values might no longer be the most appropriate thing to do.

An implicit assumption in our considerations is that the kinetic constants are fixed throughout the model dynamics. This may not be the case if some of them depend on temperature and the model includes, e.g., some exothermic reactions. It seems an interesting question whether our approach can be extended to such models.

Our result also provides a partial answer to the open question of [4] on the existence of rate constant values that can turn a structural refinement (possibly with some of the values fixed in advance based on existing literature) into a fit-preserving one with respect to the original model. The answer that follows from Theorem 1 is that such an assignment is guaranteed to exist, as long as the already fixed values do not lead to a violation of the fit-preservation constraints (this can happen for example if the sum of the fixed values already exceeds the required value, since the rate constants can only take non-negative values).

One possible use of fit-preserving refinement that we have mentioned in the introduction of this paper is the construction of hierarchical models to capture different levels of detail for the same biological process. Refinement in this case allows us to go from one model to a more detailed one. The fit-preservation constraints, on the other hand, enable us to go backward, i.e. obtain a more general model from a detailed one. In this context, it is also interesting to investigate whether it is also possible to retrieve the more general model after the detailed one has been refitted (and the fit-preservation constraints may not hold anymore).

The proof of our result relies on the interplay between multinomial expansion and the formulation of mass-action dynamics to yield very simple constraints. On the other hand, it would be interesting to investigate fit-preserving data refinement for other kinetic models as well. Another problem that remains open is to find necessary conditions for the numerical setup of refined models, thus aiming for a full solution of Problem 1.

References

1. Back, R.J., von Wright, J.: Refinement Calculus. Graduate Texts in Computer Science. Springer (1998)
2. Craciun, G., Pantea, C.: Identifiability of chemical reaction networks. Journal of Mathematical Chemistry 44(1), 244–259 (2008)
3. Czeizler, E., Czeizler, E., Iancu, B., Petre, I.: Quantitative model refinement as a solution to the combinatorial size explosion of biomodels. Electronic Notes in Theoretical Computer Science 284, 35–53 (2012)
4. Czeizler, E., Rogojin, V., Petre, I.: The phosphorylation of the heat shock factor as a modulator for the heat shock response. IEEE/ACM Transactions on Computational Biology and Bioinformatics 9(5), 1326–1337 (2012)
5. Danos, V., Feret, J., Fontana, W., Harmer, R., Krivine, J.: Rule-based modelling, symmetries, refinements. In: Fisher, J. (ed.) FMSB 2008. LNCS (LNBI), vol. 5054, pp. 103–122. Springer, Heidelberg (2008)
6. Gratie, C., Petre, I.: Fit-preserving data refinement of mass-action reaction networks. Technical report, TUCS (2014)
7. Gratie, D.E., Iancu, B., Azimi, S., Petre, I.: Quantitative model refinement in four different frameworks, with applications to the heat shock response (2013) (submitted)
8. Hirsch, M.W., Smale, S., Devaney, R.L.: Differential equations, dynamical systems and an introduction to chaos, 2nd edn. Pure and Applied Mathematics, vol. 60. Academic Press (2004)
9. Iancu, B., Czeizler, E., Czeizler, E., Petre, I.: Quantitative refinement of reaction models. International Journal of Unconventional Computing 8(5-6), 529–550 (2012)
10. Klipp, E., Herwig, R., Kowald, A., Wierling, C., Lehrach, H.: Systems Biology in Practice. WILEY-VCH Verlag GmbH & Co. KGaA, Weinheim (2005)
11. Volterra, V.: Variation and fluctuations of the number of individuals of animal species living together. In: Animal Ecology, pp. 409–448. McGraw-Hill (1931)

On Maximal Block Functions
of Computable η-like Linear Orderings

Charles M. Harris

Department of Mathematics,
University of Bristol,
Bristol BS8 1TW, U.K.

Abstract. We prove the existence of a computable η-like linear ordering $\mathscr{L}$ such that, for any Π_2^0 function $G : \mathbb{Q} \to \mathbb{N} \setminus \{0\}$ and linear ordering $\mathscr{B} \cong \mathscr{L}$, $\mathscr{B}$ does not have order type $\tau = \sum \{\, G(q) \mid q \in \mathbb{Q} \,\}$.

Keywords: Computable, linear ordering, η-like, Π_2^0, maximal block.

1   Introduction

Given their relative simplicity, the study of η-like linear orderings has attracted attention as a preliminary test case for obtaining general results for computable linear orderings. An example of this is Kierstead's [Kie87] construction of a computable linear ordering of order type $2 \cdot \eta$ with no nontrivial Π_1^0 automorphism, and subsequent conjecture that every computable copy of a computable linear ordering $\mathscr{L}$ has a strongly nontrivial[1] Π_1^0 automorphism if and only if the order type τ of $\mathscr{L}$ contains an interval of order type η. This conjecture is supported by the Theorem in [DM89] that every computable discrete linear ordering $\mathscr{L}$ has a computable copy with no strongly nontrivial Π_1^0 self embedding. In the context of η-like linear orderings, Downey and Moses deduced that Kierstead's result for the order type $2 \cdot \eta$ can be generalised to the case of any η-like order type[2] τ

---

[1] Kierstead defines an automorphism f of a linear ordering $\mathscr{L}$ to be *fairly trivial* if it is nontrivial but maps every element x to an element y with $[x, y]$ finite and f to be *strongly nontrivial* if it is neither trivial nor fairly nontrivial. Note that if $\mathscr{L}$ is η-like then any nontrivial automorphism of $\mathscr{L}$ is strongly nontrivial.

[2] The powerful *choice set* method—a choice set of a linear ordering is a set containing precisely one element from each maximal block—used by Moses and Downey in the context of embeddings of discrete linear orderings [DM89], would need to be modified in order to be applicable to *any* such order type τ. Indeed suppose that $\mathscr{L}$ is a computable η-like linear ordering with a strongly η-like interval. Choose elements $a <_{\mathscr{L}} b$ in one such interval such that a is the rightmost and b the leftmost element of its respective maximal block, and such that, for some n, the interval (a, b) contains infinitely many maximal blocks of size n and no maximal block of size $m > n$. Then the set of leftmost elements (and, in fact, for any $1 \leq i \leq n$, the set of i to *leftmost* elements) of the maximal blocks of size n in the interval (a, b) forms an infinite Σ_2^0 set. Hence any construction that diagonalises against Σ_2^0 subsets of a choice set containing, for example, the leftmost element in each maximal block, will not be applicable if $\mathscr{L}$ contains a strongly η-like interval. (Note that a proof based on similar techniques to those used in [Kie87] can be applied in this context—see [LHC14].)

A. Beckmann, E. Csuhaj-Varjú, and K. Meer (Eds.): CiE 2014, LNCS 8493, pp. 214–223, 2014.
© Springer International Publishing Switzerland 2014

provided that τ has a Π_2^0 maximal block function and no interval of order type η. This last result is the starting point of the present paper as it prompts the question of whether it can be applied to the whole class of computable η-like linear orderings and hence, in particular, of whether every computable η-like linear ordering $\mathscr{L}$ has a copy $\mathscr{B}$ with a Π_2^0 maximal block function. We answer this question in the negative in Theorem 2 by constructing a counterexample $\mathscr{B}$ via a diagonalisation argument applied using the basic properties of isomorphisms of linear orderings in this context. We note that this solves a question mentioned by several authors including Fellner [Fel76], Lerman and Rosenstein [LR82] and Downey and Moses [DM89].

2 Preliminaries

We assume $\{W_e\}_{e \in \mathbb{N}}$ to be a standard listing of c.e. sets with associated c.e. approximation $\{W_{e,s}\}_{e,s \in \mathbb{N}}$. $\emptyset'$ denotes the standard halting set for Turing machines in this context, i.e. the set $\{c \mid e \in W_e\}$ and $\mathbf{0}'$ denotes the Turing degree of $\emptyset'$. We suppose $q_0, q_1, q_2, \ldots$ to be a fixed computable listing of $\mathbb{Q}$. We also assume $\langle x, y \rangle$ to be a standard computable pairing function over $\mathbb{N}$ extended to use over $\mathbb{Q}$ via the above listing. $\{D_n\}_{n \in \mathbb{N}}$ denotes the canonical computable listing of all finite sets of nonnegative integers. Note that under this listing, for any $m, n \in \mathbb{N}$, if $D_m \subseteq D_n$ then $m \leq n$.

For any set X, we use $|X|$ to denote the cardinality of X. For any function[3] F with domain and range in $\mathbb{N}$ or $\mathbb{Q}$ we use $G(F)$ to denote the set $\{\langle x, y \rangle \mid F(x) \downarrow = y\}$, i.e. the graph of F coded into $\mathbb{N}$ via the pairing function $\langle \cdot, \cdot \rangle$. (Note that in this context we identify a pair (x, y) with its code $\langle x, y \rangle$ so that, for example, the shorthand $G(F) \subseteq \mathbb{Q} \times \mathbb{N}$ makes sense.) We define F to be Γ, for some predicate of sets Γ, if $G(F) \in \Gamma$.

Note 1. Any[4] Σ_2^0 function F with domain $\mathbb{Q}$ and codomain $\mathbb{N}$ is Δ_2^0. Indeed using a $\mathbf{0}'$ oracle we can compute $F(q)$, for any $q \in \mathbb{Q}$, as the number n found by enumerating $G(F)$ until we find $\langle x, n \rangle$ with $x = q$.

Let $\mathscr{L} = \langle L, <_{\mathscr{L}} \rangle$ be a linear ordering. We call $S \subseteq L$ an *interval* if, for all $a, b \in S$, and any c that lies $<_{\mathscr{L}}$ between a and b, c is also in S. Notice that S does not necessarily have endpoints. Note that we also use the term *interval* in direct reference to the order type of $\mathscr{L}$ with obvious meaning. For any $a, b \in L$, we say that a, b are *finitely far apart*—written $a \sim^* b$—if the interval S of elements lying between a and b is finite. (By definition $S = \emptyset$ if $a = b$.) Note that $\sim^*$ is an equivalence relation. If $\mathscr{L}$ is countably infinite we define $\mathscr{L}$ to be η-*like* if $\{c \mid c \sim^* a\}$ is finite for all $a \in L$ or, equivalently, if $\mathscr{L}$ has order type $\tau = \sum\{F(q) \mid q \in \mathbb{Q}\}$ for some function $F : \mathbb{Q} \to \mathbb{N} \setminus \{0\}$. We call any finite

---

[3] We use the convention here and in further work that maximal block functions are usually denoted using upper case letters whereas automorphisms of linear orderings are usually denoted using lower case letters.

[4] This is a particular case of the same (standard) observation generalised from $n = 2$ to any $n \geq 1$, when the domain (and codomain) of F are computable.

interval in $\mathscr{L}$ a *block* and we call the equivalence classes under $\sim^*$ *maximal* blocks. If $\mathscr{L}$ is η-like we call such a function F a *maximal block function* of $\mathscr{L}$. We say that $\mathscr{L}$ is *strongly η-like* if in addition F has finite range (i.e. the maximal block size is bounded). For any maximal block I of size $p \geq 1$ (written $|I| = p$) we use terminology of the form $I = k_1 <_{\mathscr{L}} \cdots <_{\mathscr{L}} k_p$ to denote I and we call k_1 (k_p) the *leftmost* (*rightmost*) element of I. If $\mathscr{A} = \langle A, <_{\mathscr{A}} \rangle$ is a countably infinite linear ordering we assume that $A = \mathbb{N}$ and derive a listing $a_0, a_1, a_2, \ldots$ of A computable in $<_{\mathscr{A}}$. We say that $\mathscr{A}$ is *computable* if $<_{\mathscr{A}}$ is computable.

We assume the reader to be conversant with the Arithmetical Hiearchy and Turing reducibility ($\leq_{\mathrm{T}}$). We refer the reader to [Odi89] for further background and notation in computability theory and to [Dow98] for a review of computability thoeretic results in the context of linear orders.

3   The Complexity of Maximal Block Functions

Fellner determined a bound for the arithmetical complexity of maximal block functions of a computable η-like linear ordering.

Theorem 1 ([Fel76]). *If $\mathscr{B}$ is a computable η-like linear ordering then there is a Δ_3^0 function F such that $\mathscr{B}$ has order type $\tau = \sum \{\, F(q) \mid q \in \mathbb{Q} \,\}$.*

Our present concern is with the extent to which the bound in Theorem 1 can be tightened. However before proceeding we need to take into account that care is needed when dealing with the notion of maximal block functions for η-like linear orderings.

Note 2. Let $\mathscr{A}$ be an η-like linear ordering. Then $\mathscr{A}$ may have many different maximal block functions. For example, if $\mathscr{A}$ contains maximal blocks of size $n + 1$ for all $n \geq 0$ then, for each $n \geq 0$ we can define a distinct maximal block function F_n for $\mathscr{A}$ such that $F_n(q_0) = n + 1$.

Note 3. If $\mathscr{A}$ is an η-like linear ordering and F is a maximal block function of $\mathscr{A}$ we say that a listing $I(0), I(1), I(2), \ldots$ of maximal blocks of $\mathscr{A}$ is an *assignment* of F to $\mathscr{A}$ if $F(q_n) = |I(n)|$ for all $n \geq 0$. Note that there may be many different such assignments of F to $\mathscr{A}$. For example[5], suppose that $\mathscr{A}$ is made up of sets of maximal blocks of size 2 and 3 and that each of these sets is dense (in the standard sense) in $\mathscr{A}$. Let $I(0), I(1), I(2), \ldots$ be some listing of the maximal blocks of $\mathscr{A}$ and let $I(i_0), I(i_1), I(i_2), \ldots$ be a sublisting of maximal blocks of size 2. Then we can define a block function F of $\mathscr{A}$ with $F(q_0) = 2$ such that for every $k \geq 0$ there is a distinct assignment $I_k(0), I_k(1), I_k(2), \ldots$ of F to $\mathscr{A}$ such that $I_k(0) = I(i_k)$.

Our next Lemma restates a well known property of the class of $\Pi_2^0 sets$, originally proved by Jockusch [Joc68], in a form directly applicable to our main Theorem.

---

[5] An even easier but less interesting example of this phenomemon is when $\mathscr{A}$ has order type $n \cdot \eta$ for some $n \geq 1$.

Lemma 1. *There exists a computable listing $\{U_e\}_{e \in \mathbb{N}}$ of the Π_2^0 sets with associated computable approximation $\{U_{e,s}\}_{e,s \in \mathbb{N}}$ satisfying, for all $e \geq 0$, $U_e = \{x \mid \forall t(\exists s \geq t)[x \in U_{e,s}]\}$ and such that, for any finite sets $E_0, \ldots, E_e$ with $E_i \subseteq U_i$ for all $0 \leq i \leq e$, there exist infinitely many stages s such that $E_i \subseteq U_{i,s}$ for all $0 \leq i \leq e$.*

We now proceed to our main Theorem.

Theorem 2. *There exists a computable linear ordering $\mathscr{L}$ of order type $\kappa = \sum\{F(q) \mid q \in \mathbb{Q}\}$ such that $F : \mathbb{Q} \to \mathbb{N} \setminus \{0\}$, and such that, for any Π_2^0 function $G : \mathbb{Q} \to \mathbb{N} \setminus \{0\}$ and linear ordering $\mathscr{B} \cong \mathscr{L}$, $\mathscr{B}$ does not have order type $\tau = \sum\{G(q) \mid q \in \mathbb{Q}\}$.*

Note. By Note 1 we can replace "Π_2^0" by "$\Sigma_2^0 \cup \Pi_2^0$" in the statement of Theorem 2. Notice that, taken in conjunction with Theorem 1, this implies that any *computable* $\mathscr{B} \cong \mathscr{L}$ has a properly Δ_3^0 maximal block function. In particular we will see that this is the case for the function F constructed below.

Proof. Assume $\{U_e\}_{e \in \mathbb{N}}$ to be a standard listing of the class of Π_2^0 sets with associated computable Π_2^0 approximation $\{U_{e,s}\}_{e,s \in \mathbb{N}}$ as prescribed by Lemma 1. The construction aims to construct $\mathscr{L}$ of order type $\sum\{F(q) \mid q \in \mathbb{Q}\}$ such that $F : \mathbb{Q} \to \mathbb{N} \setminus \{0\}$ and such that F satisfies, for all $e \in \mathbb{N}$, the following requirements:

$$R_e : (\forall k, j \leq e)[\langle q_k, F(e) \rangle \notin U_j \vee \exists m \exists l[m \neq l \ \& \ \langle q_k, m \rangle \in U_j \ \& \ \langle q_k, l \rangle \in U_j]].$$

We shall see in the course of the verification below that satisfaction of $\{R_e\}_{e \in \mathbb{N}}$ ensures that, for any $j \in \mathbb{N}$, if U_j is the graph of a maximal block function G_j and $\mathscr{B}$ is a linear ordering of order type $\gamma = \sum\{G_j(q) \mid q \in \mathbb{Q}\}$, then $\mathscr{B} \not\cong \mathscr{L}$.

For clarity, we use $<_\mathbb{Q}$ and $<_\mathbb{N}$ when we need to differentiate between the respective standard orderings of $\mathbb{Q}$ and $\mathbb{N}$. Our aim is to construct a computable linear ordering $\mathscr{L} = \langle L, <_\mathscr{L} \rangle$ with domain $L = \mathbb{N}$ arranged in a set of maximal blocks $\{I(n) \mid n \in \mathbb{N}\}$ such that, for all $n \geq 0$, $F(q_n) = |I(n)|$ and also such that $I(n)$ is ordered relative to $\{I(k) \mid k \neq n\}$ as q_n is ordered relative to $\{q_k \mid k \neq n\}$; i.e. under our present terminology, such that the listing $I(0), I(1), I(2), \ldots$ is an assignment of F to $\mathscr{L}$.

We will proceed by stages $s \geq 0$ defining a finite linear ordering $\mathscr{L}_s = \langle L_s, <_\mathscr{L}^s \rangle$ at stage s such that, for some $n_s, r_s \geq 0$, $L_s = \mathbb{N} \restriction n_s$ and such that $\mathscr{L}_s$ is arranged as a finite set of blocks $\{I(n, s) \mid n < r_s\}$ where, for all $n < r_s$, $I(n, s)$ is the s stage approximation to maximal block $I(n)$. We say that n is the *label* of $I(n, s)$ and use this terminology quite generally in order to distinguish this use of $\mathbb{N}$ from our use of $\mathbb{N}$ as the domain of the linear ordering. The ordering $<_\mathscr{L}^s$ is defined by the internal ordering applied within each block and the ordering between blocks dictated by $<_\mathbb{Q}$ over $\{q_n \mid n < r_s\}$. Note that, in the construction, at any stage $s \geq 0$, if $I(n, s) \neq \emptyset$, then for any elements $k, m \in I(n, s)$, $k <_\mathscr{L}^s m \Leftrightarrow k <_\mathbb{N} m$. In other words the internal ordering of blocks always coincides with the natural ordering of $\mathbb{N}$. During the construction, for any stage s, and $k, m \in L_s$, if $k <_\mathscr{L}^s m$ then $k <_\mathscr{L}^t m$ for all $t \geq s$. Hence we will in general use $<_\mathscr{L}$ as shorthand for $<_\mathscr{L}^s$.

We choose some $d > 0$ as a default maximal block size for the construction.

Notation. During the construction we use the term *new* to refer to any finite set of numbers S which has not yet been enumerated into L at the present point in the construction and which is the minimal such set of cardinality $|S|$.

Block Rebuilding. At stage $s + 1$ we may want to rebuild block $I(n, s)$ for some $n < r_s$. This means that there are distinct integers[6] $\hat{n} \geq 0$ and $\hat{m} > 0$ such that $|I(n, s)| = \hat{n}$ whereas we need $|I(n, s + 1)| = \hat{m}$. We then proceed as follows according to whichever of the two cases below applies.

(a) $\hat{n} > \hat{m}$. In this case we search for the least labels $b_1, \ldots, b_{\hat{n}-\hat{m}} \geq r_s$ such that $q_n <_{\mathbb{Q}} q_{b_1} <_{\mathbb{Q}} \cdots <_{\mathbb{Q}} q_{b_{\hat{n}-\hat{m}}} <_{\mathbb{Q}} q_a$ where $I(a, s)$ is the successor block to $I(n, s)$ in $\mathscr{L}_s$ (so that $q_n <_{\mathbb{Q}} q_a$) or q_a is simply any rational to the right of q_n if no such successor block exists. We define $S = \{ b_j \mid 1 \leq j \leq \hat{n} - \hat{m} \}$ and $T = \{ d \mid r_s \leq d \leq \max S \}$. Suppose that $k_1 <_{\mathscr{L}} \cdots <_{\mathscr{L}} k_{\hat{n}-\hat{m}}$ are the $\hat{n} - \hat{m}$ rightmost elements in $I(n, s)$. We remove $\{k_1, \ldots, k_{\hat{n}-\hat{m}}\}$ from $I(n, s)$ to obtain $|I(n, s + 1)| = \hat{m}$ and proceed as follows. We firstly construct $I(b_1, s + 1)$ by constructing it as the singleton block consisting of k_1 if $d = 1$ and otherwise we define it as $k_1 <_{\mathscr{L}} \hat{p} <_{\mathscr{L}} \cdots <_{\mathscr{L}} \hat{p} + d - 2$ (i.e. as a block of d elements) with $\{\hat{p}, \ldots, \hat{p} + d - 2\}$ a set of $d - 1$ new numbers. We then proceed for each k_j such that $1 < j \leq \hat{n} - \hat{m}$ by constructing $I(b_j, s + 1)$ in a similar fashion. Finally, for all $b \in T \setminus S$ we construct $I(b, s + 1)$ using d new numbers. (Note that each $I(b, s + 1)$ is inserted into $\mathscr{L}_{s+1}$ according to q_b's position under $<_{\mathbb{Q}}$ relative to $\{ q_n \mid n < r_s \} \cup \{ q_m \mid m \in T \} \setminus \{q_b\}$.)

(b) $\hat{n} < \hat{m}$. In this case, supposing that $I(n, s) = k_1 <_{\mathscr{L}} \cdots <_{\mathscr{L}} k_{\hat{n}}$ we choose a new set of $\hat{m} - \hat{n}$ numbers $\{\hat{p}, \ldots, \hat{p} + r\}$ where $r = \hat{m} - \hat{n} - 1$ and define $I(n, s + 1) = k_1 <_{\mathscr{L}} \cdots <_{\mathscr{L}} k_{\hat{n}} <_{\mathscr{L}} \hat{p} <_{\mathscr{L}} \cdots <_{\mathscr{L}} \hat{p} + r$.

Note that, as this is the only rebuilding process applied during the construction we will be able to see, by inspection of the construction, that the following two conditions hold.

(i) For any $n \geq 0$, $\hat{m} > 0$ and stages $0 \leq s \leq t$, such that $|I(n, r)| \geq \hat{m}$ for all $s \leq r \leq t$, the block consisting of the $\hat{m}$ leftmost (i.e. least) elements in $I(n, t)$ is the same as that in $I(n, s)$.

(ii) For any $n, b, k \geq 0$ and stages $t > s \geq 0$, if k is removed from $I(n, s)$ at stage $s + 1$ due to rebuilding and inserted into $I(b, s + 1)$ as described above, then $k \in I(b, t)$. Note that this follows from (i) as k is the least number in $I(b, s + 1)$. In other words any number can move from one block to another at most once.

The Diagonalisation Witness and Domain for R_e. For $s \geq e$, witness $m(e, s)$ for R_e is the construction's guess as to a number such that $\langle q_k, m(e, s) \rangle \notin U_j$, for all $0 \leq k, j \leq e$ such that U_j is the graph of a function $\mathbb{Q} \to \mathbb{N} \setminus \{0\}$. For all

---

[6] In fact $\hat{n} > 0$ for any such $n \leq s$ with the possible exception of $n = s$ at early stages of the construction.

stages $s > e$, $m(e,s) = |I(e,s)| = F_s(e)$, where F_s is the s stage approximation to F. The set of *diagonalisation pairs* for index e is defined to be

$$P^e = \{(i,j) \mid 0 \le i,j \le e\}.$$

Thus $|P^e| = (1+e)^2$. Letting $x_0^e, \ldots, x_{(1+e)^2-1}^e$ be the computable ordering of P^e induced by the standard pairing function $\langle \cdot, \cdot \rangle$ we have the computable listing

$$D_0^e, \ldots, D_{2^{(1+e)^2}-1}^e$$

of all subsets of P^e defined using the canonical listing of finite sets $\{D_i\}_{i \in \mathbb{N}}$ specified above. (Note that $D_0^e = \emptyset$ under this listing.) It is important to reiterate here that this means (by definition of the latter listing) that, for all $i, j \ge 0$,

$$D_i^e \subseteq D_j^e \quad \Rightarrow \quad i \le j.$$

We now define the *diagonalisation domain* for R_e to be:

$$Z_e = X_0^e \cup \ldots \cup X_{2^{(1+e)^2}-1}^e$$

where, for each $0 \le i \le 2^{(1+e)^2} - 1$, X_i^e is an interval of numbers associated with D_i^e such that (i) $|X_i^e| \ge |D_i^e| + 1$ (for reasons explained below) and (ii) for $i \ne 0$, $\min X_i^e > \max X_{i-1}^e$. To do this, for simplicity we define X_i^e such that $|X_i^e| = (1+e)^2 + 1$ for every $0 \le i \le 2^{(1+e)^2} - 1$ by defining:

$$X_i^e = \{i(1+e)^2 + (i+1), \ldots, (i+1)(1+e)^2 + (i+1)\}.$$

Accordingly Z_e is the interval $\{1, \ldots, |Z_e|\}$ partitioned by the X_i^e and having cardinality $|Z_e| = 2^{(1+e)^2}((1+e)^2 + 1)$.

The point here is that, at stage s we choose an index $i(e,s)$ for e in such a way that, $(k,j) \in D_{i(e,s)}$ if and only if there exists at most one number $n \in Z_e$ such that $\langle q_k, n \rangle \in U_{j,s}$. Since $|X_{i(e,s)}^e| \ge |D_{i(e,s)}| + 1$ we know that

$$X_{i(e,s)} \setminus \{r \mid r \in Z_e \ \& \ (\exists(k,j) \in D_{i(e,s)})[\langle q_k, r \rangle \in U_{j,s}]\} \ne \emptyset$$

so that we can define the s stage witness $m(e,s)$ to be a number in this set.

The Construction

Stage $s = 0$.

Set $\mathscr{L}_0 = \langle L_0, <_{\mathscr{L}} \rangle$ with $L_0 = <_{\mathscr{L}} = \emptyset$. Define $I(n,0) = \emptyset$, $m(n,0) = 0$ and let $i(n,0)$ be undefined for all $n \ge 0$. Set $n_0 = r_0 = 0$.

Stage $s + 1$.

We suppose that n_s and r_s are such that $L_s = \mathbb{N} \restriction n_s$ and $\{n \mid I(n,s) \ne \emptyset\} = \mathbb{N} \restriction r_s$. There are now $s + 1$ substages $0 \le e \le s$ as follows.

Substage e.

Process requirement R_e as follows. Define $i(e, s+1) \in \mathbb{N}{\upharpoonright}2^{(1+e)^2}$ to be the index l satisfying

$$(k, j) \in D_l^e \Leftrightarrow |\{\langle q_k, r \rangle \mid r \in Z_e\} \cap U_{j,s+1}| \le 1$$

for all $0 \le k, j \le e$. Let $i = i(e, s+1)$. If $i = 0$—i.e. if $D_i^e = \emptyset$—define $M^e(s+1) = X_i^e$. Otherwise define $M^e(s+1)$ as follows.

Notation. The *individual out-age* of $m \in X_i^e$ relative to any pair $(k, j) \in D_i^e$ at stage $s+1$—denoted $b^e(m, (k, j), s+1)$—is defined to be 0 if $\langle q_k, m \rangle \in U_{j,s+1}$ and otherwise is defined to be the greatest $0 < r \le s+1$ such that $\langle q_k, m \rangle \notin U_{j,t}$ for all $(s+1) - r < t \le s+1$. The *out-age* of $m \in X_i^e$ relative to D_i^e at stage $s+1$ is defined to be

$$a^e(m, s+1) = \min\{b^e(m, (k, j), s+1) \mid (k, j) \in D_i^e\}.$$

Define

$$M^e(s+1) = \{m \mid m \in X_i^e \ \& \ (\forall n \in X_i^e)[a^e(m, s+1) \ge a^e(n, s+1)]\}, \quad (1)$$

i.e. $M^e(s+1)$ contains all $m \in X_i^e$ of maximal out-age relative to D_i^e. (Note that, by construction, $a^e(m, s+1) = a^e(n, s+1) > 0$ for any $n, m \in M^e(s+1)$. Notice also that $M^e(s+1)$ contains precisely those numbers $m \in X_i^e$ for which it appears most likely that $\langle q_k, m \rangle \notin U_j$ for all $(k, j) \in D_i^e$.)

Define the $s+1$ stage witness $m(e, s+1) = \min M^e(s+1)$. If $m(e, s+1) = m(e, s)$ do nothing (so that $I(e, s+1) = I(e, s)$). Otherwise rebuild $I(e, s+1)$ from $I(e, s)$ as described under the *Block Building* above with[7] $\hat{n} = |I(e, s)|$ and $\hat{m} = m(e, s+1)$—so that $|I(e, s+1)| = m(e, s+1)$.

Ending Substage e.

If $e < s$ proceed to substage $e+1$. If $e = s$ define $\mathscr{L}_{s+1} = \langle L_{s+1}, <_{\mathscr{L}} \rangle$ as follows. Let $I(n, s+1) = I(n, s)$ for any labels $s < n < r_s$, i.e. for n such that $I(n, s)$ was a block in $\mathscr{L}_s$ but such that the block $I(n, s)$ was not rebuilt during one of the substages $0 \le e \le s$ during this stage. Set $r_{s+1} = \max\{n \mid I(n, s) \ne \emptyset\} + 1$, and $n_{s+1} = \max\{m \mid (\exists n < r_{s+1})[m \in I(n, s+1)]\} + 1$. Define $L_{s+1} = \mathbb{N}{\upharpoonright}n_{s+1}$ and define $<_{\mathscr{L}}$ as dictated by the arrangement of the blocks $\{I(n, s+1) \mid n < r_{s+1}\}$ in $\mathscr{L}_{s+1}$ (and $<_{\mathscr{L}}$'s coincidence with the natural ordering inside each block). Proceed to stage $s+2$ in this case.

Verification

Define $\mathscr{L} = \langle L, <_{\mathscr{L}} \rangle$ with $L = \bigcup_{s \ge 0} L_s$. Let

$$U = \{\langle e, m \rangle \mid \forall t(\exists s \ge t)[m(e, s) = m]\}$$

and notice that U is a Π_2^0 set (as $m(e, s)$ is computable). Define

$$F(q_e) = \mu m[\langle e, m \rangle \in U]. \quad (2)$$

---

[7] For $e < s$, $|I(e, s)| = m(e, s)$. However for $e = s$ we have $m(e, s) = 0$ whereas it may be that $|I(e, s)| \ne 0$ due to previous rebuilding activity for the sake of some R_i with $i < e$.

Note that $F(q_e)$ is defined for all e as $m(e, s)$ is defined as an element of the finite set Z_e for all $s > e$. Notice also that $F : \mathbb{Q} \to \mathbb{N} \setminus \{0\}$ (and that the construction of F is Δ_3^0 as witnessed by (2)).

We see by inspection that $L = \mathbb{N}$ and that $\mathcal{L}$ has order type $\sum \{ F(q) \mid q \in \mathbb{Q} \}$. Indeed by construction $n \in L_{n+1} \subseteq L$. Moreover n can be moved from one block $I(b, s)$ into another block $I(a, s + 1)$ at stage $s + 1$ only via the *Block Rebuilding* process. However in this case $n \in I(a, t)$ for all $t \geq s+1$ as explained in remark (ii) on page 218. Thus n changes blocks at most once. Now consider $e \geq 0$. Let s_e be a stage such that $m(e, s_e) = F(q_e)$ and such that $m(e, s) \geq m(e, s_e)$ for all $s \geq s_e$. Then $|I(e, s_e)| = F(q_e)$ and $\{ s \mid |I(e, s)| = F(q_e) \}$ is infinite. Moreover, as stated in observation (i) on page 218, the leftmost block of $F(q_e)$ is preserved in $I(e, t)$ for all $t \geq s_e$. I.e. $I(e)$ is well defined as a maximal block with cardinality $F(q_e)$.

For $e \geq 0$ as above, define $i(e)$ to be the index satisfying

$$(k, j) \in D_{i(e)}^e \Leftrightarrow |\{ \langle q_k, r \rangle \mid r \in Z_e \} \cap U_j| \leq 1$$

for all $0 \leq k, j \leq e$. Let $t_e > s_e$ be a stage such that

$$|\{ \langle q_k, r \rangle \mid r \in Z_e \} \cap U_{j,s}| \leq 1$$

for all $(k, j) \in D_i^e$ and $s \geq t_e$. Then, by definition, at any such stage s, $D_{i(e)}^e \subseteq D_{i(e,s)}^e$ and so $i(e) \leq i(e, s)$. For each $0 \leq j \leq e$ define[8]

$$E_j = \{ \langle q_k, r \rangle \mid r \in Z_e \ \& \ k \leq e \ \& \ (k, j) \notin D_{i(e)}^e \} \cap U_j \qquad (3)$$

By Lemma 1, there are infinitely many stages s such that $E_j \subseteq U_{j,s}$ for all $0 \leq j \leq e$. Moreover, at each such stage $s \geq t_e$, $i(e) = i(e, s)$ by definition of the construction. On the other hand, as $|X_{i(e)}| > |D_{i(e)}|$, we see that $X_{i(e)} = S \cup T$ with $S \neq \emptyset$ and $S \cap T = \emptyset$ where

$$T = \{ r \mid (\exists (k, j) \in D_{i(e)}^e)[\langle q_k, r \rangle \in U_j] \} \cap X_{i(e)}$$

and $S = X_{i(e)} \setminus T$. By definition of S there is a stage $\hat{t}_e \geq t_e$ such that, for all $s \geq \hat{t}_e$, and for every $r \in S$, there is no $(k, j) \in D_{i(e)}$ such that $\langle q_k, r \rangle \in U_{j,s}$. For each $r \in S$, let the *in-age* of r relative to $D_{i(e)}$ be the greatest stage s such that $\langle q_k, r \rangle \in U_{j,s}$ for some $(k, j) \in D_{i(e)}$ if such a stage exists, and otherwise define the *in-age* of r to be 0. Define $M \subseteq S$ to be the elements of S of least in-age relative to $D_{i(e)}$, and choose m to be the least number in M. (Note that by definition $M = X_{i(e)}$ if $i(e) = 0$.) Now, for each $0 \leq j \leq e$ define

$$\hat{E}_j = \{ \langle q_k, r \rangle \mid r \in X_{i(e)} \ \& \ k \leq e \ \& \ (k, j) \in D_{i(e)}^e \} \cap U_j.$$

By Lemma 1 we know that there exists a stage $u_e \geq \hat{t}_e$ such that, not only $E_j \subseteq U_{j,u_e}$ for all $0 \leq j \leq e$ (with E_j defined as in (3)), so that $i(e, u_e) = i(e)$,

---

[8] We could also simply define $E_j = \{ \langle q_k, r \rangle \mid r \in Z_e \ \& \ k \leq e \} \cap U_j$ with the same result.

but also $\widehat{E}_j \subseteq U_j$ for each such j so that, by definition $m(e, u_e) = m$ and moreover, $m(e, s) = m$ for all stages $s \geq u_e$ such that $i(e, s) = i(e)$. Accordingly, letting $m(e) = m$ we see that $m(e)$ is the witness for R_e at every such stage s and $|I(e, s)| = m(e)$. Moreover, at every stage $s \geq s_e$ such that $i(e, s) \neq i(e)$, we know that $i(e, s) > i(e)$ as $u_e \geq s_e$. However this implies that $m(e, s) > m(e)$ at every such stage s as $m(e, s) \in X_{i(e,s)}$ and $\min X_{i(e,s)} > \max X_{i(e)} \geq m(e)$. It follows that $F(q_e) = |I(e)| = m(e)$.

Note 4. Suppose that $\mathscr{B}$ is a linear ordering and $\iota : \mathscr{B} \cong \mathscr{L}$ is an isomorphism. Suppose also that $\widehat{F} : \mathbb{Q} \to \mathbb{N} \setminus \{0\}$ is a maximal block function of $\mathscr{B}$ and that $\widehat{I}(0), \widehat{I}(1), \widehat{I}(2), \ldots$ is an assignment of $\widehat{F}$ to $\mathscr{B}$. Now note that we have a listing of labels of $\mathscr{L}$,

$$m_0, m_1, m_2, \ldots$$

such that

$$\iota : \widehat{I}(j) \cong I(m_j)$$

(i.e. $I(m_j)$ is the isomorphic image of $\widehat{I}(j)$ under ι) for all $j \geq 0$. Moreover there must be infinitely many labels j of $\mathscr{B}$ such that $m_j \geq j$. Indeed, suppose otherwise so that for some l, for all $j \geq l$ we have $m_j < j$. Choose $m = \max \{ m_j \mid j < l \} \cup \{l\} + 1$. Then, under our assumption,

$$\iota^*(\{ n \mid n \leq m \}) \subseteq \{ n \mid n < m \}$$

where ι^* is the map over labels induced by ι. Thus ι^* is not one-one. This contradicts the fact that ι is an isomorphism. We therefore conclude that there are infinitely many pairs of labels (k, e) with $k \leq e$ such that $\iota : \widehat{I}(k) \cong I(e)$.

Choose any $\mathscr{B}$, ι, $\widehat{F}$ and assignment $\widehat{I}(0), \widehat{I}(1), \widehat{I}(2), \ldots$ as in Note 4. Consider any index $j \geq 0$ and suppose that U_j is the graph of a function G_j with domain $\mathbb{Q}$. As above, choose $k \geq j$ such that $\iota : \widehat{I}(k) \cong I(e)$ for some $e \geq k$. Now, by definition of the construction, $(k, j) \in D_{i(e)}$. However this implies that

$$G_j(q_k) \neq m(e) = F(q_e) = |I(e)| = |\widehat{I}(k)|.$$

Note in the above argument that the choice of $\widehat{F}$, and of its assignment to $\mathscr{B}$, as also of the isomorphism $\iota : \mathscr{B} \cong \mathscr{L}$, was in each case arbitrary. Notice also that the same observation holds for the choice of the index $j \geq 0$, and of the linear ordering $\mathscr{B} \cong \mathscr{L}$. We can thus conclude that, for *any* Π_2^0 function $G : \mathbb{Q} \to \mathbb{N} \setminus \{0\}$ and *any* $\mathscr{B} \cong \mathscr{L}$, $\mathscr{B}$ does not have order type $\tau = \sum \{ G(q) \mid q \in \mathbb{Q} \}$. $\qquad\square$

Note 5. We can choose any $p \geq 0$ and replace $F : \mathbb{Q} \to \mathbb{N} \setminus \{0\}$ by $F : \mathbb{Q} \to \mathbb{N} \setminus \{0, \ldots, p\}$ in the statement of Theorem 2, by a simple adjustment of the proof, so ensuring that $\mathscr{L}$ contains no maximal blocks of size p or less. We can also clearly force F to be injective (so making $\mathcal{L}$ rigid). For example if we define each Z_e as before but such that $\min Z_{e+1} > \max Z_e$ we obtain F strictly increasing.

We conclude by noting another application of our proof technique and how this yields an alternative proof[9] of Theorem 2 via the work of either Kach or (Kenneth) Harris. Indeed, a straightforward adaptation of the framework of the proof of Theorem 2 can be applied to show that there exists a **0**-*limitwise mono-tonic* set[10] $S \subseteq \mathbb{N} \setminus \{0, 1\}$ such that S is the range of *no* Π_1^0 function[11] G (with domain $\mathbb{N}$). Relativising this result we obtain a **0'**-*limitwise monotonic* set S such that the *shuffle sum* of S derived via the proof of Proposition 2.1 of [Kac08] and the η-*representation* of S derived via the proof of Theorem 3.3 of [Har08] are both examples of η-like computable linear orderings having no isomorphic copy with Π_2^0 maximal block function[12].

References

[DM89]   Downey, R.G., Moses, M.F.: On choice sets and strongly non-trivial self-embeddings of recursive linear orders. Zeitshrift Math. Logik Grundlagen Math. 35, 237–246 (1989)

[Dow98] Downey, R.: Computability theory and linear orderings. In: Ershov, Y.L., Goncharov, S.S., Nerode, A., Remmel, J.B. (eds.) Handbook of Recursive Mathematics. Recursive Algebra, Analysis and Combinatorics, Studies in Logic and the Foundations of Mathematics, vol. 2, pp. 823–976. North Holland (1998)

[Fel76] Fellner, S.: Recursiveness and Finite Axiomatizability of Linear Orderings. PhD thesis, Rutgers University, New Brunswick, New Jersey (1976)

[Har08] Harris, K.: η-representations of sets and degrees. Journal of Symbolic Logic 73(4), 1097–1121 (2008)

[Joc68] Jockusch, C.G.: Semirecursive Sets and Positive Reducibility. Trans. Amer. Math. Soc. 131, 420–436 (1968)

[Kac08] Kach, A.M.: Computable shuffle sums of ordinals. Archive for Mathematical Logic 47(3), 211–219 (2008)

[Kie87] Kierstead, H.A.: On Π_1-automorphisms of recursive linear orders. Journal of Symbolic Logic 52, 681–688 (1987)

[LHC14] Harris, C.M., Lee, K.I., Cooper, S.B.: Automorphisms of computable linear orders and the Ershov hierarchy (in preparation, 2014)

[LR82]   Lerman, M., Rosenstein, J.G.: Recursive linear orderings. In: Metakides, G. (ed.) Patras Logic Symposium (Proc. Logic Sympos. Patras). Studies in Logic and the Foundations of Mathematics, Greece, August 18-22, vol. 109, pp. 123–136. Elsevier Science B.V. (1982)

[Odi89] Odifreddi, P.G.: Classical Recursion Theory. Elsevier Science B.V., Amsterdam (1989)

---

[9] The author is grateful to the anonymous referees for pointing this out.

[10] A function H is a-*limitwise monotonic* if there is an a-computable function h such that, for all x, (i) $H(x) = \lim_{s \to \infty} h(x, s)$, and (ii) for all s, $h(x, s) \le h(x, s+1)$. A set S is a-*limitwise monotonic* if it is the range of an a-*limitwise monotonic* function.

[11] A $\emptyset'$ priority argument shows that, if G has infinite range, then there exists injective Π_1^0 G with the same range. We use this to apply an adapted version of Note 4.

[12] The existence of such a function F' (viewed as having domain $\mathbb{N}$) would yield an obvious contradiction—in the shuffle sum case witnessed by F' itself and in the η-representation case witnessed by the function with graph $G(F') \setminus \{\langle n, 1 \rangle \mid n \in \mathbb{N}\}$.

Lossiness of Communication Channels Modeled by Transducers

Oscar H. Ibarra[1,*], Cewei Cui[2], Zhe Dang[2], and Thomas R. Fischer[2]

[1] Department of Computer Science
University of California, Santa Barbara, CA 93106, USA
ibarra@cs.ucsb.edu
[2] School of Electrical Engineering & Computer Science
Washington State University, Pullman, WA 99164, USA
{ccui,zdang,fischer}@eecs.wsu.edu

Abstract. We provide an automata-theoretic approach to analyzing an abstract channel modeled by a transducer and to characterizing its lossy rates. In particular, we look at related decision problems and show the boundaries between the decidable and undecidable cases.

Keywords: automata, transducers, Shannon information.

1 Introduction

Modern digital communications are realized through channels. A communication system is modeled as a sender, a channel, and a receiver. The channel input is generated by the sender as an encoding of source input information. This process is referred to as channel encoding. The channel output is delivered to a receiver for decoding. Traditional analysis of such a system uses probability and random processes to model channel behavior. In the view of automata theory, the channel is a transducer, which is an automaton (not necessarily of finite states) having both input instructions and output instructions. In automata theory, textbook results [11] focus on formal language aspects of the input-output relationship exhibited in a transducer, without formulation of any probabilistic description of a transducer's behavior. It would be interesting to see if automata theory can be used to investigate certain key characteristics in a communication channel. In this paper, we use an automata-theoretic approach in studying the lossy rate of a channel modeled by a transducer.

A communication channel can be noisy. That is, the input symbols during transmission can be dropped or altered, or unwanted symbols added. As a result, the output of the channel may not be uniquely decoded back to the input. We abstract the problem as an automata theory problem: Given a transducer T, determine whether T is L-lossy. (That is, are there distinct words in L that are translated into the same word with T?) In the paper, the problem is shown decidable for nondeterministic finite state transducers (NFTs) as well as some

---

* Supported in part by NSF Grants CCF-1143892 and CCF-1117708.

A. Beckmann, E. Csuhaj-Varjú, and K. Meer (Eds.): CiE 2014, LNCS 8493, pp. 224–233, 2014.
© Springer International Publishing Switzerland 2014

NFTs augmented with reversal-bounded counters and their variations, while L is a regular language or in a certain class of nonregular languages. On the other hand, the problem is undecidable in general. Indeed, as shown in the paper, the undecidability remains even under a very restricted case: the T is a deterministic finite state transducer (DFT) augmented with the capability of making one turn on its input and the L is the universe. Hence, the decidability/undecidability boundary of the problem is subtle.

We also study the lossy rate of a channel modeled by a transducer. In the paper, we define the lossy rate based on a notion introduced by Shannon [19], which we call information rate. Using this definition, the input lossy rate (the output lossy, defined accordingly in the paper, as well) of the transducer T can be computed through computing the information rates of the input language, the output language, as well the language of input-output pairs, of T, without, as in traditional communication engineering analysis, explicitly introducing a probabilistic or stochastic model. Later in the paper, among other results, we show that the lossy rates are computable for NFT.

Because of space limitation, we omit most of the proofs here. Complete proofs will appear in the journal version of this paper.

2 Decision Problems: Decidable and Undecidable Cases

We first recall reversal-bounded nondeterministic counter machines [12] used subsequently in this paper. A counter is a nonnegative integer variable that can be incremented by 1, decremented by 1, or stay unchanged. In addition, a counter can be tested against 0. Let k be a nonnegative integer. A *nondeterministic k-counter machine (NCM)* is a one-way nondeterministic finite automaton, with input alphabet Σ, augmented with k counters. For a nonnegative integer r, we use NCM(k,r) to denote the class of k-counter machines where each counter is *r-reversal-bounded*; i.e., it makes at most r alternations between nondecreasing and nonincreasing modes in any computation; e.g., the following counter value sequence '0 0 1 2 2 3 $\underline{3\,2}$ 1 0 $\underline{0\,1}$ 1' is of 2-reversal, where the reversals are underlined. For convenience, we sometimes refer to a machine M in the class as an NCM(k,r). In particular, when k and r are implicitly given, we call M a *reversal-bounded NCM*. When M is deterministic, we use 'D' in place of 'N'; e.g., DCM. As usual, $L(M)$ denotes the language that M accepts. If M is augmented with a pushdown stack, we call it a reversal-bounded NPCM (resp., DPCM in the deterministic case).

Reversal-bounded NCMs and NPCMs have been extensively studied since their introduction in 1978 [12], and many generalizations have been identified, e.g., ones equipped with multiple tapes, with two-way tapes, etc. In particular, reversal-bounded NCMs and NPCMs have found applications in areas like Alur and Dill's [1] time-automata [8,7], Paun's [17] membrane computing systems [13], and Diophantine equations [24].

Two fundamental results in the theory of reversal-bounded NCMs and NPCMs are the following [12].

Theorem 1. *It is decidable to determine, given a reversal-bounded NPCM M, whether $L(M)$ is empty (resp., infinite).*

A two-way reversal-bounded NCM M is *finite-turn* if, for a given nonnegative integer c, M makes at most c turns on its two-way input tape.

Theorem 2. *It is decidable to determine, given a finite-turn two-way reversal-bounded NCM M, whether $L(M)$ is empty (resp., infinite).*

We now formalize the problem under study. A *transducer* T is a nondeterministic automaton that accepts pairs of words; i.e., the set of pairs accepted by T is $L(T) \subseteq \Sigma^* \times \Delta^*$. For a pair $(u, w) \in L(T)$, u is an input word and w is an output word. Suppose that L is the language from which an input u is drawn. T is *L-lossy* if there are u, v, w such that $u \neq v \in L$, and, both (u, w) and (v, w) are in $L(T)$. That is, a lossy transducer can translate distinct input words into the same output word. T is *L-lossless* if it is not L-lossy. If $L = \Sigma^*$ (i.e., the set of all finite-length input strings), then we will just use the terms lossless and lossy (omitting Σ^*). We are interested in algorithmic solutions to the problem of deciding whether a transducer is L-lossy:

Given: A transducer T and an input word language L.

Question: Is T L-lossy?

Clearly, like most decision problems in automata theory, the decidability relies on the exact classes of languages and automata to which L and T, respectively, belong.

Consider a nondeterministic finite transducer (NFT) T, which is an NFA with outputs. An instruction of T is of the form $(p, a) \rightarrow (q, b)$, where q, p are states, and a, b are in $\Sigma \cup \{\varepsilon\}$. The instruction means that M in state p reads a, outputs b, and enters state q. (Notice that the instruction can be an ε-instruction; i.e., when a or b is the null symbol ε.) As usual, $L(T)$ denotes the set of pairs (u, w) such that T enters an accepting state after it reads the input word u while it outputs w. It is fairly well known that it is decidable to determine, given an NFT T and a regular language accepted by an NFA M, whether T is L-lossy. We will generalize this.

In the results below, "augmented with reversal-bounded counters" will mean "augmented with a finite number of reversal-bounded counters".

Theorem 3. *It is decidable to determine, given an NFT T augmented with reversal-bounded counters and a language L accepted by a reversal-bounded NCM M, whether T is L-lossy.*

Proof. We construct a finite-turn two-way (with end markers on the input) reversal-bounded NCM M' to simulate T on $L = L(M)$. The idea is for M' to accept some string w if there are two distinct strings u and v in L such that they are mapped into w by T.

M' has one new 1-reversal counter, C. M', when given input w, makes two sweeps on the input. On the first sweep, M' nondeterministically guesses the symbols comprising some string $u = a_1 ... a_k$ (but not writing them) and checking that, at the end of the sweep, u is in $L(M)$. Also during the sweep, M' checks that the outputs of T match the symbols in w. Furthermore, M' uses counter C to store a nondeterministically chosen $1 \leq i \leq k$ (by incrementing the counter) and remembering in its finite control the guessed symbol a_i.

When M and T accept, M' returns to the left end marker and executes the same process as above, but this time guessing the symbols comprising $v = b_1...b_n$. Now, it decrements counter C for every symbol that it guesses. When C becomes zero and the symbol b_i it has guessed is different from a_i and M and T accept, M' accepts w.

Note that the case when u (resp., v) is a proper prefix of v (resp., u) and hence different is taken care of in the above process. Clearly, $L(M')$ is not empty if and only if T is L-lossy. The result follows, since the emptiness problem for finite-turn two-way reversal-bounded NCMs is decidable by Theorem 2.   □

We can further generalize Theorem 3. A two-way reversal-bounded NCM M is *finite-crossing* if for a given nonnegative integer c, M crosses the boundary between any adjacent cells of the input at most c times.

Theorem 4. *It is decidable to determine, given an NFT T augmented with reversal-bounded counters and a language L accepted by a two-way finite-crossing reversal-bounded NCM M, whether T is L-lossy.*

A question arises whether Theorem 4 still holds when T has a two-way input. We will show that the answer is no, even when T is deterministic and makes only one turn on its input tape: a left-to-right sweep and then a right-to-left sweep (the output is one-way). In the proof, we use the undecidability of the Post Correspondence Problem (PCP).

Theorem 5. *It is undecidable to determine, given a 1-turn DFT T, whether T is lossy.*

A transducer T is *single-valued* on a language L if for every u in L, there is at most one w such that (u, w) is in $L(T)$. In contrast to Theorem 5, it is known that it is decidable, given a finite-crossing two-way NFT M augmented with reversal-bounded counters and a language L accepted by a reversal-bounded NCM, whether T is single-valued on L [10].

A transducer T is *k-lossy* if for any word w, there are at most k words that are mapped by T into w. T is *finite-lossy* if it is k-lossy for some k. A related notion that has been extensively studied in automata theory is the notion of k-valuedness of transducers (see, e.g., [18], for an early reference). We say that a transducer T is *k-valued* if, for every input word u, there are at most k output words w such that $(u, w) \in T$. That is, T cannot have more than k outputs on any input word. T is *finite-valued* on L if it is k-valued for some k. Given an NFT T, we can construct another NFT T' such that $L(T') = \{(w, u) : (u, w) \in L(T)\}$. Clearly, T is lossless (resp., finite-lossy, k-lossy for a given k) if and only if T' is single-valued (resp., finite-valued, k-valued). The converse is also true.

The case when T is finite-lossy (resp., k-lossy for a given k) is interesting. It implies that for some k (resp., for the given k), *every* output word w received has at most k possible choices of decoded input words (no matter how long w is). Hence, this number k can also be used as an indicator on how lossy the transducer is.

It is decidable to determine, given an NFT T, whether it is finite-valued (i.e., it is k-valued for some k) [22]. It is also decidable to determine whether it is k-valued for a given k [10]. Hence, we have:

Theorem 6. *It is decidable to determine, given an NFT T and a regular language L, whether T is finite-lossy on L. In the affirmative case, the minimal k_0 such that T is k_0-lossy on L is computable.*

Currently, we do not know if the first part of Theorem 6 holds when M is an NFT augmented with an infinite memory (e.g., a reversal-bounded counter). However, we can prove the following.

Theorem 7. *It is decidable to determine, given an NFT T augmented with reversal-bounded counters, a language L accepted by a two-way finite-crossing reversal-bounded NCM M, and an integer $k \geq 1$, whether T is k-lossy on L.*

For deterministic pushdown transducers (DPDTs), the following result can be shown:

Theorem 8. *It is undecidable to determine, given a 1-reversal DPDT (i.e., the stack makes exactly one reversal: once it pops it can no longer push), whether T is·lossless (resp., k-lossy for a given k, finite-lossy).*

For the case when the NPDT's input is bounded, we have:

Theorem 9. *It is decidable to determine, given an NPDT T augmented with reversal-bounded counters whose input comes from $x_1^* \cdots x_k^*$ (where $x_1, \ldots, x_k$ are not necessarily distinct words) and a language L accepted by a reversal-bounded NCM M, whether T is L-lossy.*

Again, the theorem above generalizes to the case when L is accepted by a finite-crossing two-way reversal-bounded NCM.

Next, we investigate the subtle relationship between ambiguity in automata and lossiness in transducers. Let M be a (one-way) acceptor, e.g., DFA, NFA, DPDA, NPDA, etc. We say that a transducer T is of the same type as M, if when T's output is suppressed, it reduces to an acceptor in the class where M belongs. So a DFT (resp., NFT, DPDT, NPDT, etc,) is of the same type as DFA (resp., NFA, DPDA, NPDA, etc.) We assume that in an acceptor or transducer, an accepting state is a halting (i.e., the device has no move when it enters an accepting state).

Theorem 10. *The following statements are equivalent, where M and T are of the same type:*

1. *It is undecidable, given a nondeterministic acceptor M, whether M is unambiguous (resp., k-ambiguous for a given k, or finitely-ambiguous).*

2. *It is undecidable, given a deterministic transducer T, whether T is lossless (resp., k-lossy for a given k, or finite-lossy).*

The above result is interesting because it relates the ambiguity question of a *nondeterministic* acceptor to the lossiness question of a *deterministic* transducer of the same type as the acceptor. For example, it is undecidable, given a 1-reversal NPDA (which is equivalent to a linear context free grammar), whether it is unambiguous (resp., k-ambiguous for a given k, unboundedly ambiguous) [23]. Hence, it is also undecidable, given a 1-reversal DPDT (deterministic 1-reversal pushdown transducer), whether it is lossless (resp., k-lossy for a given k, finite-lossy).

Clearly, Theorem 10 is not valid if M is deterministic. This is because such an acceptor is always unambiguous. Hence the unambiguity question is trivially decidable (since the acceptor is always unambiguous). However, from Theorem 8, the losslessness question for 1-reversal DPDT is undecidable.

Similarly, Theorem 10 is not valid if T is nondeterministic. Consider the following example: Let $\mathcal{P}$ be the class of 1-reversal NPDAs M, where M always starts in initial state q_0 and on input ε goes to state q_{01} and q_{02}, and in the next step, the next state from q_{01} or q_{02} are the same. Clearly, any 1-reversal NPDA can be simulated by a machine in $\mathcal{P}$ and, hence, any machine in $\mathcal{P}$ is ambiguous (because, by definition of the class $\mathcal{P}$, any input accepted by the machine has at least two distinct accepting computations). It follows that the unambiguity question for $\mathcal{P}$ is decidable. Now let $\mathcal{T}$ be the class of 1-reversal NPDTs of the type defined in class $\mathcal{P}$. Clearly, any 1-reversal DPDT can be simulated by a transducer in $\mathcal{T}$. Hence, from Theorem 8, the losslessness problem for $\mathcal{T}$ is undecidable.

The next result shows that undecidability of losslessness implies undecidability of k-lossiness for any k.

Theorem 11. *Let $\mathcal{T}$ be a class of deterministic transducers. Then losslessness for $\mathcal{T}$ is undecidable if and only if k-lossiness for $\mathcal{T}$ is undecidable for any given $k \geq 1$.*

We now define a form of transducers that are Shannon channels mentioned in the Introduction. Let T be a transducer of any given type. Suppose that (u, w) is in $L(T)$. Thus, on input u, T outputs w. However, if we observe the behavior of T, i.e., we look at exactly the way that we feed T with symbols in u and we observe symbols in w to be sent out, we obtain an observed sequence which is a shuffle of the pair (u, w). For instance, if $u = ABC$ and $w = deffg$, an observed sequence could be $ABdefCfg$. That is, on input A, T runs but emits no output. Then on input B, we have output def. Finally, on input C, we have output fg. The input distance of the sequence is 3 (the length of def), that is the maximal number of output symbols between two consecutive input symbols (B and C).

Formally, define the input distance (resp., output distance) of T on (u, w) to be the maximal number of output (resp., input) symbols between two consecutive input (resp., output) symbols in the shuffled input/output behavior sequence. The input (resp., output) distance of T is the maximal input distance for all

(u, w) in $L(T)$. T is k-input Shannon (resp., k-output Shannon) if its input distance (resp., output distance) is at most k. T is finite-input (resp. finite-output) Shannon if it is k-input Shannon (resp., k-output Shannon) for some k.

Theorem 12. *The following are decidable, given a reversal-bounded NPCMT T (NPCMT is an NPCM with output):*

1. *Given $k \geq 1$, is T k-input Shannon (resp., k-output Shannon)?*
2. *Is T finite-input Shannon (resp. finite-output Shannon)?*

3   Lossy Rates of Transducers

The previous section focuses on the problem of deciding whether a channel modeled as a transducer T is L-lossy for a given input language L. Suppose that T is L-lossy. Without introducing probabilities into T, can we still define a notion that characterizes how lossy T is? Before we proceed further, we first illustrate the intuition behind the definitions.

Consider a pair (u, w) of an input word u and an output word w produced by T. The "information" contained in (u, w) is composed of the information in u and the information in w. However, since u and w are not necessarily independent, there is certain amount of mutual information shared between u and w.

The input lossy rate measures the "number" of inputs to which an average output can be decoded. Intuitively, the input lossy rate, using the classic Venn diagram of Shannon information theory, should be the information contained in the input u, given the output w. Notice that the lengths of the input and the output are in general unbounded and hence, a more scientific measurement would be information rate (in bits per symbol) instead of information (in bits). However, there is a problem. In computing the aforementioned information/mutual information, one usually needs a probability distribution which, unfortunately, the transducer T does not have and which, in practice, would be very hard to obtain.

Without an explicit probabilistic model, can we still define an information rate? There has already been a fundamental notion shown below, proposed by Shannon [19] and later Chomsky and Miller [3], that we have evaluated through experiments over C programs [25,16,9,4,5], fitting our need for the aforementioned complexity. For a number n, we use $S_n(L)$ to denote the number of words in a language L whose length is n. The *information rate* λ_L of L is defined as $\lambda_L = \lim \frac{\log S_n(L)}{n}$. Where the limit does not exist, we take the upper limit, which always exists for a finite alphabet.

The following result is fundamental.

Theorem 13. *The information rate of a regular language L is computable [3].*

The case when L is non-regular (e.g., L is the external behavior set of a software system containing (unbounded) integer variables like counters and clocks) is more interesting, considering the fact that a complex software system nowadays is almost always of infinite state, yet the notion of information rate has been applied to software testing [4,21]. However, in such a case, computing the information rate is difficult (sometimes even not computable [14]) in general. Existing results (such as unambiguous context-free languages [15], Lukasiewicz-languages [20], and regular timed languages [2]) are limited and mostly rely on Mandelbrot generating functions and the theory of complex/real functions, which are also difficult to generalize. A recent important result, using a complex loop analysis technique, is as follows.

Theorem 14. *The information rate of the language accepted by a reversal-bounded DCM is computable [6].*

Note that the case for a reversal-bounded NCM is open.

We now return to our definitions. Assume that T is *length-preserving*. That is, for all $(u, w) \in L(T)$, we have $|u| = |w|$. Example channels modeled by such transducers are binary channels that can alter a bit but never drop one. We now consider $L(T, L) = \{(u, w) : (u, w) \in L(T), u \in L\}$. Recall that the information rate $\lambda_{L(T,L)}$ is the average bit rate (number of bits per symbol) of (the string encoding of) a pair $(u, w) \in L(T, L)$. We use a simple shuffle encoding $[u, w]$ of (u, w); e.g., $[aaa, bbb] = ccc$, where c is a symbol representing the pair (a, b). Hence, the length of $[u, w]$ is the same as $|u|$ (as well as $|w|$). It is not hard to imagine that the bit rate $\lambda_{L(T,L)}$ of $[u, w]$ is "contributed" by the average bit rate λ_L in u and the average bit rate $\lambda_{T(L)}$ in w. Herein, $T(L) = \{w : (u, w) \in L(T, L)\}$. Notice that u and w are not completely independent, since $(u, w) \in L(T, L)$. What is the meaning of the bit rate amount $\lambda_{L(T,L)} - \lambda_{T(L)}$? It characterizes, for $(u, w) \in L(T, L)$, the average bit rate amount in u that is independent of w. Notice that, if T is L-lossless, the amount is simply zero. This is because, in this case, the output w completely decides the input u. Now, we define the *input lossy rate* $\lambda_{in}(L, T)$ to be $\lambda_{L(T,L)} - \lambda_{T(L)}$. Symmetrically, we define the *output lossy rate* $\lambda_{out}(L, T)$ to be $\lambda_{L(T,L)} - \lambda_L$. Notice that $\lambda_{in}(L, T) = \lambda_{out}(T(L), T^{-1})$, where T^{-1} is the inverse of T. Hence, for theoretical purposes, it suffices for us to consider only the input lossy rate in many cases.

We first consider the case when T is a length-preserving NFT (i.e., without ε-instructions).

Theorem 15. *The input and output lossy rates are computable when T is an NFT without ε-instructions and L is a regular language.*

We now consider a DFT T augmented with reversal-bounded counters. In every instruction of T, if the instruction reads a non-null inout symbol, it will also output a non-null symbol and vice versa. We call such a T *non-null* and obviously it is length-preserving. The following result uses Theorem 14.

Theorem 16. *The output lossy rate is computable when T is a non-null DFT T augmented with reversal-bounded counters and L is the language accepted by a reversal-bounded DCM.*

We currently do not know if Theorem 16 can be generalized to the input lossy rate. This is because in computing the input lossy rate, one needs $\lambda_{T(L)}$, where $T(L)$ can be accepted by a reversal-bounded NCM (instead of a DCM) and hence Theorem 14 is not applicable.

Currently, we are not clear on how to generalize the definitions of input and output lossy rates to the case when T is not necessarily length-preserving. The difficulty is that, in this case, T can map a low (resp., high) bit rate input to a high (resp., low) one, even when T is one-to-one. Hence, it is not obvious how information rates used in the definitions can faithfully catch the intuitive meaning of lossy rates. We leave this generalization for future work.

Acknowledgement. We would like to thank Klaus Wich for pointing out to us that the finite-ambiguity problem for linear context-free grammars was shown undecidable in his PhD thesis, and Eric Wang for comments that improved the presentation of our results.

References

1. Alur, R., Dill, D.L.: A theory of timed automata. Theoretical Computer Science 126(2), 183–235 (1994)
2. Asarin, E., Degorre, A.: Volume and entropy of regular timed languages: Discretization approach. In: Bravetti, M., Zavattaro, G. (eds.) CONCUR 2009. LNCS, vol. 5710, pp. 69–83. Springer, Heidelberg (2009)
3. Chomsky, N., Miller, G.A.: Finite state languages. Information and Control 1, 91–112 (1958)
4. Cui, C., Dang, Z., Fischer, T.R.: Bit rate of programs (2013) (submitted)
5. Cui, C., Dang, Z., Fischer, T.R., Ibarra, O.H.: Similarity in languages and programs. Theor. Comput. Sci. 498, 58–75 (2013)
6. Cui, C., Dang, Z., Fischer, T.R., Ibarra, O.H.: Information rate of some classes of non-regular languages: An automata-theoretic approach (submitted, 2014)
7. Dang, Z.: Pushdown timed automata: a binary reachability characterization and safety verification. Theor. Comput. Sci. 302(1-3), 93–121 (2003)
8. Dang, Z., Ibarra, O.H., Bultan, T., Kemmerer, R.A., Su, J.: Binary reachability analysis of discrete pushdown timed automata. In: Emerson, E.A., Sistla, A.P. (eds.) CAV 2000. LNCS, vol. 1855, pp. 69–84. Springer, Heidelberg (2000)
9. Dang, Z., Ibarra, O.H., Li, Q.: Sampling a two-way finite automaton (submitted, 2014)
10. Gurari, E.M., Ibarra, O.H.: A note on finite-valued and finitely ambiguous transducers. Math. Systems Theory 16, 61–66 (1983)
11. Hopcroft, J.E., Motwani, R., Ullman, J.D.: Introduction to Automata Theory, Languages, and Computation, 1st edn. Addison-Wesley (1979)
12. Ibarra, O.H.: Reversal-bounded multicounter machines and their decision problems. Journal of the ACM 25(1), 116–133 (1978)
13. Ibarra, O.H., Dang, Z., Egecioglu, O., Saxena, G.: Characterizations of Catalytic Membrane Computing Systems. In: Rovan, B., Vojtáš, P. (eds.) MFCS 2003. LNCS, vol. 2747, pp. 480–489. Springer, Heidelberg (2003)
14. Kaminger, F.P.: The noncomputability of the channel capacity of context-senstitive languages. Inf. Comput. 17(2), 175–182 (1970)
15. Kuich, W.: On the entropy of context-free languages. Information and Control 16(2), 173–200 (1970)

16. Li, Q., Dang, Z.: Sampling automata and programs (2013) (submitted)
17. Paun, G.: Membrane Computing, An Introduction. Springer (2002)
18. Schützenberger, M.P.: Sur les relations rationnelles. In: Brakhage, H. (ed.) GI-Fachtagung 1975. LNCS, vol. 33, pp. 209–213 (1975)
19. Shannon, C.E., Weaver, W.: The Mathematical Theory of Communication. University of Illinois Press (1949)
20. Staiger, L.: The entropy of lukasiewicz-languages. In: Kuich, W., Rozenberg, G., Salomaa, A. (eds.) DLT 2001. LNCS, vol. 2295, pp. 155–165. Springer, Heidelberg (2002)
21. Wang, E., Cui, C., Dang, Z., Fischer, T.R., Yang, L.: Zero-knowledge blackbox testing: Where are the faults. Int'l J. Foundations of Computer Science (to appear)
22. Weber, A.: On the valuedness of finite transducers. Acta Inf. 27(9), 749–780 (1990)
23. Wich, K.: Ambiguity functions of context-free grammars and languages. PhD thesis (2004)
24. Xie, G., Dang, Z., Ibarra, O.H.: A solvable class of quadratic Diophantine equations with applications to verification of infinite state systems. In: Baeten, J.C.M., Lenstra, J.K., Parrow, J., Woeginger, G.J. (eds.) ICALP 2003. LNCS, vol. 2719, pp. 668–680. Springer, Heidelberg (2003)
25. Yang, L., Cui, C., Dang, Z., Fischer, T.R.: An information-theoretic complexity metric for labeled graphs (2011) (in review)

Predicate Characterizations
in the Polynomial-Size Hierarchy

Christos A. Kapoutsis

Carnegie Mellon University in Qatar

Abstract. The *polynomial-size hierarchy* is the hierarchy of 'minicomplexity' classes which correspond to *two-way alternating finite automata* with polynomially many states and finitely many alternations. It is defined by analogy to the *polynomial-time hierarchy* of standard complexity theory, and it has recently been shown to be strict above its first level.

It is well-known that, apart from their definition in terms of polynomial-time *alternating Turing machines*, the classes of the polynomial-time hierarchy can also be characterized in terms of polynomial-time *predicates*, polynomial-time *oracle Turing machines*, and *formulas* of second-order logic. It is natural to ask whether analogous alternative characterizations are possible for the polynomial-size hierarchy, as well.

Here, we answer this question affirmatively for predicates. Starting with the first level of the hierarchy, we experiment with several natural ways of defining what a 'polynomial-size predicate' should be, so that existentially quantified predicates of this kind correspond to polynomial-size *two-way nondeterministic finite automata*. After reaching an appropriate definition, we generalize to every level of the hierarchy.

1 Introduction

The k-th level of the *polynomial-size hierarchy* consists of the classes $2\Sigma_k$ and $2\Pi_k$ of all (families of) regular languages which are decided by (families of) *two-way alternating finite automata* (2AFAs) with polynomially many states (i.e., of polynomial 'size'), where the start state is respectively existential or universal and every computation path on any input alternates $< k$ times between existential and universal steps, if $k > 0$; or uses only deterministic steps, if $k = 0$. The question whether this hierarchy is strict was raised in [6] and answered in the affirmative by Geffert [3] for all levels above the lowest two: for all $k \geq 1$,

$$2\Sigma_k \subsetneq 2\Sigma_{k+1} \quad \text{and} \quad 2\Sigma_k \not\subseteq 2\Pi_k \ \& \ 2\Sigma_k \not\supseteq 2\Pi_k \quad \text{and} \quad 2\Pi_k \subsetneq 2\Pi_{k+1} \,.$$

For $k = 0$, the question is still open: the classes $2\Sigma_0$ and $2\Sigma_1$ are respectively the classes 2D and 2N of all (families of) regular languages decided by (families of) *deterministic* and *nondeterministic two-way finite automata* (2DFAs and 2NFAs) with polynomially many states; hence, proving that $2\Sigma_0 \subsetneq 2\Sigma_1$ is equivalent to confirming the long-standing *Sakoda-Sipser conjecture* that $2D \subsetneq 2N$ [11,6].

The hierarchy is defined by analogy to the *polynomial-time hierarchy* of standard complexity theory, whose k-th level consists of the classes $\Sigma_k P$ and $\Pi_k P$

A. Beckmann, E. Csuhaj-Varjú, and K. Meer (Eds.): CiE 2014, LNCS 8493, pp. 234–244, 2014.
© Springer International Publishing Switzerland 2014

of languages decided by polynomial-time *alternating Turing machines* (ATMs) where the number of alternations is bounded as above [13,12]. The question whether this hierarchy is strict is, of course, a well-studied open problem, also hosting on its lowest two levels the famous question whether $P = NP$.

An important feature of the polynomial-time hierarchy, highlighting its robustness, is that its classes can be defined in several equivalent ways, which are all quite natural but also quite different from each other conceptually. Indeed, apart from their standard definition in terms of polynomial-time ATMs, these classes can also be defined in terms of:

- *Polynomial-time predicates.* For example, a language is in class $\Sigma_1 P = NP$ iff it consists of every string which can, together with a suitable 'certificate', satisfy a binary predicate which is decided by a *deterministic Turing machine* (DTM) in time polynomial in the length of the string [12].
- *Polynomial-time oracle Turing machines.* For example, a language is in class $\Sigma_2 P = NP^{NP}$ iff it is decided by a polynomial-time *nondeterministic Turing machine* (NTM) which has access to an oracle for a language of NP [10,12].
- *Logical formulas.* For example, a language is in $PH = \bigcup_{k \geq 0} \Sigma_k P$ iff it consists of every string which satisfies a formula in *second-order logic* [2,4].

It is natural to ask whether the classes of the polynomial-size hierarchy also admit analogous alternative definitions, next to their original one in terms of polynomial-size 2AFAs. That is, what kind of (i) 'polynomial-size predicates', (ii) 'polynomial-size oracle two-way finite automata', and (iii) logical formulas match 2AFAs with polynomially many states and finitely many alternations?

In this article we study (i). We identify a proper definition for *polynomial-size predicates* such that suitably quantified predicates of this kind characterize the classes $2\Sigma_k$ and $2\Pi_k$, for all k. Starting with the case $k = 1$, we experiment with several natural ways of defining predicates which characterize $2\Sigma_1 = 2N$, namely the (families of) languages decided by polynomial-size 2NFAs. After we reach the correct definition for this class, we generalize for all classes of the hierarchy.

This settles part (i). Part (ii) remains open: We know of no model of 'oracle two-way finite automaton' for characterizing the classes of the polynomial-size hierarchy. As for (iii), a partial answer was given in [8], where a class of suitably structured formulas of *monadic second-order logic with successor* were proven equivalent to polynomial-size *sweeping* 2NFAs (i.e., 2NFAs which turn their head only on the endmarkers) when the length is polynomial and certain structural parameters are appropriately bounded; the full answer involves suitably structured formulas of *first-order logic with successor and transitive closure* [9].

1.1 Preparation

If $n \geq 0$, then $[n] := \{0, 1, \ldots, n-1\}$. If Σ is an *alphabet* and the symbols $\vdash, \dashv \notin \Sigma$ are *endmarkers*, then $\Sigma_e := \Sigma \cup \{\vdash, \dashv\}$. If $z \in \Sigma^*$ is a string over Σ, then $|z|$ is its length and z_i is its i-th symbol, if $1 \leq i \leq |z|$; or $\vdash$, if $i = 0$; or $\dashv$, if $i = |z|+1$. A language $L \subseteq \Sigma^*$ is *decided* (or *solved*) by a machine M if M accepts exactly the strings in L. A language family[1] $(L_h)_{h\geq 1}$ is *decided* (or *solved*) by a family of machines $(M_h)_{h\geq 1}$ if every M_h solves L_h. A family of automata $(M_h)_{h\geq 1}$ is *polynomial-size* if M_h has $\leq p(h)$ states, for some polynomial p and all h.

A *two-way alternating finite automaton* (2AFA) is a tuple $M = (Q, U, \Sigma, \delta, q_s)$, where Q is a set of *states*, Σ is an *alphabet*, and $\delta \subseteq Q \times \Sigma_e \times Q \times \{L,R\}$ is the *transition relation*, for L,R two direction-indicating tags; one state q_s is *special* (start/accept) and each state is *universal*, if in $U \subseteq Q$, or *existential*, if in $Q \setminus U$.

An input $z \in \Sigma^*$ is presented on the tape between the endmarkers, as $\vdash z \dashv$. The automaton starts at q_s and on $\vdash$. Whenever at a state p and on a symbol a, it switches to state q and moves its head in direction d, for every q and d such that $(p, a, q, d) \in \delta$ —never violating an endmarker, except to move off $\dashv$ into q_s. The result is a tree of *configurations*, i.e., state-position pairs from $Q \times \{0, \ldots, |z|+2\}$, with $(q_s, 0)$ as root; we call this tree the *computation of M on z*, $\mathrm{COMP}_M(z)$.

The unique *accepting* configuration is $(q_s, |z|+2)$. A *rejecting* configuration is any (p, i) where $i \leq |z|+1$ and δ contains no tuple of the form $(p, z_i, ., .)$. The accepting and rejecting configurations are called *halting*. A non-halting configuration (p, i) is *existential* or *universal*, according to what p is; it is also called *deterministic*, if δ contains exactly 1 tuple of the form $(p, z_i, ., .)$.

A *full computation path* in $\mathrm{COMP}_M(z)$ is any path π which starts at the root and is infinite (*looping*) or ends at a leaf (*halting*); in the latter case, π is either *accepting* or *rejecting*, according to what the leaf is. A *full computation tree* in $\mathrm{COMP}_M(z)$ is any subtree τ such that (1) τ contains the root, (2) each existential configuration in τ has exactly 1 of its children in τ, and (3) each universal configuration in τ has all of its children in τ. We call τ *looping*, if it is infinite; *accepting*, if it is finite and all its leaves are accepting; and *rejecting*, otherwise. If $\mathrm{COMP}_M(z)$ contains an accepting full computation tree, then M *accepts* z.

Let $k \geq 1$. If every full computation path in $\mathrm{COMP}_M(z)$ for any z switches $< k$ times between existential and universal configurations, we say M is a $2\Sigma_k$FA, if $q_s \notin U$, or a $2\Pi_k$FA, if $q_s \in U$ —a $2\Sigma_1$FA is also called *nondeterministic* (a 2NFA). If every non-halting configuration ever exhibited by M is actually deterministic, we say M is a $2\Sigma_0$FA or a $2\Pi_0$FA or simply *deterministic* (a 2DFA). If δ never uses the L tag, we say M is *one-way* (1AFA, 1NFA, 1DFA).

Let $k \geq 0$. The class $2\Sigma_k$ (respectively, $2\Pi_k$) consists of every language family which is solved by a polynomial-size family of $2\Sigma_k$FAs (respectively, $2\Pi_k$FAs):

$$2\Sigma_k := \left\{ (L_h)_{h\geq 1} \;\middle|\; \begin{array}{l} \text{there exists a } 2\Sigma_k\text{FAs family } (M_h)_{h\geq 1} \text{ and a polynomial } p \\ \text{such that every } M_h \text{ solves } L_h \text{ with } \leq p(h) \text{ states.} \end{array} \right\},$$

and similarly for $2\Pi_k$. Easily, $2\Sigma_k, 2\Pi_k \subseteq 2\Sigma_{k+1}, 2\Pi_{k+1}$ for all k. We also write 2D for $2\Sigma_0 = 2\Pi_0$; 2N for $2\Sigma_1$; and 2H for $\cup_{k\geq 0} 2\Sigma_k = \cup_{k\geq 0} 2\Pi_k$.

2   The Case of 2N

The class 2N is the minicomplexity analogue of NP. The predicate characterization of NP is given by the following well-known fact (which uses Def. 1):

---

[1] For an example of a language family, see TWL $= (\mathrm{TWL}_h)_{h\geq 1}$ on page 237.

Theorem 1. *A language L is in* NP *iff there exists a polynomial-time binary predicate R such that, for all x: $x \in L \Longleftrightarrow (\exists y)R(x,y)$.* [2]

Definition 1. *A binary predicate R is* polynomial-time *if there is a* DTM *M and a polynomial p such that, for all x,y: $R(x,y) \Longleftrightarrow M$ accepts $\langle x,y \rangle$ in time $p(|x|)$.*

E.g., if L is SAT (the *satisfiability problem* [12]), then R is the predicate which is true whenever x is a Boolean formula (the *instance*) and y is a truth-assignment which satisfies it (the *certificate*); M is the DTM which computes the value of x under y and accepts iff the result is "true"; and p is the small polynomial which bounds the time spent by M as a function of the length of x.

Our goal is to replicate this setting for 2N. That is, we want a characterization of 2N as captured by the following statement and definition:

Theorem 2. *A language family $(L_h)_{h \geq 1}$ is in* 2N *iff there exists a polynomial-size binary predicate family $(R_h)_{h \geq 1}$ such that, for all h and all x:*

$$x \in L_h \iff (\exists y)R_h(x,y).$$

Definition 2. *A binary predicate family $(R_h)_{h \geq 1}$ is* polynomial-size *if there exists a family of 'deterministic finite-state acceptors' $(M_h)_{h \geq 1}$ and a polynomial p such that, for all h and all x,y:*

$$M_h \text{ has } \leq p(h) \text{ states} \qquad \& \qquad R_h(x,y) \iff M_h \text{ accepts } \langle x,y \rangle.$$

E.g., if L_h is TWL$_h$ (the *two-way liveness problem* on h-tall graphs [6,7]), then R_h should be the predicate which is true whenever x is a string of h-tall two-column graphs and y is a path from the leftmost to the rightmost column of the respective multi-column graph; M_h should be some kind of a deterministic finite-state machine which scans the arrows of y and accepts iff they are all present in the graph of x, the first one departs from the leftmost column, and the last one arrives at the rightmost column; and p should be a polynomial bounding the number of states needed to perform these checks.

All we need to do, in order to complete this setting, is to clarify what type of acceptors we should use in Def. 2 so that Th. 2 holds. We explore our options in the next sections. We start with two naive attempts, and explain why they fail. We then continue with a more educated guess which, although it fails, too, it captures a different minicomplexity class. The correct choice is given in Sect. 2.3.

2.1   Two Naive Attempts

The straightforward attempt is to simply have each M_h be a 2DFA which receives the pair $\langle x,y \rangle$ on its input tape as the #-delimited concatenation $x\#y$. But this model is

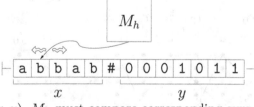

too weak. Intuitively, to check $R_h(x,y)$, M_h must compare corresponding symbols of x and y (i.e., symbols around x_i with symbols around y_i), a task which is impossible for a finite-state machine when x and y become arbitrarily long.

---

[2] Note that y need never be more than polynomially long, as R is polynomial-time.

To enable M_h to compare corresponding symbols of x and y, we may place x and y on different tapes, each with its own, independent, two-way head. Formally, $M_h = (Q, \Sigma, \Delta, \delta, q_s)$, where Σ and Δ are the alphabets for instances and certificates, respectively, and the transition function has the form $\delta : Q \times \Sigma_e \times \Delta_e \longrightarrow Q \times \{L,R\} \times \{L,R\}$. But now the model is too strong: M_h can use the distance between $\vdash$ and the head on the second tape as counter to solve problems that are even non-regular.

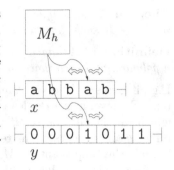

2.2   A Better Attempt

To fix our problems, we must prevent M_h from using its second head as counter. One way to do this, is to first require that x and y are (almost) equally long, then remove the ability of the heads to move independently. Formally, we require that $|y| = |x| + 2$ and $\delta : Q \times \Sigma_e \times \Delta \longrightarrow Q \times \{L,R\}$. Let us call this type of machine a *synchronous two-way deterministic finite verifier* (2DFV$_*$). It looks promising.

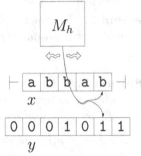

For one, we can now prove the forward direction of Th. 2. It follows from the next lemma, when we apply it to every member of a family $(L_h)_{h \geq 1} \in$ 2N.

Lemma 1. *If L is solved by an s-state 2NFA, then some binary predicate R is solved by an s-state 2DFV$_*$ and is such that, for all x: $x \in L \Longleftrightarrow (\exists y)R(x,y)$.*

Proof. Let $N = (Q, \Sigma, \delta, q_s)$ be the 2NFA which solves L.

To motivate R, consider any $x \in L$. Let $n := |x|$. Consider any accepting computation of N on x. Remove all cycles from it, to get the corresponding *minimal* accepting computation —call it c. Because c is minimal, its representation in the configuration graph of N on x (i.e., the graph with all configurations in $Q \times \{0, \ldots, n+2\}$ as vertices, and all computation steps allowed by δ as arrows) is a path where no two arrows have a common endpoint. Split this $(n+3)$-column representation into $n+2$ three-column graphs $f_0, f_1, \ldots, f_{n+1}$, one for each column but the last one, where each f_i represents only the steps performed on x_i.

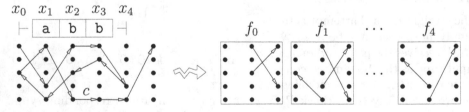

Since no two arrows have a common endpoint, each f_i is really a partial injection from Q to $Q \times \{L,R\}$. Let $\Delta := (Q \to Q \times \{L,R\})$ be the alphabet of all such partial injections. Then, we can use $y := f_0 f_1 \ldots f_{n+1} \in \Delta^*$ as a certificate for x.

Indeed, define $R \subseteq \Sigma^* \times \Delta^*$ so that $R(x,y)$ holds iff (1) $|x|+2 = |y|$; (2) the $(|x|+3)$-column graph derived from y (by viewing each y_i as a three-column graph; then identifying the last two columns of each y_i with the first two columns of y_{i+1}; then dropping the first column of the leftmost y_i) contains a path from the top of the leftmost column to the top of the rightmost one; and (3) every arrow (p,q,d) of every y_i is a legal step of N on x_i: $(p,q,d) \in y_i \Longrightarrow (p,x_i,q,d) \in \delta$. Then the argument of the previous paragraph proves that $x \in L \Longrightarrow (\exists y)R(x,y)$. Conversely, if $R(x,y)$, then (3) means that the path guaranteed by (2) is an accepting computation of N on x, and thus $x \in L$.

Finally, R is solved by the s-state 2DFV$_*$ $M = (Q, \Sigma, \Delta, \delta', q_s)$ which, on input $\langle x,y \rangle$, interprets y as a $(|x|+3)$-column graph as above and follows the unique path out of q_s of the leftmost column, verifying that all arrows in the graph are consistent with δ and that the path terminates at q_s of the rightmost column. Formally, every $\delta'(p,a,f)$ is either $f(p)$, if $f(p)$ is defined and all arrows in f are consistent with δ; or undefined, otherwise.   □

To complete the proof of Th. 2, we would need the converse lemma: *If a binary predicate R is solved by an s-state 2DFV$_*$, then $L := \{x \mid (\exists y)R(x,y)\}$ is solved by a poly(s)-state 2NFA.* However, in trying to prove this claim, one would find it hard to build the desired 2NFA N for L from the given 2DFV$_*$ for R: the natural approach, where N simply guesses y symbol-by-symbol, fails because, upon returning to an input symbol x_i that has been visited before, N would need to re-guess the corresponding y_i identically as in all previous visits.

As a matter of fact, the backward direction of Th. 2 is false:

Lemma 2. *There exists a polynomial-size binary predicate family $(R_h)_{h \geq 1}$ such that the language family $(L_h)_{h \geq 1}$ where $L_h := \{x \mid (\exists y)R_h(x,y)\}$ is not in* 2N.

Proof. For every h, let $R_h \subseteq \{0\}^* \times [2^h]^*$ be a binary predicate such that $R_h(x,y)$ holds only when $x = 0^{2^h-2}$ and y is the ordered string of all symbols of $[2^h]$:

$$y := \boxed{0} \; \boxed{1} \; \boxed{2} \; \boxed{3} \; \ldots \; \boxed{2^h-2} \; \boxed{2^h-1}$$

A 2DFV$_*$ M_h can solve R_h by focusing on y and checking that (1) it starts with 0; (2) each of the other symbols is derived from its previous one by adding 1; and (3) the last symbol is 2^h-1. To check (2), M_h goes through every pair of successive symbols, y_i and y_{i+1}, and checks that $y_i+1 = y_{i+1}$ by zig-zagging h times between the two positions and comparing the binary representations of y_i and y_{i+1} bit-by-bit. Easily, this requires $O(h)$ states, so $(R_h)_{h \geq 1}$ is polynomial-size.

Finally, the only x admitting a certificate under R_h is 0^{2^h-2}, so $L_h = \{0^{2^h-2}\}$, which needs $\geq 2^h-2$ states on a 2NFA [1, Fact 5.2]. Hence, $(L_h)_{h \geq 1} \notin$ 2N.   □

Overall, our current definitions led us to a strict superset of 2N (Lemmas 1, 2). Before modifying them, let us see which class they really capture. The next two lemmas show that it is the class 2^{1N} corresponding to exponential-size 1NFAs [6].

Lemma 3. *If L is solved by an s-state 1NFA, then some binary predicate R is solved by a $O(\log s)$-state 2DFV$_*$ and satisfies $x \in L \Longleftrightarrow (\exists y)R(x,y)$, for all x.*

Proof. Let $N = (Q, \Sigma, \delta, q_s)$ be the 1NFA which solves L, with $|Q| = s$. Without loss of generality, assume that $Q = [s]$ and that $q_s = 0$. Let $t := \lceil \log_2 s \rceil$.

To motivate R, consider any $x \in L$. Let $n := |x|$. Pick any accepting computation of N on x. This is a list $p_0, p_1, \ldots, p_{n+2} \in Q$ such that $p_0 = q_s = p_{n+2}$ and $(p_i, x_i, p_{i+1}, \text{R}) \in \delta$ for all i. Recast this $(n+3)$-item list into the list of $n+2$ successive pairs $\pi_0, \pi_1, \ldots, \pi_{n+1}$, where $\pi_i := (p_i, p_{i+1})$.

Now, letting $\Delta := Q \times Q$ be the alphabet of all pairs of states, we can use the string of pairs $y := \pi_0 \pi_1 \ldots \pi_{n+1} \in \Delta^*$ as a certificate for x.

Therefore, we define $R \subseteq \Sigma^* \times \Delta^*$ so that $R(x, y)$ holds iff (1) $|x|+2 = |y|$; (2) y is really a sequence of states (i.e., every two successive symbols are of the form $(., p)$ and $(p, .)$ for some p) from q_s to q_s (the first and last symbols are of the form $(q_s, .)$ and $(., q_s)$, respectively); and (3) this sequence of states is a computation of N on x (i.e., every symbol $y_i = (p, q)$ is a legal step of N on x_i, namely $(p, x_i, q, \text{R}) \in \delta$). Then the argument of the last paragraph shows that $x \in L \Longrightarrow (\exists y) R(x, y)$. Conversely, if $R(x, y)$, then (3) means that the sequence guaranteed by (2) is an accepting computation of N on x, and thus $x \in L$.

Finally, R is solved by a 2DFV$_*$ M which, on input $\langle x, y \rangle$, works as follows. It scans y and, on every two successive symbols $y_i = (p_i, q_i)$ and $y_{i+1} = (p_{i+1}, q_{i+1})$, checks that $q_i = p_{i+1}$ by zig-zagging t times between y_i and y_{i+1} to test that the corresponding bits of $q_i, p_{i+1} \in [s]$ are identical. At the start and end of the scan, M also checks that the first and last symbols of y have respectively the form $(0, .)$ and $(., 0)$. This confirms condition (2). Condition (3) is checked in the same scan: whenever M reads a new symbol $y_i = (p_i, q_i)$, it also verifies that $(p_i, x_i, q_i, \text{R}) \in \delta$. Easily, M needs no more than $O(t) = O(\log s)$ states. □

Lemma 4. *If a binary predicate R is solved by an s-state 2DFV$_*$, then the language $L := \{x \mid (\exists y) R(x, y)\}$ is solved by a $2^{O(s)}$-state 1NFA.*

Proof. Let $M = (Q, \Sigma, \Delta, \delta, q_s)$ be the 2DFV$_*$ which solves R, with $|Q| = s$.

Pick any $x \in \Sigma^*$. Let $n := |x|$. To check whether $x \in L$, a 1NFA N guesses a $(n+2)$-long $y \in \Delta^*$ and an accepting computation of M on x and the guessed y. The certificate is guessed one symbol per step, as N scans x on its tape; likewise, the accepting computation is guessed one *frontier* per step [5, p. 547].

Formally, $N := (Q', \Sigma, \delta', F_s)$ for $Q' := \{(U, V) \mid U, V \subseteq Q \ \& \ |U|+1 = |V|\}$ the set of all frontiers of M and $F_s := (\emptyset, \{q_s\})$. When at a state (U, V) reading an input symbol $a \in \Sigma_e$, the automaton guesses a corresponding certificate symbol $b \in \Delta$, together with a frontier (U', V') such that (U, V) is (a, b)-*compatible* to it (with respect to δ [5, Def. 2]), and moves to state (U', V'):

$$((U, V), a, (U', V'), \text{R}) \in \delta' \iff (\exists b \in \Delta)[(U, V) \text{ is } (a, b)\text{-compatible to } (U', V')].$$

Therefore, N accepts x iff there exists a sequence of guesses $b_i, (U_{i+1}, V_{i+1})$ for $i = 0, 1, \ldots, n+1$ such that the sequence of frontiers $F_s = (U_0, V_0), (U_1, V_1), \ldots,$

$(U_{n+1}, V_{n+1}), (U_{n+2}, V_{n+2}) = F_s$ *fits* the string $(\vdash, b_0)(x_1, b_1) \cdots (x_n, b_n)(\dashv, b_{n+1})$ of symbols over $\Sigma_e \times \Delta$ [5, Def. 3], and thus contains an accepting computation of M on $\langle x, b_0 b_1 \cdots b_{n+1} \rangle$ [5, Lemma 2 and converse]. Hence, N accepts x iff there exists $y \in \Delta^*$ and an accepting computation of M on $\langle x, y \rangle$; i.e., iff $(\exists y) R(x, y)$.

Finally, the number of states of N is $\binom{2s}{s+1} = 2^{O(s)}$ [5, p. 552]. □

Theorem 3. *A language family* $(L_h)_{h \geq 1}$ *is in* 2^{1N} *iff there exists a binary predicate family* $(R_h)_{h \geq 1}$ *which is solved by a polynomial-size family of* 2DFV$_*$s *and is such that, for all h and all x:* $x \in L_h \iff (\exists y) R_h(x, y)$.

By similar arguments, we can also characterize the class 1N corresponding to polynomial-size 1NFAs in terms of *synchronous one-way deterministic finite verifiers* (1DFV$_*$s), the restriction of 2DFV$_*$s where the heads move only forward.

Theorem 4. *A language family* $(L_h)_{h \geq 1}$ *is in* 1N *iff there exists a binary predicate family* $(R_h)_{h \geq 1}$ *which is solved by a polynomial-size family of* 1DFV$_*$s *and is such that, for all h and all x:* $x \in L_h \iff (\exists y) R_h(x, y)$.

2.3   The Right Choice

To fix our problems, we must restore M_h's ability to move its heads independently, but still prevent the use of the second head as counter. One way to do this, is to have the second head be one-way. Formally, $\delta : Q \times \Sigma_e \times \Delta \longrightarrow Q \times \{\text{L,R}\}$ again, but now L,R indicate only the first head's motion; the second head moves always right. Let us call this a *two-way deterministic finite verifier* (2DFV).

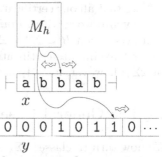

Now we can finally prove Th. 2. It follows from the next two lemmas.

Lemma 5. *If L is solved by an s-state* 2NFA, *then some binary predicate R is solved by an s-state* 2DFV *and is such that, for all x:* $x \in L \iff (\exists y) R(x, y)$.

Proof. Let $N = (Q, \Sigma, \delta, q_s)$ be the 2NFA which solves L. To motivate R, pick any $x \in L$. Pick any accepting computation c of N on x. Let m be its length. The 'instructions' followed by N along c are the pairs $\iota_1, \dots, \iota_m \in Q \times \{\text{L,R}\}$, where $\iota_i := (q, d)$ iff in the i-th step N switched to q and moved its head towards d.

$$x_0 \quad x_1 \quad x_2 \quad x_3 \quad x_4$$

| | | | | | | ι_1 | ι_2 | ι_3 | ι_4 | ι_5 | ι_6 | ι_7 |

$\vdash$ a b b $\dashv$

$q_s \quad q_3 \quad q_7$

$q_2 \quad q_4 \quad q_2 \quad q_1 \quad q_s$

| q_3 R | q_7 R | q_2 L | q_4 R | q_2 R | q_1 R | q_s R |

Hence, letting $\Delta := Q \times \{\text{L,R}\}$, we can use $y := \iota_1 \cdots \iota_m \in \Delta^*$ as certificate for x.

So, we define $R \subseteq \Sigma^* \times \Delta^*$ so that $R(x, y)$ holds iff the list of state-position pairs derived from $(q_s, 0)$ by following the instructions $y_1, \cdots, y_m \in \Delta$ is an accepting computation of N on x. It should be clear that $x \in L \iff (\exists y) R(x, y)$.

Moreover, R is solved by the 2DFV $M = (Q, \Sigma, \Delta, \delta', q_s)$ which, on input $\langle x, y \rangle$, simply follows the instructions in y and accepts iff they lead it off $\dashv$ into q_s and never violate δ: when at state p reading $a \in \Sigma_e$ and $(q, d) \in \Delta$, it checks that $(p, a, q, d) \in \delta$ and, if so, switches to q and moves towards d. Easily, M accepts $\langle x, y \rangle$ iff y causes an accepting computation of N on x; i.e., iff $R(x, y)$.   □

Lemma 6. *If a binary predicate R is solved by an s-state 2DFV, then the language $L := \{ x \mid (\exists y) R(x, y) \}$ is solved by an s-state 2NFA.*

Proof. Let $M = (Q, \Sigma, \Delta, \delta, q_s)$ be the 2DFV which solves R. Pick any $x \in \Sigma^*$. To check that $x \in L$, a 2NFA $N := (Q, \Sigma, \delta', q_s)$ simulates M on $\langle x, y \rangle$, for $y \in \Delta^*$ a certificate which is guessed on the fly, symbol-by-symbol. When at state p reading symbol $a \in \Sigma_e$, the automaton guesses the next symbol $b \in \Delta$ on the certificate tape, then switches to q and moves towards d, where $(q, d) = \delta(p, a, b)$. Formally, $(p, a, q, d) \in \delta' \iff (\exists b \in \Delta)[(q, d) = \delta(p, a, b)]$.

Easily, N accepts x iff there is a sequence of guesses $b_0, b_1, \ldots, b_m$ such that M accepts $\langle x, b_0 b_1 \cdots b_m \rangle$; namely, iff there exists $y \in \Delta^*$ such that $R(x, y)$.   □

Note that all our certificates are *finite* strings, which makes sense for $2\Sigma_1$. But we may also work with *infinite* certificates: easily, Lemmas 5 and 6 (and Th. 2) hold even when $R \subseteq \Sigma^* \times \Delta^\omega$, where $\Delta^\omega := \{$all infinite strings over $\Delta\}$, and 2DFVs have infinite certificate tape. This variation of our definitions is optional for $2\Sigma_1$; however, for $2\Pi_1$ and for general $2\Sigma_k, 2\Pi_k$ it is essential.

3   The General Case

We now turn to classes $2\Sigma_k$ and $2\Pi_k$ for arbitrary k. For concreteness, we treat only $2\Sigma_3$. (Our proof does generalize to $2\Sigma_k$, it is straightforward but tedious; then, $2\Pi_k$ is handled by a dual argument.) So, our goal is to prove the following.

Theorem 5. *A language family $(L_h)_{h \geq 1}$ is in $2\Sigma_3$ iff there is a polynomial-size quaternary predicate family $(R_h)_{h \geq 1}$ such that, for all h and all x:*

$$x \in L_h \iff (\exists z_1)(\forall z_2)(\exists z_3) R_h(x, z_1, z_2, z_3).$$

Definition 3. *A quaternary predicate family $(R_h)_{h \geq 1}$ is* polynomial-size *if some family of 2DFVs $(M_h)_{h \geq 1}$ and polynomial p are such that, for all h and x, z_1, z_2, z_3:*

$$M_h \text{ has } \leq p(h) \text{ states} \quad \& \quad R_h(x, z_1, z_2, z_3) \iff M_h \text{ accepts } \langle x, z_1, z_2, z_3 \rangle.$$

Now, each predicate relates a *finite* string x with three *infinite* strings z_1, z_2, z_3. Accordingly, a 2DFV M has three *infinite* certificate tapes, one per z_j, with its own head h_j. Crucially, the heads are *used in order*: first, M reads from h_1, keeping h_2, h_3 stationary; later, it deactivates h_1 and starts reading from h_2, keeping h_3 stationary; eventually, it deactivates h_2 too, and starts reading from h_3. Formally, $M = (Q, J, \Sigma, \Delta, \delta, q_s)$, where again $\delta : Q \times \Sigma_e \times \Delta \longrightarrow Q \times \{L, R\}$ but now the single certificate symbol always comes from the currently active head; and $J \subseteq Q$ is the states which cause a jump to the next certificate tape: entering $q \in J$ causes M to deactivate the currently active head h_j and activate h_{j+1}.

As usual, the proof consists of two lemmas, each for a single direction and h.

Lemma 7. *If L is solved by an s-state $2\Sigma_3\text{FA}$, then some quaternary predicate R is solved by an $O(s)$-state 2DFV and is such that, for all x:*

$$x \in L \iff (\exists z_1)(\forall z_2)(\exists z_3)R(x, z_1, z_2, z_3).$$

Proof idea. Let $A = (Q, ., \Sigma, ., .)$ be a $2\Sigma_3\text{FA}$ for L. Then A follows 'instructions' from $\Delta := Q \times \{\text{L},\text{R}\}$ and we define $R \subseteq \Sigma^* \times (\Delta^\omega)^3$ so that $R(x, z_1, z_2, z_3)$ iff *either* $z_1 z_2 z_3$ is the list of instructions followed along an accepting full computation path on x, and z_2 consists of those followed from universal configurations; *or* $z_1 z_2 z_3$ starts as such a list, but contains an invalid instruction in z_2. $\qquad\square$

Lemma 8. *If a quaternary predicate R is solved by an s-state 2DFV, then the language $L := \{x \mid (\exists z_1)(\forall z_2)(\exists z_3)R(x, z_1, z_2, z_3)\}$ is solved by a $3s$-state $2\Sigma_3\text{FA}$.*

Proof idea. Let $M = (Q, ., \Sigma, \Delta, ., q_s)$ be a 2DFV for R. Then a $2\Sigma_3\text{FA}$ $A := (Q_1 \cup Q_2 \cup Q_3, Q_2, \Sigma, ., q_s^1)$, where each $Q_j := \{p^j \mid p \in Q\}$ is a copy of Q, checks whether an input $x \in \Sigma^*$ is in L by simulating M on $\langle x, z_1, z_2, z_3 \rangle$, for some strings $z_1, z_2, z_3 \in \Delta^\omega$ which are respectively guessed, universally selected, and guessed, each of them up to some prefix and on the fly. This works in three phases, where each phase j uses states exclusively from Q_j. $\qquad\square$

References

1. Birget, J.-C.: Two-way automata and length-preserving homomorphisms. Mathematical Systems Theory 29, 191–226 (1996)
2. Fagin, R.: Generalized first-order spectra and polynomial-time recognizable sets. In: Karp, R.M. (ed.) Complexity of Computation. AMS-SIAM Symposia in Applied Mathematics, vol. VII, pp. 43–73 (1974)
3. Geffert, V.: An alternating hierarchy for finite automata. Theoretical Computer Science 445, 1–24 (2012)
4. Immerman, N.: Descriptive complexity. Springer (1998)
5. Kapoutsis, C.: Removing bidirectionality from nondeterministic finite automata. In: Jedrzejowicz, J., Szepietowski, A. (eds.) MFCS 2005. LNCS, vol. 3618, pp. 544–555. Springer, Heidelberg (2005)
6. Kapoutsis, C.: Size complexity of two-way finite automata. In: Diekert, V., Nowotka, D. (eds.) DLT 2009. LNCS, vol. 5583, pp. 47–66. Springer, Heidelberg (2009)
7. Kapoutsis, C.A.: Minicomplexity. In: Kutrib, M., Moreira, N., Reis, R. (eds.) DCFS 2012. LNCS, vol. 7386, pp. 20–42. Springer, Heidelberg (2012)
8. Kapoutsis, C., Lefebvre, N.: Analogs of Fagin's Theorem for small nondeterministic finite automata. In: Yen, H.-C., Ibarra, O.H. (eds.) DLT 2012. LNCS, vol. 7410, pp. 202–213. Springer, Heidelberg (2012)

9. Kapoutsis, C., Mulaffer, L.: A descriptive characterization of the power of small 2NFAs (in preparation, 2014)
10. Meyer, A.R., Stockmeyer, L.J.: The equivalence problem for regular expressions with squaring requires exponential space. In: Proceedings of the Symposium on Switching and Automata Theory, pp. 125–129 (1972)
11. Sakoda, W.J., Sipser, M.: Nondeterminism and the size of two-way finite automata. In: Proceedings of STOC, pp. 275–286 (1978)
12. Sipser, M.: Introduction to the theory of computation, 3rd edn. Cengage Learning (2012)
13. Stockmeyer, L.J.: The polynomial-time hierarchy. Theoretical Computer Science 3, 1–22 (1976)

Function Spaces
for Second-Order Polynomial Time[*]

Akitoshi Kawamura and Arno Pauly

[1] Department of Computer Science,
University of Tokyo, Japan
kawamura@is.s.u-tokyo.ac.jp
[2] Clare College, University of Cambridge
United Kingdom
Arno.Pauly@cl.cam.ac.uk

Abstract. In the context of second-order polynomial-time computability, we prove that there is no general function space construction. We proceed to identify restrictions on the domain or the codomain that do provide a function space with polynomial-time function evaluation containing all polynomial-time computable functions of that type.

As side results we show that a polynomial-time counterpart to admissibility of a representation is not a suitable criterion for natural representations, and that the Weihrauch degrees embed into the polynomial-time Weihrauch degrees.

Keywords: cartesian closed, computational complexity, higher order, computable analysis, admissible representation, Weihrauch reducibility.

1   Introduction

Computable analysis (e.g. [32]) deals with computability questions for operators from analysis such as integration, differentiation, Fourier transformation, etc.. In general, the actual computation is envisioned to be performed on infinite sequences over some finite or countable alphabet, this model is then lifted to the spaces of interest by means of representations. Thus, an adequate choice of representations for the various relevant spaces is the crucial foundation for any investigation in computable analysis.

At first, the search for good representations proceeded in a very ad-hoc fashion, exemplified by TURING's original definition of a computable real number as one with computable decimal expansion [29] and later correction to one with a computable sequence of nested rational intervals collapsing to the number [30][1].

The development of more systematic techniques to identify good representations had two interlocked main components: One, the identification of *admissibility* as the central criterion whenever the space in question already carries a

---

[*] A full version containing the omitted proofs is available on the arXiv as [18].
[1] This choice of a representation, which is indeed a *correct* one, is credited to BROUWER by TURING.

A. Beckmann, E. Csuhaj-Varjú, and K. Meer (Eds.): CiE 2014, LNCS 8493, pp. 245–254, 2014.
© Springer International Publishing Switzerland 2014

natural topology by KREITZ and WEIHRAUCH [20] and later SCHRÖDER [28]. Two, the observation that one can form function spaces in the category of represented spaces (e.g. [31], [2]). Using the ideas of synthetic topology [7], this suffices to obtain good representations of spaces just from their basic structure[2] (demonstrated in [23]).

While computable analysis has obtained a plethora of results, for a long time the aspect of computational complexity has largely been confined to restricted settings (e.g. [33]) or non-uniform results (e.g. [19]). This was due to the absence of a sufficiently general theory of second-order polynomial-time computability – a gap which was filled by COOK and the first author in [15]. This theory can be considered as a refinement of the computability theory. In particular, this means that for doing complexity theory, one has to choose well-behaved representations for polynomial-time computation out of the equivalence classes w.r.t. computable translations.

Various results on individual operators have been obtained in this new framework [13,16,17,26], leaving the field at a very similar state as the early investigation of computability in analysis: While some indicators are available what good choices of representations are, an overall theory of representations for computational complexity is missing. Our goal here is to provide the first steps towards such a theory by investigating the role of admissibility and the presence of function spaces for polynomial-time computability.

2   Background on Second-Order Polynomial-Time Computability

We will use (a certain class of) string functions to encode the objects of interest. We fix some alphabet Σ. We say that a (total) function $\varphi\colon \Sigma^* \to \Sigma^*$ is *regular* if it preserves relative lengths of strings in the sense that $|\varphi(u)| \leq |\varphi(v)|$ whenever $|u| \leq |v|$. We write **Reg** for the set of all regular functions. We restrict attention to regular functions (rather than using all functions from Σ^* to Σ^*) to keep the notion of their *size* (to be defined shortly) simple.

We use an oracle Turing machine (henceforth just "machine") to convert regular functions to regular functions.

Definition 1. *A machine M computes a partial function $F \colon\subseteq \mathbf{Reg} \to \mathbf{Reg}$ if for any $\varphi \in \mathrm{dom}F$, the machine M on oracle φ and any string u outputs $F(\varphi)(u)$ and halts.*

Remark 2. For computability, this is equivalent to the model where a Turing machine converts infinite strings to infinite strings. For the discussion of polynomial-time computability, however, we really need to use strings functions in order to encode information efficiently and to measure the input size, as we will see below.

---

[2] The concept of structure here goes beyond topologies, as witnessed e.g. by the treatment of hyperspaces of measurable sets and functions in [24].

Regular functions map strings of equal length to strings of equal length. Therefore it makes sense to define the *size* $|\varphi|: \mathbb{N} \to \mathbb{N}$ of a regular function φ to be the (non-decreasing) function $|\varphi|(|u|) = |\varphi(u)|$. We will use **Mon** to denote the strictly monotone functions from $\mathbb{N}$ to $\mathbb{N}$. For technical reasons, we will tacitly restrict ourselves to those regular functions φ with $|\varphi| \in$ **Mon**, this does not impede generality[3].

We will make use of a polynomial-time computable pairing function $\langle , \rangle$: $\Sigma^* \times \Sigma^* \to \Sigma^*$, which we want[4] to satisfy $|\langle u, v \rangle| = |u| \times |v|$. This is then lifted to a pairing function on **Reg** via $\langle \varphi, \phi \rangle(u) = \langle \varphi(u), \psi(u) \rangle$, and to a mixed pairing function for $\langle -, - \rangle : \Sigma^* \times$ **Reg** $\to$ **Reg**.

Now we want to define what it means for a machine to run in polynomial time. Since $|\varphi|$ is a function, we begin by defining polynomials in a function, following the idea of Kapron and Cook [12]. *Second-order polynomials* (in type-1 variable L and type-0 variable n) are defined inductively as follows: a positive integer is a second-order polynomial; the variable **n** is also a second-order polynomial; if P and Q are second-order polynomials, then so are $P + Q$, $P \cdot Q$ and $L(P)$. An example is

$$L\big(L(\mathbf{n} \cdot \mathbf{n})\big) + L\big(L(\mathbf{n}) \cdot L(\mathbf{n})\big) + L(\mathbf{n}) + 4. \tag{1}$$

A second-order polynomial P specifies a function, which we also denote by P, that takes functions $L \in$ **Mon** to another function $P(L) \in$ **Mon** in the obvious way. For example, if P is the above second-order polynomial (1) and $L(n) = n^2$, then $P(L)$ is given by

$$P(L)(n) = \big((n \cdot n)^2\big)^2 + (n^2 \cdot n^2)^2 + n^2 + 4 = 2 \cdot n^8 + n^2 + 4. \tag{2}$$

As in this example, $P(L)$ is a (usual first-order) polynomial if L is.

Definition 3. *A machine M runs in* polynomial time *if there is a second-order polynomial P such that, given any $\varphi \in$ **Reg** as oracle and any $u \in \Sigma^*$ as input, M halts within $P(|\varphi|)(|u|)$ steps.*

This defines the class of (polynomial-time) computable functions from **Reg** to **Reg**. We can suitably define some other complexity classes related to nondeterminism or space complexity, as well as the notions of reduction and hardness [15].

A *representation* δ of a set X is formally a partial function from **Reg** to X that is surjective—that is, for each $x \in X$, there is at least one $\varphi \in$ **Reg** with $\delta(\varphi) = x$. We say that φ is a δ-*name* of x. A *represented space* is a pair $\mathbf{X} = (X, \delta_X)$ of a set X together with a representation δ_X of it. For a function $f :\subseteq \mathbf{X} \to \mathbf{Y}$ between represented spaces $\mathbf{X}, \mathbf{Y}$ and $F :\subseteq$ **Reg** $\to$ **Reg**, we call F a realizer of f (notation $F \vdash f$), iff $\delta_Y(F(p)) = f(\delta_X(p))$ for all $p \in \mathrm{dom}(f\delta_X)$.

---

[3] Given some $\varphi \in$ **Reg**, let φ' be defined by $\varphi'(v) = v\varphi(v)$. Then the function $\cdot' :$ **Reg** $\to$ **Reg** is polynomial-time computable, and has a polynomial-time computable inverse. Moreover, $|\varphi'| \in$ **Mon** for all $\varphi \in$ **Reg**.

[4] While this choice is a bit wasteful, it is useful for technical reasons, and ultimately does not matter for polynomial-time computability.

A map between represented spaces is called (polynomial-time) computable, iff it has a (polynomial-time) computable realizer.

Type-2 complexity theory generalizes classical complexity theory, as we can regard the objects of the latter as special **Reg**-represented spaces. In the following, we will in particular understand $\mathbb{N}$ to be represented via $\delta_\mathbb{N}(\varphi) = |\varphi(0)|$, i.e. using an adaption of the unary representation (although not much would change if the binary representation were used instead).

3   Some Properties of Second-Order Polynomials

We will establish some properties of second-order polynomials as the foundation for our further investigations. For this, we first introduce the notion of the second-order degree of a second-order polynomial by $\deg(1) = 0$, $\deg(n) = 0$, $\deg(P + Q) = \max\{\deg(P), \deg(Q)\}$, $\deg(P \times Q) = \max\{\deg(P), \deg(Q)\} + 1$ for $P, Q \neq 0$ and $\deg(\mathrm{L}(P)) = \deg(P) + 1$. Just as the degree of an ordinary polynomial uniquely determines its $\mathcal{O}$-notation equivalence class, we find a similar result for the second-order degree and second-order polynomials. The role of the monomials x^n are taken by the second-order polynomials P_n defined via $P_0(p)(k) = k$ and $P_{n+1}(p)(k) = p(P_n(p)(k))$.

Lemma 4. *For any second-order polynomial Q there are $q \in \mathbf{Mon}$ and $n \in \mathbb{N}$ such that $Q(p)(k) \leq P_{\max\{\deg(Q),1\}}(p \times q)((k + 1)^n)$ for all $p \in \mathbf{Mon}$, $k \in \mathbb{N}$.*

Proof. Omitted.

Lemma 5. *For no $q \in \mathbf{Mon}$, $n, m \in \mathbb{N}$ we have $P_{n+1}(p)(k) \leq P_n(p \times q)((k + 1)^m)$ for all $p \in \mathbf{Mon}$, $k \in \mathbb{N}$.*

Proof. Omitted.

4   Failure of Cartesian Closure

We shall show that the category of **Reg**-represented spaces and polynomial-time computable functions is not cartesian closed. For this we define the functions $\Phi_n : \mathbf{Reg} \to \mathbf{Reg}$ via $\Phi_0(\varphi)(w) = w$ and $\Phi_{n+1}(\varphi)(w) = \varphi(\Phi_n(\varphi)(w))$. Then computing $\Phi_n(\varphi)(w)$ takes time $\Omega(P_n(|\varphi|)(|w|))$, as already the length of the output provides a lower bound.

Theorem 6. *Let the second-order polynomial P witness polynomial-time computability of the function $F :\subseteq \mathbf{Reg} \times \mathbf{Reg} \to \mathbf{Reg}$. For no $\psi \in \mathbf{Reg}$ we may have $F(\psi, \varphi) = \Phi_{\deg(P)+1}(\varphi)$ for all $\varphi \in \mathbf{Reg}$.*

Proof. If one considers the runtime bounds available for F by assumption, and for $\Phi_{\deg(P)+1}$ as above, the claim becomes an immediate consequence of Lemma 5.

Corollary 7. *There cannot be an exponential in the category of **Reg**-represented spaces and polynomial-time computable functions.*

Proof. Any realizer of the evaluation operation would violate Theorem 6.

5 Clocked Type-Two Machines

Despite the negative result above, we can identify spaces of functions with some of the desired properties of exponentials. The required technical tool is a type-two version of clocked Turing machines. We pick a Universal Turing Machine (UTM) M which simulates efficiently, meaning that on input n, φ, w the time M needs to compute the output of the nth Oracle Turing machine on input w with oracle φ is bounded by a quadratic polynomial in n and the time T needed by the nth Turing machine itself to compute the output on w with oracle φ [5]. Then M is extended by a clock evaluating the standard second-order polynomial[6] P_m on $|\langle n, \varphi \rangle|, |w|^l$ for fixed m and some $l \in \mathbb{N}$ encoded as $(x \mapsto x^l) \in \mathbf{Mon}$ and aborts the computation of M once the runtime exceeds the value of P_m. Denote the resulting machine with $M^{T=P_m}$. The runtime of $M^{T=P_m}$ can be bounded by $KP_{m+1}^2 + K$ for some constant $K \in \mathbb{N}$. In particular we find that the second-order degree of the runtime of $M^{T=P_m}$ is $m + 1$.

Theorem 8. *For any partial function $f :\subseteq \mathbf{Reg} \to \mathbf{Reg}$ computable in polynomial time P with $\deg(P) \leq m$ there are some $\psi \in \mathbf{Reg}$, $n, l \in \mathbb{N}$ such that for any $\varphi \in dom(f)$ we find $f(\varphi) = M^{T=P_m}(\langle n, \langle \varphi, \psi \rangle, x^l \rangle)$.*

Proof. Omitted.

Based on the preceding theorem, we see that rather than a single function space, we obtain a family of function spaces indexed by a natural number corresponding to the second-order degree. Given two $\mathbf{Reg}$-represented spaces $\mathbf{X}$, $\mathbf{Y}$ we define the function space $\mathcal{C}^{T=P_m}(\mathbf{X}, \mathbf{Y})$ by letting $\langle n, \psi, x^l \rangle \in \mathbf{Reg}$ be a name for $f : \mathbf{X} \to \mathbf{Y}$ if $\varphi \mapsto M^{T=P_m}(\langle n, \langle \varphi, \psi \rangle, x^l \rangle)$ is a realizer of f. This definition just enforces that $\mathrm{Eval} : \mathcal{C}^{T=P_m}(\mathbf{X}, \mathbf{Y}) \times \mathbf{X} \to \mathbf{Y}$ is computable with polynomial time bound $KP_{m+1}^2 + K$.

We can then reformulate Theorem 6 as $\mathcal{C}^{T=P_m}(\mathbf{Reg}, \mathbf{Reg}) \subsetneq \mathcal{C}^{T=P_{m+1}}(\mathbf{Reg}, \mathbf{Reg})$ and Theorem 8 as $f \in \mathcal{C}^{T=P_m}(\mathbf{X}, \mathbf{Y})$ for any $f : \mathbf{X} \to \mathbf{Y}$ computable in a polynomial time-bound of $\deg \leq m$. We can easily obtain an even stronger version of the latter by adapting the proof:

Corollary 9. *For a function $f : \mathbf{X} \to \mathbf{Y}$ the following properties are equivalent:*

1. *f is computable in polynomial time P with $\deg(P) \leq m$.*
2. *$f \in \mathcal{C}^{T=P_m}(\mathbf{X}, \mathbf{Y})$ has a polynomial time computable name.*

6 Effectively Polynomial-Bounded Spaces

Our next goal is to investigate restrictions we can employ on $\mathbf{X}$ (and later on $\mathbf{Y}$) in order to force the collapse of the time hierarchy $\mathcal{C}^{T=P_m}(\mathbf{X}, \mathbf{Y}) \subseteq$

---

[5] A straight-forward adaption of the classical result by HENNIE and STEARNS [10] provides the existence of such a universal machine.

[6] More generally, we could use an arbitrary time-constructible function in place of P_m. That P_m actually is time-constructible is witnessed by Φ_m.

$\mathcal{C}^{T=P_{m+1}}(\mathbf{X}, \mathbf{Y})$. The collapse will only occur at the second level, as this is the minimal level where a query to the second-order input may depend on the result of another such query, which is required in order to fully utilize the function-argument depending on the input-argument.

Definition 10. *We call $\mathbf{X}$ effectively polynomially bounded (epb)[7] , iff it admits a **Reg**-representation $\delta_{\mathbf{X}}$ such that there is a constant $c \in \mathbb{N}$ and a monotone polynomial $Q : \mathbb{N} \to \mathbb{N}$ s.t.:*

$$\forall \varphi \in dom(\delta_{\mathbf{X}}) \; \forall i \in \mathbb{N} \quad |\varphi|(i) \le c|\varphi|(c)^c Q(i)$$

Theorem 11. *Let $\mathbf{X}$ be epb. Then for any $m \ge 2$ we find $\mathcal{C}^{T=P_2}(\mathbf{X}, \mathbf{Y}) \cong \mathcal{C}^{T=P_m}(\mathbf{X}, \mathbf{Y})$ where $\cong$ denotes polytime isomorphic.*

Proof. It suffices to show only the direction $\subseteq: \mathcal{C}^{T=P_m}(\mathbf{X}, \mathbf{Y}) \to \mathcal{C}^{T=P_2}(\mathbf{X}, \mathbf{Y})$. Let M be the UTM used in the definition of $\mathcal{C}^{T=P_m}(\mathbf{X}, \mathbf{Y})$, let M' behave with the oracle $\langle \varphi, \langle \psi, \psi' \rangle \rangle$ in exactly the same way as M does with $\langle \varphi, \psi \rangle$, and then finally, use M' to define $\mathcal{C}^{T=P_2}(\mathbf{X}, \mathbf{Y})$.

The assumption that $\mathbf{X}$ is epb allows us to estimate:

$$
\begin{aligned}
P_m(|\langle \varphi, \psi \rangle|)(k) &= |\langle \varphi, \psi \rangle|(P_{m-1}(|\langle \varphi, \psi \rangle|)(k)) \\
&\le c|\varphi|(c)^c Q(P_{m-1}(|\langle \varphi, \psi \rangle|)(k)) \times |\psi|(P_{m-1}(|\langle \varphi, \psi \rangle|)(k)) \\
&\le (cQ^c \times |\psi|)(P_{m-1}(|\langle \varphi, \psi \rangle|)((k+1)^c)) \\
&\le (cQ^c \times |\psi|) \left((cQ^c + |\psi|)(P_{m-2}(|\langle \varphi, \psi \rangle|)((k+1)^{c^2})) \right) \\
&\le (cQ^c \times |\psi|)^{(m)}(|\langle \varphi, \psi \rangle|((k+1)^{c^m})) \\
&\le P_2(|\langle \varphi, \psi \rangle| \times (cQ^c \times |\psi|)^{(m)})((k+1)^{c^m})
\end{aligned}
$$

Now given ψ, we can compute some ψ' with $|\langle \varphi, \psi \rangle| \times (cQ^c \times |\psi|)^{(m)} \le |\langle\langle \varphi, \psi \rangle, \psi' \rangle|$ in polynomial time (note that Q, c and m are all constants here). The l in the original name is replaced by lc^m.

It is worthwhile pointing out that the function spaces for computability do not only contain the computable functions as elements, but comprise exactly the continuous functions as discussed very well in [1], yielding a structure dubbed *category extension* in [23,22]. This is due to the fact that the (partial) functions $f :\subseteq \mathbb{N}^{\mathbb{N}} \to \mathbb{N}^{\mathbb{N}}$ arising as sections of computable (partial) functions $F :\subseteq \mathbb{N}^{\mathbb{N}} \times \mathbb{N}^{\mathbb{N}} \to \mathbb{N}^{\mathbb{N}}$ are just the continuous functions.

In a similar way, we shall investigate which functions appear in a space $\mathcal{C}^{T=P_2}(\mathbf{X}, \mathbf{Y})$ for epb $\mathbf{X}$. It turns out that (a modification of) uniform continuity plays a central role. A connection between run-time bounds and the modulus of continuity was also found for multivalued functions in [25].

---

[7] Note that the epb-condition acts on the domain of the representation only, it does not relate to any hypothetical additional structure available on $\mathbf{X}$ (such as a metric). In particular, this condition is unrelated to the notion of a concise representation introduced by WEIHRAUCH in [33].

Definition 12. *We call a partial function $f :\subseteq \mathbf{Reg} \to \mathbf{Reg}$ polytime-locally uniformly continuous, if there is a polynomial-time computable function $\chi :\subseteq \mathbf{Reg} \to \mathbb{N}$, such that $dom(f) \subseteq dom(\chi)$ and any $f|_{\chi^{-1}(\{n\})}$ is uniformly continuous.*

Theorem 13. *Let $\mathbf{X} \subseteq \mathbf{Reg}$ be epb. Then for $f : \mathbf{X} \to \mathbf{Reg}$ the following are equivalent:*

1. *f is polytime-locally uniformly continuous*
2. *$f \in \mathcal{C}^{T=P_2}(\mathbf{X}, \mathbf{Reg})$*

Proof. 1. $\Rightarrow$ 2. Given Theorem 11 and Corollary 9, it suffices to show that such an f is polynomial-time computable relative to some oracle ψ. We start by some $\varLambda \in \mathbf{Mon}$ such that $i \mapsto \varLambda(\langle n, i \rangle)$ is a modulus of continuity of $f|_{\chi^{-1}(\{n\})}$. Then $f(\varphi)(u)$ depends only on values $\varphi(w)$ with $|w| \leq \varLambda(\langle \chi(\varphi), |u| \rangle)$, and we may encode this dependency in some table ψ. In order to write the query to ψ, the machine needs time $2^{\varLambda(\langle \chi(\varphi), |u| \rangle)}$. By providing $\langle 2^{\varLambda}, \psi \rangle$ as an oracle, this time is made available.

2. $\Rightarrow$ 1. Omitted.

Note that the same argument used for 1. $\Rightarrow$ 2. in the preceding proof also establishes that $\mathcal{C}^{T=P_2}(\mathbb{R}, \mathbb{R})$ contains all the continuous functions, where $\mathbb{R}$ is represented as suggested in [15], as observed by the first author in [14]. In particular, $\mathbb{R}$ as defined there is an epb space – and the best example of an epb space available to us.

Observation 14. *If $\mathbf{X}$ and $\mathbf{Y}$ are epb, then so are $\mathbf{X} + \mathbf{Y}$ and $\mathbf{X} \times \mathbf{Y}$. Any subspace of an epb-space is epb itself. However, $\mathcal{C}^{T=P_2}(\mathbf{X}, \mathbf{Y})$ is not necessarily epb. If $\mathbf{X} \cong \mathbf{X}'$, we also cannot conclude that $\mathbf{X}'$ is epb, as $\mathbf{X}'$ may have superfluous fast-growing names[8].*

7 Padding and Polytime Admissibility

In this section we shall explore two distinct but similar arguments based on using padding-like concepts on the codomain of a function in order to make time bounds irrelevant. This technique both reveals polynomial-time admissibility as a far too restrictive concept (as opposed to computable admissibility) and allows us to draw some conclusions about degree structures.

We define a $\mathbf{Reg}$-representation π of Cantor space via $dom(\pi) = \{\varphi \in \mathbf{Reg} \mid range(|\varphi|) = \mathbb{N}\}$ and $\pi(\varphi)(i) = \varphi(0^n)(i)$ where $n = \min\{j \in \mathbb{N} \mid |\varphi(0^j)| = i\}$. Now any Cantor-representation δ can be turned into a $\mathbf{Reg}$-representation by composing with π, and by this we obtain a strong correspondence between computability and polynomial-time computability.

---

[8] This aspect raises the question whether there is a convenient characterization of representations that are polynomial-time equivalent to an epb representation.

Proposition 15. *A function* $f : \mathbf{X} \to (Y, \delta_\mathbf{Y})$ *is computable if and only if* $f : \mathbf{X} \to (Y, \delta_\mathbf{Y} \circ \pi)$ *is polynomial-time computable.*

Proof. Omitted.

Weihrauch reducibility (e.g. [6,5,4,11]) is a computable many-one reduction between multivalued functions that serves as the basis of a metamathematical research programme. Likewise, a reduction that could be called polynomial-time Weihrauch reducibility has been investigated by some authors (e.g. [3,15]). In [21,22] abstract principles were demonstrated that provide a very similar degree structure for both. Let $(\mathfrak{W}, \oplus, +, \times)$ and $(\mathfrak{P}, \oplus, +, \times)$ be the corresponding degree structures for Weihrauch reducibility and polynomial-time Weihrauch reducibility. We then find:

Corollary 16. $(\mathfrak{W}, \oplus, +, \times)$ *embeds as a substructure into* $(\mathfrak{P}, \oplus, +, \times)$.

The characterization of admissibility that admits a translation into the setting of computational complexity is due to SCHRÖDER [27] (see also [23]). Given the Sierpiński space $\mathbb{S}$ and the function space $\mathcal{C}(-, -)$, we find that there is a canonic map $\kappa_\mathbf{X} : \mathbf{X} \to \mathcal{C}(\mathcal{C}(\mathbf{X}, \mathbb{S}), \mathbb{S})$ with $\kappa(x)(f) = f(x)$. A space $\mathbf{X}$ is called computably admissible, if $\kappa_\mathbf{X}$ admits a computable partial inverse.

The space $\mathbb{S}$ has the underlying set $\{\top, \bot\}$, and the representation $\delta_\mathbb{S} : \mathbf{Reg} \to \mathbb{S}$ defined by $\delta_\mathbb{S}(\varphi) = \top$ iff $\exists w . \varphi(w) = 1$. By the same argument as Proposition 15, any computable function into $\mathbb{S}$ is computable in polynomial time – in fact, even linear time suffices. Thus, just as in Section 6 we can use the space $\mathcal{C}^{T=P_1}(\mathbf{X}, \mathbb{S})$ as a function space and subsequently obtain a definition of polynomial-time admissibility by calling $\mathbf{X}$ polynomial-time admissible iff the (polynomial-time computable) map $\kappa_\mathbf{X} : \mathbf{X} \to \mathcal{C}^{T=P_1}(\mathcal{C}^{T=P_1}(\mathbf{X}, \mathbb{S}), \mathbb{S})$ has a polynomial-time computable partial inverse. However, this notion is of limited use:

Proposition 17. *If* $x \in \mathbf{X}$ *for polynomial-time admissible* $\mathbf{X}$ *has a computable name, then it has a polynomial-time computable name.*

Proof. Omitted.

Note that this implies that all the representations suggested in [15] fail to be polynomial-time admissible, despite appearing to be very reasonable choices[9].

8   Conclusions

The trusted techniques developed for the theory of represented spaces and computable functions are insufficient to fully comprehend polynomial-time computability. Function spaces are not always available, and even where they are,

---

[9] Nevertheless, there are non-trivial polynomial-time admissible spaces. In particular, any space $\mathcal{C}^{T=P_1}(\mathbf{X}, \mathbb{S})$ will be polynomial-time admissible. Consequently, we find that there is a polynomial-time admissible space in any equivalence class regarding computable translations that is computably admissible – but for these spaces, the formally defined polynomial-time computability actually is just computability, without any complexity-theoretic flavour to it.

they might differ from the familiar one of the continuous functions[10]. Instead, some form of uniform continuity will be appear as the central notion.

What can be used as a guiding principle for the choice of representations is the epb property. If compatible with other criteria, choosing a representation that makes a space epb also makes function spaces well-behaved. For example, separable metric spaces are traditionally represented by encoding points by fast converging sequences of basic elements. For computability theory it does not matter what *fast* means – for complexity theory it does. A sensible choice could be: As fast as possible while retaining the epb property. Whether this already determines a representation up to polynomial-time equivalence is open, though.

Acknowledgements. The second author is grateful to Anuj Dawar, Carsten Rösnick and Martin Ziegler for valuable discussions pertaining to the topic of the paper. This work is supported in part by the Japanese Grant-in-Aid for Scientific Research (Kakenhi) 24106002, and the Marie Curie International Research Staff Exchange Scheme *Computable Analysis*, PIRSES-GA-2011- 294962.

References

1. Bauer, A.: Realizability as the connection between computable and constructive mathematics. Tutorial at CCA 2004, notes (2004)
2. Bauer, A.: A relationship between equilogical spaces and type two effectivity. Mathematical Logic Quarterly 48(1), 1–15 (2002)
3. Beame, P., Cook, S., Edmonds, J., Impagliazzo, R., Pitassi, T.: The relative complexity of NP search problems. Journal of Computer and System Science 57, 3–19 (1998)
4. Brattka, V., de Brecht, M., Pauly, A.: Closed choice and a uniform low basis theorem. Annals of Pure and Applied Logic 163(8), 968–1008 (2012)
5. Brattka, V., Gherardi, G.: Effective choice and boundedness principles in computable analysis. Bulletin of Symbolic Logic 1, 73–117 (2011), arXiv:0905.4685
6. Brattka, V., Gherardi, G.: Weihrauch degrees, omniscience principles and weak computability. Journal of Symbolic Logic 76, 143–176 (2011), arXiv:0905.4679
7. Escardó, M.: Synthetic topology of datatypes and classical spaces. Electronic Notes in Theoretical Computer Science 87 (2004)
8. Férée, H., Gomaa, W., Hoyrup, M.: Analytical properties of resource-bounded real functionals. Journal of Complexity (to appear)
9. Férée, H., Hoyrup, M.: Higher-order complexity in analysis. In: CCA 2013 (2013)
10. Hennie, F.C., Stearns, R.E.: Two-tape simulation of multitape turing machines. J. ACM 13(4), 533–546 (1966)
11. Higuchi, K., Pauly, A.: The degree-structure of Weihrauch-reducibility. Logical Methods in Computer Science 9(2) (2013)
12. Kapron, B.M., Cook, S.A.: A new characterization of type-2 feasibility. SIAM Journal on Computing 25(1), 117–132 (1996)

---

[10] This observation was also made by FÉRÉE and Hoyrup [9] (see also [8]), and they suggested to use higher-order functionals on the machine level to retain spaces of continuous function with efficient evaluation. However, as shown by SCHRÖDER (personal communication), this would change the notion of computability, too.

13. Kawamura, A.: Lipschitz continuous ordinary differential equations are polynomial-space complete. Computational Complexity 19(2), 305–332 (2010)
14. Kawamura, A.: On function spaces and polynomial-time computability. Dagstuhl Seminar 11411 (2011)
15. Kawamura, A., Cook, S.: Complexity theory for operators in analysis. ACM Transactions on Computation Theory 4(2), Article 5 (2012)
16. Kawamura, A., Müller, N., Rösnick, C., Ziegler, M.: Parameterized Uniform Complexity in Numerics: from Smooth to Analytic, from NP-hard to Polytime. arXiv 1211.4974 (2012)
17. Kawamura, A., Ota, H., Rösnick, C., Ziegler, M.: Computational complexity of smooth differential equations. Logical Methods in Computer Science 10(1), Paper 6 (2014)
18. Kawamura, A., Pauly, A.: Function spaces for second-order polynomial time. arXiv 1401.2861 (2014)
19. Ko, K.I.: Polynomial-time computability in analysis. Birkhäuser (1991)
20. Kreitz, C., Weihrauch, K.: Theory of representations. Theoretical Computer Science 38, 35–53 (1985)
21. Pauly, A.: Many-one reductions between search problems. arXiv 1102.3151 (2011), http://arxiv.org/abs/1102.3151
22. Pauly, A.: Multi-valued functions in computability theory. In: Cooper, S.B., Dawar, A., Löwe, B. (eds.) CiE 2012. LNCS, vol. 7318, pp. 571–580. Springer, Heidelberg (2012)
23. Pauly, A.: A new introduction to the theory of represented spaces (2012), http://arxiv.org/abs/1204.3763
24. Pauly, A., de Brecht, M.: Towards synthetic descriptive set theory: An instantiation with represented spaces. arXiv 1307.1850
25. Pauly, A., Ziegler, M.: Relative computability and uniform continuity of relations. Journal of Logic and Analysis 5 (2013)
26. Rösnick, C.: Closed sets and operators thereon. In: CCA 2013 (2013)
27. Schröder, M.: Admissible Representations for Continuous Computations. Ph.D. thesis, FernUniversität Hagen (2002)
28. Schröder, M.: Extended admissibility. Theoretical Computer Science 284(2), 519–538 (2002)
29. Turing, A.: On computable numbers, with an application to the Entscheidungsproblem. Proceedings of the LMS 2(42), 230–265 (1936)
30. Turing, A.: On computable numbers, with an application to the Entscheidungsproblem: Corrections. Proceedings of the LMS 2(43), 544–546 (1937)
31. Weihrauch, K.: Type 2 recursion theory. Theoretical Computer Science 38, 17–33 (1985)
32. Weihrauch, K.: Computable Analysis. Springer (2000)
33. Weihrauch, K.: Computational complexity on computable metric spaces. Mathematical Logic Quarterly 49(1), 3–21 (2003)

Complexity of Operation Problems

Martin Kutrib

Institut für Informatik, Universität Giessen
Arndtstr. 2, 35392 Giessen, Germany
kutrib@informatik.uni-giessen.de

Abstract. The operation problem for several classes of automata and other language descriptors is addressed: Fix an operation on formal languages. Given a class of automata (or other language descriptors), is the application of this operation to the given class still a language represented by a device of that class? In particular, several aspects of complexity in connection with these problems are considered. Is the problem decidable or not? What is the computational complexity of the decision procedure, or what is its precise level in the arithmetic hierarchy? What is the blow-up of the size of the resulting device, if it exists, in terms of the sizes of the given ones? Otherwise, is there a so-called non-recursive trade-off between the representation by devices combined with the operation and the representation by just one device? We present some selected results on the computational and descriptional complexity of operation problems and draw attention to the overall picture and some of the main ideas involved.

1 Introduction

From an implementation point of view, the operation problem is related to the question whether, for example, a parser or acceptor for a given language can be decomposed into several simpler parsers. Advantages of simpler parsers, whose combination according to the operation is equivalent to the given device, are obvious. For example, the total size of the simpler devices could be smaller than the given parser, the verification is easier, etc. So, there is a natural interest in efficient decomposition algorithms. From this point of view, the complexity of the converse question, whether the composition of languages yields a given language, is interesting. The operation problem can be seen as a weaker class of such problems.

Here we address several aspects of complexity in connection with these problems. If a class is closed under a certain operation, the decidability of the problem is trivial. In this case the descriptional complexity is of particular interest: What is the blow-up of the size of the resulting device in terms of the sizes of the given ones? If a class is not closed under a certain operation, an immediate question asks for the decidability of the problem whether or not the result still belongs to the class. At this point the operation problem has strong relations to computational complexity. What is the complexity of the decidability process?

A. Beckmann, E. Csuhaj-Varjú, and K. Meer (Eds.): CiE 2014, LNCS 8493, pp. 255–264, 2014.
© Springer International Publishing Switzerland 2014

Or else, what is its precise level in the arithmetic hierarchy? Also in these cases descriptional complexity issues are interesting. Is there a so-called non-recursive trade-off between the representation by devices combined with the operation and the representation by just one device, if the latter is possible at all? The phenomenon of non-recursive trade-offs means that the trade-offs are not bounded by any recursive function. With other words, the gain in economy of description can be arbitrary. At this point, the operation problem has strong relations to computability.

The reader is assumed to be familiar with the basic notions of automata theory as contained, for example, in [22]. In the present paper we will use the following notational conventions. An alphabet Σ is a non-empty finite set, its elements are called letters or symbols. We write Σ^* for the set of all words over the finite alphabet Σ. The complement of a language $L \subseteq \Sigma^*$ is denoted by $\overline{L}$. We use $\subseteq$ for *inclusions* and $\subset$ for *strict inclusions*.

2   Closed Classes

We turn to the operation problem for classes of automata that are closed under the operation. Seemingly, this scenario is very simple, since the decision problem becomes trivial, just answer *yes*, and its computational complexity is constant. However, one can ask for the sizes of the representations of the language by one automaton as opposed to one or more automata combined with the operation. Here the descriptional complexity turns out to be a finer apparatus compared with computational complexity. In the following we exemplarily consider the classes of deterministic (DFA), nondeterministic (NFA), alternating (AFA), and Boolean finite automata (BFA).

Now, the problem for DFAs and a regularity preserving binary operation ∘ reads as follows:

− Given two DFAs A and B of sizes m and n.
− Which size is sufficient and necessary in the worst case (in terms of m and n) for a DFA to accept the language $L(A) \circ L(B)$?

Clearly, this problem generalizes as well to unary language operations as, for example, complementation, and to other devices such as, for example, NFA, AFA, BFA, and their two-way variants. As implied by the definition, here we deal with the language operation problem in terms of worst case complexity. In the following the notion *state complexity* is used to express that the size of the finite automata is measured by their number of states.

Concerning the main historical development of operational state complexity of finite automata, first observations for DFAs can be found in [36] without proof and in [32]. Later, the field of research was revitalized in [49]. Recent surveys of results dealing with this topic are [20,47,48]. A systematic study of language operations in connection with NFAs is [19] (cf. [23]). Tight bounds for AFAs and BFAs have recently been obtained in [24] (cf. [11]).

The bounds for some basic operations on DFAs, NFAs, AFAs, and BFAs accepting general and unary regular languages are summarized and compared in Table 1.

Table 1. DFA, NFA, AFA, and BFA state complexities, where t is the number of accepting states of the "left" automaton. The tight lower bounds for union, intersection, and concatenation of unary DFAs require m and n to be relatively prime.

	DFA		NFA		AFA	BFA
	general	unary	general	unary	general	general
$\cup$	$m+n+1$	$m+n+1$	mn	2^n	$m+n+1$	$m+n$
$\sim$	n	n	2^n	$2^{\Theta(\sqrt{n\cdot\log n})}$	n	n
$\cap$	mn	mn	mn	mn	$m+n+1$	$m+n$
R	2^n	n	$n+1$	n	$2^n \leq \cdot \leq 2^n+1$	2^n
$\cdot$	$m2^n - t2^{n-1}$	mn	$m+n$	$m+n-1 \leq \cdot \leq m+n$	$2^m+n \leq \cdot \leq 2^m+n+1$	2^m+n
$*$	$3\cdot 2^{n-2}$	$(n-1)^2+1$	$n+1$	$n+1$	$2^n \leq \cdot \leq 2^n+1$	$2^n \leq \cdot \leq 2^n+1$
$+$			n	n		

We chose the complementation from the range of possible operations and discuss it in more detail. Complementation often plays a crucial role in connection with nondeterminism. In fact, compared with DFAs the complementation of NFAs is expensive. Since the complementation operation on DFAs neither increases nor decreases the number of states (simply exchange accepting and rejecting states), we obtain the upper bounds for the NFA complementation by determinization, that is, 2^n states [35,39,40]. The story of the lower bound is a little longer: In [45] an example of languages over a growing alphabet size is given which reaches the upper bound 2^n. In [3] the result for a three-letter alphabet was claimed and later corrected to a four-letter alphabet in [4]. In [19] the lower bound 2^{n-2} is achieved for a two-letter alphabet and, finally, by a fooling set technique the bound 2^n was proven to be tight for a two-letter alphabet [23].

Interestingly, the complementation becomes cheap again, when the nondeterminism is generalized to alternations. The tight bound of n follows from a construction presented in [11].

It turned out that the unary case is different for NFAs compared with DFAs. In [8] it has been shown that for any unary n-state NFA there exists an equivalent $(2^{\Theta(\sqrt{n\cdot\log n})})$-state deterministic finite automaton, and in [19] it is shown that this is a tight bound in the order of magnitude for the unary NFA complementation. More detailed results on the relation between the sizes of unary NFAs and their complements are obtained in [38]. In particular, if a unary language L has a succinct NFA, then nondeterminism cannot help to recognize its complement, namely, the smallest NFA accepting the complement of L has as many states as the minimal DFA accepting it.

The operational state complexity for two-way deterministic finite automata (2DFA) has recently been investigated in [25]. Before, in [13] it has been shown that the complement of any n-state 2DFA can be accepted by a $4n$-state 2DFA that always halts. In the same paper, the polynomial upper bound of $O(n^8)$

has been obtained for *unary* two-way *nondeterministic* finite automata. Here we have a station where one can change the train to computational complexity. The challenging open question about the costs for simulating 2NFAs by 2DFAs raised by Sakoda and Sipser in [45] is unanswered for decades. The best known upper bound is of order $2^{O(\ln^2 n)}$ [12]. As shown in [14], a tightness proof of such bound (or any other superpolynomial lower bound) would imply the separation between the classes L and NL, thus solving another long-standing open problem. Now let $s(n)$ be the state complexity for the simulation of 2NFAs by 2DFAs. Then for any given n-state 2NFA there is an equivalent $s(n)$-state 2DFA and, hence, a $4s(n)$-state 2DFA (and thus 2NFA) accepting the complement of the language accepted by the given 2NFA. So, if the answer to the open question of the operational state complexity of 2NFA and complementation yields a superpolynomial lower bound, then L and NL would be separated. Other interesting connections between the problem of Sakoda and Sipser and the question whether L equals NL have been proved in [2] and, recently, in [26,27].

3   Classes with Closed Subclasses

Here we focus on the operation problem for classes of automata that are not closed under the operation, but some non-trivial subclasses are closed. Clearly, only subclasses that are decidable make sense from this point of view. Now the scenario is twofold. For the subclasses the decidability of the operation problem is again trivial, while it may be undecidable in general. Here we mention exemplarily the classes of deterministic pushdown automata, nondeterministic one-turn pushdown automata (that accept the linear context-free languages), and one-way multi-head finite automata. These classes have in common that their *unary* subclasses accept only regular languages. So, for unary devices the operation problem for regularity preserving operations becomes decidable. Though deep results on descriptional complexity issues of these important subclasses have been obtained (see, for example, [30,31,42,43]), a systematic study of operation problems is a still open and challenging task.

In the following, we turn to the undecidability of operation problems for the devices in question and, in particular, their level in the arithmetic hierarchy [44]. In [9] the recursively enumerable one-one Turing degrees were investigated and it was shown that various unsolvable problems, including some operation problems, of formal languages are in these degrees. A definition of the *arithmetic hierarchy* can be given as follows:

$$\Sigma_1 = \{\, L \mid L \text{ is recursively enumerable} \,\} \text{ and}$$

$$\Sigma_{n+1} = \{\, L \mid L \text{ is recursively enumerable in some } A \in \Sigma_n \,\}, \text{for } n \geq 1,$$

where a language L is said to be recursively enumerable in some A if there is a Turing machine with oracle A that semi-decides L. The complement of Σ_n is denoted by Π_n, that is, $\Pi_n = \{\, L \mid \overline{L} \text{ is in } \Sigma_n \,\}$. Notice, that $\Sigma_1 \cap \Pi_1$ is the class of all recursive sets. Completeness and hardness are always meant with respect to many-one reducibilities.

In [5] the following theorem is shown. It gives an upper bound on the unsolvability of the operation problem, if some easy properties are met.

Let $M_1, M_2, \ldots$ be an effective enumeration of machines of a certain type, so that the induced language class $\{ L(M_i) \mid i \geq 1 \}$ is effectively included in the class NSPACE(lin). Further let $\circ$ be a k-ary operation, $k \geq 1$, under which NSPACE(lin) is effectively closed. Then the $\circ$ operation problem for the induced language class is contained in Σ_2.

Notice, that the theorem applies to the devices in question and, for example, to the operations union, intersection, complementation, (marked) concatenation, (marked) Kleene star, non-erasing homomorphism, non-erasing substitution, shuffle, and root. In [5] the Σ_2-completeness for deterministic pushdown automata and nondeterministic one-turn pushdown automata and all these operations has been shown, if the corresponding class is not closed under the operation. The proofs use the major technique of (in)valid computations of Turing machines. For Boolean operations the result reads as follows.

1. Given two nondeterministic one-turn pushdown automata, the problems whether the intersection of both accepted languages, or whether the complement of an accepted language is again accepted by a nondeterministic one-turn pushdown automaton is Σ_2-complete.
2. Given two deterministic pushdown automata, the problems whether the intersection of both accepted languages, or whether their union is again accepted by a deterministic pushdown automaton is Σ_2-complete.

For example, the Σ_2 hardness of the intersection problems can be derived as follows. Basically, a *valid computation* of a deterministic Turing machine with one single tape and one single read-write head is a word built from a sequence of configurations passed through in an accepting computation. In [1] it has been shown that such a valid computation can be represented by the intersection of two deterministic linear context-free languages and that the corresponding acceptors can effectively be constructed. Another result in [1] says that the language accepted by an arbitrary Turing machine is finite if and only if the valid computations of that Turing machine are context free. So, the finiteness problem for Turing machines reduces to the intersection problem for the pushdown automata. Since the finiteness problem for Turing machines is Σ_2-complete [44], the Σ_2-hardness follows. Together with the above mentioned Σ_2-containment we obtain the Σ_2-completeness.

4   Classes with Decidable Subclasses

In this section we consider a scenario which is slightly different from the above. Here we are interested in classes of languages and operations so that the operation problem is undecidable in general, but is decidable for some non-trivial subclasses not closed under the operation. In this setting, the level of undecidability as well as the decidability procedure and its computational complexity are of natural interest.

In order to discuss briefly another aspect of operation problems, here we stick with the decidability procedure itself. As classes of languages we chose Lindenmayer systems (L systems for short). These systems have been introduced to describe the development of lower organisms [33,34]. L systems can be seen as a parallel counterpart to sequential rewriting mechanisms, they can be considered as finite substitutions over a free monoid, which are iteratively applied to a designated element of the monoid, the so-called axiom of the system. Basically, in every derivation step, all symbols in the sentential form have to be rewritten (in parallel), and there is no distinction between terminal and nonterminal symbols.

Formally, a T0L system is a tuple $G = \langle \Sigma, P_1, P_2, \ldots, P_r, \omega \rangle$, where r is a positive integer, Σ is an alphabet, $\omega \in \Sigma^+$ is the axiom, and P_i, for $1 \leq i \leq r$, is a finite subset of $\Sigma \times \Sigma^*$ such that for every $a \in \Sigma$, there is a word $v \in \Sigma^*$ with $(a, v) \in P_i$. The sets P_i are called the tables of G. A T0L system is *propagating* (a PT0L system) if all tables of G are finite subsets of $\Sigma \times \Sigma^+$. A 0L system is a T0L system with only one table, that is, $r = 1$. A 0L system is *propagating* (a P0L system) if the only table of G is a finite subset of $\Sigma \times \Sigma^+$.

The elements of the tables are called rules and define how a symbol of the current sentential form may be rewritten. In a single step of a T0L system, all symbols are rewritten in parallel according to one of its tables. More precisely, with every table P we associate the finite substitution σ_P defined by $\sigma_P(a) = \{ v \mid (a, v) \in P \}$. Now, the language generated by a T0L system is defined as follows. A word $x \in \Sigma^+$ directly derives a word $y \in \Sigma^*$ if there is i with $1 \leq i \leq r$, such that $y \in \sigma_{P_i}(x)$. We write $x \Rightarrow y$ in this case. The language $L(G)$ *generated* by G is defined to be the set $L(G) = \{ w \in \Sigma^* \mid \omega \Rightarrow^* w \}$, where $\Rightarrow^*$ refers to the reflexive, transitive closure of the derivation relation $\Rightarrow$.

By definition, every P0L language is also a 0L, PT0L as well as a T0L language, and both every 0L and every PT0L language is also a T0L language.

The operation problem for families of languages generated by L systems has been investigated only for the union of 0L and propagating 0L languages [10], where also the unary variant for 0L was studied. It was shown that in general the union problem for 0L languages is undecidable, while for the restricted variants of PD0L and unary 0L languages the problem becomes decidable. In [6] the operation problem for the families of 0L and T0L languages and their propagating variants are investigated to a large extent. For intersection, substitution, and all AFL operations but Kleene star, the problem turned out to be non-semi-decidable, that is, Π_1 is a lower bound. The proof is by reduction of the Post's Correspondence Problem.

Further results in [6] show the decidability of the operation problems for Kleene star, complementation, and intersection with regular sets for *unary* L systems. The proofs utilize the following characterization [18]: If L is a unary 0L language over alphabet $\{a\}$, then either L is regular, that is, there is a finite set F and integers $d \geq 1$ and $1 \leq i_1 < i_2 < \cdots < i_k$, for some $k \geq 0$, such that $L = F \cup \{a^{i_1}, a^{i_2}, \ldots, a^{i_k}\}\{a^d\}^*$, or there are integers $i \geq 1$ and $k \geq 2$ such that $L = \{ a^{i \cdot k^n} \mid n \geq 0 \}$. Moreover, given a 0L system, there is an algorithm to determine the parameters of the language. A simple application of this result yields the following theorem.

Given a unary 0L language L over a unary alphabet, it is decidable, whether or not the Kleene star of L is a 0L language.

In order to give evidence of the assertion, recall that the Kleene star of any unary language is regular. Due to the effective characterization, for a given unary 0L system G a finite automaton accepting $L(G)^*$ can effectively be constructed. Since it is decidable whether a given regular language is a 0L language [46], it is decidable whether or not $L(G)^*$ is a 0L language.

5   Classes Defined by Operations

In this section we consider the operation problem from a slightly different point of view. We use an operation to define a class of languages. For example, given two (incomparable) classes of languages where each has its own neat properties, the *intersection* of both classes should – to some extend – have the best properties of both classes, it can be the best of the two worlds. Again exemplarily, we discuss the intersection of context-free languages (CFL) and Church-Rosser languages (CRL). The known upper bound on the time complexity for the membership problem of CFL still exceeds $O(n^2)$. Church-Rosser languages have been introduced in [37]. They are defined via finite, confluent, and length-reducing Thue systems. Church-Rosser languages are incomparable to the context-free languages [7]. They parse rapidly in linear time, contain non-semilinear as well as inherently unambiguous languages [37]. Moreover, they are characterized by deterministic automata models [7,41] and contain the deterministic context-free languages as well as their reversals properly [37]. The intersection CFL ∩ CRL has been studied in [29].

Any language from the intersection CFL ∩ CRL has two representations, say in terms of context-free grammars and length-reducing Thue systems (or in terms of the equivalent automata types). One immediate question is about the succinctness of these two equivalent representations, which one is more succinct? As it turns out this question is closely related to the decidability problem whether or not a given context-free language is Church-Rosser, or vice versa. The answer to the questions reveals an interesting phenomenon, there are so-called non-recursive trade-offs between context-free and Church-Rosser languages *and vice versa*. This means, given a language from the intersection which is represented by one of the mechanism, say of size n, there is no recursive function bounding from above the size of the other representation in terms of n. We now turn to discuss this phenomenon in more detail.

For our purposes, the set of context-free grammars as well as the set of finite, confluent, and length-reducing Thue systems are called *descriptional systems*. The descriptional complexity of elements from the systems is measured by the lengths of their representations, by their *sizes* (over some fixed alphabet). A total function $f : \mathbb{N} \to \mathbb{N}$ is an *upper bound* for the increase in complexity when changing from a representation r_1 in one system to an equivalent representation in the other system, if there is always a representation r_2 in the other system so that the size of r_2 is at most f applied to the size of r_1. However, for the systems in question this function is not effectively computable. One can choose an arbitrarily large recursive function f but the gain in economy of description eventually exceeds f when changing the representation. For establishing non-recursive trade-offs the following general result from [28] is useful which is a slightly generalized and unified form of a result in [17]. In particular, it emphasizes the relation to decidability problems (see, for example, [15,16,21,28] for more on non-recursive trade-offs).

Let S_1 and S_2 be two descriptional systems for recursive languages such that any descriptor D in S_1 and S_2 can effectively be converted into a Turing machine that decides $L(D)$. If there exists a descriptional system S_3 and a property P that is not semi-decidable for descriptors from S_3, such that, given an arbitrary $D_3 \in S_3$, (i) there exists an effective procedure to construct a descriptor D_1 in S_1, and (ii) D_1 has an equivalent descriptor in S_2 if and only if D_3 does not have property P, then the trade-off between S_1 and S_2 is non-recursive.

The following levels of unsolvability have been shown in [29].

1. *Given a Church-Rosser language L, the problem whether L is context free is Σ_2-complete.*
2. *Given a context-free language L, the problem whether L is Church-Rosser is Σ_2-complete.*

In order to apply the above technique to derive non-recursive trade-offs, let S_1 be the class of nondeterministc pushdown automata and S_2 be the class of finite, confluent, and length-reducing Thue systems. The descriptional system S_3 is set to be S_1, and property P is to have no equivalent descriptor in S_2. The converse non-recursive trade-off follows symmetrically.

References

1. Baker, B.S., Book, R.V.: Reversal-bounded multipushdown machines. J. Comput. System Sci. 8, 315–332 (1974)
2. Berman, P., Lingas, A.: On the complexity of regular languages in terms of finite automata. Tech. Rep. 304, Polish Academy of Sciences (1977)
3. Birget, J.C.: Partial orders on words, minimal elements of regular languages and state complexity. Theoret. Comput. Sci. 119, 267–291 (1993)
4. Birget, J.C.: Erratum: Partial orders on words, minimal elements of regular languages and state complexity (2002),
http://clam.rutgers.edu/~birget/papers.html
5. Bordihn, H., Holzer, M., Kutrib, M.: Unsolvability levels of operation problems for subclasses of context-free languages. Int. J. Found. Comput. Sci. 16, 423–440 (2005)
6. Bordihn, H., Holzer, M., Kutrib, M.: Decidability of operation problems for t0l languages and subclasses. Inform. Comput. 209, 344–352 (2011)
7. Buntrock, G., Otto, F.: Growing context-sensitive languages and Church-Rosser languages. Inform. Comput. 141, 1–36 (1998)

8. Chrobak, M.: Finite automata and unary languages. Theoret. Comput. Sci. 47(2), 149–158 (1986); errata: Theoret. Comput. Sci. 302, 497–498 (2003)
9. Cudia, D.F.: The degree hierarchy of undecidable problems of formal grammars. In: Symposium on Theory of Computing (STOC 1970), pp. 10–21. ACM Press (1970)
10. Dassow, J., Păun, G., Salomaa, A.: On the union of 0l languages. Inform. Process. Lett. 47, 59–63 (1993)
11. Fellah, A., Jürgensen, H., Yu, S.: Constructions for alternating finite automata. Internat. J. Comput. Math. 35, 117–132 (1990)
12. Geffert, V., Mereghetti, C., Pighizzini, G.: Converting two-way nondeterministic unary automata into simpler automata. Theoret. Comput. Sci. 295, 189–203 (2003)
13. Geffert, V., Mereghetti, C., Pighizzini, G.: Complementing two-way finite automata. Inform. Comput. 205(8), 1173–1187 (2007)
14. Geffert, V., Pighizzini, G.: Two-way unary automata versus logarithmic space. Inform. Comput. 209, 1016–1025 (2011)
15. Goldstine, J., Kappes, M., Kintala, C.M.R., Leung, H., Malcher, A., Wotschke, D.: Descriptional complexity of machines with limited resources. J. UCS 8, 193–234 (2002)
16. Gruber, H., Holzer, M., Kutrib, M.: On measuring non-recursive trade-offs. J. Autom., Lang. Comb. 15, 107–120 (2010)
17. Hartmanis, J.: On Gödel speed-up and succinctness of language representations. Theoret. Comput. Sci. 26, 335–342 (1983)
18. Herman, G.T., Lee, P., van Leeuwen, J., Rozenberg, G.: Characterization of unary developmental languages. Discrete Mathematics 6, 235–247 (1973)
19. Holzer, M., Kutrib, M.: Nondeterministic descriptional complexity of regular languages. Int. J. Found. Comput. Sci. 14, 1087–1102 (2003)
20. Holzer, M., Kutrib, M.: Nondeterministic finite automata – Recent results on the descriptional and computational complexity. Int. J. Found. Comput. Sci. 20, 563–580 (2009)
21. Holzer, M., Kutrib, M.: Descriptional complexity – An introductory survey. In: Scientific Applications of Language Methods, pp. 1–58. Imperial College Press (2010)
22. Hopcroft, J.E., Ullman, J.D.: Introduction to Automata Theory, Languages, and Computation. Addison-Wesley (1979)
23. Jirásková, G.: State complexity of some operations on binary regular languages. Theoret. Comput. Sci. 330(2), 287–298 (2005)
24. Jirásková, G.: Descriptional complexity of operations on alternating and Boolean automata. In: Hirsch, E.A., Karhumäki, J., Lepistö, A., Prilutskii, M. (eds.) CSR 2012. LNCS, vol. 7353, pp. 196–204. Springer, Heidelberg (2012)
25. Jirásková, G., Okhotin, A.: On the state complexity of operations on two-way finite automata. In: Ito, M., Toyama, M. (eds.) DLT 2008. LNCS, vol. 5257, pp. 443–454. Springer, Heidelberg (2008)
26. Kapoutsis, C.A.: Two-way automata versus logarithmic space. In: Kulikov, A., Vereshchagin, N. (eds.) CSR 2011. LNCS, vol. 6651, pp. 359–372. Springer, Heidelberg (2011)
27. Kapoutsis, C.A., Pighizzini, G.: Two-way automata characterizations of l/poly versus nl. In: Hirsch, E.A., Karhumäki, J., Lepistö, A., Prilutskii, M. (eds.) CSR 2012. LNCS, vol. 7353, pp. 217–228. Springer, Heidelberg (2012)
28. Kutrib, M.: The phenomenon of non-recursive trade-offs. Int. J. Found. Comput. Sci. 16, 957–973 (2005)

29. Kutrib, M., Malcher, A.: When Church-Rosser becomes context free. Int. J. Found. Comput. Sci. 18, 1293–1302 (2007)
30. Kutrib, M., Malcher, A., Wendlandt, M.: States and heads do count for unary multi-head finite automata. In: Yen, H.-C., Ibarra, O.H. (eds.) DLT 2012. LNCS, vol. 7410, pp. 214–225. Springer, Heidelberg (2012)
31. Kutrib, M., Malcher, A., Wendlandt, M.: Size of unary one-way multi-head finite automata. In: Jurgensen, H., Reis, R. (eds.) DCFS 2013. LNCS, vol. 8031, pp. 148–159. Springer, Heidelberg (2013)
32. Leiss, E.L.: Succinct representation of regular languages by Boolean automata. Theoret. Comput. Sci. 13, 323–330 (1981)
33. Lindenmayer, A.: Mathematical models for cellular interactions in development I. Filaments with one-sided inputs. J. Theor. Biol. 18, 280–299 (1968)
34. Lindenmayer, A.: Mathematical models for cellular interactions in development II. Simple and branching filaments with two-sided inputs. J. Theor. Biol. 18, 300–315 (1968)
35. Lupanov, O.B.: A comparison of two types of finite sources. Problemy Kybernetiki 9, 328–335 (1963) (in Russian); German translation: Über den Vergleich zweier Typen endlicher Quellen. Probleme der Kybernetik 6, 328–335 (1966)
36. Maslov, A.N.: Estimates of the number of states of finite automata. Soviet Math. Dokl. 11, 1373–1375 (1970); (English translation); Dokl. Akad. Nauk SSSR 194, 1266–1268 (1970) (in Russian)
37. McNaughton, R., Narendran, P., Otto, F.: Church-Rosser Thue systems and formal languages. J. ACM 35, 324–344 (1988)
38. Mera, F., Pighizzini, G.: Complementing unary nondeterministic automata. Theoret. Comput. Sci. 330, 349–360 (2005)
39. Meyer, A.R., Fischer, M.J.: Economy of description by automata, grammars, and formal systems. In: Symposium on Switching and Automata Theory (SWAT 1971), pp. 188–191. IEEE (1971)
40. Moore, F.R.: On the bounds for state-set size in the proofs of equivalence between deterministic, nondeterministic, and two-way finite automata. IEEE Trans. Comput. 20, 1211–1214 (1971)
41. Niemann, G., Otto, F.: The Church-Rosser languages are the deterministic variants of the growing context-sensitive languages. Inform. Comput. 197, 1–21 (2005)
42. Pighizzini, G.: Deterministic pushdown automata and unary languages. Int. J. Found. Comput. Sci. 20, 629–645 (2009)
43. Pighizzini, G., Shallit, J., Wang, M.W.: Unary context-free grammars and pushdown automata, descriptional complexity and auxiliary space lower bounds. J. Comput. System Sci. 65, 393–414 (2002)
44. Rogers, H.: Theory of Recursive Functions and Effective Computability. McGraw-Hill (1967)
45. Sakoda, W.J., Sipser, M.: Nondeterminism and the size of two way finite automata. In: Symposium on Theory of Computing (STOC 1978), pp. 275–286. ACM Press (1978)
46. Salomaa, A.: Solutions of a decision problem concerning unary Lindenmayer systems. Discrete Mathematics 9, 71–77 (1974)
47. Yu, S.: State complexity of regular languages. J. Autom., Lang. Comb. 6, 221–234 (2001)
48. Yu, S.: State complexity of finite and infinite regular languages. Bull. EATCS 76, 142–152 (2002)
49. Yu, S., Zhuang, Q., Salomaa, K.: The state complexities of some basic operations on regular languages. Theoret. Comput. Sci. 125, 315–328 (1994)

A Computational Model of XACML–Based Access Control Management in Distributed Networks

Katalin Anna Lázár

Faculty of Informatics, Eötvös Loránd University
Pázmány Péter sétány 1/C, 1117 – Budapest, Hungary
kati@elte.hu

Abstract. In this paper, we propose a novel approach to enforcing eXtensible Access Control Markup Language (XACML) policy specifications in distributed environments. Our approach is based on a formal language theoretic construction, a variant of networks of parallel language processors. The language processors form teams, send and receive information through component and team level filters. The hierarchical nature of the network supports multiple levels of nesting. Consequently, different security needs can be defined at varying levels of granularity. We use various context conditions for filtering information, thus controlling information flow. Our theoretical contributions include establishing the connection between the growth of the number of strings at the components of the networks and the growth functions of developmental systems.

Keywords: distributed access control enforcement, XACML, rule combining algorithms, networks of parallel multiset string processors, information dynamics, developmental systems.

1 Introduction

Distributed computing systems have led to the development of new technologies, such as peer–to–peer (P2P) networks, Service–Oriented Architecture (SOA), web services and cloud computing. With the increase in the amount of exchanged and published information and the rapid growth in the number of computing resources such as sensors, smart phones, desktop and portable computers and virtual (cloud–based) resources, the need to integrate these computing resources of many types into ongoing computations, has become an increasingly difficult task to manage. Furthermore, the protection of sensitive information has to be guaranteed. Several models have been developed to express access control requirements in distributed systems (e.g. [5], [13]). These models, however, do not fully embrace the dynamic nature of the open environment, where participants join and leave a network in an unpredictable manner. Minsky et al. presented a decentralized coordination and control mechanism, called Law–Governed Interaction (LGI) for distributed systems [10]. LGI enables a distributed, often large, heterogeneous and open group of actors or agents to engage in interaction governed

A. Beckmann, E. Csuhaj-Varjú, and K. Meer (Eds.): CiE 2014, LNCS 8493, pp. 265–274, 2014.

by an explicitly specified policy, called the *interaction law* of the group. This law is enforced, turning a disparate group of actors into a community with members relying on each other to comply with the given law. Rather than building a centralized security policy, LGI supports the concept of conformance based on the hierarchical structure of the system. Our work is closely related to LGI, since it supports LGI–type information flow control. In addition, our model is able to incorporate concepts specified by powerful access control policy languages, such as the eXtensible Access Control Markup Language (XACML) [1]. The XACML policy language model has three main components: rules, policies and policy sets. The rule is the most elementary unit of the policy, each rule either allows (Permit) or denies (Deny) an access control request. A policy comprises a set of rules, a policy set a set of policies or other policy sets. The rule combining algorithm (RCA) determines how to combine the results of evaluating the component rules when evaluating the policy. The policy combining algorithm (PCA) specifies how to combine the results of evaluating the component policies when evaluating the policy set. In this paper, we focus on RCAs. In particular, we build formalisms for the Deny–overrides and Permit–overrides RCAs. The result of the Deny–overrides RCA is Deny, if any decision is Deny. The result of the Permit–overrides RCA is Permit, if any decision is Permit.

Our approach is based on a formal language theoretic construction, a variant of networks of parallel language processors [4]. The language processors form teams, send and receive information through component and team level filters. We extend the model proposed in [6] and [7] by developing a framework that can accommodate a more versatile access control model. The hierarchical nature of our network supports multiple levels of nesting, and as a consequence different security needs can be defined at varying levels of granularity (e.g. individual, unit, organization, etc.). We use various context conditions for filtering information, thus controlling information flow. Our theoretical contributions include establishing the connection between the growth of the number of strings at the components of the networks and the growth functions of developmental systems.

The organization of this article is as follows. In Sect. 2, we overview the notations and definitions used throughout this paper. In Sect. 3, we provide a formal specification of RCAs in XACML in our distributed networks. In Sect. 4, we characterize the dynamics of information in these networks. Finally, in Sect. 5, we summarize our achievements and propose some further research directions.

2   Preliminaries

In the followings, V^* will denote the set of words over V for an alphabet V, and $V^+ = V^* \setminus \{\lambda\}$, the set of all nonempty words, where λ is the empty string. $length(x)$ denotes the length of $x \in V^*$ and $alph(x)$ the set of symbols occurring in $x \in V^*$. For $L \subseteq V^*$, let $alph(L) = \bigcup_{x \in L} alph(x)$. For $V' \subseteq V$, $|x|_{V'}$ is the number of occurrences of letters V' in $x \in V^*$. If $V' = \{a\}$, then we simply write $|x|_a$. For a finite set A, $card(A)$ stands for the number of elements of A. $\mathbb{N}$ denotes the set of natural numbers and $\mathbb{N}_0 = \mathbb{N} \cup \{0\}$. For further notions from formal language theory, the reader is referred to [11] and [12].

A multiset is a pair $M = (V, f)$, where V is an arbitrary (not necessarily finite) set (the universe) of objects of U and $f : U \to \mathbb{N}_0$ is a mapping that assigns the multiplicity to each object, such that if $a \notin V$, then $f(a) = 0$. We define the support of $M = (V, f)$ by $supp(M) = \{a \in V \mid f(a) \geq 1\}$. M is a finite multiset, if $supp(M)$ is finite. The set of all finite multisets over the set V is denoted by V°. $card(M) = \sum_{a \in V} f(a)$ defines the cardinality of a finite multiset $M = (V, f)$. $a \in M = (V, f)$, if $a \in supp(M)$, and $M_1 = (V_1, f_1) \subseteq M_2 = (V_2, f_2)$, if $supp(M_1) \subseteq supp(M_2)$ and for all $a \in V_1, f_1(a) \leq f_2(a)$. $(M_1 \cup M_2) = (V_1 \cup V_2, f')$, where for all $a \in V_1 \cup V_2, f'(a) = f_1(a) + f_2(a)$, defines the union of two multisets. M is an empty multiset, denoted by ϵ, if $supp(M) = \emptyset$. A multiset M over the finite set of objects V can be represented by a string ω over the alphabet V with $|w|_a = f(a)$, $a \in V$, and λ represents the empty multiset ϵ. The finite multiset of objects with the word ω over V representing M is identified by $M = (V, f)$, hence $\omega \in V^\circ$ is written.

By a context condition ϱ over V^*, where V is an alphabet, we mean a computable mapping $\varrho : V^* \to \{true, false\}$. ϱ is of type **reg** (regular context condition), given by a regular language $L \subseteq V^*$, if $\varrho(\omega) = true$ for any $\omega \in V^*$, where $\omega \in L$, and $\varrho(\omega) = false$ otherwise. ϱ is of type **rc** (random context condition), given by a pair (Q, R), where $Q, R \subseteq V$, if $\varrho(\omega) = true$ for any $\omega \in V^*$ that contains each element of Q, but no element of R, and $\varrho(\omega) = false$ otherwise. We call Q the permitting and R the forbidding context condition.

A 0L system (an *interactionless Lindenmayer system*) is a triplet $G = (V, \omega, P)$, where V is an alphabet, $\omega \in V^+$ is the axiom, and P is a finite set of context–free rewriting rules over V, such that for each $a \in V$, there is a rule $a \to x$ in P (we say that P is *complete*). A deterministic 0L system is called a D0L system. If the axiom is replaced by a finite language, then we have an F0L (FD0L) system. Since the production set P of a D0L system $G = (V, \omega, P)$ defines a homomorphism $h : V \to V^*$, $G = (V, \omega, h)$ is often used instead of the first notation. By a word sequence of a D0L system $G = (V, \omega, h)$, we mean the following sequence: $h^0(\omega) = \omega, h(\omega), h^2(\omega), h^3(\omega), \ldots$. Function $f : \mathbb{N}_0 \to \mathbb{N}_0$ defined by $f(t) = length(h^t(\omega))$, $t \geq 0$, is called the growth function of G, and sequence $length(h^t(\omega))$ for $t = 0, 1, 2, \ldots$, is its growth sequence.

3 Formal Specification of Rule Combination in XACML

We introduce the notion of hierarchical networks of parallel multiset string processors with component and team level filtering. We show how our model can support rule combination strategies expressed by the XACML policy specifications.

Definition 1. *A hierarchical network of parallel multiset string processors with component and team level filtering (an* $\mathrm{H_{ct}NPMP_{F0L}}$ *system) is a construct* $\Gamma = (V, \{(t_i, \theta_i, \xi_i) \mid 1 \leq i \leq n\}, \{(c_j, \psi_j, \upsilon_j) \mid 1 \leq j \leq m\})$, *where*
- V *is an alphabet, the alphabet of the system,*
- t_i, $1 \leq i \leq n$, *is the* i*–th team,* $t_i = \{t_{i_1}, \ldots, t_{i_{s_i}}, c_{\bar{i}_1}, \ldots, c_{\bar{i}_{s_{\bar{i}}}}\} \neq \emptyset$, *where*
  $\{i_1, \ldots, i_{s_i}\} \subseteq \{1, \ldots, n\} \setminus \{i\}, \{\bar{i}_1, \ldots, \bar{i}_{s_{\bar{i}}}\} \subseteq \{1, \ldots, m\}, 0 \leq s_i \leq n, 0 \leq$

$s_{\bar{i}} \leq m$, $\{t_{i_1}, \ldots, t_{i_{s_i}}\} \cap \{t_{j_1}, \ldots, t_{j_{s_j}}\} = \emptyset$, if $1 \leq i \neq j \leq n$, *(teams comprise other teams and/or components of the network; a component may belong to different teams simultaneously, but teams are not allowed to have common team members)*,

- θ_i, ξ_i, $1 \leq i \leq n$, *are context conditions over V^*, the exit filter and the entrance filter of the i–th team (these filters limit access to strings at team level information filtering)*,
- $c_j = (P_j, F_j)$, $1 \leq j \leq m$, *is the j–th component (entity), where*
  - P_j *is a finite and complete set of pure context–free rules over V (i.e. rules of the form $A \to \alpha$ with $A \in V$, $\alpha \in V^*$ and for each $A \in V$, there is a rule $A \to \alpha$ in P_j), the production set of the j–th component,*
  - $F_j \in V^\circ$ *is a non–empty finite multiset of strings (e.g. advertisements, messages), the multiset of axioms of the j–th component, and*
- ψ_j, υ_j, $1 \leq j \leq m$, *are contexts conditions over V^*, the exit filter and the entrance filter of the j–th component (these filters limit access to strings at component level information filtering)*.

According to the type of the filters and the type of the production sets we distinguish different classes of H_{ct}NPMP systems. We denote by H_{cxt_Y}NPMP$_Z$ the class of H_{ct}NPMP systems with (X)–type component and (Y)–type team level filters, where $X, Y \in \{reg, rc\}$ and $Z \in \{\text{0L, D0L, F0L}, \ldots\}$. The H_{ct}NPMP$_{F0L}$ system functions by changing its states.

Definition 2. *By a state (or a configuration) of an H_{ct}NPMP$_{F0L}$ system $\Gamma = (V, \{(t_i, \theta_i, \xi_i) \mid 1 \leq i \leq n\}, \{(c_j, \psi_j, \upsilon_j) \mid 1 \leq j \leq m\})$, we mean a tuple $s_t = (M_1^{(t)}, \ldots, M_m^{(t)})$, where $M_j^{(t)} \in V^\circ$, $1 \leq j \leq m$, $t \in \mathbb{N}_0$, is the state of the j–th component at step t and it represents the multiset of strings present at component j at that step. $s_0 = (F_1, \ldots, F_m)$ is the initial state of the system.*

For the sake of legibility, we introduce the following notations. Let $\Gamma = (V, \{(t_i, \theta_i, \xi_i) \mid 1 \leq i \leq n\}, \{(c_j, \psi_j, \upsilon_j) \mid 1 \leq j \leq m\})$ be an H_{ct}NPMP$_{F0L}$ system as it is defined in Definition 1. We say that $t_p \prec t_q$ (t_p precedes t_q / t_q succeeds t_p), $1 \leq p \neq q \leq n$, if $t_p \in t_q$ and $t_q \notin t_p$. t_p and t_p are incomparable, if neither $t_p \prec t_q$, nor $t_q \prec t_p$ is true. If $t_p \prec t_q$ and $t_q \prec t_s$, then $t_p \prec t_s$. Furthermore, let us suppose that $c_j \in t_{k_1}$, for some j, $1 \leq j \leq m$, and that $t_{k_1} \prec t_{k_2} \ldots \prec t_{k_{s_k}}$, $\{k_1, \ldots, k_{s_k}\} \subseteq \{1, \ldots, n\}$, $0 \leq s_k \leq n$. We say that t_{k_1} is the minimal team with respect to c_j. Let us denote this fact by $c_j \prec_{min} t_{k_1}$.

Definition 3. *(Configuration transition.)* Let $\Gamma = (V, \{(t_i, \theta_i, \xi_i) \mid 1 \leq i \leq n\}, \{(c_j, \psi_j, \upsilon_j) \mid 1 \leq j \leq m\})$ be an H_{ct}NPMP$_{F0L}$ system. Let $s_t = (M_1^{(t)}, \ldots, M_m^{(t)})$, $s_{t+1} = (M_1^{(t+1)}, \ldots, M_m^{(t+1)})$ be two states of Γ at step t and $t+1$, respectively, $t \in \mathbb{N}_0$. We say that

1. s_{t+1} is derived from s_t by a rewriting step in Γ, written as

$$(M_1^{(t)}, \ldots, M_m^{(t)}) \Rightarrow (M_1^{(t+1)}, \ldots, M_m^{(t+1)}),$$

if $M_j^{(t)} = \{\{\alpha_{j_1}, \ldots, \alpha_{j_{g_j}}\}\}$, $M_j^{(t+1)} = \{\{\beta_{j_1}, \ldots, \beta_{j_{g_j}}\}\}$, where $\alpha_{j_k}, \beta_{j_k} \in V^*$, $\alpha_{j_k} \Rightarrow \beta_{j_k}$ in P_j, $1 \leq k \leq g_j$, $1 \leq j \leq m$.

2. s_{t+1} is derived from s_t by a communication step in Γ, written as

$$(M_1^{(t)}, \ldots, M_m^{(t)}) \vdash (M_1^{(t+1)}, \ldots, M_m^{(t+1)}),$$

(a) through the application of the Deny–overrides rule combining algorithm, if for every $1 \leq j \leq m$, $M_j^{(t+1)} = M_j^{(t)} \cup C_j \cup I_j$, where

i. $C_j = \{\{\gamma \mid \gamma \in M_k^{(t)}, \theta_e(\gamma) = true, \xi_d(\gamma) = true, c_j \prec_{min} t_d, c_k \prec_{min} t_e, 1 \leq d \neq e \leq n, 1 \leq k \leq m, k \neq j$, and for all $d', e', 1 \leq d' \neq e' \leq n, t_d \prec t_{d'}, t_e \prec t_{e'} : \theta_{e'}(\gamma) = true, \xi_{d'}(\gamma) = true\}\}$, and

ii. $I_j = \{\{\gamma \mid \gamma \in M_k^{(t)}, \psi_k(\gamma) = true, \upsilon_j(\gamma) = true, c_j \prec_{min} t_d, c_k \prec_{min} t_e, 1 \leq d, e \leq n, 1 \leq k \leq m, k \neq j$, for all $d', e', 1 \leq d', e' \leq n, t_d \prec t_{d'}, t_e \prec t_{e'} : \theta_{e'}(\gamma) = true, \xi_{d'}(\gamma) = true$, and there exists $f, 1 \leq f < n, t_d \prec t_f, t_e \prec t_f : \theta_f(\gamma) = true, \xi_f(\gamma) = true\}\}$,

(b) through the use of the Permit–overrides rule combining algorithm, if for every $1 \leq j \leq m$, $M_j^{(t+1)} = M_j^{(t)} \cup C_j \cup I_j$, where

i. $C_j = \{\{\gamma \mid \gamma \in M_k^{(t)}, c_j \prec_{min} t_d, c_k \prec_{min} t_e, 1 \leq d \neq e \leq n, 1 \leq k \leq m, k \neq j, 1 \leq d' \neq e' \leq n, t_d \prec t_{d'}, t_e \prec t_{e'}$, and there exist $e'' \in \{e, e' \mid t_e \prec t_{e'}\}, d'' \in \{d, d' \mid t_d \prec t_{d'}\} : \theta_{e''}(\gamma) = true, \xi_{d''}(\gamma) = true\}\}$, and

ii. $I_j = \{\{\gamma \mid \gamma \in M_k^{(t)}, c_j \prec_{min} t_d, c_k \prec_{min} t_e, 1 \leq d, e \leq n, 1 \leq k \leq m, k \neq j, 1 \leq d', e' \leq n, t_d \prec t_{d'}, t_e \prec t_{e'}$, and $\psi_k(\gamma) = true, \upsilon_j(\gamma) = true$, or there exist $e'' \in \{e, e' \mid t_e \prec t_{e'}\}, d'' \in \{d, d' \mid t_d \prec t_{d'}\} : \theta_{e''}(\gamma) = true, \xi_{d''}(\gamma) = true\}\}$.

In Cond.1 of Def. 3, as a result of the rewriting step, only one new string can be derived from each string through the application of 0L productions. Moreover, some of these strings may be identical. The components communicate the copies of the strings at their disposal. In accordance with the XACML policy specifications, we interpret the allowance and the denial of access control decisions as true and false logical values, respectively, yielded by the filters in our framework. In more details, in Cond. 2(a)i, C_j consists of all those strings that are allowed to penetrate the exit filters of the minimal team of c_k (the sender) and the teams that succeed the minimal team of c_k, the entrance filters of the minimal team of c_j (the receiver) and the teams that succeed the minimal team of c_j. In Cond. 2(a)ii, I_j comprises all those strings that are permitted to pass the exit filters of c_k (the sender), the minimal team of c_k and the teams that succeed the minimal team of c_k, the entrance filters of c_j (the receiver), the minimal team of c_j and the teams that succeed the minimal team of c_j. In Cond. 2(b)i, C_j consists of all those strings that are allowed to penetrate the exit and the entrance filters of the minimal teams of c_k (the sender) and c_j (the receiver), or at least one of the exit filters of the teams that succeed the minimal team of c_k and at least one of the entrance filters of the teams that succeed the minimal team of c_j. In Cond. 2(b)ii, I_j comprises all those strings that are permitted to pass the exit and the entrance filters of c_k (the sender) and c_j (the receiver), or at least one of the exit filters of the minimal team of c_k and the teams that succeed the minimal team of c_k and at least one of the entrance filters of the minimal team of c_j and the teams that succeed the minimal team of c_j.

A sequence of subsequent states determines a computation in Γ. Let $\Gamma = (V, \{(t_i, \theta_i, \xi_i) \mid 1 \leq i \leq n\}, \{(c_j, \psi_j, \upsilon_j) \mid 1 \leq j \leq m\})$ be an $H_{ct}NPMP_{F0L}$ system. By a computation C in Γ we mean a sequence of states $s_0, s_1, \ldots$, where $s_t \Rightarrow s_{t+1}$, if $t = 2j + 1, j \geq 0$, and $s_t \vdash s_{t+1}$, if $t = 2j, j \geq 1$.

4   Information Dynamics

In the followings, we characterize the dynamics of information in $H_{crctrc}NPMP_{FD0L}$ systems.

Definition 4. *Let* $\Gamma = (V, \{(t_i, \theta_i, \xi_i) \mid 1 \leq i \leq n\}, \{(c_j, \psi_j, \upsilon_j) \mid 1 \leq j \leq m\})$, *be an* $H_{crctrc}NPMP_{FD0L}$ *system and let* $(M_1^{(t)}, \ldots, M_m^{(t)})$ *be the state of* Γ *at step* t *during the computation in* Γ, *where* $t \geq 0$. *We define*

1. *the population growth function of* Γ *by* $m : \mathbb{N}_0 \rightarrow \mathbb{N}_0$, $m(t) = \sum_{j=1}^{m} card(M_j^{(t)})$, *for* $t \geq 0$;
2. *the population growth function of* Γ *at node* j, $1 \leq j \leq m$, *by* $m_j : \mathbb{N}_0 \rightarrow \mathbb{N}_0$, $m_j(t) = card(M_j^{(t)})$, *for* $t \geq 0$;
3. *the communication functions of* Γ *from node* k *to node* l *(cases 3(a)i and 3(b)i: the minimal teams of nodes* k *and* l *have no common successor, cases 3(a)ii and 3(b)ii: the minimal teams of nodes* k *and* l *have a common successor) by*
   (a) *(if we apply the Deny-overrides rule combining algorithm:)*
      i. $f_{k,l}^{DENY} : \mathbb{N}_0 \rightarrow \mathbb{N}_0$, $f_{k,l}^{DENY}(t) = card(\{\{\gamma \in M_k^{(t-1)} \mid \theta_e(\gamma) = true, \xi_d(\gamma) = true, c_l \prec_{min} t_d, c_k \prec_{min} t_e, 1 \leq d \neq e \leq n, 1 \leq l \leq m, k \neq l$, *and for all* $d', e', 1 \leq d' \neq e' \leq n, t_d \prec t_{d'}, t_e \prec t_{e'}$: $\theta_{e'}(\gamma) = true, \xi_{d'}(\gamma) = true\}\})$, *for* $t = 2k', k' \geq 1$, *and* $f_{k,l}^{DENY}(t) = 0$ *otherwise;*
      ii. $\bar{f}_{k,l}^{DENY} : \mathbb{N}_0 \rightarrow \mathbb{N}_0$, $\bar{f}_{k,l}^{DENY}(t) = card(\{\{\gamma \in M_k^{(t-1)} \mid \psi_k(\gamma) = true, \upsilon_l(\gamma) = true, c_l \prec_{min} t_d, c_k \prec_{min} t_e, 1 \leq d, e \leq n, 1 \leq l \leq m, k \neq l$, *for all* $d', e', 1 \leq d', e' \leq n, t_d \prec t_{d'}, t_e \prec t_{e'}$: $\theta_{e'}(\gamma) = true, \xi_{d'}(\gamma) = true$, *and there exists* $f, 1 \leq f \leq n, t_d \prec t_f, t_e \prec t_f$: $\theta_f(\gamma) = true, \xi_f(\gamma) = true\}\})$, *for* $t = 2k', k' \geq 1$, *and* $\bar{f}_{k,l}^{DENY}(t) = 0$ *otherwise;*
   (b) *(if we use the Permit-overrides rule combining algorithm:)*
      i. $f_{k,l}^{PERMIT} : \mathbb{N}_0 \rightarrow \mathbb{N}_0$, $f_{k,l}^{PERMIT}(t) = card(\{\{\gamma \in M_k^{(t-1)} \mid c_l \prec_{min} t_d, c_k \prec_{min} t_e, 1 \leq d \neq e \leq n, 1 \leq l \leq m, k \neq l, 1 \leq d' \neq e' \leq n, t_d \prec t_{d'}, t_e \prec t_{e'}$, *and there exist* $e'' \in \{ e, e' \mid t_e \prec t_{e'}\}, d'' \in \{ d, d' \mid t_d \prec t_{d'} \}$: $\theta_{e''}(\gamma) = true, \xi_{d''}(\gamma) = true\}\})$, *for* $t = 2k', k' \geq 1$, *and* $f_{k,l}^{PERMIT}(t) = 0$ *otherwise;*
      ii. $\bar{f}_{k,l}^{PERMIT} : \mathbb{N}_0 \rightarrow \mathbb{N}_0$, $\bar{f}_{k,l}^{PERMIT}(t) = card(\{\{\gamma \in M_k^{(t-1)} \mid c_l \prec_{min} t_d, c_k \prec_{min} t_e, 1 \leq d, e \leq n, 1 \leq l \leq m, k \neq l, 1 \leq d', e' \leq n, t_d \prec t_{d'}, t_e \prec t_{e'}$, *and* $\psi_k(\gamma) = true, \upsilon_j(\gamma) = true$, *or there exist* $e'' \in$

$$\{e, e' \mid t_e \prec t_{e'}\}, d'' \in \{d, d' \mid t_d \prec t_{d'}\} : \theta_{e''}(\gamma) = true, \xi_{d''}(\gamma) =$$
$$true\}\}), for\ t = 2k', k' \geq 1,\ and\ \bar{f}_{k,l}^{PERMIT}(t) = 0\ otherwise.$$

The population growth function of Γ, m, describes the increase in the number of pieces of information in the network, the population growth function of Γ at node j, m_j, the increase in the number of pieces of information at node j, and the communication functions of Γ from node k to node l, f^{DENY}, $\bar{f}^{DENY}$, f^{PERMIT} and $\bar{f}^{PERMIT}$, the increase in the number of pieces of information at a given time step during the communication between node k and node l, respectively.

We demonstrate that the change of the rewritten and the communicated string collections using random context filters can be described by developmental systems.

Theorem 1. *Let* $\Gamma = (V, \{(t_i, \theta_i, \xi_i) \mid 1 \leq i \leq n\}, \{(c_j, \psi_j, \upsilon_j) \mid 1 \leq j \leq m\})$, *be an* $H_{c_{rc}t_{rc}}$NPMP$_{FD0L}$ *system. Let us suppose that we apply the Deny–overrides (Permit–overrides) RCA throughout the function of the network. Then a D0L system* $H = (\Sigma, \omega, h)$ *can be constructed, such that*

1. $m(t) = f(t)$, *where* m *is the population growth function of* Γ *and* f *is the growth function of* H;
2. $m_j(t) = card(\bar{h}_j(h^t(\omega)))$ *for some erasing homomorphism* $\bar{h}_j : \Sigma \to \Sigma$, *where* m_j *is the population growth function of* Γ *at node* j;
3. *(Communication functions.)*
   (a) $f_{j,k}^{DENY}(t) = card(\bar{h}_{j,k}(h^t(\omega)))$ $(f_{j,k}^{PERMIT}(t) = card(\bar{h}'_{j,k}(h^t(\omega))))$ *for some erasing homomorphism* $\bar{h}_{j,k} : \Sigma \to \Sigma$ $(\bar{h}'_{j,k} : \Sigma \to \Sigma)$, *where* $f_{j,k}^{DENY}$ $(f_{j,k}^{PERMIT})$ *is the communication function of* Γ *from node* j *to node* k, $t \geq 0$, $c_j \prec_{min} t_r, c_k \prec_{min} t_q$, $1 < r \neq q \leq n$, $1 \leq j \neq k \leq m$, $t_r \prec t_{r_x}, t_q \prec t_{q_y}$, $1 \leq x \leq g_j, 1 \leq y \leq h_k$, $0 \leq g_j, h_k \leq n$, $\{t_{r_1}, \ldots, t_{r_{g_j}}\} \cap \{t_{q_1}, \ldots, t_{q_{h_k}}\} = \emptyset$.
   (b) $\bar{f}_{j,k}^{DENY}(t) = card(\bar{h}_{j,k}(h^t(\omega)))$ $(\bar{f}_{j,k}^{PERMIT}(t) = card(\bar{h}'_{j,k}(h^t(\omega))))$ *for some erasing homomorphism* $\bar{h}_{j,k} : \Sigma \to \Sigma$ $(\bar{h}'_{j,k} : \Sigma \to \Sigma)$, *where* $\bar{f}_{j,k}^{DENY}$ $(\bar{f}_{j,k}^{PERMIT})$ *is the communication function of* Γ *from node* j *to node* k, $t \geq 0$, $c_j \prec_{min} t_r, c_k \prec_{min} t_q$, $1 \leq r, q \leq n$, $1 \leq j \neq k \leq m$, $t_r \prec t_{r_x}, t_q \prec t_{q_y}$, $1 \leq x \leq g_j, 1 \leq y \leq h_k$, $0 \leq g_j, h_k \leq n$, $\{t_r, t_{r_1}, \ldots, t_{r_{g_j}}\} \cap \{t_q, t_{q_1}, \ldots, t_{q_{h_k}}\} \neq \emptyset$.

Proof. D0L systems define homomorphism and the number of strings with a fixed minimal alphabet at a node is known, therefore the number of strings with the same minimal alphabet at the nodes can be calculated after we have performed the rewriting step. Through the context conditions we can check the presence and/or absence of various symbols in the string. Since the (minimal) alphabet of the string is known, we can decide whether the string satisfies the given context condition. As a consequence, we can represent any multiset of strings present at some stage of computation in Γ by a multiset of symbols. These multisets of symbols identify the different alphabets in a unique manner.

Theorem 1 describes how to construct a communication graph by means of communication functions, since the sequence of communication functions with respect to a given time step defines a sequence of communication graphs.

By the theory of D0L systems (see [11]), we obtain the following corollaries:

Corollary 1. *Let* $\Gamma = (V, \{(t_i, \theta_i, \xi_i) \mid 1 \leq i \leq n\}, \{(c_j, \psi_j, \upsilon_j) \mid 1 \leq j \leq m\})$, *be an* $H_{crctrc}NPMP_{FD0L}$ *system. Let us assume that we apply the Deny–overrides (Permit–overrides) RCA throughout the function of the network. Then the population growth function of an* $H_{crctrc}NPMP_{FD0L}$ *system is either exponential or polynomially bounded, which is decidable.*

Corollary 1 guarantees that the access control model reaches a decision.

Corollary 2. *Let* $\Gamma = (V, \{(t_i, \theta_i, \xi_i) \mid 1 \leq i \leq n\}, \{(c_j, \psi_j, \upsilon_j) \mid 1 \leq j \leq m\})$, *be an* $H_{crctrc}NPMP_{FD0L}$ *system. Let us assume that we apply the Deny–overrides (Permit–overrides) RCA throughout the function of the network. Suppose that* $H = (\Sigma, \omega, h)$ *is the D0L system for which conditions 1, 2 and 3 of Theorem 1 hold. Let* $\omega = \omega_0, \omega_1, \omega_2, \ldots$, *be the word sequence generated by the D0L system H. Then the sets* $\Sigma_i = alph(\omega_i), i \geq 0$, *form an almost periodic sequence, i.e. there are numbers* $p > 0$ *and* $q \geq 0$, *such that* $\Sigma_i = \Sigma_{i+p}$ *holds for every* $i \geq q$. *If a letter* $a \in \Sigma$ *occurs in some* Σ_i, *then it also appears in some* Σ_j, *with* $j \leq card(\Sigma) - 1$.

According to Corollary 2, after some time the function of these $H_{crctrc}NPMP_{FD0L}$ systems results in the saturation of information. When the system reaches the state of saturation, we can evaluate all disclosed information, i.e. the components cannot disclose any additional data under the current access control specification.

Corollary 3. *Let* $\Gamma_z = (V_z, \{(t_{i_z}, \theta_{i_z}, \xi_{i_z}) \mid 1 \leq i \leq n\}, \{(c_{j_z}, \psi_{j_z}, \upsilon_{j_z}) \mid 1 \leq j \leq m\})$, *be an* $H_{crctrc}NPMP_{FD0L}$ *system for* $z = 1, 2$. *Let us suppose that we apply the Deny–overrides (Permit–overrides) RCA throughout the function of the network. Then the sequence and language equivalence problems are decidable for the D0L systems* $H_z = (\Sigma_z, \omega_z, h_z), z = 1, 2$, *constructed for* $\Gamma_z, z = 1, 2$, *and satisfying conditions 1, 2 and 3 of Theorem 1.*

Corollary 3 implies that it is decidable for two $H_{crctrc}NPMP_{FD0L}$ systems whether they function in the same manner concerning the dynamics of information. Informally, it means that given two networks, we can determine whether they accumulate/transmit information in the same order.

5   Conclusions and Future Work

The transition from centralized and monolithic systems to open and distributed architecture made it necessary to develop access control models that are suitable to accommodate the new, dynamic environments. XACML has become the de facto standard for specifying access control policies for distributed networks.

Existing policy specification languages usually define a fixed set of rule and policy combining algorithms. Extensibility and flexibility of RCAs and PCAs, however, are desirable to meet the needs of distributed applications. Despite the fact that XACML explicitly allows additional user–defined combining algorithms, it does not provide a formal language for specifying new RCAs and PCAs [8]. Without such a specification language, automated processing is impossible, and each new RCA and PCA must be hard–coded in policy evaluation engines, which makes the wide deployment of RCAs and PCAs infeasible in practice. Thus formal language theoretic–based approaches to XACML are justifiable by all means.

In this paper, we used a variant of networks of parallel multiset string processors to present an approach to XACML–based access control management in distributed networks. In our model, the multiset string processors form teams, send and receive information through filters. The teams are organized into a hierarchical structure. Our approach allows the participants to specify their own local policies as well as to adopt the policies of any organizational units as needed. Our model supports autonomy of the individual components while guarantees the satisfaction of team–level policies. We established the connection between the growth of the number of strings at the components of networks of parallel multiset string processors employing the Deny–overrides and Permit–overrides RCAs and the growth functions of developmental systems.

In our future work, we plan to refine the model presented herein and construct new computational models, examine how our tools can be employed to give an appropriate description of additional features in XACML, study properties such as computational and descriptional complexity, patterns and dynamics of behaviour, the dependence of these properties on the structure, the organization and the functioning of the system. We aim at investigating other normative RCAs and PCAs [1] (e.g. Ordered deny–overrides, Ordered–permit–overrides, First–applicable and Only–one–applicable) of XACML and extending the PCAs proposed in [8] (e.g. Weak–consensus, Strong–consensus, Weak–majority, Strong–majority, Super–majority–permit). Besides the combining algorithms, we plan to deal with policy integration algorithms [9] (to define a common policy for resources jointly owned by multiple parties, i.e. to compose different or even conflicting access control policies into a coherent policy), as well. To this end, we will incorporate different measures into our model to describe the different types of rule and policy similarity (convergence, divergence, restriction, extension and shuffle).

Though XACML is a powerful access control specification language, it still lacks some features to fully support regulations mandating how private information collected by organizations can be used or disclosed [3]. The missing features include event history and obligations. Privacy rules may restrict a request for disclosure or usage of protected information based on some past events. These rules may impose obligations associated with deadlines on the network entities in conjunction with the enforcement of an authorization decision. In our model, we can keep track of the event history in the network on the basis of the configuration transitions, however, the study of how changing the enforcement of an authorization decision influences the behaviour of the system, is a subject of further research. We will also examine how the incorporation of time constraints into the model impacts on its properties.

Our future research targets applications that would benefit early disease detection, medical treatment (e.g. controlled drug delivery) and at–a–distance diagnosis (personalized health care) built upon new technologies (P2P, SOA, web services and cloud computing) in distributed computing. In these systems, biological hardware offers an alternative to silicon hardware and ensures energy–efficiency. Furthermore, the secure access control to life–critical data has to be guaranteed. We will consider bio–inspired operations (e.g. operations used in the existing models of DNA and cellular computing [2]) to describe access control requirements. We anticipate that our approach will simplify the policy modification process. The architectures using bio–inspired operations give us insights into the limits of these operations. We will explore the limits of these operations.

References

1. eXtensible Access Control Markup Language (XACML) Version 3.0., OASIS Standard (2013), http://docs.oasis-open.org/xacml/3.0/xacml-3.0-core-spec-os-en.html
2. Rozenberg, G., Back, T., Kok, J. (eds.): Handbook of Natural Computing. Springer, Heidelberg (2012)
3. Chowdhury, O., Chen, H., Niu, J., Li, N., Bertino, E.: On XACML's Adequacy to Specify and to Enforce HIPAA. In: Proceedings of the 3rd USENIX Conference on Health Security and Privacy, p. 11. USENIX, Berkeley (2012)
4. Csuhaj–Varjú, E., Salomaa, A.: Networks of Parallel Language Processors. In: Păun, G., Salomaa, A. (eds.) New Trends in Formal Languages. LNCS, vol. 1218, pp. 299–318. Springer, Heidelberg (1997)
5. Jajodia, S., Samarati, P., Sapino, M.L., Subrahmanian, V.S.: Flexible Support for Multiple Access Control Policies. ACM Trans. Database Syst. 26(2), 214–260 (2001)
6. Lázár, K., Csuhaj–Varjú, E., Lőrincz, A.: Peer–to–Peer Networks: a Language Theoretic Approach. Computing and Informatics 27(3), 403–422 (2008)
7. Lázár, K.A., Farkas, C.S.: Security in T_{ci}NMP Systems. In: Proceedings of the 5th International Workshop on Security in Information Systems (WOSIS 2007), pp. 95–104. INSTICC Press, Portugal (2007)
8. Li, N., Wang, Q., Qardaji, W., Bertino, E., Rao, P., Lobo, J., Lin, D.: Access Control Policy Combining: Theory Meets Practice. In: Proceedings of the 14th ACM Symposium on Access Control Models and Technologies (SACMAT 2009), pp. 135–144. ACM, New York (2009)
9. Mazzoleni, P., Crispo, B., Sivasubramanian, S., Bertino, E.: XACML Policy Integration Algorithms. ACM Trans. Inf. Syst. Secur. 11(1), 4:1–4:29 (2008)
10. Minsky, N.H., Ungureanu, V.: Law–Governed Interaction: a Coordination and Control Mechanism for Heterogeneous Distributed Systems. ACM Trans. Softw. Eng. Methodol. 9(3), 273–305 (2000)
11. Rozenberg, G., Salomaa, A.: The Mathematical Theory of L Systems. Academic Press, New York (1980)
12. Salomaa, A.: Formal Languages. Academic Press, New York (1973)
13. Sandhu, R.S., Samarati, P.: Access Control: Principles and Practice. IEEE Communications 32(9), 40–48 (1994)

Early Machine Translation

Integration and Transfers between Computing and the Language Sciences

Jacqueline Léon

CNRS, Université Paris Diderot, Université Sorbonne Nouvelle,
Sorbonne Paris Cité, Paris, France
jleon@linguist.univ-paris-diderot.fr

Abstract. Early Machine Translation was devised as a war technology originating in war sciences, and was intended to provide mass translations for the strategic purposes of the Cold war. Linguistics, which did not belong to war sciences, did not play any role at the beginning of Machine Translation. However, thanks to machine translation, the language sciences have been engaged in the process of the second mathematization of language which can be called the computational mathematization of language. In my paper, I propose to examine how linguistics integrated such a technology and entered into the second mathematization by doing a comparative study of two European traditions, the British tradition and the Russian tradition.

Keywords: Machine Translation, mathematization of language, universal languages, history.

1   Introduction

Early experiments in Machine Translation started in 1949 and marked the beginning of the application of computers to the language sciences. Machine translation, although one of the most difficult task in Natural Language Processing, was the first non digital application of computers. It was devised as a war technology, originating in the war sciences [7] which were characterized by the intertwining of engineering with fundamental research prevailing during the 2nd World War. They were devised at MIT (Massachussetts Institute of Technology), which was the very place of the new scientifico-technological configuration. In that context, Machine Translation was intended to provide mass translations for the strategic purposes of the Cold war.

It should be said that linguistics, which did not belong to the war sciences, did not play any role at the beginning of Machine Translation. However, thanks to Machine Translation, the language sciences have been engaged in the second mathematization of language which can be called the computational mathematization of language. The question remains to know how linguistics integrated such a technology which dealt with language, especially with the translation of one language into another, without using its methods and theoretical insights. We assume that, in order to perform that

A. Beckmann, E. Csuhaj-Varjú, and K. Meer (Eds.): CiE 2014, LNCS 8493, pp. 275–282, 2014.

integration, the language sciences rested on specific intellectual and cultural traditions of knowledge. Thus, the computerization of the language sciences took various forms according to the traditions of the fourth main protagonists of the Cold War, the USA, the URSS, Great Britain and France, especially according to their respective anchorage in the first mathematization of language of the 1930s [11].

In my paper, I will first focuse on early Machine Translation. Second, I will address two European traditions comparatively, the British tradition and the Russian tradition, and examine how they dealt with the integration of machine translation into the language sciences. I will examine which theoretical and methodological transfers had been necessary to carry out such an integration.

2   Early Machine Translation

Early Machine translation was mainly anchored in "the first mathematization of language" of the 1930s which was characterized by the rise of formalization promoted by the School of Vienna as a common objective for every science. Thanks to computer programming, the interaction of syntax, logical mathematics, formal languages and algorithms became dynamic and formed the second mathematization of language. This framework was the common background of the two main personalities who promoted Machine Translation, Warren Weaver (1894-1978) and Yehoshua Bar-Hillel (1915-1975). In addition to those personalities, three reports guided the setting of Machine Translation on the institutional level:

(i) 1949 [17] : *Translation*, Warren Weaver.

(ii) 1960 [6]: *The present Status of Automatic Translation of Languages*, Yehoshua Bar-Hillel.

(iii) 1966 [9] : *Language and Machines. Computers in translation and linguistics*, Automatic Language Processing Advisory Committee (ALPAC) of the National Research Council.

Warren Weaver (1894–1978), trained as a mathematician and the co-author of Shannon and Weaver's Mathematical Theory of Communication [16] was a most influential personality. Before and after the second World War, he was the director of The Natural Sciences Division of the Rockefeller foundation, and spent many years promoting scientific programs in the USA and Europe. He was an active player of the War Sciences[1]. He managed the D2 "fire control" section at MIT during the war, and was one of the founder of the Rand Corporation after the war. Machine translation was thus conceived as a war technology, suitably adapted to Culture (Cold) War objectives: the intertwining of engineering with fundamental research; the belief that science is able to resolve every problem; huge financial and human means provided mainly by the State; the development of Operational Research, inspired by the military, advocating to deal with one unique and measurable objective; finally a strong tendency to the automation of every human task.

---

[1] Mathematics, logics, physics, electronics, neurosciences, cryptography and computering belonged to war sciences.

Weaver's Memorandum "Translation", he sent to a hundred prominent personalities in 1949, boosted the creation of centers for Machine Translation, mainly in the universities. The objective was to produce mass translations of scientific texts from Russian into English. His view was that the translations thus produced did not need to be perfect. Actually tolerance for errors was permitted provided translated texts be understandable by scientists. In order to provide such rough translations, only "machine linguistics" was necessary, in other words grammatical rules which did not need to meet linguists' descriptive requirements were sufficient. In that view, Machine Translation was a flawed technology which denied any legitimacy to linguistics.

Yehoshua Bar-Hillel (1915-1975), an Israeli philosopher of language and one of Carnap's followers, was the second key player of the second mathematization of language. In 1951 he was appointed as full-time researcher in Machine Translation at MIT, where he developed an operational syntax for Machine Translation, "A Quasi-arithmetical notation for syntactic description" [5] based on Harris's distributional method and Ajdukiewicz's logical notation. In 1958, he was appointed by the National Science Foundation to assess the Machine Translation groups at work in the USA and elsewhere in the world. His critical report, published in 1960, rested on several arguments, one of them concerned more specifically the feasability of Machine Translation. For him, « Fully Automatic High Quality Translation" as a goal cannot be achieved. The only reasonable objective was machine-aided translation or human-aided translation. Actually, the results were very poor given the amounts of human and financial means invested in Machine Translation. They did not meet the expectations in the public mainly encouraged by the press. In the face of such a situation a second assessment report was commissioned by the NSF which was published in 1966. The report Language and Machines. Computers in translation and linguistics written by the Automatic Language Processing Advisory Committee (ALPAC) was published in 1966 and led to the end of Machine Translation research in the USA, as well as elsewhere in the world2. In so doing, the ALPAC report promoted Computational Linguistics to replace Machine Translation which has become totally infamous. Its disgrace remained till the end of the 1970s. Computational Linguistics was essentially based on Bar-Hillel's early views, the primacy of logical syntax and formal languages, and was dominated by syntactic parsing. For years, those views have been at the core of natural language processing.

3   Machine Translation and the Language Sciences. Horizon of Retrospection and Horizon of Projection

Linguistics did not belong to the War sciences, and Machine Translation experimenters ignored linguists' works completely. In other words, neither Machine Translation nor its successor, Computational Linguistics, belonged to the "horizon of retrospection" of linguists or could be part of their "horizon of projection" [3,4].

---

[2] Except in France, where Machine Translation, supported by the Centre National de la Recherche Scientifique, did not need to be a profitable business.

« Because it is a social (and not a biological) phenomenon, scientific activity cannot be carried out without the transmission of knowledge, without the institutionalization of training and of knowledge accumulation (in libraries for instance), and without individual memory. The scientific text comprises two types of expressions essentially. The first ones refer to the sole domain of phenomena which can be dealt with concepts belonging to common knowledge… the second ones refer to other works, that is earlier works by definition. Let us call « horizon of retrospection », this set of references … the structure of the horizon of retrospection is involved in the momentary scientific production…Conversely, the structure of scientific systems determines the structure of horizons of retrospection » [3 p.29].

« Because it is limited, the act of knowing has a temporal thickness by definition, a horizon of retrospection, as well as a horizon of projection. Knowledge (or the instances that implement it) does not destroy its past, as is often mistakenly believed. It organizes it, elects it, forgets it, imagines it or idealizes it, in the same way that it anticipates the future by dreaming it when building it. Without memory and without projects, there simply is no knowledge » [4 p.49].

It could be said that the second mathematization of language, that is the interaction between formal languages, syntax, algorithms and programming, set up a new horizon of retrospection for the language sciences. The question remains to know how the language sciences integrated the new horizon of restrospection, in other words how the computerization of the language sciences was carried out.

4   Early Machine Translation in Britain and in the Soviet Union

In the last section of my paper, I will examine how British and Russian Machine Translation experimenters dealt with the new horizons imposed on them with the second mathematization of language. Most surprisingly, early machine translation experiments in Britain and in the USSR showed similarities. Both began research on Machine Translation in the wake of the first demonstration on (IBM) computer which took place in New York in 1954. While working completely separately, both the British and the Russians, while most of the time[3] ignoring completely their respective work, developed Machine Translation methods based on semantic intermediary languages, anchored in their linguistic and philosophic specific traditions.

Contrary to the Americans, who privileged syntactic analysis, the British and the Russians, for different reasons, gave priority to meaning and the transfer of meaning in the translation process. However they were not equally familiar with the first mathematization. Thanks to Russell's, Turing's and Wittgenstein's works, the British were well aware of the first mathematization. Besides, they had their own school of

---

[3] Some Russians knew the British works but it was one-sided and the British never mentioned the Russian ones.

Information Theory led by Colin Cherry, Dennis Gabor and David McKay. The situation was quite different in the former Soviet Union. The researchers in Machine Translation had only access to the first mathematization of language through the debates on cybernetics and the complex process of the reception of Wiener's work [14].

The most significative point is that Semantic intermediary languages were, in both cases, anchored in strong traditions of international languages and universal language schemes. Thus, the earliest Machine Translation projects were based on Esperanto, showing the importance of universal languages in researchers's horizon of retrospection. This was the case of the machine translation project designed by the Soviet engineer Smirnov-Trojanskij (1894-1950) in 1931 – thus before the apparition of computer machines [2]. This was also the case for the project called 'Pidgin English' proposed by Andrew D. Booth (1918-2009) and Richard H. Richens (1919-1984). Pidgin English was a mixture of the source language, the target language4 and inflectional endings borrowed from Esperanto [15].

British early machine translation was essentially carried out at the Cambridge Language Research Unit (CLRU). Led by a Wittgensteinian philosopher, Margaret Masterman (1910–1986), the center was not, contrary to the Americans, under political and economic pressure to provide swift and cost-effective results. The group consisting of linguists (Martin Kay and M.A.K. Halliday), mathematicians and computer scientists (A.F. Parker-Rhodes, Yorick Wilks and Karen Spark-Jones) ... and botanists (Richard Richens) did not comprise any engineer.

As a method for machine translation, they designed a semantically-based intermediary language at the crossroads of two trends of the British tradition : XVIIth century universal language schemes and contextual theories of meaning of the 1930s [10]. It should be noted that universal languages schemes and intermediary language projects were both anchored in a strong social demand for interlingual means of communication respectively in the XVIIth century and in the XXth century.

Three projects were developed : Nude, Nude II, and templates. Richens' Nude was inspired by Dalgarno's Ars Signorum (1661) and Wilkins' Essay towards a real character and a philosophical language (1668). As one of Wilkins' descendants was the botanist Linné, it may be assumed that Richens was well aware of this hierarchical type of classification. The intermediary language was a semantic network composed of fifty 'naked ideas' where the source language structural characteristics had been deleted. Composed of semantic primitives and syntactic operators, the semantic network was what remained invariant during the translation process. As such it was used for machine translation. Margaret Masterman designed a second project which was based on three distinct sources of inspiration : Nude, Roget's thesaurus and Wittgenstein's conception of meaning by usage [12]. The simultaneous recourse to universal languages and to meaning by usage, which might seem contradictory at first sight, was justified by Masterman's use of a thesaurus. As Wittgenstein's pupil, Margaret Masterman could not develop Nude primitives as universal concepts. She agreed with Wittgenstein that the logic unit for studying language should not be word or proposition but word context, in particular word use. Because of its structure, based on the

---

4 In their paper, Richens and Booth gave examples of twenty languages translated into English.

classification of words organized by a set of contexts, Masterman chose the thesaurus organization to create a new intermediary language, " a thesauric interlingua ".

As a third project, Yorick Wilks (b. 1939) adapted Nude in order to resolve semantic ambiguities. In the line of Wittgenstein, Firth and Masterman, he envisaged solving semantic ambiguities by implementing context of usage. Wilks modified Nude in order to resolve semantic ambiguities in texts. Contrary to the common view of Machine Translation experimenters, according to whom ambiguities should be defined with reference to dictionaries, Wilks chose to define them within a text. He was then led to develop what he called " preferential semantics ": for a given text, a specific meaning is chosen preferably over another, so that no definitive choice should be made. He devised a system of semantic representation of texts based on templates [18].

The Soviets undertook research on machine translation in 1954, just after Stalin's death and once cybernetics was back in favor. Contrary to the USA, Machine Translation did not play a strategic part in the Soviet conception of the Cold War [2], so that computers were reserved for secret services and military objectives. Machine Translation, at least during the first years, remained confined to speculative research projects. The researchers making Machine Translation belonged to domains such as mathematics and linguistics which were considered ideologically harmless. In the USSR, the aim was less to translate English into Russian than Russian into the other languages of the Union. Finally, translation was not the only objective : Machine Translation was considered the first step of a more ambitious and general program aiming at information retrieval from texts written in various languages.

Russian researchers conceived translation as the transfer of meaning from one language to another. That view led them to give more importance to synthesis than to analysis and to work out a semantic approach based on intermediary languages. Two main projects were devised. Nikolaj Dmitrivic Andreev (1920-1997), who led the translation laboratory at the University of Leningrad, conceived a model of intermediary language based on the idea of unequivocal information language promoted by 1930s language planners such as Drezen. His method rested on the statistical treatment of linguistic invariants and on a conception of languages as social facts [1].

Igor Mel'čuk (b.1932) chose a method based on a semantic intermediary language when facing languages with completely different word order systems, such as Hungarian, where pure syntactic treatment was impossible. The intermediary language he developed was based on a protolanguage, originated from the Russian tradition of comparative linguistics, more specifically from Vjačeslav Ivanov's protolanguage scheme [8]. Such an intermediary language was neither totally an artificial language, nor a natural language. It was based on properties extracted from natural languages. For this type of Machine Translation process, synthesis should be preferred over analysis. The reason was that synthesis needs only linguistic knowledge while analysis often requires disambiguisation which cannot be solved without the recourse to extra-linguistic context.

5   Conclusion

As a conclusion, let us say that, for the British as for the Russians, translation from one language into another involved cultural issues and long-term views. What was imported in the translation process were models stemming from universal language schemes, philosophy of language, comparative linguistics and linguistic planning pertaining to their respective intellectual and cultural traditions. More than transferring concepts and methods between theories of meaning and language sciences, computering acted as a catalyst. Semantic models were implemented in computer systems in order to achieve practical tasks. Finally, the computerization of translation gave way to new models for the language sciences which can be regarded as the horizon of projection of Machine Translation projects. The CLRU works gave way to semantic networks and artificial intelligence projects, raising debates on non referential semantics. Mel'čuk's intermediary language led to his meaning-text project still at work in Natural Language Processing.

References

1. Andreev, N.D.: The Intermediary Language as the Focal Point of Machine Translation. In: Booth, A.D. (ed.) Machine Translation, pp. 3–27. North Holland Publishing Company, Amsterdam (1967)
2. Archaimbault, S., Léon, J.: La langue intermédiaire dans la Traduction Automatique en URSS (1954-1960). Filiations et modèles. Histoire Epistémologie Langage 19-2, 105-132 (1997)
3. Auroux, S.: Histoire des sciences et entropie des systèmes scientifiques. Les horizons de retrospection. In: Schmitter, P. (ed.) Zur Theorie und Methode der Geschichtsschreibung der Linguistik, pp. 20–42. Narr, Tübingen (1987)
4. Auroux, S.: L'histoire des sciences du langage et le paradoxe historiographique. Le Gré des Langues 8, 40–63 (1995)
5. Bar-Hillel, Y.: A Quasi-Arithmetic Notation for Syntactic Description. Language 29, 47–58 (1953)
6. Bar-Hillel, Y.: The present Status of Automatic Translation of Languages (1). In: Alt, F.C. (ed.) Advances in Computers, vol. 1, pp. 91–141. Academic Press, N.Y. (1960)
7. Dahan, A., Pestre, D. (eds.) : Les sciences pour la guerre (1940-1960). Editions de l'EHESS, Paris (2004)
8. Kulagina, O.S., Mel'čuk, I.A.: Automatic Translation : some Theoretical Aspects and the Design of a Translation System. In: Booth, A.D. (ed.) Machine Translation, pp. 137–173. North Holland Publishing Company, Amsterdam (1967)
9. Language and Machines. Computers in Translation and Linguistics, A report by the Automatic Language Processing Advisory Committee (ALPAC). National Academy of Sciences, National Research Council (1966)
10. Léon, J.: Histoire de l'automatisation du langage, ENS Editions, Lyon (in press)
11. Léon, J.: From Universal Languages to Intermediary Languages in Machine Translation: the Work of the Cambridge Language Research Unit (1955-1970). In: Guimaraes, E., Pessoa de Barros, D.L. (eds.) History of Linguistics 2002, pp. 123–132. John Benjamins Publishing Company, Amsterdam (2007)
12. Masterman, M.: What is a Thesaurus? Essays on and in Machine Translation by the Cambridge Language Research Unit (1959). In: Wilks, Y. (ed.) Margaret Masterman. Language, Cohesion and Form, pp. 107–145. Cambridge University Press, Cambridge (2005)
13. Mel'čuk, I.A., Zholkovskij, A.K.: Sur la synthèse sémantique. T.A. Informations 2, 1–85 (1970)

14. Mindell, D., Segal, J., Gerovitch, S.: Cybernetics and Information Theory in the United States, France and the Soviet Union. In: Walker, M. (ed.) Science and Ideology: A Comparative History, pp. 66–95. Routledge, London (2003)
15. Richens, R., Booth, A.D.: Some Methods of Mechanized Translation. In: Locke, W.N., Booth, A.D. (eds.) Machine Translation of Languages, 14 Essays, pp. 24–46. MIT, John Wiley, Cambridge, New York (1955)
16. Shannon, C.E., Weaver, W.: The Mathematical Theory of Communication. Univ. of Illinois Press, Urbana-Champaign (1949)
17. Weaver, W.: Translation. In: Locke, W.N., Booth, A.D. (eds.) Machine Translation of Languages, 14 Essays, pp. 15–23. MIT, John Wiley, Cambridge, New York (1955)
18. Wilks, Y.: On line Semantic Analysis of English Texts. Mechanical Translation 11(3-4), 59–72 (1968)

Lines Missing Every Random Point

Jack H. Lutz[1,*] and Neil Lutz[2,**]

[1] Department of Computer Science, Iowa State University, Ames, IA 50011, USA
lutz@cs.iastate.edu
[2] Department of Computer Science, Rutgers University, Piscataway, NJ 08854, USA
njlutz@cs.rutgers.edu

Abstract. We prove that there is, in every direction in Euclidean space, a line that misses every computably random point. We also prove that there exist, in every direction in Euclidean space, arbitrarily long line segments missing every double exponential time random point.

Keywords: randomness, algorithmic geometric measure theory, computable analysis.

1 Introduction

One objective of the theory of computing is to investigate the fine-scale geometry of algorithmic information in Euclidean space. Recent work along these lines has included algorithmic classifications of points lying on computable curves and arcs [6, 14, 22, 25, 31] and in more exotic sets [8, 18, 21].

This paper concerns a simple, fundamental question: Can the direction of a line in Euclidean space force the line to meet at least one random point? That is, can the set of Martin-Löf random points, which is everywhere dense and contains almost every point in Euclidean space, be avoided by lines in every direction? For example, it is reasonable to conjecture that every line of random slope in $\mathbb{R}^2$ contains a random point. We show here that this conjecture is false, and in fact that—regardless of slope—every line can be translated so that it contains no Martin-Löf random point. Moreover, the line can miss the larger class of all computably random points.

Our solution of this problem builds on a very old—and ongoing—line of research in geometric measure theory. In 1917 Fujiwara and Kakeya [13, 16] posed the question of the minimum area of a plane set in which a unit segment can be continuously reversed without leaving the set, a *Kakeya needle set*. This question was resolved in 1928 by Besicovitch [2]: such a set can have arbitrarily small measure. The work made use of a construction by Besicovitch from 1919 [1] (but not widely circulated until its republication in 1928 [3]) of a plane set of area 0 containing a unit line segment in every direction, a *Kakeya set*. This set was constructed using a clever iterated process of partitioning and translating the pieces of an equilateral triangle.

* Research supported in part by National Science Foundation Grant 1247051.
** Research supported in part by National Science Foundation Grant 1101690.

A. Beckmann, E. Csuhaj-Varjú, and K. Meer (Eds.): CiE 2014, LNCS 8493, pp. 283–292, 2014.

In 1964 Besicovitch used a duality principle to construct a plane set with area 0 that contains a line in every direction, a *Besicovitch set* [5]. Falconer [10,11] used an alternative duality principle to give a somewhat simpler construction of a Besicovitch set. This latter set B, which is the point-line dual of a simply defined "fractal dust," is described in detail in Section 4. Our main result is achieved by showing that B has computable measure 0, as does its Cartesian product with $\mathbb{R}^n$, for every $n \in \mathbb{N}$. We also sketch an alternative proof suggested to us by Turetsky (personal communication) and an anonymous reviewer.

Our main result leads us to conjecture that there is, in every direction in Euclidean space, a line that misses not only every computably random point, but every point that is feasibly random (i.e., polynomial time random, as defined in Section 2). We are unable to prove this conjecture at this time, but in Section 5 we prove a weaker result along these lines. Specifically, we show that there exist, in every direction in the Euclidean plane, arbitrarily long line segments missing every point that is double exponential time random (a randomness condition defined in Section 2). Our proof of this fact uses Besicovitch's above-mentioned 1919 construction of a Kakeya set, together with later refinements of this proof by Perron [24], Schoenberg [28], and Falconer [11].

More recent work on the "sizes" of Besicovitch sets and Kakeya sets has focused on their dimensions. Davies showed that every Kakeya set in $\mathbb{R}^2$ has Hausdorff dimension 2 [7], and the famous Kakeya conjecture states that Kakeya sets in $\mathbb{R}^n$ have Hausdorff dimension n for all $n \geq 2$. For more on this history, consult [11,17].

The remainder of the paper is organized as follows. Section 2 contains preliminary information regarding computable and time-bounded measure and randomness in $\mathbb{R}^n$. In Section 3, we present a class of martingales for betting on open sets. In Section 4, we describe Falconer's Besicovitch set B and prove the main theorem in $\mathbb{R}^2$. In Section 5 we briefly describe a Kakeya set K and use it to prove our result on segments missing every double exponential time random point in $\mathbb{R}^2$. Section 6 extends our two theorems to $\mathbb{R}^n$ ($n \geq 2$). Section 7 mentions open problems. All proofs are omitted from this proceedings version of the paper.

2   Computable and Time-Bounded Randomness in $\mathbb{R}^n$

We now discuss the elements of computable measure and randomness in $\mathbb{R}^n$. For each $r \in \mathbb{N}$ and each $\mathbf{u} = (u_1, ..., u_n) \in \mathbb{Z}^n$, let

$$Q_r(\mathbf{u}) = \left[u_1 \cdot 2^{-r}, (u_1 + 1) \cdot 2^{-r}\right) \times ... \times \left[u_n \cdot 2^{-r}, (u_n + 1) \cdot 2^{-r}\right)$$

be the r-*dyadic cube* at $\mathbf{u}$. Note that each $Q_r(\mathbf{u})$ is "half-open, half-closed" in such a way that, for each $r \in \mathbb{N}$, the family

$$\mathcal{Q}_r = \left\{Q_r(\mathbf{u}) \mid \mathbf{u} \in \{0, ..., 2^r - 1\}^n\right\}$$

is a partition of the unit cube $Q_0 (\mathbf{0}) = [0, 1)^n$. The family

$$\mathcal{Q} = \bigcup_{r=0}^{\infty} \mathcal{Q}_r$$

is the set of all *dyadic cubes* in $[0, 1)^n$.

A *martingale* on $[0, 1)^n$ is a function $d : \mathcal{Q} \to [0, \infty)$ satisfying

$$d (Q_r (\mathbf{u})) = 2^{-n} \sum_{\mathbf{a} \in \{0, 1\}^n} d (Q_{r+1} (2\mathbf{u} + \mathbf{a})) \qquad (1)$$

for all $Q_r (\mathbf{u}) \in \mathcal{Q}$. Intuitively, a martingale d is a strategy for placing successive bets on the location of a point $\mathbf{x} \in [0, 1)^n$. After r bets have been placed, the bettor's capital is

$$d^{(r)} (\mathbf{x}) = d (Q_r (\mathbf{u})) \ ,$$

where $\mathbf{u}$ us the unique element of $\{0, ..., 2^r - 1\}^n$ such that $\mathbf{x} \in Q_r (\mathbf{u})$. The bettor's next bet is on which of the 2^n immediate subcubes $Q_{r+1} (2\mathbf{u} + \mathbf{a})$ of $Q_r (\mathbf{u})$ has $\mathbf{x}$ as an element. The condition (1) says that the bettor's expected capital after this bet is exactly the bettor's capital before the bet, i.e., the payoffs are fair. A martingale d *succeeds* at a point $\mathbf{x} \in [0, 1)^n$ if

$$\limsup_{r \to \infty} d^{(r)} (\mathbf{x}) = \infty \ .$$

A well known theorem of Ville [29], restated in the present setting, says that a set $E \subseteq [0, 1)^n$ has Lebesgue measure 0 if and only if there is a martingale d that succeeds at every point $\mathbf{x} \in E$. It follows easily by the countable additivity and translation invariance of Lebesgue measure that a set $E \subseteq \mathbb{R}^n$ has Lebesgue measure 0 if and only if there is a martingale d that succeeds at every point $\mathbf{x} \in E^{\#}$, where

$$E^{\#} = [0, 1)^n \cap \bigcup_{\mathbf{t} \in \mathbb{Z}^n} (E + \mathbf{t}) \ . \qquad (2)$$

Let

$$J = \left\{ (r, \mathbf{u}) \in \mathbb{N} \times \mathbb{Z}^n \mid \mathbf{u} \in \{0, ..., 2^r - 1\}^n \right\} \ .$$

Then a martingale $d : \mathcal{Q} \to [0, \infty)$ is *computable* if there is a computable function $\widehat{d} : \mathbb{N} \times J \to \mathbb{Q} \cap [0, \infty)$ such that, for all $(s, r, \mathbf{u}) \in \mathbb{N} \times J$,

$$\left| \widehat{d} (s, r, \mathbf{u}) - d (Q_r (\mathbf{u})) \right| \leq 2^{-s} \ . \qquad (3)$$

A set $E \subseteq \mathbb{R}^n$ is defined to have *computable measure* 0 if there is a computable martingale d that succeeds at every point $\mathbf{x} \in E^{\#}$, where $E^{\#}$ is defined as in (2). A point $\mathbf{x} \in \mathbb{R}^n$ is *computably random* if it is not an element of any set of computable measure 0, i.e., if there is no computable martingale that succeeds at $\mathbf{x}$. Computable randomness was introduced by Schnorr [26, 27]. It is well known [9, 23] that every random point in $\mathbb{R}^n$ (i.e., every Martin-Löf random point in $\mathbb{R}^n$) is computably random and that the converse does not hold. In particular, then, almost every point in $\mathbb{R}^n$ is computably random.

Resource-bounded measure, a complexity-theoretic generalization of Lebesgue measure that induces measure on complexity classes, has been used to define complexity-theoretic notions of randomness [20]. Adapting these notions to Euclidean space, a martingale $d : \mathcal{Q} \to [0, \infty)$ is p-*computable* (respectively, ee-*computable*) if there is a function $\widehat{d} : \mathbb{N} \times J \to \mathbb{Q} \cap [0, \infty)$ that satisfies 3 and is computable in $(s+r)^{O(1)}$ time (respectively, in $2^{2^{O(s+r)}}$ time). A point $\mathbf{x} \in \mathbb{R}^n$ is p-*random* (or *polynomial time random*, or *feasibly random*) if no p-computable martingale succeeds at $\mathbf{x}$ [20]. A point $\mathbf{x} \in \mathbb{R}^n$ is ee-*random* (or *double exponential time random*) if no ee-computable martingale succeeds at $\mathbf{x}$ [15]. It is routine to show that every computably random point is ee-random, that every ee-random point is p-random, and that the converses of these statements are false.

3   Betting on Open Sets

In this section we describe a class of martingales that are used in the proof of the main theorem in Section 4. These martingales are also likely to be useful in future investigations.

For any set $G \subseteq [0, 1)^n$ with $m(G) > 0$, define a martingale $d_G : \mathcal{Q} \to [0, \infty)$ recursively as follows.

(i) $d_G(Q_0(\mathbf{0})) = 1$.
(ii) For all $r \geq 0$, $\mathbf{u} \in \{0, ..., 2^r - 1\}^n$, and $\mathbf{a} \in \{0, 1\}^n$,

$$d_G(Q_{r+1}(2\mathbf{u} + \mathbf{a})) = \begin{cases} 0 & \text{if } d_G(Q_r(\mathbf{u})) = 0 \\ 2^n d_G(Q_r(\mathbf{u})) \frac{m(G \cap Q_{r+1}(2\mathbf{u}+\mathbf{a}))}{m(G \cap Q_r(\mathbf{u}))} & \text{otherwise} . \end{cases}$$

That is, for each cube $Q \in \mathcal{Q}_r$, the values of the martingale on the immediate subcubes of Q are proportional to the measures of the subcubes' intersections with G.

Theorem 1. *For every nonempty set G that is open as a subset of the subspace $[0, 1)^n$ of $\mathbb{R}^n$ and every $\mathbf{x} \in G$, $d_G^{(r)}(\mathbf{x}) = 1/m(G)$ for all sufficiently large r.*

When G is open, we call d_G the *open set martingale* for G.

4   Betting on a Besicovitch Set

This section reviews Falconer's construction of the Besicovitch set B mentioned in the introduction and proves that the set B in fact has computable measure 0. Hence B contains a line in every direction in $\mathbb{R}^2$, and each of these lines misses every computably random point in $\mathbb{R}^2$.

For each $m, b \in \mathbb{R}$, let $\mathcal{L}_{m,b} \subseteq \mathbb{R}^2$ be the line with slope m and y-intercept b. Falconer defined the *line set operator* $\mathcal{L} : \mathcal{P}(\mathbb{R}^2) \to \mathcal{P}(\mathbb{R}^2)$ by

$$\mathcal{L}(F) = \bigcup \{\mathcal{L}_{m,b} \mid (m, b) \in F\}$$

for all $F \subseteq \mathbb{R}^2$. We call $\mathcal{L}(F)$ the *line set* of F. It is easy to verify that the operator $\mathcal{L}$ is monotone and maps compact sets to closed sets.

We are interested in the line set of a particular self-similar fractal F, which we now define. Consider the alphabet $\Sigma = \{0, 1, 2, 3\}$. For each $i \in \Sigma$ define the contraction $S_i : \mathbb{R}^2 \to \mathbb{R}^2$ by

$$S_i(x, y) = \frac{1}{4}\left((x, y) + (i, a_i)\right),$$

where $a_0 = 2$, $a_1 = 0$, $a_2 = 3$, and $a_3 = 1$. For each $w \in \Sigma^*$ define the set $F(w) \subseteq \mathbb{R}^2$ by the recursion

$$F(\lambda) = [0, 1]^2$$
$$F(iw) = S_i(F(w)),$$

for all $i \in \Sigma$ and $w \in \Sigma^*$. For each $k \in \mathbb{N}$ let

$$F_k = \bigcup \{F(w) \mid w \in \Sigma^k\}.$$

The sets F_0 and F_1, along with their line sets, are depicted in Figure 1. We are interested in the set

$$F = \bigcap_{k=0}^{\infty} F_k.$$

This set F is an uncountable, totally disconnected set, informal called a "fractal dust." More formally it is the *attractor* of the *iterated function system* (S_0, S_1, S_2, S_3), i.e., it is a *self-similar fractal*.

Let $\mathrm{Ref}_Y : \mathbb{R}^2 \to \mathbb{R}^2$ and $\mathrm{Rot}_\theta : \mathbb{R}^2 \to \mathbb{R}^2$ denote reflection across the y-axis and rotation about the origin by the angle θ, respectively. The set

$$B = \mathcal{L}(F) \cup \mathrm{Rot}_{\frac{\pi}{2}}(\mathcal{L}(F)) \cup \mathrm{Ref}_Y(\mathcal{L}(F) \cup \mathrm{Rot}_{\frac{\pi}{2}}(\mathcal{L}(F))) \tag{4}$$

is the Besicovitch set that we use for our main theorem.

Observation 2. *The set B contains a line in every direction in $\mathbb{R}^2$.*

Using the duality principle and some nontrivial fractal geometry, Falconer also proved the following.

Lemma 3. ([10,11]) *The set B has Lebesgue measure 0.*

It is not obvious whether or how the proof of Lemma 3 can be effectivized. Nevertheless we prove the following.

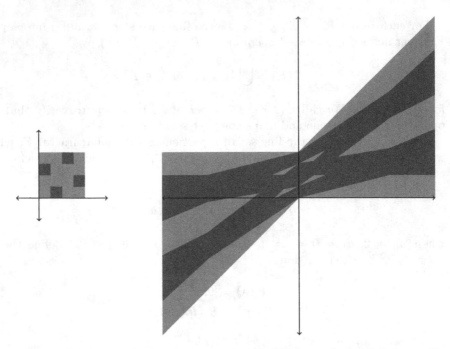

Fig. 1. F_0 and F_1, along with their line sets. F_0 and $\mathcal{L}(F_0)$ are shaded light gray; F_1 and $\mathcal{L}(F_1)$ are dark gray.

Theorem 4. (main theorem, in $\mathbb{R}^2$) *The set B has computable measure 0. Hence there is, in every direction in $\mathbb{R}^2$, a line that misses every computably random point.*

In remarks on an early draft of this paper, Turetsky and an anonymous reviewer pointed out an alternative proof of Theorem 4. The key fact, proved by Wang [9, 30], is that every computably random point $\mathbf{x}$ is *Kurtz random* (also called *weakly random* [18]), meaning that $\mathbf{x}$ is not an element of any computably closed (i.e., Π_1^0) set of measure 0. Furthermore, the above-mentioned fact that the operator $\mathcal{L}$ maps compact sets to closed sets can be extended to prove that $\mathcal{L}$ maps bounded Π_1^0 sets to Π_1^0 sets. Finally, it is routine to verify that the fractal dust F is a bounded Π_1^0 set. These things and Lemma 3 imply that $\mathcal{L}(F)$ contains no computably random point, whence Theorem 4 holds by Observation 2. This elegant proof is simpler than our martingale construction, even when Wang's proof is included. However, we believe that the direct martingale construction may help illuminate the path to results on time-bounded randomness, so we retain the martingale proof in this paper.

5   Betting in Doubly Exponential Time

In light of Theorem 4 it is natural to ask whether there is, in every direction in $\mathbb{R}^2$, a line that misses not only every computably random point, but every feasibly random point. We do not know the answer to this question at the time of this writing, but we prove a weaker result of this type in this section.

As noted in the introduction, Besicovitch constructed a *Kakeya set*, a Lebesgue measure 0 plane set containing a unit line segment in every direction, in 1919. Our objective here is to specify a Kakeya set K and prove that it has ee-measure 0 (a condition defined in Section 2). Our specification and proof take advantage of Besicovitch's original work, together with subsequent refinements by Perron [24], Schoenberg [28], and Falconer [11].

We first describe *Perron trees*, the building blocks of our set K. Let $\tau = \triangle(U, V, W)$ be a triangle oh height h with its base $\overline{UV}$ on the x-axis. We regard this triangle as *including* its interior. Note that τ contains a line segment of length h in every direction between the directions of $\overline{UW}$ and $\overline{VW}$. Given a positive integer k, cut τ into 2^k nonoverlapping triangles as indicated in Figure 2(a). (Throughout this discussion, sets in $\mathbb{R}^2$ are *nonoverlapping* if their interiors are disjoint.) Besicovitch showed that these 2^k smaller triangles can be slid horizontally along the x-axis in such a way that their union, due to high overlap, has very small area. Perron simplified Besicovitch's overlap scheme to that depicted in Figure 2(b). Note that, notwithstanding its small area, the set in Figure 2(b) still contains a line segment in every direction between the directions of $\overline{UW}$ and $\overline{VW}$. Schoenberg coined the term *Perron trees* for sets of the type depicted in Figure 2(b) and gave a simpler, recursive "sprouting construction" of the Perron tree $P_k(\tau)$ as a union of $2^{k+1} - 1$ *nonoverlapping* triangles as in Figure 2(c). This is useful for our purpose here because it simplifies the algorithm for betting on Perron trees.

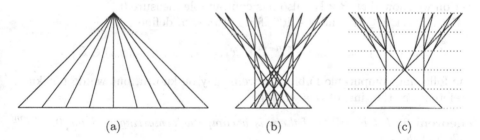

      (a)   (b)   (c)

Fig. 2. (a) a triangle cut into eight pieces; (b) a Perron tree constructed by sliding those pieces together; (c) the same Perron tree via Schoenberg's sprouting construction

It is also possible to cut τ into smaller triangles $\tau_1, ..., \tau_m$ in the manner of Figure 2(a), then divide each τ_i further to construct the Perron trees $P_k(\tau_1), ..., P_k(\tau_m)$. Since the area of a Perron tree $P_k(T)$ is proportional to the area of T, the area of the union $\bigcup_i P_k(\tau_i)$ is equal to the area of $P_k(\tau)$. Furthermore, for a triangle T with base length b, the distance from point in $P_k(T)$ to T is less than b, so the initial subdivision of τ into triangles with smaller bases makes the construction more *local*. This enables us to construct a nested sequence of open sets G_k, where each G_k contains a union of many Perron trees, such that for every $k \in \mathbb{N}$,

(a) G_k contains a unit segment in every direction $\theta \in [\pi/4, 3\pi/4]$.
(b) The area of G_k is at most 2^{-k}.
(c) $G_{k+1} \subseteq G_k$.
(d) G_k is a union of $2^{2^{O(k)}}$ fully specified nonoverlapping triangles.

Let

$$F = \bigcap_{k \in \mathbb{N}} \overline{G_k}$$

and

$$K_0 = \bigcup_{c \in \mathbb{N}} cF \, ,$$

where $cF = \{cx \mid x \in F\}$. Then

$$K = K_0 \cup \mathrm{Rot}_{\pi/4}(K_0)$$

is the Kakeya set that we use for the following.

Theorem 5. *The set K has ee-measure 0. Hence there exist, in every direction in $\mathbb{R}^2$, arbitrarily long line segments that miss every ee-random point.*

6 Higher Dimensions

For every $n \in \mathbb{N}$, the set $B \times \mathbb{R}^n$ contains a line in every direction in $\mathbb{R}^{n+2}$, and Fubini's theorem implies that this set has Lebesgue measure 0 [12]. In this section we show that $B \times \mathbb{R}^n$ also has computable measure 0.

For any set $E \subseteq \mathbb{R}^n$ and $y \in \mathbb{R}^m$, for $1 \leq m < n$, define

$$E_y = \{(x_1, ..., x_{n-m}) \in \mathbb{R}^{n-m} \mid (x_1, ..., x_{n-m}, y_1, ..., y_m) \in E\} \, .$$

The following computable Fubini theorem may be known, but we do not know a reference at the time of this writing.

Theorem 6. *Let $E \in \mathbb{R}^n$. If there is a computable martingale d on $[0,1)^{n-m}$ such that the set*

$$N_E(d) = \{y \in [0,1)^m \mid \exists x \in E_y^\# \text{ such that } d \text{ does not succeed at } x\}$$

has computable measure 0, then E has computable measure 0.

Corollary 7. *For every computable measure 0 set E and $n \in \mathbb{N}$, the set $E \times \mathbb{R}^n$ has computable measure 0.*

Theorem 8. (main theorem, in $\mathbb{R}^n$) *For every $n \geq 2$ there is, in every direction in $\mathbb{R}^n$, a line that misses every computably random point.*

It is routine to prove double exponential time versions of Theorem 6 and Corollary 7, and hence to extend Theorem 5 to $\mathbb{R}^n$ in a similar fashion.

7   Open Problems

As noted in the introduction, we conjecture that there is a line in every direction missing every feasibly random point in Euclidean space. Proving or disproving this conjecture may require a significant advance beyond current understanding of the algorithmic geometric measure theory of Besicovitch and Kakeya sets. In the meantime, more modest goals may be achieved. Can Theorem 5 be improved to singly exponential time, or to lines instead of segments?

Besicovitch's duality idea for constructing the set B came soon after, and was perhaps prompted by, the Mathematical Association of America's production of a film in which he explained his 1919 solution of the Kakeya needle problem. (The article [4] is based on this film.) Does a copy of this film still exist?

Acknowledgment. We thank Dan Turetsky and an anonymous reviewer for pointing out the alternate proof of Theorem 4 and for a useful correction.

References

[1] Besicovitch, A.S.: Sur deux questions d'intégrabilité des fonctions. Journal de la Société de physique et de mathematique de l'Universite de Perm 2, 105–123 (1919)

[2] Besicovitch, A.S.: On Kakeya's problem and a similar one. Mathematische Zeitschrift 27, 312–320 (1928)

[3] Besicovitch, A.S.: On the fundamental geometric properties of linearly measurable plane sets of points. Mathematische Annalen 98, 422–464 (1928)

[4] Besicovitch, A.S.: The Kakeya problem. American Mathematical Monthly 70, 697–706 (1963)

[5] Besicovitch, A.S.: On fundamental geometric properties of plane line sets. Journal of the London Mathematical Society 39, 441–448 (1964)

[6] Couch, P.J., Daniel, B.D., McNicholl, T.H.: Computing space-filling curves. Theory of Computing Systems 50(2), 370–386 (2012)

[7] Davies, R.O.: Some remarks on the Kakeya problem. Proceedings of the Cambridge Philosophical Society 69, 417–421 (1971)

[8] Dougherty, R., Lutz, J.H., Mauldin, R.D., Teutsch, J.: Translating the Cantor set by a random real. Transactions of the American Mathematical Society 366, 3027–3041 (2014)

[9] Downey, R., Hirschfeldt, D.: Algorithmic Randomness and Complexity. Springer (2010)

[10] Falconer, K.J.: Sections of sets of zero Lebesgue measure. Mathematika 27, 90–96 (1980)

[11] Falconer, K.J.: The Geometry of Fractal Sets. Cambridge University Press (1985)

[12] Falconer, K.J.: Fractal Geometry: Mathematical Foundations and Applications, 2nd edn. Wiley (2003)

[13] Fujiwara, M., Kakeya, S.: On some problems of maxima and minima for the curve of constant breadth and the in-revolvable curve of the equilateral triangle. Tôhoku Science Reports 11, 92–110 (1917)

[14] Gu, X., Lutz, J.H., Mayordomo, E.: Points on computable curves. In: FOCS, pp. 469–474. IEEE Computer Society (2006)

[15] Harkins, R.C., Hitchcock, J.M.: Upward separations and weaker hypotheses in resource-bounded measure. Theoretical Computer Science 389(1-2), 162–171 (2007)

[16] Kakeya, S.: Some problems on maxima and minima regarding ovals. Tôhoku Science Reports 6, 71–88 (1917)

[17] Katz, N., Tao, T.: Recent progress on the Kakeya conjecture. In: Proceedings of the 6th International Conference on Harmonic Analysis and Partial Differential Equations, pp. 161–180. Publicacions Matematiques (2002)

[18] Kurtz, S.: Randomness and Genericity in the Degrees of Unsolvability. PhD thesis, University of Illinois at Urbana-Champaign (1981)

[19] Kjos-Hanssen, B., Nerode, A.: Effective dimension of points visited by Brownian motion. Theoretical Computer Science 410(4-5), 347–354 (2009)

[20] Lutz, J.H.: Almost everywhere high nonuniform complexity. Journal of Computer and System Sciences 44(2), 220–258 (1992)

[21] Lutz, J.H., Mayordomo, E.: Dimensions of points in self-similar fractals. SIAM J. Comput. 38(3), 1080–1112 (2008)

[22] McNicholl, T.H.: The power of backtracking and the confinement of length. Proceedings of the American Mathematical Society 141(3), 1041–1053 (2013)

[23] Nies, A.: Computability and Randomness. Oxford University Press, Inc., New York (2009)

[24] Perron, O.: Über einen Satz von Besicovitch. Mathematische Zeitschrift 28, 383–386 (1928)

[25] Rettinger, R., Zheng, X.: Points on computable curves of computable lengths. In: Královič, R., Niwiński, D. (eds.) MFCS 2009. LNCS, vol. 5734, pp. 736–743. Springer, Heidelberg (2009)

[26] Schnorr, C.-P.: A unified approach to the definition of a random sequence. Mathematical Systems Theory 5, 246–258 (1971)

[27] Schnorr, C.-P.: Zufälligkeit und Wahrscheinlichkeit. Lecture Notes in Mathematics, vol. 218. Springer (1971)

[28] Schoenberg, I.J.: On the Besicovitch-Perron solution of the Kakeya problem. In: Studies in Mathematical Analysis and Related Topics, Pólya, vol. 383-386 (1962)

[29] Ville, J.: Étude Critique de la Notion de Collectif. Gauthier–Villars, Paris (1939)

[30] Wang, Y.: Randomness and Complexity. PhD thesis, University of Heidelberg (1996)

[31] Zheng, X., Rettinger, R.: Point-separable classes of simple computable planar curves. Logical Methods in Computer Science 8(3) (2012)

Natural Descriptions and Anthropic Bias: Extant Problems In Solomonoff Induction

Simon McGregor

University of Sussex

Abstract. According to some advocates, algorithmic information theory (a branch of theoretical computer science) promises to underwrite an ultimate formal theory of comprehensible patterns. The arguments have an intuitive appeal when expressed in terms of well-known computer languages, and can both inspire and explain practical results in machine learning. The theory of Solomonoff induction, which combines algorithmic information theory and Bayesian inference, has been suggested as a solution to the philosophical problem of induction and an idealisation of the scientific method; an extension of it forms part of a proposed mathematical theory of intelligence.

Unfortunately, the philosophical import of algorithmic information theory is undermined by its dependence on an arbitrary choice of language (reference machine). While the choice of reference machine is irrelevant in the infinite limit, I observe that considered over finite sets there are infinitely many reference machines which give arbitrary evaluations of simplicity. I also explain why, regardless of how much data has been observed, infinitely many reference machines will always give every conceivable "best guess" answer to finite questions in Solomonoff induction.

Finally, I argue that algorithmic information theory is philosophically incomplete because it pretends to a "God's-eye view" and ignores relevant information in the structure of the observer. This issue has been raised before, but given relatively little focus. The question of anthropic bias - how to take the existence of the reasoner into account when reasoning - is still a subject of major disagreement in Bayesian inference, and is likely to be so in algorithmic information theory as well.

Keywords: algorithmic information, Solomonoff induction, anthropic bias, anthropic principle, natural Turing machines, Kolmogorov complexity.

1 Introduction

In some sense, algorithmic information theory - the study of descriptions which are computer-interpretable - attempts to formalise a theory of everything we can possibly talk about. Such an ambitious remit means that, if well-founded, it has wide-reaching ramifications. An obvious example of such ramifications is the framework of Solomonoff induction [14], which has attracted recent discussion regarding its philosophical implications [18,23,24].

A. Beckmann, E. Csuhaj-Varjú, and K. Meer (Eds.): CiE 2014, LNCS 8493, pp. 293–302, 2014.
© Springer International Publishing Switzerland 2014

One of the outstanding problems in the theory is the identification of formally "natural" Turing machines [11], which applies to other theoretical applications of algorithmic information theory [6,15,5]. This paper argues that, contra claims in [18], the relevance of Solomonoff induction is severely limited without some theory of a "natural" descriptive language. In particular, all finite "best" predictions made by Solomonoff induction are arbitrarily dependent on the choice of reference machine, with infinitely many machines giving every possible answer. This applies at every scale, including the "inconceivably large" scale. Moreover, there are infinitely many machines which give meaningless answers for algorithmic entropy when applied to every element of a set of arbitrary finite size.

A further foundational question is raised regarding Solomonoff induction, from a radically situated/embodied perspective on intelligence: philosophically speaking, we do not have a "God's-eye view" but rather a particular subjective one. Solomonoff induction asks what rules produced the observed data; we should also ask what rules produced the process of observation itself, since data implies an observer. In other words, algorithmic induction needs to address anthropic considerations [1]. (The anthropic principle in cosmology is the observation that the fact of our own existence constrains the possible form of the Universe.)

I conclude that a well-known foundational problem for Solomonoff induction is more serious than has been previously argued, and that it is philosophically incomplete due to the omission of anthropic considerations even if the language dependence problem can be solved. However, these problems are not necessarily insuperable within the formal domain; Solomonoff induction remains a suitable framework for discussing the mathematical structure of rationality.

2   Weaknesses of Algorithmic Entropy

Kolmogorov-Chaitin complexity, also known as algorithmic entropy, essentially quantifies the degree of randomness of a string X by the length L of the shortest computer program which outputs X.

Randomness is intuitively taken to be the opposite of pattern or structure, so algorithmic entropy can also be used to measure the degree of structure in a string, and the shortest program then provides something like a semantic model of what the structure is. This intuition provides a very powerful and general approach to conceptualising structure.

From a purely theoretical standpoint, there are at least three well-known ways in which KC complexity falls short of a principled optimal method for measuring randomness in empirically observed structures.

1. It is non-computable (although it is "lower semi-computable" - approximable from below)
2. It is dependent on the choice of "reference machine" - effectively, the language programs are written in - although only up to an additive constant.
3. If the object under study is not intrinsically a binary string, it must be encoded as one. The choice of encoding will also affect the KC complexity attributed to the object.

There is also a fourth problem, which is mentioned in [19] but not explored further:

4. It does not address anthropic considerations, i.e. the question of what information the observer's existence and relationship to the data provides. Intuitively, since it does not take relevant information into account, it cannot be optimal.

This paper focuses on problems arising from the second and fourth issues: language dependence and anthropic bias[1].

It is worth commenting that all invariance results which depend on an infinite limit (such as the asymptotic equivalence of reference machines for KC complexity, or the washing out of priors in Bayesian inference) cannot be used to justify arbitrary parameter decisions when applied to problems involving finite sets of strings.

The language dependence of algorithmic entropy arises because there is currently no formulation of a "natural" or canonical reference machine. Appeals to machine minimality or simplicity are circular; in this way, the lack of a foundation resembles the "symbol grounding problem" in artificial intelligence. One way to resolve the symbol grounding problem in AI is via a situated/embodied approach [3,8], in which the intelligent agent is assumed to be part of (and in interactive contact with) the world it reasons about. I discuss the implications of this insight for Solomonoff induction, where it translates into a variant of the anthropic principle [4]. In particular, like [19], I propose that Solomonoff induction should assume that a single computational process is responsible for the observer, the data, and the interaction between the two which constitutes the observation.

Note: this paper provides an intuitive treatment of the concept of KC complexity, and blurs a few technical distinctions (for instance, the difference between Turing machines, prefix Turing machines and monotone Turing machines). A comprehensive formal treatment is given in [14].

2.1 Kolmogorov's Invariance Theorem

The theoretical class of reference machines considered for the purposes of KC complexity are those machines which are universal. If a machine P is universal, this effectively means that P can be programmed to run any modern computer language (say, Java) and that a simulation of P can be written in any standard computer language (given infinite memory).

Discussions of algorithmic complexity typically invoke the invariance theorem, which states that the Kolmogorov-Chaitin complexity of a string X evaluated under reference machines P and Q differ by no more than a constant C_{PQ} which depends on P and Q but not on X. However, this constant may be arbitrarily large: hence, larger than the largest P-complexity of any string in a finite set.

---

[1] The non-computability issue is well-known and will not be discussed here. The paper will also omit a discussion of problems associated with the third issue (encoding), although they clearly relate to both language dependence and anthropic bias.

2.2 Lookup Machines

If we consider a finite set of strings S, it is even worse than that: given a machine P, we can construct a machine Q such that the complexity $K_Q(s)$ of every string in S (according to Q) is effectively unrelated to its complexity under P. We can do this by constructing a pathological Turing-complete machine I will call a "lookup machine". A lookup machine contains an arbitrarily large (but finite) lookup table, and runs as follows:

- If its input is prefixed with a zero, it looks up the remaining input in the lookup table.
  - If an entry is found, it outputs that entry and terminates.
  - Otherwise, it outputs the empty string and terminates.
- If its input is prefixed with n ones, where n is the longest complexity we want to assign to any string in S, it runs the remaining input on P.
- Otherwise (i.e. its input is prefixed with at least one, but less than n, ones) it outputs the empty string and terminates.

Clearly, subject to the constraint that we can only assign a complexity of k to at most 2^{k-1} strings, we are free to assign arbitrary Q-complexities to each member of S.

This machine Q has a compiler under P, since it can be specified using standard computational operations, and has a compiler for P (the input consisting of n ones), so it is computationally equivalent in power to P. Hence, according to the most abstract current notion of algorithmic entropy, Q is every bit as sensible a choice of reference machine as any other. Moreover, it is trivial to construct infinitely many of these machines since n is arbitrary.

Intuitively, the problem is that for any finite string or set of strings, imposing the constraint that the strings must have been generated by a computable process gives us nothing at all, since all finite structures can be thus generated.

3   Solomonoff Induction

The framework of Solomonoff induction, perhaps best formally described in [14] and given differing recent interpretative accounts in [18,23,24], addresses a very general problem: upon observing a series of things, what should we expect to happen next?

The first step, in line with algorithmic information theory more generally, is to encode the observations as a binary string. The problem then becomes, after observing the first n digits of a binary string, what will the next digits be?

According to Solomonoff induction, we should assume that the string was generated by a computer program, determine which of all possible infinitely many computer programs would have produced the digits we have observed so far (there will still be infinitely many such programs), consider the next digits they all produce, and take a sort of vote amongst these possibilities giving exponentially more weight to the shorter programs' output.

The "method" of Solomonoff induction corresponds to Bayesian inference (typically claimed to be the optimal reasoning process under uncertainty [18,12]) based on a so-called Universal Prior. It has several attractive formal properties: for instance, it "majorises" all computable priors ([14]) and minimises both worst-case error and worst-case retractions in formal learning theory ([13]).

In the next section, we will discuss the implications of language dependence on Solomonoff induction.

3.1   Reasoning on Finite Data

In both algorithmic information theory and Bayesian inference, there are formal results corresponding to the intuition that ideal observers can begin with differences of opinion which will disappear with enough data.

In algorithmic information theory, KC complexity is invariant to the choice of reference machine (providing the reference machines are computationally equivalent) up to an additive constant. For sufficiently long (finite) strings, this constant C becomes an insignificant proportion of the string's overall complexity.

In Bayesian inference, two different priors (assuming that the KL divergence in each direction is finite) will converge on the same posterior given enough relevant data.

These elegant results unfortunately have a gaping hole when applied in practice. There is no guarantee that differences of opinion will even be reduced after observing a finite amount of data. (See [12], Chapter 5, for an example where Bayesian posteriors diverge after a finite amount of data.) The implications for Solomonoff induction are serious.

In [18], the authors claim that

> "It is worth noting that this problem of arbitrary predictions for short sequences x is largely mitigated if we use the method of prefixing x with prior knowledge y. The more prior knowledge y that is encoded, the more effective this method becomes. Taken to the extreme we could let y represent all prior (scientific) knowledge, which is possibly all relevant knowledge. This means that for any x the string yx will be long and therefore prediction will be mostly unaffected by the choice of universal reference machine."

This quote is reproduced in [23] but is unfortunately misleading. The choice of reference machine entirely determines the next prediction, regardless of the length of any finite data string.

Let's consider two different reference machines. The first, which we will call Jekyll, is some monotone machine based on the elegant combinator language described in [14]. The second, Hyde, is a monstrous lookup machine which assigns an arbitrary complexity to every data string we could possibly hope to observe in our physical Universe.

What happens when we apply Solomonoff induction using Jekyll and Hyde to perform inference based on a large (but finite) body of scientific observations?

Well, leaving formal problems of computability aside (the Universal Prior is uncomputable), it seems at least plausible following the arguments in [18] that, given a large but not physically impossible amount of data, inference based on Jekyll would make predictions which accord reasonably well with current scientific theories - and the shortest program on Jekyll to produce the data would in fact resemble the theories (or a better, more unified/accurate theory). By contrast, inference based on Hyde would make predictions which resemble neither scientific theory nor common sense. What's more, we can plausibly expect our species to be extinct before it ever gathers enough data to overcome Hyde's pathology.

This thought experiment raises a tantalising question. Given appropriate initial assumptions (i.e. a "sensible" reference machine), it looks like Solomonoff induction attacks the thorny philosophical problem of induction in an elegant and precise way. Indeed, computable approximations to Kolmogorov complexity (using "reasonable" models) have already been shown to yield practically useful results [7,20]. But given inappropriate initial assumptions, algorithmic entropy falls flat on its face - and our theory gives us no way yet to distinguish between appropriate and inappropriate assumptions.

Arbitrary Best Prediction Theorem. Let's say that a reference machine U "predicts y given x" if Solomonoff induction (under U) on a string x yields another string y as the maximum posterior estimate of the next k characters (where k is the length of y). Then, regardless of observed data x, each possible (finite) string y is predicted by infinitely many reference machines.

This is easy to prove, since given x and y we can construct infinitely many machines which produce a string prefixed with xy when run on any program of sufficiently short length. Under Solomonoff induction, the short programs' output will dominate the shape of the posterior, resulting in the desired maximum posterior prediction.

4   "Minimal" Reference Machines

A fairly intuitive approach to determining a choice of individual reference machine is to suppose that we should prefer reference machines which are "simpler" than others. For instance, researchers have tried to identify minimal Turing machines, minimal combinator machines, minimal Turing-equivalent cellular automata, and so on.

But in exactly what formal sense are these machines minimal? It seems that they are minimal in the language which humans commonly use to describe those machines. This may be a particular formulation of CAs (or Turing machines), or it may even be something as fundamental as ZF set theory. But that language itself is computationally arbitrary. Just as there are infinitely many computationally equivalent reference machines, there are infinitely many symbolic languages which are equivalent in their power to describe reference machines. Why use ZF set theory? Why not Von Neumann set theory? Or some equivalent axiomatic system which assigns totally different complexities to reference machines, which looks hideously complex from the viewpoint of a ZF set theory (and vice versa)?

Looking for the least complex reference machine is not the answer: it simply shifts the problem from finding a non-arbitrary machine, to finding a non-arbitrary descriptive foundational language for mathematics. The problem of finding a non-arbitrary base structure among infinitely many ones of identical power still remains.

5 Anthropic Bias

It is tempting to suppose that languages such as Java are more "minimal" than pathological lookup machines because they are simpler to implement mechanically; that is to say, to formalise the principle of Occam's razor for our physical universe, we should use reference machines which are easier to realise physically. Indeed, while I was able to describe how in principle to construct the monstrous Hyde machine, I would be unable to give a complete description of it (because it is based on a random lookup table larger than the Universe); this contrasts starkly with the relatively agreeable Jekyll machine.

Unfortunately, this intuition does not help on its own to resolve the philosophical problem. If algorithmic information theory offers a solution to the philosophical conundrum of scientific induction, the solution should extend "all the way down". In particular, the inductive empirical reasoning we use to judge which reference machines will prove easier to construct (or operate) in the real world will ultimately need to be justified by reference to Occam's razor. Again, the problem generates an infinite regress.

One possible solution would be to introduce what [1] has called an anthropic bias. Solomonoff induction implies a so-called "God's-eye view", where the reasoner pretends to have no subjective perspective or relationship with the data. In practice, of course, the only way we can observe any data at all is by interacting with an external world, and in order for those observations to be well-structured we will presumably need to assume that we ourselves, and our interactions with the external world, are also structured. The correct application of anthropic considerations in Bayesian reasoning is disputed, although it is relevant to such questions as "How likely are we to be the last humans on Earth?" [4]. (A detailed discussion of the problems involved is given in [1].)

In intuitive terms, we can interpret the formal framework of Solomonoff induction as based on the assumption that our observable data is generated by a computational process. Considerations from both rational materialist and phenomenological perspectives would suggest an additional, reflexive, assumption: not only is the data generated by a computational process, but the observer is also also computationally generated - and a common underlying process produces the data, the observer and their interaction. Effectively, if there are laws of physics, then they describe us as well as the data.

This intuition echoes a perspective on (artificial and natural) intelligence which stresses the importance of situated and embodied cognition [3]: the observation that cognition occurs in agents which are part of - and affect - the very world they reason about. It should be mentioned that [10] considers the interactive aspect of situated/embodied cognition, although it does not consider the anthropic question.

[19] does raise the application of the anthropic principle to algorithmic induction, where he calculates that the probability of finding yourself in a particular conceivable Universe x, given the additional information that you exist, is equal to the a priori probability of x multiplied by a normalising constant (providing that Universe x implies your existence). His treatment of anthropic considerations dismisses it by concluding that "the probability of finding yourself in universe x [given the fact of your own existence] is essentially determined by $P(x)$, the prior probability of x".

However, this statement is true only for Universes which imply our existence! For Universes which do not, the posterior probability is zero. This raises the serious problem of specifically how to formalise our existence in a way which allows us to say whether or not a particular Universe would contain us or not.

The argument does not apply if human cognition cannot be represented computationally. Some researchers (e.g. [17,2,22]) have claimed to prove that human cognition is non-computable, but computability probably remains the consensus in the Artificial Intelligence community. At the very least, it seems reasonable for rational materialists to make the assumption that when we observe data, we ourselves are products of the same sort of process which ultimately generates the data, whether or not that process is computable. Hence, an objection to anthropic Solomonoff induction on the grounds of human non-computability should arguably also be grounds for rejecting Solomonoff induction *in toto*, and the theoretical use of Kolmogorov complexity more generally by analogy.

6   Reasons for Hope

It may be possible to find a formally justified "natural" base structure. Reference machines do have individual properties which do not "wash out" in the infinite limit, and hence distinguish them from one another in a non-arbitrary way. Two examples are time complexity (as used in [21]) and the typical behaviours of random programs (as considered, unsuccessfully, in [16]).

These properties might suffice to describe a single unique "natural" machine, or more likely a measure of reasonableness which describes a unique "natural" distribution over machines. An alternative approach would be to identify some way of asking the relevant question which is genuinely invariant to the choice of reference machine.

The formalised mathematical treatment of our subjective selfhood is a particularly interesting problem, and will likely require insights from cognitive science and philosophy as well as physics, mathematics and computer science. Some viewpoints within a non-algorithmic context are provided by e.g. [25] and [9], who consider formalised versions of embodied inference.

While the task is daunting, it seems that it could in principle be possible. An adequate account should also provide insights into the encoding problem:

how to translate between the phenomenological and digital realms. If we are particularly fortunate, perhaps the anthropic principle might in the end provide a justification for using reference machines which strike us as more intuitively reasonable, although there is (at present) no obvious formal reason why that should be.

7   Conclusion

Algorithmic information theory describes the structure of binary strings in terms of a computer-interpretable descriptive language. With a few assumptions, this formalism provides an exceptionally powerful abstract way to talk about simplicity, information, and systematicity: concepts which underlie some of our most abstract notions in philosophy and cognitive science. The theory comes supplied with mathematically proven results and demonstrable practical applications.

The problem is that any real-world application or interpretation of the theory relies utterly on the choice of an arbitrary descriptive language, and the most obvious intuitive justifications for the decision (simplicity, practical utility) turn out to be circular. I have sketched a proof of finite language dependence in the Solomonoff induction case, which should hold straightforwardly for other applications of KC complexity.

As well as the need for a "natural" descriptive language, I have argued that we need to close the gap between subject and object by considering an anthropic principle: if there is to be an "algorithmic theory of everything", it needs to account for the process of embodied observation, including the observer as well as the data.

These arguments should not be seen as suggesting that algorithmic information theory has nothing to contribute to cognitive science or philosophy; on the contrary, the precision of its formalism allows difficult questions to be highlighted more clearly.

Acknowledgments. Many thanks to Nathaniel Virgo, Daniel Polani, Lionel Barnett, and Chris Thornton for helpful discussions on earlier drafts of this paper.

References

1. Bostrom, N.: Anthropic bias: Observation selection effects in science and philosophy. Psychology Press (2002)
2. Bringsjord, S., Zenzen, M.J.: Superminds: People Harness Hypercomputation, and More. Kluwer Academic Publishers, Norwell (2003)
3. Brooks, R.A.: Cambrian intelligence: the early history of the new AI. The MIT Press (1999)
4. Carter, B., McCrea, W.H.: The Anthropic Principle and its Implications for Biological Evolution [and Discussion]. Philosophical Transactions of the Royal Society of London. Series A, Mathematical and Physical Sciences 310(1512), 347–363 (1983), http://dx.doi.org/10.1098/rsta.1983.0096

5. Chater, N., Vitányi, P.: Simplicity: a unifying principle in cognitive science? Trends in Cognitive Sciences 7(1), 19–22 (2003)
6. Chrisley, R.: Natural intensions. In: Adaptation and Representation, pp. 3–11 (2007), http://interdisciplines.org/medias/confs/archives/archive_4.pdf
7. Cilibrasi, R., Vitányi, P.M.: Clustering by compression. IEEE Transactions on Information Theory 51(4), 1523–1545 (2005)
8. Clark, A.: Being there: Putting brain, body, and world together again. MIT Press (1998)
9. Friston, K.J., Daunizeau, J., Kiebel, S.J.: Reinforcement learning or active inference? PloS One 4(7), e6421 (2009)
10. Hutter, M.: Universal algorithmic intelligence: A mathematical top→down approach. In: Goertzel, B., Pennachin, C. (eds.) Artificial General Intelligence. Cognitive Technologies, pp. 227–290. Springer, Berlin (2007)
11. Hutter, M.: Open problems in universal induction & intelligence. Algorithms 2(3), 879–906 (2009)
12. Jaynes, E.T.: Probability theory: the logic of science. Cambridge University Press (2003)
13. Kelly, K.T.: Justification as truth-finding efficiency: how ockham's razor works. Minds and Machines 14(4), 485–505 (2004)
14. Li, M., Vitanyi, P.M.: An Introduction to Kolmogorov Complexity and Its Applications, 3rd edn. Springer Publishing Company, Incorporated (2008)
15. McGregor, S.: Algorithmic Information Theory and Novelty Generation. In: Proceedings of the 4th Internation Joint Workshop on Computational Creativity, pp. 109–112 (2007)
16. Müller, M.: Stationary algorithmic probability. Theoretical Computer Science 411(1), 113–130 (2010)
17. Penrose, R.: The Emperor's New Mind. Oxford University Press (1989)
18. Rathmanner, S., Hutter, M.: A philosophical treatise of universal induction. Entropy 13(6), 1076–1136 (2011)
19. Schmidhuber, J.: Algorithmic theories of everything. Tech. Rep. IDSIA-20-00, quant-ph/0011122, IDSIA, Manno (Lugano), Switzerland (2000)
20. Schmidhuber, J.: Discovering neural nets with low kolmogorov complexity and high generalization capability. Neural Networks 10(5), 857–873 (1997)
21. Schmidhuber, J.: The speed prior: a new simplicity measure yielding near-optimal computable predictions. In: Computational Learning Theory, pp. 216–228. Springer (2002)
22. Searle, J.R., et al.: Minds, brains, and programs. Behavioral and Brain Sciences 3(3), 417–457 (1980)
23. Sterkenburg, T.F.: The Foundations of Solomonoff Prediction. Master's thesis, University of Utrecht (2013)
24. Vallinder, A.: Solomonoff Induction: A Solution to the Problem of the Priors? Master's thesis, Lund University (2012)
25. Wolpert, D.H.: Physical limits of inference. Physica D: Nonlinear Phenomena 237(9), 1257–1281 (2008)

An Efficient Algorithm for the Equation Tree Automaton *via* the k-C-Continuations

Ludovic Mignot, Nadia Ouali Sebti, and Djelloul Ziadi*

Laboratoire LITIS - EA 4108 Université de Rouen,
Avenue de l'Université
76801 Saint-Étienne-du-Rouvray Cedex, France
{ludovic.mignot,nadia.ouali-sebti,djelloul.ziadi}@univ-rouen.fr

Abstract. Champarnaud and Ziadi, and Khorsi *et al.* show how to compute the equation automaton of word regular expression E *via* the k-C-Continuations. Kuske and Meinecke extend the computation of the equation automaton to a regular tree expression E over a ranked alphabet Σ and produce a $O(R \cdot |\,E\,|^2)$ time and space complexity algorithm, where R is the maximal rank of a symbol occurring in Σ and $|\,E\,|$ is the size of E. In this paper, we give a full description of the algorithm based on the acyclic minimization of Revuz. Our algorithm, which is performed in an $O(|Q|\cdot|\,E\,|)$ time and space complexity, where $|Q|$ is the number of states of the produced automaton, is more efficient than the one obtained by Kuske and Meinecke.

1 Introduction

Regular expressions, which are finite representatives of potentially infinite languages, are widely used in various application areas such as XML Schema Languages [10], logic and verification, *etc.* The concept of word regular expressions has been extended to tree regular expressions. Similarly to word expressions, one can convert them into finite recognizers, the tree automata.

The study of the different ways of conversion of regular expressions into automata and *vice versa* is a very active field. There exists a lot of techniques to transform regular expressions (resp. regular tree expressions) into finite automata [2,6,7,14] (resp. into finite tree automata [8,9]). As far as tree automata are concerned, computation algorithms are extensions of word cases. In [9], the computation of the position tree automaton from a regular tree expression has been achieved by extending the classical notions of Glushkov functions defined in [6], leading to the computation of an automaton which number of states is linear w.r.t. the number of occurrences of symbols but which number of transitions can be exponential. In the same paper, it is proved that this automaton can be reduced into a quadratic size recognizer.

On the other side, Kuske and Meinecke have extended the notion of word partial derivatives [1] into tree partial derivatives. They also present how to compute them extending from words to trees [8] the k-C-Continuation algorithm

---

* D. Ziadi was supported by the MESRS - Algeria under Project 8/U03/7015.

A. Beckmann, E. Csuhaj-Varjú, and K. Meer (Eds.): CiE 2014, LNCS 8493, pp. 303–313, 2014.

by Champarnaud and Ziadi [3]. They obtain an algorithm with $O(R \cdot |E| \cdot |E|)$ space and time complexity where R is the maximal rank of a symbol occurring in the finite ranked alphabet Σ and $|E|$ is the size of the regular expression.

In this paper, we show how to extend a notion of k-C-Continuation in order to compute from a regular tree expression its equation tree automaton with an $O(|E| + |Q| \cdot |E|)$ time and space complexity where $|Q|$ is the number of its states. This constitutes an improvement in comparison with Kuske and Meinecke algorithm [8]. The paper is organized as follows: Section 2 outlines finite tree automata over ranked trees, regular tree expressions, and linearized regular tree expressions which allows the set of positions to be defined. Next, in Section 3 the notions of derivation and partial derivative of regular expression and set of regular expressions are introduced. Thus the definitions of equation tree automaton and k-C-Continuation tree automaton associated with the regular expression E is obtained. Afterwards, in Section 4 we present our algorithm which builds the equation tree automaton with an $O(|E| + |Q| \cdot |E|)$ time and space complexity.

2 Preliminaries

Let (Σ, ar) be *a ranked alphabet*, where Σ is a finite set and ar represents the *rank* of Σ which is a mapping from Σ into $\mathbb{N}$. The set of symbols of rank n is denoted by Σ_n. The elements of rank 0 are called *constants*. A *tree* t over Σ is inductively defined as follows: $t = a$, $t = f(t_1, \ldots, t_k)$ where a is any symbol in Σ_0, k is any integer satisfying $k \geq 1$, f is any symbol in Σ_k and $t_1, \ldots, t_k$ are any k trees over Σ. We denote by T_Σ the set of trees over Σ. A *tree language* is a subset of T_Σ. Let $\Sigma_{\geq 1} = \Sigma \backslash \Sigma_0$ denote the set of *non-constant symbols* of the ranked alphabet Σ. A *Finite Tree Automaton* (FTA) [5,8] $\mathcal{A}$ is a tuple (Q, Σ, Q_T, Δ) where Q is a finite set of states, $Q_T \subset Q$ is the set of *final states* and $\Delta \subset \bigcup_{n \geq 0} (Q \times \Sigma_n \times Q^n)$ is the set of *transition rules*. This set is equivalent to the function Δ from $Q^n \times \Sigma_n \to 2^Q$ defined by $(q, f, q_1, \ldots, q_n) \in \Delta \Leftrightarrow q \in \Delta(q_1, \ldots, q_n, f)$. The domain of this function can be extended to $(2^Q)^n \times \Sigma_n \to 2^Q$ as follows: $\Delta(Q_1, \ldots, Q_n, f) = \bigcup_{(q_1, \ldots, q_n) \in Q_1 \times \cdots \times Q_n} \Delta(q_1, \ldots, q_n, f)$. Finally, we denote by Δ^* the function from $T_\Sigma \to 2^Q$ defined for any tree in T_Σ as follows:
$$\Delta^*(t) = \begin{cases} \Delta(a) & \text{if } t = a, a \in \Sigma_0 \\ \Delta(f, \Delta^*(t_1), \ldots, \Delta^*(t_n)) & \text{if } t = f(t_1, \ldots, t_n), f \in \Sigma_n, t_1, \ldots, t_n \in T_\Sigma \end{cases}$$
A tree is *accepted* by $\mathcal{A}$ if and only if $\Delta^*(t) \cap Q_T \neq \emptyset$. The language *recognized* by $\mathcal{L}(\mathcal{A})$ is the set of trees accepted by $\mathcal{A}$ i.e. $\mathcal{L}(\mathcal{A}) = \{t \in T_\Sigma \mid \Delta^*(t) \cap Q_T \neq \emptyset\}$. A state $q \in Q$ is *coaccessible* if $q \in Q_T$ or if $\exists Q' = \{q_1, \ldots, q_n\} \subset Q$, $f \in \Sigma_n$, q' a coaccessible state in Q such that $q \in Q'$ and $q' \in \Delta(f, q_1, \ldots, q_n)$. The *coaccessible part* of the automaton $\mathcal{A}$ is the tree automaton $\mathcal{A}' = (Q', \Sigma, \Delta', Q_T')$ where $Q' = \{q \in Q \mid q \text{ is coaccessible}\}$ and $\Delta' = \{(q, f, q_1, \ldots, q_n) \in \Delta \mid \{q, q_1, \ldots, q_n\} \subset Q'\}$. It is easy to show that $\mathcal{L}(\mathcal{A}) = \mathcal{L}(\mathcal{A}')$.

Let $\sim$ be an equivalence relation over Q. We denote by $[q]$ the equivalence class of any state q in Q. The *quotient* of A w.r.t. $\sim$ is the tree automaton $A_{/\sim} = (Q_{/\sim}, \Sigma, Q_{T/\sim}, \Delta_{/\sim})$ where: $Q_{/\sim} = \{[q] \mid q \in Q\}$, $Q_{T/\sim} = \{[q] \mid q \in Q_T\}$, $\Delta_{/\sim} = \{([q], f, [q_1], \ldots, [q_n]) \mid (q, f, q_1, \ldots, q_n) \in \Delta\}$.

For any integer $n \geq 0$, for any n languages $L_1, \ldots, L_n \subset T_\Sigma$, and for any symbol $f \in \Sigma_n$, $f(L_1, \ldots, L_n)$ is the tree language $\{f(t_1, \ldots, t_n) \mid t_i \in L_i\}$. The *tree substitution* of a constant c in Σ by a language $L \subset T_\Sigma$ in a tree $t \in T_\Sigma$, denoted by $t\{c \leftarrow L\}$, is the language inductively defined by L if $t = c$; $\{d\}$ if $t = d$ where $d \in \Sigma_0 \setminus \{c\}$; $f(t_1\{c \leftarrow L\}, \ldots, t_n\{c \leftarrow L\})$ if $t = f(t_1, \ldots, t_n)$ with $f \in \Sigma_n$ and $t_1, \ldots, t_n$ any n trees over Σ. Let c be a symbol in Σ_0. The *c-product* $L_1 \cdot_c L_2$ of two languages $L_1, L_2 \subset T_\Sigma$ is defined by $L_1 \cdot_c L_2 = \bigcup_{t \in L_1} \{t\{c \leftarrow L_2\}\}$. The *iterated c-product* is inductively defined for $L \subset T_\Sigma$ by: $L^{0_c} = \{c\}$ and $L^{(n+1)_c} = L^{n_c} \cup L \cdot_c L^{n_c}$. The *c-closure* of L is defined by $L^{*_c} = \bigcup_{n \geq 0} L^{n_c}$.

A *regular expression* over a ranked alphabet Σ is inductively defined by $E \in \Sigma_0$, $E = f(E_1, \cdots, E_n)$, $E = (E_1 + E_2)$, $E = (E_1 \cdot_c E_2)$, $E = (E_1^{*_c})$, where $c \in \Sigma_0$, $n \in \mathbb{N}$, $f \in \Sigma_n$ and $E_1, E_2, \ldots, E_n$ are any n regular expressions over Σ. Parenthesis can be omitted when there is no ambiguity. We write $E_1 = E_2$ if E_1 and E_2 graphically coincide. We denote by $\mathrm{RegExp}(\Sigma)$ the set of all regular expressions over Σ. Every regular expression E can be seen as a tree over the ranked alphabet $\Sigma \cup \{+, \cdot_c, *_c\}$ with $c \in \Sigma_0$ where $+$ and $\cdot_c$ can be seen as a symbol of rank 2 and $*_c$ has rank 1. This tree is the syntax-tree T_E of E. The *alphabetical width* $\|E\|$ of E is the number of occurrences of symbols of Σ in E. *The size* $|E|$ of E is the size of its syntax tree T_E. The *language* $[\![E]\!]$ *denoted by* E is inductively defined as $[\![c]\!] = \{c\}$, $[\![f(E_1, E_2, \cdots, E_n)]\!] = f([\![E_1]\!], \ldots, [\![E_n]\!])$, $[\![E_1 + E_2]\!] = [\![E_1]\!] \cup [\![E_2]\!]$, $[\![E_1 \cdot_c E_2]\!] = [\![E_1]\!] \cdot_c [\![E_2]\!]$, $[\![E_1^{*_c}]\!] = [\![E_1]\!]^{*_c}$ where $n \in \mathbb{N}$, $E_1, E_2, \ldots, E_n$ are any n regular expressions, $f \in \Sigma_n$ and $c \in \Sigma_0$. It is well known that a tree language is accepted by some tree automaton if and only if it can be denoted by a regular expression [5,8]. A regular expression E defined over Σ is *linear* if and only if every symbol of $\Sigma_{\geq 1}$ appears at most once in E. Note that any constant symbol may occur more than once. Let E be a regular expression over Σ. The *linearized regular expression* $\overline{E}^E$ in E of a regular expression E is obtained from E by marking differently all symbols of a rank greater than or equal to 1 (symbols of $\Sigma_{\geq 1}$). The set of *marked symbols* with symbols of Σ_0 is the ranked alphabet containing symbols called *positions*. We denote this set by $\mathrm{Pos_E}(E)$. When there is no ambiguity we denote by $\overline{F}$ the subexpression $\overline{F}^E$ with F is a subexpression of E. The mapping h is defined from $\mathrm{Pos_E}(E)$ to Σ with $h(\mathrm{Pos_E}(E)_m) \subset \Sigma_m$ for every $m \in \mathbb{N}$. It associates with a marked symbol $f_j \in \mathrm{Pos_E}(E)_{\geq 1}$ the symbol $f \in \Sigma_{\geq 1}$ and for a symbol $c \in \Sigma_0$ the symbol $h(c) = c$. We can extend the mapping h naturally to $\mathrm{RegExp}(\mathrm{Pos_E}(E)) \to \mathrm{RegExp}(\Sigma)$ by $h(a) = a$, $h(E_1 + E_2) = h(E_1) + h(E_2)$, $h(E_1 \cdot_c E_2) = h(E_1) \cdot_c h(E_2)$, $h(E_1^{*_c}) = h(E_1)^{*_c}$, $h(f_j(E_1, \ldots, E_n)) = f(h(E_1), \ldots, h(E_n))$, with $n \in \mathbb{N}$, $a \in \Sigma_0$, $f \in \Sigma_n$, $f_j \in \mathrm{Pos_E}(E)_n$ such that $h(f_j) = f$ and $E_1, \ldots, E_n$ any regular expressions over $\mathrm{Pos_E}(E)$.

3   Tree Automata Computations

In this section, we recall how to compute from a regular expression E a tree automaton that accepts $[\![E]\!]$. We first recall the computation of the equation automaton $\mathcal{A}_E$ of E, then we define the k-c-continuation automaton $\mathcal{C}_E$.

3.1 The Equation Tree Automaton

In [8], Kuske and Meinecke extend the notion of word partial derivatives [1] to tree partial derivatives in order to compute from a regular expression E a tree automaton recognizing $[\![E]\!]$. Due to the notion of ranked alphabet, partial derivatives are no longer sets of expressions, but sets of tuples of expressions.

Let $\mathcal{N} = (E_1, \ldots, E_n)$ be a tuple of regular expressions, F be some regular expression and $c \in \Sigma_0$. Then $\mathcal{N} \cdot_c F$ is the tuple $(E_1 \cdot_c F, \ldots, E_n \cdot_c F)$. For $\mathcal{S}$ a set of tuples of regular expressions, $\mathcal{S} \cdot_c F$ is the set $\mathcal{S} \cdot_c F = \{\mathcal{N} \cdot_c F \mid \mathcal{N} \in \mathcal{S}\}$. Finally, $\text{SET}(\mathcal{N}) = \{E_1, \cdots, E_m\}$ and $\text{SET}(\mathcal{S}) = \bigcup_{\mathcal{N} \in \mathcal{S}} \text{SET}(\mathcal{N})$.

In the following of this paper, E and F are two regular expressions over a ranked alphabet Σ, and f and g are symbols in $\Sigma_{\geq 1}$.

The set $f^{-1}(E)$ of tuples of regular expressions is defined [8] as follows:

$$f^{-1}(g(E_1, \cdots, E_n)) = \begin{cases} \{(E_1, \cdots, E_n)\} & \text{if } f = g \\ \emptyset & \text{otherwise} \end{cases}$$

$$f^{-1}(F + G) = f^{-1}(F) \cup f^{-1}(G)$$

$$f^{-1}(F \cdot_c G) = \begin{cases} f^{-1}(F) \cdot_c G & \text{if } c \notin [\![F]\!] \\ f^{-1}(F) \cdot_c G \cup f^{-1}(G) & \text{otherwise} \end{cases}$$

$$f^{-1}(F^{*_c}) = f^{-1}(F) \cdot_c F^{*_c}$$

The function f^{-1} is extended to any set S of regular expressions by $f^{-1}(S) = \bigcup_{E \in S} f^{-1}(E)$. The *partial derivative* of E w.r.t. a word $w \in \Sigma_{\geq 1}^*$, denoted by $\partial_w(E)$, is the set of regular expressions inductively defined by:

$$\partial_w(E) = \begin{cases} \{E\} & \text{if } w = \varepsilon \\ \text{SET}(f^{-1}(\partial_u(E))) & \text{if } w = uf, f \in \Sigma_{\geq 1}, u \in \Sigma_{\geq 1}^* \end{cases}$$

The partial derivation is extended to any subset U of $\Sigma_{\geq 1}^*$ as by $\partial_U(E) = \bigcup_{w \in U} \partial_w(E)$. Note that $\partial_{uf}(E) = \partial_f(\partial_u(E)) = \bigcup_{F \in \partial_u(E)} \partial_f(F)$.

Definition 1. *Let* E *be a regular expression over a ranked alphabet* Σ. *The Equation Automaton of* E *is the tree automaton* $\mathcal{A}_E = (Q, \Sigma, Q_T, \Delta)$ *defined by* $Q = \partial_{\Sigma_{\geq 1}^*}(E)$, $Q_T = \{E\}$, *and*

$$\Delta = \begin{cases} \{(F, f, G_1, \ldots, G_m) \mid F \in Q, f \in \Sigma_m, m \geq 1, (G_1, \ldots, G_m) \in f^{-1}(F)\} \\ \cup \{(F, c) \mid c \in ([\![F]\!] \cap \Sigma_0)\} \end{cases}$$

Theorem 1 ([8]). $\mathcal{L}(\mathcal{A}_E) = [\![E]\!]$.

3.2 The C-Continuation Tree Automaton

In [8], Kuske and Meinecke show how to efficiently compute the equation tree automaton of a regular expression *via* an extension of Champarnaud and Ziadi's k-C-Continuation [3,4,7]. In this section, we show how to inductively compute them. The main difference with [8] is that the k-c-continuations are here computed using alternative formulae, and not using the partial derivation. As a consequence, any symbol that appears in the expression E admits a non-empty k-c-continuation (*e.g.* in [8], there is no continuation for g in $E = a \cdot_b g(c)$).

Definition 2. *Let* E *be linear. Let* k *and* m *be two integers such that* $1 \leq k \leq m$. *Let* f *be in* $(\Sigma_E \cap \Sigma_m)$. *The* k-*C-continuation* $C_{f^k}(E)$ *of* f *in* E *is the regular expression defined by:*

$$C_{f^k}(g(E_1, \cdots, E_m)) = \begin{cases} E_k & \text{if } f = g \\ C_{f^k}(E_j) & \text{if } f \in \Sigma_{E_j} \end{cases}$$

$$C_{f^k}(F + G) = \begin{cases} C_{f^k}(F) & \text{if } f \in \Sigma_F \\ C_{f^k}(G) & \text{if } f \in \Sigma_G \end{cases}$$

$$C_{f^k}(F \cdot_c G) = \begin{cases} C_{f^k}(F) \cdot_c G & \text{if } f \in \Sigma_F \\ C_{f^k}(G) & \text{otherwise} \end{cases}$$

$$C_{f^k}(F^{*c}) = C_{f^k}(F) \cdot_c F^{*c}$$

By convention, we set $C_{\varepsilon^1}(E) = E$.

Let us first show the relation between partial derivation and k-c-continuation.

Lemma 1. *Let* E *be linear,* n, m *and* k *be three integers such that* $n, m \geq 1$, $1 \leq k \leq m$, $f \in \Sigma_n$ *and* $g \in \Sigma_m \cup \{\varepsilon\}$. *If* $f^{-1}(C_{g^k}(E)) \neq \emptyset$ *then* $f^{-1}(C_{g^k}(E)) = \{(C_{f^1}(E), \ldots, C_{f^n}(E))\}$.

Proposition 1. *Let* E *be linear and* $f \in \Sigma_n$ *with* $n \geq 1$. *Let* u *be a word in* $\Sigma_{\geq 1}^*$. *If* $f^{-1}(\partial_u(E)) \neq \emptyset$ *then* $f^{-1}(\partial_u(E)) = \{(C_{f^1}(E), \ldots, C_{f^n}(E))\}$.

Definition 3. *The automaton* $\overline{C}_E = (Q_{\overline{C}}, \mathrm{Pos}_E(E), \{C_{\varepsilon^1}(\overline{E})\}, \Delta_{\overline{C}})$ *is defined by*

$$Q_{\overline{C}} = \{C_{f_j^k}(\overline{E}) \mid f_j \in \mathrm{Pos}_E(E)_m, 1 \leq k \leq m\} \cup \{C_{\varepsilon^1}(\overline{E})\},$$

$$\Delta_{\overline{C}} = \begin{cases} \{(C_x(\overline{E}), g_i, \mathfrak{C}_{g_i}) \mid g_i \in \mathrm{Pos}_E(E)_m, m \geq 1, \mathfrak{C}_{g_i} \in g_i^{-1}(C_x(\overline{E}))\} \\ \cup \{(C_x(\overline{E}), c) \mid, c \in [\![C_x(\overline{E})]\!] \cap \Sigma_0\} \end{cases}$$

where for any symbol g_i *in* $\mathrm{Pos}_E(E)_m$, $\mathfrak{C}_{g_i} = (C_{g_i^1}(\overline{E}), \ldots, C_{g_i^m}(\overline{E}))$.

The following lemma illustrates the link between $\overline{C}_E$ and $\mathcal{A}_{\overline{E}}$.

Lemma 2. *The coaccessible part of* $\overline{C}_E$ *is equal to* $\mathcal{A}_{\overline{E}}$.

Corollary 1. *The automaton* $\overline{C}_E$ *accepts* $[\![\overline{E}]\!]$.

The *C-Continuation tree automaton* C_E associated with E is obtained by replacing each transition $(C_x(\overline{E}), g_i, C_{g_i^1}(\overline{E}), \ldots, C_{g_i^m}(\overline{E}))$ of the tree automaton $\overline{C}_E$ by $(C_x(\overline{E}), h(g_i), C_{g_i^1}(\overline{E}), \ldots, C_{g_i^m}(\overline{E}))$.

Corollary 2. $h(\mathcal{L}(\overline{C}_E)) = \mathcal{L}(C_E) = [\![E]\!]$.

In what follows, for any two trees s and t, we denote by $s \preccurlyeq t$ the relation "s is a subtree of t". Let k be an integer. We denote by $\mathrm{root}(s)$ the root of any tree s and by k-child(t), for a tree $t = f(t_1, \ldots, t_n)$, the kth child of f in t that is root of t_k if it exists.

Let $1 \leq k \leq m$ be two integers and f_j be a symbol in $\mathrm{Pos_E}(\mathrm{E})_m$. The sets $\mathrm{First}(\mathrm{E})$ is the subset of $\mathrm{Pos_E}(\mathrm{E})$ defined by $\mathrm{First}(\mathrm{E}) = \{\mathrm{root}(t) \in \mathrm{Pos_E}(\mathrm{E}) \mid t \in [\![\overline{\mathrm{E}}]\!]\}$. The set $\mathrm{Follow}(\mathrm{E}, f_j, k)$ is the subset of $\mathrm{Pos_E}(\mathrm{E})$ defined by $\mathrm{Follow}(\mathrm{E}, f_j, k) = \{g_i \in \mathrm{Pos_E}(\mathrm{E}) \mid \exists t \in [\![\overline{\mathrm{E}}]\!], \exists s \preccurlyeq t, \mathrm{root}(s) = f_j, k\text{-child}(s) = g_i\}$.

Proposition 2 ([9]). *The computation of all the sets* $(\mathrm{Follow}(\mathrm{E}, f_j, k))$ *for* $1 \leq k \leq m$ *and* $f \in \mathrm{Pos_E}(\mathrm{E})_m$ *can be done in an* $O(|\mathrm{E}|)$ *time and space complexity.*

Proposition 3. *Let* $1 \leq k \leq m$ *and* f_j *be a position in* $\mathrm{Pos_E}(\mathrm{E})_m$. *If* $\mathrm{Follow}(\mathrm{E}, f_j, k) \neq \emptyset$ *then* $\mathrm{Follow}(\mathrm{E}, f_j, k) = \mathrm{First}(C_{f_j^k}(\overline{\mathrm{E}}))$.

Proposition 4. *Let* $1 \leq k \leq m$ *be two integers,* f_j *be a symbol in* $\mathrm{Pos_E}(\mathrm{E})_m$ *and* g_i *be a symbol in* $\mathrm{Pos_E}(\mathrm{E})$. *Then* $g_i^{-1}(C_{f_j^k}(\overline{\mathrm{E}})) \neq \emptyset \Leftrightarrow g_i \in \mathrm{First}(C_{f_j^k}(\overline{\mathrm{E}}))$.

Lemma 3. *Let* $1 \leq k \leq m$ *be two integers and* f_j *be a position in* $\mathrm{Pos_E}(\mathrm{E})_m$. *If* $\mathrm{Follow}(\mathrm{E}, f_j, k) = \emptyset$ *then* $C_{f_j^k}(\overline{\mathrm{E}})$ *is not a coaccessible state in* $\mathcal{C}_\mathrm{E}$.

3.3   From k-C-Continuation Automaton to Equation Automaton

The equation automaton is a quotient of the C-Continuation one w.r.t. the equivalence relation denoted by $\sim_e$ over the set of states of $\overline{\mathcal{C}}_\mathrm{E}$ defined for any two states $q_1 = C_{f_j^k}(\mathrm{E})$ and $q_2 = C_{g_i^p}(\mathrm{E})$ by $q_1 \sim_e q_2 \Leftrightarrow h(q_1) = h(q_2)$.

Proposition 5. *The coaccessible part of the finite tree automaton* $\mathcal{C}_\mathrm{E}/_{\sim_e}$ *is isomorphic to the equation tree automaton* $\mathcal{A}_\mathrm{E}$.

4   Construction of the Equation Tree Automaton $\mathcal{A}_\mathrm{E}$

In [8], the computation of the k-C-Continuations requires a preprocessing step which is the identification of subexpression of E in $O(|\mathrm{E}|^2)$ time and space complexity. We propose an algorithm for the computation of the set of states with an $O(|\mathrm{E}|)$ time and space complexity.

4.1   Computation of the Set of States $Q_{\overline{\mathcal{C}}}/_{\sim_e}$

The main idea is to efficiently compute the quotient $\overline{\mathcal{C}}_\mathrm{E}/_{\sim_e}$ by converting the syntax tree into a finite acyclic deterministic word automaton.

Let T_E be the syntax tree associated with E. The set of nodes of T_E is written as $\mathrm{Nodes}(\mathrm{E})$. For a node ν in $\mathrm{Nodes}(\mathrm{E})$, $\mathrm{sym}(\nu)$, $\mathrm{father}(\nu)$, $\mathrm{son}(\nu)$, $\mathrm{right}(\nu)$ and $\mathrm{left}(\nu)$ denote respectively the symbol, the father, the son, the right son and the left son of the node ν if they exist. We denote by E_ν the subexpression rooted at ν; In this case we write ν_E to denote the node associated to E_ν. Let $\gamma : \mathrm{Nodes}(\mathrm{E}) \cup \{\bot\} \to \mathrm{Nodes}(\mathrm{E}) \cup \{\bot\}$ be the function defined by:

$$\gamma(\nu) = \begin{cases} \mathrm{father}(\nu) & \text{if } \mathrm{sym}(\mathrm{father}(\nu)) =^{*c} \text{ and } \nu \neq \nu_\mathrm{E} \\ \mathrm{right}(\mathrm{father}(\nu)) & \text{if } \mathrm{sym}(\mathrm{father}(\nu)) = \cdot_c \\ \bot & \text{otherwise} \end{cases}$$

where $\perp$ is an artificial node such that $\gamma(\perp) = \perp$. The ZPC-Structure is the syntax tree equipped with $\gamma(\nu)$ links. We extend the relation $\preccurlyeq$ to the set of nodes of T_E: For two nodes μ and ν we write $\nu \preccurlyeq \mu \Leftrightarrow T_{E_\nu} \preccurlyeq T_{E_\mu}$. We define the set $\Gamma_\nu(E) = \{\mu \in \text{Nodes}(E) \mid \nu \preccurlyeq \mu \wedge \gamma(\mu) \neq \perp\}$ which is totally ordered by $\preccurlyeq$.

Proposition 6. *Let* E *be linear,* $1 \leq k \leq n$ *be two integers and* f *be in* $\Sigma_E \cap \Sigma_n$. *Then* $C_{f^k}(E) = ((((E_{\nu_0} \cdot_{op(\nu_1)} E_{\gamma(\nu_1)}) \cdot_{op(\nu_2)} E_{\gamma(\nu_2)}) \cdots \cdot_{op(\nu_m)} E_{\gamma(\nu_m)})$ *where* ν_f *is the node of* T_E *labelled by* f, ν_0 *is the* k-*child*(ν_f), $\Gamma_{\nu_f}(E) = \{\nu_1, \ldots, \nu_m\}$ *and for* $1 \leq i \leq m$, $op(\nu_i) = c$ *such that* $\text{sym}(\text{father}(\nu_i)) \in \{\cdot_c, *_c\}$.

Corollary 3. *Let* E *be linear,* $f \in (\Sigma_E)_m$ *and* $k \leq m$. *Then* $|C_{f_j^k}(E)| \leq |E|^2$.

Example 1. Let Σ be the ranked alphabet such that $\Sigma_0 = \{a, b\}$, $\Sigma_1 = \{h\}$ and $\Sigma_2 = \{f\}$. Let $E = (f(a,a) + f(a,a))^{*_a} \cdot_a h(b)$. Then $\overline{E} = (f_1(a,a) + f_2(a,a))^{*_a} \cdot_a h_3(b)$. The ZPC-Structure associated with $\overline{E}$ is represented in Figure 1 restricted to some γ links. As stated in Proposition 6, $C_{f_1^1}(\overline{E}) = ((a \cdot_a (f_1(a,a) + f_2(a,a))^{*_a}) \cdot_a h_3(b)) = ((E_{\nu_0} \cdot_a E_{\gamma_{\nu_1}}) \cdot_a E_{\gamma_{\nu_2}})$.

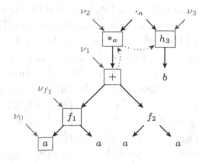

In order to identify the equivalent k-C-Continuations, we can sort them in lexicographic order. This be done in $O(|E|^3)$ time and space complexity using Paige and Tarjan's Algorithm [12]. This is due to the fact that the size of k-C-Continuations is in $O(|E|^2)$ (by Corollary 3). This complexity has been improved by using k-*Pseudo-Continuations* instead of k-C-Continuations [3,7].

Fig. 1. ZPC-Structure of E

A k-*Pseudo-Continuation* $l_{f_j^k}(E)$ of f_j in E is obtained from the k-C-Continuation $C_{f_j^k}(\overline{E})$ by replacing some subexpression $\overline{F}$ of $\overline{E}$ by a symbol $\psi(h(\overline{F}))$ such that for two subexpressions F and G of E: $\psi(F) = \psi(G) \Leftrightarrow F = G$.

Definition 4. *Let* H *be a regular expression over* Σ *and* ψ *be a bijection that associates to each subexpression of* E *a symbol in an alphabet* Ψ. *We define the word* $\psi'(H)$ *over the alphabet* $\Psi \cup \{\cdot_a \mid a \in \Sigma_0\}$ *inductively as follows:*

$$\psi'(H) = \begin{cases} \psi'(F) \cdot_c \psi'(G) & \text{if } H = F \cdot_c G \text{ and } G \text{ a subexpression of } E \\ \psi(H) & \text{if } H \neq F \cdot_c G \text{ and } H \text{ a subexpression of } E \\ \varepsilon & \text{otherwise.} \end{cases}$$

The function ψ' *is said to be an* (E, Ψ)-*encoding.*

Definition 5. *Let* n *and* k *be two integers such that* $1 \leq k \leq n$, f_j *be a symbol in* $\text{Pos}_E(E)$ *and* ψ' *an* (E, Ψ)-*encoding for some alphabet* Ψ. *The* k-*Pseudo-Continuation of* f_j *in* E, *denoted by* $l_{f_j^k}(E)$, *is the word over* $\Psi \cup \{\cdot_a \mid a \in \Sigma_0\}$ *defined by* $l_{f_j^k}(E) = \psi'(h(C_{f_j^k}(\overline{E})))$.

In the following, we consider that the pseudo-continuations of E are defined over Ψ a finite subset of $\mathbb{N}$, bounded by the number of subexpressions of E.

Proposition 7. *The two following propositions hold:*

1. $|l_{f_j^k}(\mathrm{E})|$ *is at most linear w.r.t.* $|\mathrm{E}|$,
2. $\sum_{f_j \in \mathrm{Pos}_\mathrm{E}(\mathrm{E})_n, 1 \leq k \leq n} |\psi'(\mathrm{E}_{k\text{-child}(\nu_{f_j})})|$ *is at most linear w.r.t.* $|\mathrm{E}|$.

Proposition 8. *Let* $f_j \in \mathrm{Pos}_\mathrm{E}(\mathrm{E})_n$, $g_i \in \mathrm{Pos}_\mathrm{E}(\mathrm{E})_m$, $k \leq n$ *and* $p \leq m$ *be two integers. Then* $h(C_{f_j^k}(\overline{\mathrm{E}})) = h(C_{g_i^p}(\overline{\mathrm{E}})) \Leftrightarrow l_{f_j^k}(\mathrm{E}) = l_{g_i^p}(\mathrm{E})$.

From Proposition 8 we can deduce that the k-C-Continuations identification can be achieved by considering the k-Pseudo-Continuations. In the following we show that this identification step (computation of $\sim_e$) can be done without the computation of the k-Pseudo-Continuations and that it amounts to the minimization of a word acyclic deterministic automaton. Before seeing how the identification of k-Pseudo-Continuations $l_{f_j^k}(\mathrm{E})$ is performed, we prove that the computation of the function ψ can be done in a linear time in the size of E.

Let us consider the syntax tree T_E associated with E. This syntax tree contains all the subexpressions of E. Each node ν in T_E corresponds to the subexpression E_ν of E. The equivalence relation $\sim$ over the nodes of the tree T_E is defined by $\nu_1 \sim \nu_2 \Leftrightarrow \mathrm{E}_{\nu_1} = \mathrm{E}_{\nu_2}$. We show that the computation of the equivalence relation $\sim$ amounts to the minimization of the word acyclic deterministic automaton $\mathcal{A}_{T_\mathrm{E}} = (Q, \Sigma_\mathcal{A}, \{\nu_\mathrm{E}\}, \{\nu_T\}, \delta)$, where ν_E is the node associated to the root of E, $Q = \mathrm{Nodes}(\mathrm{E}) \cup \{\nu_T\} \cup \{\bot\}$ with $\nu_T, \bot \notin \mathrm{Nodes}(\mathrm{E})$, $\Sigma_\mathcal{A} = \Sigma_0 \cup \{g_+, d_+\} \cup \{*_a, g_{\cdot_a}, d_{\cdot_a} \mid a \in \Sigma_0\} \cup \{f^1, \ldots, f^n \mid f \in \Sigma_n, n \geq 1\}$, and δ is defined by $\delta(\nu, *_a) = \mathrm{son}(\nu)$ if $\mathrm{sym}(\nu) = *_a$, $\delta(\nu, g_{\mathrm{sym}(\nu)}) = \mathrm{left}(\nu)$ and $\delta(\nu, d_{\mathrm{sym}(\nu)}) = \mathrm{right}(\nu)$ if $\mathrm{sym}(\nu) \in \{+, \cdot_a, a \in \Sigma_0\}$, $\delta(\nu, \mathrm{sym}(\nu)) = \nu_T$ if $\mathrm{sym}(\nu) \in \Sigma_0$, $delta(\nu, f^k) = k\text{-child}(\nu)$ if $\mathrm{sym}(\nu) = f \in \Sigma_{\geq 1}$, and $\delta(\nu, x) = \bot$ in all otherwise.

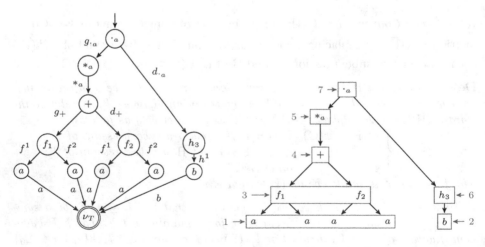

Fig. 2. The automaton $\mathcal{A}_{T_\mathrm{E}}$ **Fig. 3.** The Equivalence Classes

Lemma 4. $E = F \Leftrightarrow \mathcal{L}(\mathcal{A}_{T_E}) = \mathcal{L}(\mathcal{A}_{T_F})$.

According to Lemma 4, $\nu_1 \sim \nu_2 \Leftrightarrow \mathcal{L}(\mathcal{A}_{T_{E_{\nu_1}}}) = \mathcal{L}(\mathcal{A}_{T_{E_{\nu_2}}})$, that is the equivalence relation $\sim$ coincides with Myhill-Nerode equivalence [11] over the states of the automaton $\mathcal{A}_{T_E}$, that can be computed in $O(|E|)$ time and space complexity using Revuz Algorithm [13].

Lemma 5. *The computation of $\psi(F)$ for all subexpression F of E can be done in $O(|E|)$ time and space complexity.*

Example 2. Let us consider the regular expression $E = (f(a,a) + f(a,a))^{*a} \cdot_a h(b)$ of the Example 1. Applying Myhill-Nerode equivalence [11] to the states of the automaton $\mathcal{A}_{T_E}$ (Figure 2) results in 7 equivalence classes labeled by $\Psi = \{1, 2, \ldots, 7\}$. For example $\psi(f(a,a)) = 3$ and $\psi(E) = 7$ (Figure 3). Finally, $l_{f_1^1}(E) = 1 \cdot_a 6 \cdot_a 5$.

Recall that the k-Pseudo-Continuation identification can be achieved in $O(|E|^2)$ [4,8] using Paige and Tarjan's sorting algorithm [12]. In what follows we show that this step amounts to the minimization of the acyclic deterministic word automaton $\mathcal{B}_{T_E} = (Q_\mathcal{B}, \Sigma_\mathcal{B}, \{\nu_T\}, \{\nu_{\overline{E}}\}, \delta_\mathcal{B})$ defined with $\nu_T \notin \text{Nodes}(\overline{E})$ and $\mathfrak{F} = \{f_j^k \mid 1 \leq k \leq m, f_j \in \text{Pos}_E(E)_m\}$ by $Q_\mathcal{B} = (\text{Nodes}(\overline{E}) \setminus \Sigma_0) \cup \mathfrak{F} \cup \{\nu_T, \perp\}$, $\Sigma_\mathcal{B} = \{\psi(\nu) \mid \nu \in \text{Nodes}(\overline{E}) \cap Q_\mathcal{B}\} \cup \mathfrak{F} \cup \{\cdot_a \mid a \in \Sigma_0\} \cup \{\varepsilon\}$, and $\delta_\mathcal{B}$ is defined as follows: $\delta(\nu_T, f_j^i) = f_j^i$ for all $f_j^i \in \mathfrak{F}$, $\delta(f_j^i, \psi'(h(E_{\nu_k}))) = f_j$ if ν_k is the kth child of f_j $\delta(\nu, \cdot_a \psi(E_{\gamma(\nu)})) = \text{father}(\nu)$ if $\text{sym}(\text{father}(\nu)) \ in \{\cdot_a, *_a\}$ and $\gamma(\nu) \neq \perp$, $\delta(\nu, \varepsilon) = \text{father}(\nu)$ and if $\gamma(\nu) = \perp$ and $\delta(\nu, x) = \perp$ in all otherwise.

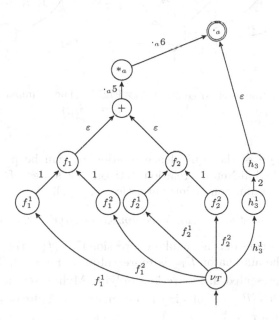

Fig. 4. The automaton $\mathcal{B}_{T_E}$

Proposition 9. $\mathcal{L}(\mathcal{B}_{T_{\overline{E}}}) = \{f_j^k \cdot l_{f_j^k}(E) \mid f_j \in \mathrm{Pos}_E(E)_m, k \leq m\}$

Let f_j and g_i be two positions in $\mathrm{Pos}_E(E)$. As a direct consequence of Proposition 9, $C_{f_j^k}(\overline{E}) \sim_e C_{g_i^p}(\overline{E})$ if and only if the states f_j^k and g_i^p of $\mathcal{B}_{T_{\overline{E}}}$ are equivalent. We eliminate the ε-transitions from the automaton $\mathcal{B}_{T_{\overline{E}}}$. Since it has no ε-transitions cycles, this elimination can be performed in a linear time in the size of E. Hence, we obtain a more compacted but equivalent structure, which we denote by ε-free$(\mathcal{B}_{T_{\overline{E}}})$.

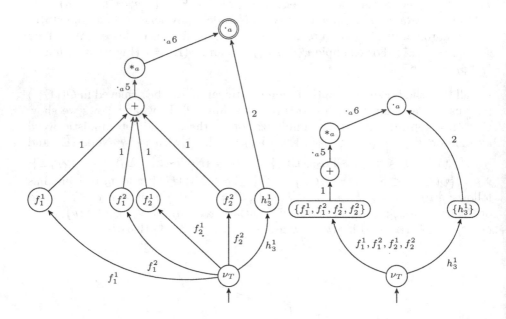

Fig. 5. The automaton ε-free$(\mathcal{B}_{T_{\overline{E}}})$ **Fig. 6.** The Minimal Automaton of ε-free$(\mathcal{B}_{T_{\overline{E}}})$

The computation of the equivalence relation $\sim_e$ can be performed by the computation of Myhill-Nerode relation [11] on the states of the automaton ε-free$(\mathcal{B}_{T_{\overline{E}}})$. This automaton is deterministic and acyclic.

Theorem 2. *The relation $\sim_e$ can be computed in $O(|E|)$ time complexity.*

Example 3. Let us consider the regular expression $E = (f(a,a) + f(a,a))^{*a} \cdot_a h(b)$ of Example 1. The automaton $\mathcal{B}_{T_{\overline{E}}}$ is represented by Figure 4. The automaton ε-free$(\mathcal{B}_{T_{\overline{E}}})$ is represented in Figure 5. Applying Myhill-Nerode equivalence to the automaton ε-free$(\mathcal{B}_{T_{\overline{E}}})$ results in the automaton in Figure 6. We deduce from this automaton that $C_{f_1^1}(\overline{E}) \sim_e C_{f_1^2}(\overline{E}) \sim_e C_{f_2^1}(\overline{E}) \sim_e C_{f_2^2}(\overline{E})$. Consequently the set of states of $\mathcal{C}_E / \sim_e$ is $\{[C_{\varepsilon^1}(\overline{E})], [C_{f_1^1}(\overline{E})], [C_{h_3^1}(E)]\}$.

4.2   Computation of the Set of Transition Rules

Using Proposition 3 and Proposition4, we can show that the computation of the set of transitions of the equation tree automaton is performed by computing the function Follow. The computation of a transition rule using Proposition 3 requires a linear time, according to Proposition 2. Then for all transition rules we get an $O(|Q/_{\sim_e}| \times |E|)$ time and space complexity where Q is the set of k–C-Continuations of $\overline{E}$. The computation of the set of states $Q_{\overline{C}}/_{\sim_e}$ make possible the creation of non-coaccessible states. Removing these states requires an $O(|Q_{\overline{C}}/_{\sim_e}| \cdot |E|)$ time complexity.

Theorem 3. *The equation tree automaton $\mathcal{A}_E$ of E can be computed in $O(|Q| \cdot |E|)$ time and space complexity with Q the set of states of $\mathcal{A}_E$.*

References

1. Antimirov, V.M.: Partial derivatives of regular expressions and finite automaton constructions. Theor. Comput. Sci. 155(2), 291–319 (1996)
2. Brüggemann-Klein, A.: Regular expressions into finite automata. Theor. Comput. Sci. 120(2), 197–213 (1993)
3. Champarnaud, J.M., Ziadi, D.: From c-continuations to new quadratic algorithms for automaton synthesis. IJAC 11(6), 707–736 (2001)
4. Champarnaud, J.M., Ziadi, D.: Canonical derivatives, partial derivatives and finite automaton constructions. Theor. Comput. Sci. 289(1), 137–163 (2002)
5. Comon, H., Dauchet, M., Gilleron, R., Jacquemard, F., Lugiez, D., Loding, C., Tison, S., Tommasi, M.: Tree automata techniques and applications (October 2007), http://www.grappa.univ-lille3.fr/tata
6. Glushkov, V.M.: The abstract theory of automata. Russian Mathematical Surveys 16, 1–53 (1961)
7. Khorsi, A., Ouardi, F., Ziadi, D.: Fast equation automaton computation. J. Discrete Algorithms 6(3), 433–448 (2008)
8. Kuske, D., Meinecke, I.: Construction of tree automata from regular expressions. RAIRO - Theor. Inf. and Applic. 45(3), 347–370 (2011)
9. Laugerotte, É., Sebti, N.O., Ziadi, D.: From regular tree expression to position tree automaton. In: Dediu, A.-H., Martín-Vide, C., Truthe, B. (eds.) LATA 2013. LNCS, vol. 7810, pp. 395–406. Springer, Heidelberg (2013)
10. Murata, M.: Hedge automata: a formal model for xml schemata (2000), http://www.xml.gr.jp/relax/hedge_nice.html
11. Nerode, A.: Linear automata transformation. Proc. Amer. Math. Soc. 9, 541–544 (1958)
12. Paige, R., Tarjan, R.E.: Three partition refinement algorithms. SIAM J. Comput. 16(6), 973–989 (1987)
13. Revuz, D.: Minimisation of acyclic deterministic automata in linear time. Theor. Comput. Sci. 92(1), 181–189 (1992)
14. Ziadi, D., Ponty, J.L., Champarnaud, J.M.: Passage d'une expression rationnelle a un automate fini non deterministe. Bulletin of the Belgian Mathematical Society - Simon Stevin 4, 177–203 (1997)

On the Effectiveness of Symmetry Breaking

Russell Miller[1], Reed Solomon[2], and Rebecca M. Steiner[3]

[1] Queens College and the Graduate Center of the City University of New York
Flushing NY 11367
[2] University of Connecticut
Storrs CT 06269
[3] Vanderbilt University
Nashville TN 37240
r.m.steiner@vanderbilt.edu

Abstract. Symmetry breaking involves coloring the elements of a structure so that the only automorphism which respects the coloring is the identity. We investigate how much information we would need to be able to compute a 2-coloring of a computable finite-branching tree under the predecessor function which eliminates all automorphisms except the trivial one; we also generalize to n-colorings for fixed n and for variable n.

1 Introduction

Symmetry has always been a crucial concept in mathematics. We think of symmetry as a geometric property, but in fact, symmetries appear in many other branches of math as well. The symmetries of a mathematical structure are precisely its automorphisms – the bijections from the structure onto itself which preserve the essential properties of the structure. Some structures have many symmetries, and others have only one (the identity, or trivial automorphism).

Symmetry breaking involves coloring the elements of a structure in such a way that the only automorphism which respects the coloring is the trivial one; "breaking" symmetries can be thought of as "killing off" automorphisms.

Definition 11. An *n-coloring* of a structure is a function from the domain of the structure into a set of size n. It is said to *distinguish* the structure if there are no nontrivial automorphisms of the structure which respect the equivalence relation defined by the coloring. If a distinguishing n-coloring exists, then the structure is said to be *n-distinguishable*.

Definition 12. The *distinguishing number* of a structure is the smallest $n \in \omega$ such that the structure has a distinguishing n-coloring. If it exists, then the structure is *finitely distinguishable*.

As an example, consider a graph that looks like the integers, as shown here. As a graph, it has infinitely many automorphisms. But there is a way to color

A. Beckmann, E. Csuhaj-Varjú, and K. Meer (Eds.): CiE 2014, LNCS 8493, pp. 314–323, 2014.
© Springer International Publishing Switzerland 2014

the elements of this graph with just two colors so that the only automorphism which respects the coloring is the trivial one. A certain three elements are given the "solid" color, as in the figure here, while all the rest are given the "striped" color, and the only symmetry of this graph which respects this coloring is the identity. So we would say that this graph has distinguishing number 2.

Symmetry breaking has been studied extensively by combinatorists, with very recent results detailed in [1], [5], and [6]. In fact, the work found in the next section on symmetry breaking from a computability-theoretic perspective was inspired by a result from one of these articles:

Theorem 13 ([1], Theorem 3.1). *The countable random graph has distinguishing number* 2.

This is extremely surprising. The random graph has continuum-many automorphisms, and is ultrahomogeneous: every finite partial automorphism extends to an automorphism of the entire graph. (This says that, in a certain sense, its automorphisms are dense, within the finite partial maps respecting its edge relation.) Moreover, while the result of Theorem 13 was not at all intended to be an effectiveness result, the construction of the distinguishing coloring of the countable random graph in [1] is indeed effective in the edge relation of the graph. In other words, a computable copy of the random graph has a *computable* distinguishing 2-coloring.

Knowing that this holds for the random graph inspired us to investigate the same question for other structures: what kinds of computable structures have *computable* distinguishing n-colorings? It was this question which led to our study of effective symmetry breaking in computable finite-branching predecessor trees, which form a natural first step in the subject, mainly because the automorphisms of such structures are readily understood.

Definition 14. A *tree* is a partial order $\prec$ on a set $\mathcal{T}$ of nodes, with a least element r (the *root*) under $\prec$, such that for every $x \in T$, the set $\{y \in T : y \prec x\}$ is well-ordered by $\prec$. If every chain under $\prec$ has order type $\leq \omega$ (that is, if the tree has *height* $\leq \omega$), then each $x \in T$ has an immediate predecessor under $\prec$. A *predecessor tree* is a tree of height $\leq \omega$ in a language with equality and one unary function P, the *predecessor function*, for which $P(r) = r$ and $P(x)$ is the immediate predecessor of x whenever $x \neq r$. If $\mathcal{T}$ has domain ω and P is computable, we call $\mathcal{T}$ a *computable predecessor tree*; such $\mathcal{T}$ correspond precisely to computable subtrees of the tree $\omega^{<\omega}$ of finite strings from ω. (The underlying partial order on such a T is computable from P, although not definable by finitary formulas using P.) A predecessor tree is *finite-branching* if, for every y, there are only finitely many $x \in T$ with $P(x) = y$.

Trees which are computable as partial orders (but for which the predecessor function is not necessarily computable) are considered in a different context in [3,4], which may provide useful background for readers interested in investigating these questions. In such a tree, it is not generally possible to compute the level of a node, and this makes it substantially more difficult to determine which nodes lie in the same orbit under automorphisms of the tree. It would be natural to attempt to extend the results of this article to computable trees under $\prec$. Some previous effectiveness results about predecessor trees appear in [7], while for more general effectiveness results about symmetries as automorphisms, we refer the reader to [2].

2   The Effectiveness Results

The most natural question to address first is whether a computable finite-branching predecessor tree with distinguishing number 2 must have a *computable* distinguishing 2-coloring.

Theorem 21. *There is a computable finite-branching predecessor tree which is distinguished by a 2-coloring but not by any computable 2-coloring.*

Proof. We will build our tree $\mathcal{T}$ in such a way that no (partial) computable function φ_e can be a distinguishing 2-coloring of the tree. We start by describing the basic module – the strategy which will guarantee that for some fixed e, φ_e is not a distinguishing 2-coloring of the tree. We build a finite tree $\mathcal{T}_e$ beginning with a root, five immediate successors of the root, and one immediate successor each for four of those five immediate successors of the root, as shown here.

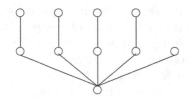

We wait for φ_e to converge on each of these ten inputs. If φ_e doesn't converge on all the inputs or converges outside the set $\{0, 1\}$, then we do nothing further, because φ_e is not a 2-coloring of $\mathcal{T}_e$. If φ_e converges on all ten nodes to values in $\{0, 1\}$ in such a way that there is already a nontrivial automorphism of $\mathcal{T}_e$ which preserves the coloring, then we do nothing further, because φ_e is not a distinguishing coloring of $\mathcal{T}_e$. So, without loss of generality, suppose φ_e converges on all ten nodes to values in $\{0, 1\}$ and colors them as shown.

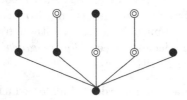

We do not want φ_e to be a distinguishing coloring of $\mathcal{T}_e$, so we respond by adding three more nodes to $\mathcal{T}_e$, as seen below. (For convenience, nodes are now identified as in $\omega^{<\omega}$.)

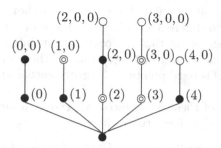

We have now made it impossible for φ_e to be a distinguishing 2-coloring of $\mathcal{T}_e$. φ_e must color the new node $(4, 0)$ at level 2, and whichever color it chooses, there will be a nontrivial automorphism of $\mathcal{T}_e$ which respects that coloring. However, there does exist another 2-coloring f_e which distinguishes $\mathcal{T}_e$: change (4) to a striped node, while keeping the other colors and coloring the remaining three nodes with either color. Under this coloring the tree is rigid.

We put these basic modules together to build one big tree $\mathcal{T}$ as follows. Start with a *spine*: a single path $d_0 < d_1 < d_2 < \cdots$ of nodes, among which d_0 will be the root of $\mathcal{T}$. Above each d_{2e+1}, in addition to d_{2e+2}, we place a node r_e. Then we build a copy of the tree $\mathcal{T}_e$ with r_e as its root. Thus no φ_c is a distinguishing 2-coloring of $\mathcal{T}$, but $\mathcal{T}$ is distinguishable by combining the 2-colorings f_e for each $\mathcal{T}_e$ into a single f and coloring every node on the spine striped. The spine is fixed by every automorphism of T, as is each r_e, and this (noncomputable) f then ensures that no automorphism except the identity can respect f. □

In the tree constructed in Theorem 21, the branching function is not computable: it was not decidable which of the $(4, 0)$ nodes (in all the different finite subtrees $\mathcal{T}_e$) have successors and which do not. So one naturally asks whether a computable finite-branching tree with distinguishing number 2 and with computable branching function would necessarily have a computable distinguishing 2-coloring. However, the answer is still no: below, in Theorem 23, we construct such a tree with no computable distinguishing 2-coloring.

We start by describing the *balanced 2-coloring* of the complete binary tree $2^{<\omega}$. Two nodes in this tree are called *siblings* if they are of the form $\sigma^\frown 0$ and $\sigma^\frown 1$, that is, if they have the same immediate predecessor σ. Thus every node except the root has exactly one sibling. A 2-coloring is *balanced* if it colors each node differently from its sibling: for instance, every node $\sigma^\frown 0$ is solid and every node $\sigma^\frown 1$ is striped. This example is isomorphic to every other balanced 2-coloring of $2^{<\omega}$, except for the color of the root. We will therefore speak of the *balanced 2-coloring with striped root* or the *balanced 2-coloring with solid root*. It is clear, by induction on the lengths of nodes, that each balanced 2-coloring distinguishes $2^{<\omega}$, i.e., no automorphism of $2^{<\omega}$ except the identity respects this coloring. Likewise, we speak of balanced 2-colorings of finite binary trees 2^n.

However, it is also possible for an *unbalanced* 2-coloring to distinguish $2^{<\omega}$. As an example, color the nodes so that every node 1^n is striped, and also every node 1^n0 is striped, with all other nodes colored according to the balanced coloring with striped root. The two siblings 0 and 1 at level 1 are both striped, but the two successors of 0 are two different colors, while those of 1 are both striped. Therefore no automorphism respecting the coloring can interchange 0 with 1, and one then uses this same argument to go upwards through all levels of the tree and see that each level is fixed pointwise by every automorphism respecting the coloring.

There are in fact many of these unbalanced colorings. However, once an imbalance has been introduced, it perpetuates itself.

Lemma 1. *In a distinguishing 2-coloring of $2^{<\omega}$, if some siblings $\sigma\hat{\ }0$ and $\sigma\hat{\ }1$ share a color, then either some two siblings extending $\sigma\hat{\ }0$ share a color as well, or else some two siblings extending $\sigma\hat{\ }1$ share a color.*

Proof. If not, then the coloring would restrict to the balanced coloring on the tree above $\sigma\hat{\ }0$, and also on the tree above $\sigma\hat{\ }1$, with the same colored root in both. Therefore, there would be an automorphism interchanging $\sigma\hat{\ }0$ with $\sigma\hat{\ }1$ and respecting the coloring.   □

Corollary 22. *For every finite binary tree $2^{<n}$, no unbalanced 2-coloring is distinguishing.*

Proof. The reasoning is the same as in the lemma: any imbalance forces there to be another imbalance above itself. However, now this yields a pair of siblings with the same color at the very top level of the tree, and the automorphism which interchanges this pair and fixes all other nodes respects the unbalanced coloring.   □

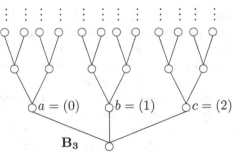

With these unbalanced colorings, we see that the tree B_3 in the figure above is 2-distinguishable:

$$B_3 = \{\sigma \in \omega^{<\omega} : \sigma(0) < 3 \ \& \ \sigma(n) < 2 \text{ for } 0 < |\sigma| \leq n\}.$$

Indeed, B_3 has a computable distinguishing 2-coloring: just give the balanced coloring with solid root on the binary tree above a, the balanced coloring with striped root on the binary tree above b, and any unbalanced coloring (say with striped root) on the binary tree above c. We also have a distinguishing 2-coloring of each tree $B_{3,n}$:

$$B_{3,n} = B_3 - \{\sigma \in B_3 : \sigma(0) \neq 0 \ \& \ |\sigma| \geq n\}.$$

This $B_{3,n}$ is the tree gotten by chopping off (at level n) two of the three binary trees in B_3. To get a distinguishing 2-coloring of $B_{3,n}$, however, one is forced by the corollary above to use balanced colorings, with roots of different colors, on each of the finite trees above b and c. The binary tree above a is still complete, however: one can color it using the balanced 2-coloring with either color for the root, or using any unbalanced distinguishing 2-coloring.

Theorem 23. *There is a computable finite-branching predecessor tree with computable branching function which is distinguished by a 2-coloring but not by any computable 2-coloring.*

Proof. With the above observations, we can produce a tree $\mathcal{T}_e$ with distinguishing number 2 for which a given partial computable function φ_e is not a distinguishing 2-coloring. Moreover, $\mathcal{T}_e$ will have computable branching (uniformly in e). This will form the basic module of the construction below. To build $\mathcal{T}_e$, start with three distinct immediate successors a, b, c of the root node r, and begin building a copy of $2^{<\omega}$ above each, exactly as in the tree B_3. When we add level s to these binary trees, we check to see whether $\varphi_{e,s}$ has converged yet on all three nodes at level 1. If it never does so, or if it gives values $\notin \{0, 1\}$ for any of them, then we simply keep building a copy of B_3. However, if it does output values in $\{0, 1\}$ for all three, then we change our strategy. Without loss of generality, say that $\varphi_e(b) = \varphi_e(c)$. Once we see this, we end the construction of the binary trees above b and c: they are complete up to level s, but contain no nodes at all above level s. (Above a, we continue to build the complete binary tree, although in fact putting a single node at level $n + 1$ above a would suffice.)

The point is that, having committed to the same color for both b and c, φ_e is now trapped into giving a non-distinguishing 2-coloring of $\mathcal{T}_e$. By the Corollary, the only way to give a distinguishing 2-coloring above b is to give the balanced 2-coloring, up to level n; and the same above c. However, then there will be an automorphism of $\mathcal{T}_e$ interchanging b with c and respecting this coloring, so φ_e failed to distinguish $\mathcal{T}_e$ by its coloring.

On the other hand, $\mathcal{T}_e$ is 2-distinguishable, exactly as above: just color b and c different colors, and then use the balanced coloring above each of them, while coloring the complete binary tree above a with any distinguishing 2-coloring of $2^{<\omega}$. (Clearly no automorphism of this $\mathcal{T}_e$ can avoid fixing a, so the choice between balanced and unbalanced above a is irrelevant.)

Finally, we wish to combine these basic modules to build a single computable tree $\mathcal{T}$, with computable finite branching, which has distinguishing number 2 but has no computable distinguishing 2-coloring. This is straightforward. Start with a spine $d_0 < d_1 < d_2 < \cdots$, among which d_0 will be the root of $\mathcal{T}$. Above each d_{2e+1}, in addition to d_{2e+2}, we place a node r_e. Then we build a copy of the tree $\mathcal{T}_e$ with r_e as its root, diagonalizing against the possible coloring φ_e exactly as above in Theorem 21, using the three successors a_e, b_e, and c_e of r_e. No φ_e can be a distinguishing 2-coloring of the entire tree $\mathcal{T}$, because, assuming φ_e is total, there will be some nontrivial automorphism of $\mathcal{T}_e$ which respects the coloring φ_e, and this automorphism extends to an automorphism of all of $\mathcal{T}$ just by fixing the rest of $\mathcal{T}$ pointwise. $\square$

The branching function of a computable finite-branching predecessor tree is always $\mathbf{0}'$-computable. However, it turns out that even with a $\mathbf{0}'$-oracle, we could not necessarily compute a distinguishing 2-coloring of such a tree with distinguishing number 2.

Theorem 24. *There is a computable finite-branching predecessor tree which is distinguished by a 2-coloring but not by any $\mathbf{0}'$-computable 2-coloring.*

Proof. This proof uses a simple modification to the trees $\mathcal{T}_e$ used to build $\mathcal{T}$ in Theorem 23. Now that the branching is allowed to be noncomputable, we may temporarily stop building the tree $\mathcal{T}_e$ at levels $> n$ above b and c (when φ_e has given the same color to b and c), and then resume building the complete binary tree above b and c when/if φ_e "changes its mind" about its coloring of b and c. This makes the branching (above these nodes at level n) noncomputable, since we do not know whether we will ever add nodes at level $n+1$. However, it enables us to satisfy the following requirement.

$$\mathcal{R}_e : \lim_t \varphi_e(x, t) \text{ is not a distinguishing 2-coloring of } \mathcal{T}_e.$$

These requirements together will show that $\mathcal{T}$ has no $\mathbf{0}'$-computable distinguishing 2-coloring, where $\mathcal{T}$ is built from the trees $\mathcal{T}_e$ exactly as before.

The alteration to the construction of $\mathcal{T}_e$ is simple. As before, wait for $\varphi_{e,s}(a, 0)$, $\varphi_{e,s}(b, 0)$, and $\varphi_{e,s}(c, 0)$ to halt with values in $\{0, 1\}$. Pick two of them which have the same value, and stop building the binary trees above those two nodes (while continuing to build a binary tree above the third node). Meanwhile, wait for $\varphi_{e,s}(a, 1)$, $\varphi_{e,s}(b, 1)$, and $\varphi_{e,s}(c, 1)$ to halt with values in $\{0, 1\}$. When and if this happens, these values supersede those from before: for example, if previously we had halted construction above b and c (as in the original description of $\mathcal{T}_e$), but now $\varphi_e(a, 1) = \varphi_e(b, 1) = 0 \neq \varphi_e(c, 1)$, then we build up the trees above b and c until all three have the same height, then continue building the tree above c but stop building the ones above a and b. On the other hand, if $\varphi_{e,s}(a, 1) = \varphi_e(a, 0)$, $\varphi_{e,s}(b, 1) = \varphi_e(b, 0)$, and $\varphi_{e,s}(c, 1) = \varphi_e(c, 0)$, then φ_e has not changed its mind, and we do not resume construction above the two nodes above which it was stopped. We then continue on to consider $\varphi_e(a, 2)$, etc., using the same program relative to the values $\varphi_e(a, 1)$, etc., and so on for all t.

If $\lim_t \varphi_e(x, t)$ exists for all three values $x \in \{a, b, c\}$, then we wind up in the same situation as in the previous proof, showing that this limit cannot be a distinguishing 2-coloring of $\mathcal{T}_e$, so that $\mathcal{R}_e$ holds. On the other hand, if the limit fails to exist (but φ_e is total with range $\subseteq \{0, 1\}$), then we simply built B_3 above r_e, and B_3 is indeed 2-distinguishable (although not by $\lim_t \varphi_e(x, t)$). Finally, if φ_e is not total or assumes values > 1, then $\mathcal{R}_e$ will hold, and the $\mathcal{T}_e$ we build is either a copy of B_3 or a copy of some $B_{3,n}$, both of which are 2-distinguishable. (Technically, even if range(φ_e) contains some values > 1, the limit could still have values 0 and 1 only. However, if this holds, then some other $\varphi_{e'}$ would have the same limit and would have range $\subseteq \{0, 1\}$, so that $\mathcal{R}_{e'}$ would have taken care of showing that $\lim_t \varphi_e$ was not a distinguishing 2-coloring.)   □

So how much information would we need to compute a distinguishing 2-coloring of a computable finite-branching predecessor tree with distinguishing number 2? We answer this below in Theorem 25, but we begin by defining an *extendible* node of a tree to be a node which lies on an infinite path.

Lemma 2. *If all nodes of a computable predecessor tree T are extendible, then T is 2-distinguishable.*

Proof. This is true even if the tree is infinite-branching, so we prove it for this more general case. Suppose we have a computable predecessor tree with every node extendible. Label the nodes at level 1 of the tree $x_1, x_2, \ldots$ in order of their enumeration into the tree. Color x_1 striped. Color x_2 solid, and color every node at level 2 above x_2 striped. Color x_3 solid, color every node at level 2 above x_3 solid, and color every node at level 3 above x_3 striped, and so on for $x_4, x_5, \ldots$. This procedure distinguishes each node at level 1 from every other node at level 1.

Fix an $n > 0$, and consider the immediate (level-2) successors $y_1, y_2, \ldots$ of x_n. These may have already been colored by the previous instructions; in fact, their successors up to level n will already be colored. Color the level-$(n+1)$ successors of y_1 striped, the level-$(n+1)$ successors of y_2 solid and its level-$(n+2)$ successors striped, then the same above y_3 with solid-solid-striped, and so on. When we do this for every n, each node at level 2 is distinguished from all its siblings at level 2.

We continue in this vein to distinguish each node at level k from every other node at level k, for every k.

Here is the algorithm for determining the color of an arbitrary node on such a tree if we've used the above coloring:

Choose a node on the tree. Call it n. Call the root r.

$(\star)$ Use the predecessor function to determine the level of n above r. Call this level l.

Label the immediate successors of r with $x_1, x_2, \ldots$.

If n sits above x_l, then n is red.

Else, if n sits above x_i for some $i > l$, then n is blue.

Else, n sits above x_i for some $i < l$. Let this x_i be the new r, and go back to $(\star)$. $\qquad\square$

Theorem 25. *If a computable finite-branching predecessor tree has distinguishing number 2, then it has a $\mathbf{0}''$-computable distinguishing 2-coloring.*

Proof. Because the tree is finite-branching, with a $\mathbf{0}''$-oracle we can determine, for each immediate successor y of a given node x, whether y is extendible or not: König's Lemma states that a non-extendible node must have only finitely many nodes extending it. Above the extendible ones we use the process illustrated above in Lemma 2. For each extendible x, consider the (finite) subtree containing x, the non-extendible immediate successors of x and all of their successors. There must be a way to distinguish this subtree with a 2-coloring, since the tree has a distinguishing 2-coloring. Each non-extendible node has only finitely many nodes above it. With the $\mathbf{0}''$-oracle we can find them all, and we can try out all the possible colorings until we find one which admits no non-trivial automorphism. Having done all of this, we know that every automorphism of the tree which respects our coloring and which fixes x must also fix each immediate successor y (and all the nodes above each non-extendible y). By induction on levels, this means that an automorphism which respects the coloring must fix each single node, i.e. must be the identity. $\square$

Thus the existence of a distinguishing 2-coloring is equivalent to the existence of such a coloring computable in $\mathbf{0}''$. This significantly reduces the complexity of the property (for computable finite-branching trees $\mathcal{T}$ under predecessor) of being 2-distinguishable. On its face, that property was Σ_2^1: it said that there exists a function (the coloring) such that every automorphism of $\mathcal{T}$ either fixes all nodes or disrespects the coloring. Of course, this complexity can quickly be reduced, since the complexity of an orbit in a finite-branching computable predecessor tree is at most Π_2^0. Nevertheless, 2-distinguishability still could have been Σ_1^1-hard for these trees, up until we established Theorem 25, which showed that we need only quantify over $\mathbf{0}''$-computable functions, not over all functions, to define 2-distinguishability.

We now investigate how much further we can lower the complexity of the property of 2-distinguishability, and whether the same complexity level holds for n-distinguishability. Theorem 26 answers these questions. Subsequently we will consider distinguishability by finite colorings, i.e., colorings with finitely many colors, but with no fixed bound on the number of colors.

Theorem 26. *For each fixed n, the property of having a distinguishing n-coloring is Π_2^0-complete within the class of finite-branching predecessor trees.*

Proof. We will show that having a distinguishing 2-coloring is Π_2^0-complete, and we will explain how the argument extends to an n-coloring for fixed n.

A finite-branching predecessor tree has **no** distinguishing 2-coloring if and only if

$$\exists \sigma_1, \ldots, \sigma_k \left[\begin{array}{l} \sigma_1, \ldots, \sigma_k \text{ are not extendible \& have a common predecessor } \tau \ \& \\ \text{the tree } \{\tau\} \cup \bigcup_{i=1}^k \{\delta : \delta \supseteq \sigma_i\} \text{ has no distinguishing 2-coloring} \end{array} \right].$$

Non-extendibility is a Σ_2^0 property, and the other two conjuncts inside the brackets are each computable. So, **not** having a distinguishing 2-coloring is Σ_2^0. Thus, having a distinguishing 2-coloring is Π_2^0. Notice that if "2-coloring" in the above argument were replaced with "n-coloring," the result would still hold; so, for fixed n, having a distinguishing n-coloring is Π_2^0.

To show completeness, we start by building a copy of the tree B_3 as follows. Begin with the root and the three nodes at level 1. Then, whenever a new element is enumerated into the e-th c.e. set W_e, we add the whole next level of B_3 to our tree. If W_e turns out to be finite, our copy of B_3 will only have finitely many levels, and thus will not be 2-distinguishable. If W_e turns out to be infinite, then our B_3 will likewise be infinite, and thus 2-distinguishable. So the tree has a distinguishing 2-coloring just if W_e is infinite.

If we define B_{n+1} to be the tree whose root has exactly $(n + 1)$ immediate successors and every other node has exactly n immediate successors, then if we replace B_3 with B_{n+1} in the immediately preceding paragraph, we show that having a distinguishing n-coloring for some fixed n is Π_2^0-complete as well.   □

It remains to consider "having a distinguishing n-coloring for some (arbitrary) n." This is the property of being *finitely distinguishable*. Expressing this property takes an extra $\exists$ quantifier, so it is plausible that having a finite distinguishing coloring is Σ_3^0-complete.

Theorem 27. *The property of having a distinguishing finite coloring is Σ_3^0-complete within the class of finite-branching predecessor trees.*

Proof. Define S_∞ to be the following tree: There is an infinite "spine," and the node α_n on the spine at level n has exactly n additional immediate successors, all of which are terminal. We will show, using the tree S_∞, that being finitely distinguishable is Σ_3^0-complete by giving a 1-reduction from the set of indices for finitely distinguishable trees to the set Cof.

We start by building the tree S_∞. Let A_k be the subtree consisting of α_k and all its non-extendible immediate successors. We wait for elements to be enumerated into W_e. At stage s, suppose $m \in W_{e,s} - W_{e,s-1}$. Then we make A_m rigid by adding paths of distinct finite lengths above the immediate successors of α_m.

We claim that the resulting tree is finitely distinguishable if and only if W_e is cofinite. Suppose $\overline{W_e}$ is finite and nonempty. (If $\overline{W_e}$ is empty, then the tree is rigid, i.e., 1-distinguishable.) Then every tree A_k with $k > \max(\overline{W_e})$ is rigid. Thus the whole tree is $\max(\overline{W_e})$-distinguishable. Now suppose $\overline{W_e}$ is infinite. Then, given any n, there is A_k with $k > n$ for which A_k is k-distinguishable but not n-distinguishable. Thus, for each n, the whole tree is not n-distinguishable. So the tree has a finite distinguishing coloring just if W_e is cofinite.   □

References

1. Imrich, W., Klavžar, S., Trofimov, V.: Distinguishing infinite graphs. Electronic Journal of Combinatorics 14, #R36 (2007)
2. Harizanov, V., Miller, R., Morozov, A.: Simple structures with complex symmetry. Algebra and Logic 49, 51–67 (2010)
3. Lempp, S., McCoy, C., Miller, R.G., Solomon, R.: Computable categoricity of trees of finite height. Journal of Symbolic Logic 70, 151–215 (2005)
4. Miller, R.G.: The computable dimension of trees of infinite height. Journal of Symbolic Logic 70, 111–141 (2005)
5. Smith, S.M., Watkins, M.E.: Bounding the distinguishing number of infinite graphs (submitted for publication)
6. Smith, S.M., Tucker, T.W., Watkins, M.E.: Distinguishability of infinite groups and graphs. Electronic Journal of Combinatorics 19, #R27 (2012)
7. Steiner, R.M.: Effective algebraicity. Archive for Mathematical Logic 52, 91–112 (2013)

On the Ramseyan Factorization Theorem

Shota Murakami[1], Takeshi Yamazaki[1], and Keita Yokoyama[2,⋆]

[1] Mathematical Institute, Tohoku University, Japan
{sb0m33,yamazaki}@math.tohoku.ac.jp
[2] Japan Advanced Institute of Science and Technology, Japan
y-keita@jaist.ac.jp

Abstract. We study, in the context of reverse mathematics, the strength of Ramseyan factorization theorem (RF_k^s), a Ramsey-type theorem used in automata theory. We prove that RF_k^s is equivalent to RT_2^2 for all $s, k \geq 2$, $k \in \omega$ over RCA_0. We also consider a weak version of Ramseyan factorization theorem and prove that it is in between ADS and CAC.

1 Introduction

In the current study of reverse mathematics, deciding the strength of Ramsey's theorem for pairs (RT_2^2) is one of the most important topics (see e.g., Cholak/Jockusch/Slaman[1] and Hirschfeldt[5], and for the study of reverse mathematics, Simpson[9] is the standard reference). In this paper, we study, in the context of reverse mathematics, the strength of a Ramsey-type theorem which is called Ramseyan factorization theorem. Ramseyan factorization theorem is used in the theory of automata (see, for example, [8]). We show that some kinds of Ramseyan factorization theorem are equivalent to RT_2^2. We also study a weak version of Ramseyan factorization theorem. We discuss it in section 3, and show that a weak version is in between ADS and CAC. Note that ADS and CAC are just separated by Lerman/Solomon/Towsner[7]. Thus, it must be strictly stronger than ADS or strictly weaker than CAC. We also consider other variations of Ramseyan factorization theorem in section 5.

Notations and Definitions

Let A be a set. Then $A^{<\mathbb{N}}$ (resp. $A^{\mathbb{N}}$) denotes the set of all finite (resp. infinite) sequences of elements from A. If $u, v \in A^{<\mathbb{N}}$, u_i denotes the i-th element of u, $u^\frown v$ (and uv for short) denotes the concatenation of u and v, and $|u|$ denotes the length of u. The Ramseyan factorization theorem is the following statement.

Definition 1 (Ramseyan factorization theorem). *For any $A \subseteq \mathbb{N}$ and finite $B \subseteq \mathbb{N}$, the following statement (RF_B^A) holds:*

⋆ The third author is partially supported by JSPS Grant-in-Aid for Research Activity Start-up grant number 25887026.

A. Beckmann, E. Csuhaj-Varjú, and K. Meer (Eds.): CiE 2014, LNCS 8493, pp. 324–332, 2014.
© Springer International Publishing Switzerland 2014

For any $u \in A^{\mathbb{N}}$ and $f : A^{<\mathbb{N}} \to B$, there exists $v \in (A^{<\mathbb{N}})^{\mathbb{N}}$ such that $u = v_0 \frown v_1 \frown \cdots$ and for any $j \geq i > 0$ and $j' \geq i' > 0$, $f(v_i \frown v_{i+1} \frown \cdots \frown v_j) = f(v_{i'} \frown v_{i'+1} \frown \cdots \frown v_{j'})$.

If u, f and v satisfy the above condition, we call v a Ramseyan factorization for u and f. In this paper, we aim to study $\mathrm{RF}_k^{\mathbb{N}}$ and RF_k^s for $s, k \in \mathbb{N}$.

2 Ramseyan Factorization Theorem and Ramsey's Theorem for Pairs

In this section, we see the relation between Ramsey's theorem (RT_k^n) and Ramseyan factorization theorem (RF_k^s).

Proposition 2 (RCA$_0$). *For any $k \in \mathbb{N}$, $\mathrm{RF}_k^{\mathbb{N}} \Rightarrow \cdots \Rightarrow \mathrm{RF}_k^2 \Rightarrow \mathrm{RF}_k^1$.*

Proof. Trivial from the definition.

Theorem 3 (RCA$_0$). *For any $k \in \mathbb{N}$, RT_k^2 implies $\mathrm{RF}_k^{\mathbb{N}}$.*

Proof. Let $u \in \mathbb{N}^{\mathbb{N}}$ and $f : \mathbb{N}^{<\mathbb{N}} \to k$. Define $P : [\mathbb{N}]^2 \to k$ as follows:

$$P(i, j) = f(u_i u_{i+1} \ldots u_{j-1}).$$

Let X be an infinite homogeneous set for P. Define $l \in \mathbb{N}^{\mathbb{N}}$ by setting l_i to be the i-th smallest element in X and define $v \in (\mathbb{N}^{<\mathbb{N}})^{\mathbb{N}}$ by setting $v_0 = u_0 \ldots u_{l_0-1}$ and $v_i = u_{l_{i-1}} \ldots u_{l_i-1}$ for all $i \geq 1$. Then clearly v is a Ramseyan factorization for u and f.

Theorem 4 (RCA$_0$). *For any $k \in \mathbb{N}$, RF_k^2 implies RT_k^2.*

Proof. Let $P : [\mathbb{N}]^2 \to k$. We will find an infinite homogeneous set for P. Define $u \in 2^{\mathbb{N}}$ and $f : 2^{<\mathbb{N}} \to k$ as follows:

$$u = 1010010001 \ldots 10^{n-1}10^n10^{n+1}1 \ldots$$

$$f(\sigma) = \begin{cases} P(m, n+2) & \text{if } \sigma = 0^k10^m1\tau10^n10^l \text{ for some } k, l, m, n \geq 0 \text{ and } \tau \in 2^{<\mathbb{N}}, \\ 0 & \text{otherwise.} \end{cases}$$

Let v be a Ramseyan factorization for u and f. By combining v_i's if necessary, we may assume that each v_i contains at least four 1's, i.e., v_i is of the form $0^k10^m1\tau10^n10^l$. Let $H = \{m \in \mathbb{N} \mid 1 \leq \exists i \leq m \ v_i = 0^k10^m1\tau10^n10^l\}$. We can easily check that this H is an infinite homogeneous set for P.

From the above proposition and theorems, we can show that RF_k^s is equivalent to RT_2^2 for all $s, k \geq 2, k \in \omega$.

Corollary 5. *The following are equivalent over* RCA$_0$.

1. RT$^2_2$.
2. RF$^{\mathbb{N}}_k$ *(k $\geq$ 2, k $\in \omega$).*
3. RF^2_k *(k $\geq$ 2, k $\in \omega$).*

Proof. This is clear from the previous theorems and the fact that RCA$_0$ proves RT$^2_k \Rightarrow$ RT$^2_{k+1}$ for all $k \geq 2$.

Corollary 6. *The following are equivalent over* RCA$_0$.

1. RT$^2_{<\infty}$.
2. $\forall k$RF$^{\mathbb{N}}_k$.
3. $\forall k$RF^2_k.

Next, we consider the remaining case, i.e. the strength of RF^1_k. In order to study RF^1_k, we consider the following version of Ramsey's theorem.

Definition 7. *For a given function $f : [\mathbb{N}]^n \to \mathbb{N}$, RT$^f_k$ is the following statement:*

For any $P : \mathbb{N} \to k$, there exists an infinite set $H \subseteq \mathbb{N}$ such that for any $u, v \in [H]^n$, $P(f(u)) = P(f(v))$.

If f is a bijection, we can prove the following.

Proposition 8 (RCA$_0$). *For any $n \in \mathbb{N}$ and any bijection $f : [\mathbb{N}]^n \to \mathbb{N}$, RT$^f_k$ is equivalent to* RT^n_k.

The full version of RF^f_k, i.e. $\forall f : [\mathbb{N}]^n \to \mathbb{N}$ RF^f_k, is still equivalent to RT^n_k.

Proposition 9 (RCA$_0$). *RT^n_k is equivalent to $\forall f : [\mathbb{N}]^n \to \mathbb{N}$ RT^f_k.*

Proof. From left to right is trivial, because $P \circ f$ is a function from $[\mathbb{N}]^n$ to k when $P : \mathbb{N} \to k$. From right to left is proved from the above proposition.

If f is not a bijection, RT^f_k may not be equivalent to RT^n_k. In case f is the subtraction Subt$(a, b) = b - a$, RT^f_k is equivalent to RF^1_k. (The function Subt is considered as a function of $[\mathbb{N}]^2$.)

Proposition 10 (RCA$_0$). *For any $k \in \mathbb{N}$, RF^1_k is equivalent to* RT$^{\mathrm{Subt}}_k$.

Proof. We first prove RF$^1_k \Rightarrow$ RT$^{\mathrm{Subt}}_k$. Assume RF^1_k and let $P : \mathbb{N} \to k$. Define $f : 1^{<\mathbb{N}} \to k$ by $f(0^n) = P(n)$ and let v be a Ramseyan factorization for $0^{\mathbb{N}}$ and f. Let $X = \{\sum_{j \leq i} |v_j| \mid i \in \mathbb{N}\}$. Then X is an infinite homogeneous set for $P \circ$ Subt.

Next, we prove RT$^{\mathrm{Subt}}_k \Rightarrow$ RF^1_k. Assume RT$^{\mathrm{Subt}}_k$ and let $f : 1^{<\mathbb{N}} \to k$. Define $P : \mathbb{N} \to k$ by $P(n) = f(0^n)$. Then there exists an infinite homogeneous set $H := \{l_0 < l_1 < \cdots\} \subseteq \mathbb{N}$ for P. Define $v \in (1^{<\mathbb{N}})^{\mathbb{N}}$ by $v_0 = 0^{l_0}$ and $v_i = 0^{l_i - l_{i-1}}$ for all $i \geq 1$. Then v is a Ramseyan factorization for $0^{\mathbb{N}}$ and f.

From the above, we can show that $\forall k \mathrm{RF}^1_k$ is strong enough to prove the bounding principle for Σ^0_2 formulas.

Corollary 11 (RCA$_0$). $\forall k \mathrm{RF}^1_k$ *implies* $\mathrm{B}\Sigma^0_2$.

Proof. Because of the above and the equivalence of $\mathrm{B}\Sigma^0_2$ and $\mathrm{RT}^1_{<\infty}$, it's enough to prove $\mathrm{RT}^{\mathrm{Subt}}_k \Rightarrow \mathrm{RT}^1_k$ for all $k \in \mathbb{N}$. Assume $\mathrm{RT}^{\mathrm{Subt}}_k$ and let $P : \mathbb{N} \to k$. Then there exists an infinite set $H \subseteq \mathbb{N}$ such that for any $u, v \in [H]^2$, $P(u_1 - u_0) = P(v_1 - v_0)$. Then $X = \{h - \min H \mid h \in H \setminus \{\min H\}\}$ is an infinite homogeneous set for P.

Question 12. *Is* RF^1_k *equivalent to* RT^2_2 *or* RT^1_k?

3   Weak Factorization

In this section, we consider a weaker version of Ramseyan factorization theorem. For applications in automata theory, the following weaker version of Ramseyan factorization theorem is usually good enough.

Definition 13. *For given sets* $A, B \subseteq \mathbb{N}$, *weak Ramseyan factorization theorem for* A *and* B *(WRF^A_B) is the following statement:*

> *For any* $u \in A^{\mathbb{N}}$ *and* $f : A^{<\mathbb{N}} \to B$, *there exists* $v \in (\mathbb{N}^{<\mathbb{N}})^{\mathbb{N}}$ *such that* $u = v_0 \frown v_1 \frown \dots$ *and for any* $i, j > 0$, $f(v_i) = f(v_j)$.

Here, such v *is said to be a weak Ramseyan factorization for* u *and* f.

Similarly, we consider a weaker version of Ramsey's theorem as follows.

Definition 14. *Pseudo Ramsey's Theorem* $psRT^n_k$ *is the following statement:*

> *For any coloring* $P : [\mathbb{N}]^n \to k$, *there exists an infinite set* $H = \{a_0 < a_1 < \dots\}$ *such that for any* $i, j \in \mathbb{N}$, $P(a_i, \dots, a_{i+n-1}) = P(a_j, \dots, a_{j+n-1})$.

Such H *is called pseudo homogeneous set for* P. [1]

Remark 15. *In general, a subset of a weak homogeneous set might not be weak homogeneous again.*

Question 16. *Does* WRT^n_k *imply* WRT^n_{k+1} *over* RCA$_0$?

Proposition 17 (RCA$_0$). *For any* $m \in \mathbb{N}$, $\mathrm{WRF}^{\mathbb{N}}_m \Leftrightarrow \mathrm{WRT}^2_m$. *In particular,* $\mathrm{WRF}^{\mathbb{N}}_2$ *is equivalent to* WRT^2_2.

Proof. We first show for a given $m \in \mathbb{N}$ that $\mathrm{WRF}^{\mathbb{N}}_m \Rightarrow \mathrm{WRT}^2_m$. Fix $u = \langle i \mid i \in \mathbb{N} \rangle \in \mathbb{N}^{\mathbb{N}}$. For a given coloring $P : [\mathbb{N}]^2 \to m$, define $f : \mathbb{N}^{<\mathbb{N}} \to m$ by $f(\sigma) = P(a, a + k)$ if $\sigma = \langle a + i \mid i < k \rangle$ for some $a, k \in \mathbb{N}$, $k \geq 1$, and $f(\sigma) = 0$ otherwise. Now, let v be a weak Ramseyan factorization for u and f. Then, one

---

[1] In Friedman/Pelupessy[4], this set is called adjacent homogeneous.

can easily check that the set $H = \{\sum_{j \le i} |v_j| \mid i \in \mathbb{N}\}$ is a weak homogeneous set for P.

Next, we show $m \in \mathbb{N}$, $\mathrm{WRT}_m^2 \Rightarrow \mathrm{WRF}_m^\mathbb{N}$. Let $u \in \mathbb{N}^\mathbb{N}$, and let $f : \mathbb{N}^{<\mathbb{N}} \to m$. Then, define a coloring $P : [\mathbb{N}]^2 \to m$ by $P(a,b) = f(\langle u_i \mid a \le i < b\rangle)$. Let $H = \{a_0 < a_1 < \ldots\}$ be an infinite weak homogeneous set for P. Define $v_0 = \langle u_i \mid 0 \le j < a_0\rangle$ and $v_{i+1} = \langle u_j \mid a_i \le j < a_{i+1}\rangle$. Then, v is a weak Ramseyan factorization for u and f.

Question 18. *Is* WRF_2^2 *equivalent to* WRT_2^2 *over* RCA_0?

Proposition 19 (RCA_0). *For any* $k \in \mathbb{N}$, WRF_{k+5}^2 *implies* WRT_k^2.

Proof. Let $w_i = 10^i \in 2^{<\mathbb{N}}$, and let $u = w_0 ^\frown w_1 ^\frown \ldots$. For a given coloring $P : [\mathbb{N}]^2 \to k$, we define a function $f : 2^{<\mathbb{N}} \to k + 5$ as follows:

$$f(\sigma) = \begin{cases} P(m, n+2) & \text{if } \sigma = 0^i {}^\frown w_m ^\frown \ldots ^\frown w_n ^\frown 10^j \text{ for some } i, j \ge 0 \text{ and } 1 \le m \le n, \\ k & \text{if } \sigma = 0^i 10^j \text{ for some } i, j \ge 0 \text{ such that } i \text{ and } j \text{ are both even}, \\ k+1 & \text{if } \sigma = 0^i 10^j \text{ for some } i, j \ge 0 \text{ such that } i \text{ is odd and } j \text{ is even}, \\ k+2 & \text{if } \sigma = 0^i 10^j \text{ for some } i, j \ge 0 \text{ such that } i \text{ is even and } j \text{ is odd}, \\ k+3 & \text{if } \sigma = 0^i 10^j \text{ for some } i, j \ge 0 \text{ such that } i \text{ and } j \text{ are both odd}, \\ k+4 & \text{otherwise}. \end{cases}$$

Take a weak Ramseyan factorization v for u and f, and let $f(v_i) = d$ for all $i \ge 1$. If v_i contains at least one '1', then $f(v_i) \ne k+4$. Thus, $d \ne k+4$. If $k \le d < k+4$, then each v_i contains only one '1'. However, one can easily check that this is impossible. Therefore, for any $i \ge 1$, $f(v_i) = d$ for some $d < k$. This means that $H = \{m \in \mathbb{N} \mid v_l = 0^i {}^\frown w_m ^\frown \cdots ^\frown w_n ^\frown 10^j \text{ for some } i, j \ge 0, 1 \le m \le n, \text{ and } l \ge 1\}$ is a weak homogeneous set for P.

Question 20. *Is it possible to reduce the number of colorings in the above proof?*

One of reviewers told us that if we change the color "$k + 4$" to "0", the above proof still works without changing the weak Ramseyan factorization v. Therefore, thank to him or her, we can prove the following.

Proposition 21 (RCA_0). *For any* $k \in \mathbb{N}$, WRF_{k+4}^2 *implies* WRT_k^2.

4   The Strength of $\mathrm{WRF}_k^\mathbb{N}$, or Equivalently WRT_k^2

Our main goal in this section is to prove that $\mathrm{WRF}_2^\mathbb{N}$, or equivalently WRT_2^2, is in between CAC and ADS. In order to show it, we use the facts that ADS is equivalent to trRT_2^2, transitive Ramsey's theorem for pairs, and CAC is equivalent to strRT_2^2, semi-transitive Ramsey's theorem for pairs, which were both proved in Hirschfeldt/Shore[6].

Definition 22 (Transitive and semi-transitive colorings [6])

1. *A k-coloring $P : [\mathbb{N}]^2 \to k$ is said to be transitive if $P(a,b) = P(b,c) = i \Rightarrow$ $P(a,c) = i$.*
2. *A k-coloring $P : [\mathbb{N}]^2 \to k$ is said to be semi-transitive if $P(a,b) = P(b,c) = i > 0 \Rightarrow P(a,c) = i$.*

Now, we consider the following variations of Ramsey's theorem for pairs.

Definition 23. 1. *Transitive Ramsey's theorem trRT_k^2: Any transitive k-coloring $P : [\mathbb{N}]^2 \to k$ has an infinite homogeneous set.*
2. *Semi-transitive Ramsey's theorem strRT_k^2: Any semi-transitive k-coloring $P : [\mathbb{N}]^2 \to k$ has an infinite homogeneous set.*
3. *Semi-weak Ramsey's theorem sWRT_k^2: Any k-coloring $P : [\mathbb{N}]^2 \to k$ has an infinite homogeneous set H such that $P([H]^2) = \{0\}$ or an infinite weak homogeneous set $H' = \{h_0 < h_1 < \ldots\}$ such that $P(h_i, h_{i+1}) > 0$.*

Clearly, sWRT_k^2 is a stronger version of WRT_k^2. First, we show the lower bound for WRT_2^2.

Theorem 24 (RCA$_0$). *For any $m \in \mathbb{N}$, WRT_m^2 implies trRT_m^2.*

Proof. If P is a transitive coloring, a weak homogeneous set for P is actually a homogeneous set for P.

Next, we consider the upper bound for WRT_2^2

Lemma 25 (RCA$_0$). *For any $m \in \mathbb{N}$, sWRT_m^2 implies strRT_m^2.*

Proof. If P is a semi-transitive coloring, a weak homogeneous set H for P with $P([H]^2) \neq \{0\}$ is actually a homogeneous set for P.

The converse is true for the case $m = 2$.

Lemma 26 (RCA$_0$). strRT_2^2 *implies* sWRT_2^2.

Proof. Let $P : [\mathbb{N}]^2 \to 2$. We want to find a homogeneous set for 0, or a weak homogeneous set for 1. Define $\bar{P} : [\mathbb{N}]^2 \to 2$ as follows: $\bar{P}(a,b) = 1$ if there exists a sequence $a = a_0 < \cdots < a_l = b$ such that $P(a_i, a_{i+1}) = 1$ for any $i < l$, and $\bar{P}(a,b) = 0$ otherwise. Then, $\bar{P}$ is a semi-transitive coloring. Thus, by strRT_2^2, take an infinite homogeneous set H for $\bar{P}$. If $\bar{P}([H]^2) = \{0\}$, then we have $P([H]^2) = \{0\}$ and we have done. If $\bar{P}([H]^2) = \{1\}$, then for any $a, b \in H$, we can (effectively) find a sequence $a = a_0 < \cdots < a_l = b$ such that $P(a_i, a_{i+1}) = 1$ for every $i < l$. Thus, we can construct a set $H' \supseteq H$ which is a weak homogeneous set for P with the value 1.

Question 27. *Over RCA$_0$, does $\text{strRT}_{<\infty}^2$ imply $\text{sWRT}_{<\infty}^2$ or $\text{WRT}_{<\infty}^2$?*

Although WRT_k^2 might not prove WRT_{k+1}^2, we can show the following.

Lemma 28 (RCA$_0$). *For any $m \geq 2$, sWRT_m^2 implies sWRT_{m+1}^2.*

Proof. Let $P : [\mathbb{N}]^2 \to m + 1$. Define $\bar{P} : [\mathbb{N}]^2 \to m$ by $\bar{P}(a, b) = 0$ if $P(a, b) \in \{0, 1\}$ and $\bar{P}(a, b) = P(a, b) - 1$ if $P(a, b) \geq 2$. If $\bar{P}$ has a weak homogeneous set with the value $d \geq 1$, then it is a weak homogeneous set for P. Otherwise, $\bar{P}$ has a homogeneous set H with the value 0. Then, $P \upharpoonright [H]^2$ is a 2-coloring, thus we can apply sWRT$^2_2$ again, and we have done. [2]

Combining the above, we have the following.

Theorem 29. *The following are equivalent over* RCA$_0$.

1. sWRT$^2_2$.
2. strRT$^2_2$.
3. sWRT^2_k *for any* $k \in \omega$, $k \geq 2$.
4. strRT^2_k *for any* $k \in \omega$, $k \geq 2$.

Thus, within RCA$_0$, WRT$^2_2$ *is provable from any one of the above.*

Corollary 30 (RCA$_0$). WRT$^2_2$ *is stronger than* ADS *and weaker than* CAC.

Proof. By Hirschfeldt/Shore[6], ADS is equivalent to trRT$^2_2$ and CAC is equivalent to strRT$^2_2$.

Question 31. *Is* WRT$^2_2$ *equivalent to* ADS *or* CAC *over* RCA$_0$?

Corollary 32 (RCA$_0$). SRT$^2_2$ *does not imply* WRT$^2_2$.

Proof. By Chong/Slaman/Yang[2], SRT$^2_2$ does not imply COH. On the other hand, by Hirschfeldt/Shore[6], ADS implies COH, and thus WRT$^2_2$ implies COH.

Corollary 33 (RCA$_0$). WRT$^2_2$ *does not imply* DNR.

Proof. By Hirschfeldt/Shore[6], CAC does not imply DNR, thus WRT$^2_2$ does not, either.

Question 34. *Does* P$^2_2$ *or* RWKL$^{0'}$ *imply* WRT$^2_2$? *(See, e.g., Flood[3] for the definitions of these statements. Note that* RWKL$^{0'}$ *is introduced as* RKL$^{(1)}$ *in [3].)*

5   Other Topics

In this section, we focus on some other versions of Ramseyan factorization theorem.

---

[2] Note that this argument still works for any n-tuples.

5.1   Stable Versions

We can consider stable versions of RF or WRF. For given $u \in \mathbb{N}^{<\mathbb{N}}$ and $f : \mathbb{N}^{<\mathbb{N}} \to k$, f is said to be *stable* on u if for any $m \in \mathbb{N}$, there exists $n > m$ such that for any $l > n$, $f(\langle u_i \mid m \le i < n \rangle) = f(\langle u_i \mid m \le i < l \rangle)$. Then, SRF_k^A and SWRF_k^A are the following statements:

Definition 35. 1. SRF_k^A: *For any $u \in A^{\mathbb{N}}$ and $f : A^{<\mathbb{N}} \to k$ such that f is stable on u, there exists a Ramseyan factorization for u and f.*
2. SWRF_k^A: *For any $u \in A^{<\mathbb{N}}$ and $f : A^{<\mathbb{N}} \to k$ such that f is stable on u, there exists a weak Ramseyan factorization for u and f.*

As in Theorems 3 and 4, we can show the following.

Theorem 36. *Within RCA_0, the following are equivalent for any $m \in \mathbb{N}$.*

1. SRT_m^2.
2. $\mathrm{SRF}_m^{\mathbb{N}}$.
3. SRF_m^2.

Theorem 37. *Within RCA_0, the following are equivalent for any $m \in \mathbb{N}$.*

1. SWRT_m^2: *Any stable coloring $P : [\mathbb{N}]^2 \to m$ has an infinite weak homogeneous set.*
2. $\mathrm{SWRF}_m^{\mathbb{N}}$.

5.2   Tree Versions

In this subsection, we consider a slightly stronger version of RF_m^2. For given two trees $T, S \subseteq 2^{<\mathbb{N}}$, a tree embedding is an injective function $\pi : S \to T$ such that for any $\sigma, \tau \in S$, $\pi(\sigma) \cap \pi(\tau) = \pi(\sigma \cap \tau)$. For a given tree embedding $\pi : S \to T$, and for any $\sigma, \tau \in S$ such that $\sigma \subsetneq \tau$, the edge between $\pi(\sigma)$ and $\pi(\tau)$, denoted by $E_\pi(\sigma, \tau)$, is the sequence $\rho \in 2^{<\mathbb{N}}$ such that $\pi(\sigma) ^\frown \rho = \pi(\tau)$. Then, we consider the following tree version of Ramseyan factorization theorem.

Definition 38. *Ramseyan factorization theorem for trees TRF_k^2 is the following statement:*

> *For any infinite tree $T \subseteq 2^{<\mathbb{N}}$ and a coloring $f : 2^{<\mathbb{N}} \to k$, there exists an infinite tree $S \subseteq 2^{<\mathbb{N}}$ and a tree embedding $\pi : S \to T$ such that for any $\sigma \subsetneq \tau \in S$ and $\sigma' \subsetneq \tau' \in S$, $f(E_\pi(\sigma, \tau)) = f(E_\pi(\sigma', \tau'))$.*

Proposition 39 (RCA_0). TRF_k^2 *implies* RF_k^2 *for all $k \in \mathbb{N}$. In particular, TRF_2^2 implies RF_2^2 (and, equivalently, RT_2^2).*

Proof. Assume TRF_k^2 and let $u \in 2^{\mathbb{N}}$ and $f : 2^{<\mathbb{N}} \to k$. Define a tree $T \subseteq 2^{<\mathbb{N}}$ by $T = \{u_0 u_1 \ldots u_{i-1} \mid i \in \mathbb{N}\}$. By TRF_k^2, there exist $S = \{s_0 < s_1 < \cdots\} \subseteq 2^{<\mathbb{N}}$ and an embedding $\pi : S \to T$ such that for all $\sigma \subsetneq \tau \in S$ and $\sigma' \subsetneq \tau' \in S$, $f(E_\pi(\sigma, \tau)) = f(E_\pi(\sigma', \tau'))$. Define $v \in (2^{<\mathbb{N}})^{\mathbb{N}}$ by setting $v_0 = \pi(s_0)$ and $v_i = E_\pi(s_{i-1}, s_i)$ for all $i \ge 1$. Then, v is a Ramseyan factorization for u and f.

We can also show that TRF_2^2 is weaker than $\mathsf{WKL}_0 + \mathrm{RT}_2^2$.

Proposition 40. $\mathsf{WKL}_0 + \mathrm{RT}_2^2$ *implies* TRF_2^2.

Proof. Let $T \subseteq 2^{<\mathbb{N}}$ be an infinite tree and $f : 2^{<\mathbb{N}} \to 2$. By WKL_0, there is an infinite path $u \in 2^{\mathbb{N}}$ through T. By RF_2^2, which is equivalent to RT_2^2, there is a Ramseyan factorization $v \in (2^{<\mathbb{N}})^{\mathbb{N}}$ for u and f. Define $S \subseteq 2^{<\mathbb{N}}$ and $\pi : S \to T$ by $S = \{0^i \mid i \in \mathbb{N}\}$ and $\pi(0^i) = v_0 \frown v_1 \frown \cdots \frown v_i$ for all $i \in \mathbb{N}$. Then S and π satisfy the condition.

Therefore, TRF_2^2 is in between $\mathsf{WKL}_0 + \mathrm{RT}_2^2$ and RT_2^2.

Question 41. *Does* TRF_2^2 *imply* WKL_0 *over* RCA_0?

Remark 42. TRF_2^2 *may be equivalent to the following stronger version of* RT_2^2:

RT_2^{2+} : *If* $\mathcal{P}$ *be a class of colorings* $P : [F_P]^2 \to 2$ *where* $F_P = \{0, 1, \ldots, l\}$ *for some* $l \in \mathbb{N}$, *then there exists an infinite set* $H \subseteq \mathbb{N}$ *such that there exist infinitely many* $P \in \mathcal{P}$ *such that* P *is constant on* $[H \cap F_P]^2$.

We think that the equivalence should hold, but we do not know either $\mathrm{TRF}_2^2 \Rightarrow \mathrm{RT}_2^{2+}$ *or* $\mathrm{RT}_2^{2+} \Rightarrow \mathrm{TRF}_2^2$. *This kind of strengthened Ramsey's theorem is studied in [10].*

Acknowledgment. We would like to thank Florian Pelupessy and anonymous reviewers for their helpful comments and suggestions on this paper.

References

1. Cholak, P.A., Jockusch, C.G., Slaman, T.A.: On the strength of Ramsey's theorem for pairs. Journal of Symbolic Logic 66(1), 1–55 (2001)
2. Chong, C.-T., Slaman, T.A., Yang, Y.: The metamathematics of stable Ramsey's theorem for pairs (to appear)
3. Flood, S.: Reverse mathematics and a Ramsey-type König's lemma. Journal of Symbolic Logic 77(4), 1272–1280 (2012)
4. H. Friedman, F. Pelupessy.: Independence of Ramsey theorem variants using ϵ_0. draft.
5. Hirschfeldt, D.R.: Slicing the truth: On the computability theoretic and reverse mathematical analysis of combinatorial principles (to appear)
6. Hirschfeldt, D.R., Shore, R.A.: Combinatorial principles weaker than Ramsey's theorem for pairs. Journal of Symbolic Logic 72(1), 171–206 (2007)
7. Lerman, M., Solomon, R., Towsner, H.: Separating principles below ramsey's theorem for pairs. Journal of Mathematical Logic 13(2), 1350007 (2013)
8. Perrin, D., Pin, J.-É.: Infinite Words: Automata, Semigroups, Logic and Games, vol. 141. Academic Press (2004)
9. Simpson, S.G.: Subsystems of Second Order Arithmetic. Perspectives in Mathematical Logic, 2nd edn., pp. XIV + 445 pages. Springer (1999); Perspectives in Logic, Association for Symbolic Logic. Cambridge University Press, pp. XVI+ 444 pages (2009)
10. Yokoyama, K.: Finite iterations of infinite and finite Ramsey's theorem (in preparation)

The Complexity of Satisfaction Problems
in Reverse Mathematics

Ludovic Patey

Laboratoire PPS, Université Paris Diderot, Paris, France
ludovic.patey@computability.fr

Abstract. Satisfiability problems play a central role in computer science and engineering as a general framework for studying the complexity of various problems. Schaefer proved in 1978 that truth satisfaction of propositional formulas given a language of relations is either NP-complete or tractable. We classify the corresponding satisfying assignment construction problems in the framework of Reverse Mathematics and show that the principles are either provable over RCA_0 or equivalent to WKL_0. We formulate also a Ramseyan version of the problems and state a different dichotomy theorem. However, the different classes arising from this classification are not known to be distinct.

1 Introduction

A common way to solve a constrained problem in industry consists of reducing it to a satisfaction problem over propositional logic and using a SAT solver. The generality of the framework and its multiple applications make it a natural subject of interest for the scientific community and constraint satisfaction problems remains an active field of research.

In 1978, Schaefer [9] gave a great insight in the understanding of the complexity of satisfiability problems by studying a parameterized class of problems and showing they admit a dichotomy between NP-completeness and tractability. Many other dichotomy theorems have been proven since, about refinements to AC^0 reductions [1], variants about counting, optimization, 3-valued domains and many others [4,7,3]. The existence of dichotomies for n-valued domains with $n > 3$ remains open.

Reverse Mathematics is a vast program of classification of the strength of mathematical theorems by emphasizing on their computational content. This study has led to the main observation that many theorems are computationally equivalent to one of four axioms. On particular axiom is Weak König's lemma (WKL_0) which allows formalization of many compactness arguments and the solution of many satisfiability problems. We believe that studying constraint satisfaction problems within this framework can lead to insights in both fields: in Reverse Mathematics, we can exploit the generality of constraint satisfaction problems to compare existing principles by reducing them to satisfaction problems. In CSP, Reverse Mathematics can yield a better understanding of the computational strength of satisfiability problems for particular classes of formulas. In particular we answer to the question of Marek & Remmel [8] whether there exists dichotomy theorems for infinite recursive versions of constraint satisfaction problems.

A. Beckmann, E. Csuhaj-Varjú, and K. Meer (Eds.): CiE 2014, LNCS 8493, pp. 333–342, 2014.

Definition 1. *As set of Boolean formulas C is* satisfiable *if every conjunction of a finite set of formulas in C is satisfiable.* SAT *is the statement "for every satisfiable set C of Boolean formulas over an infinite set of variables V there is an infinite assignment $v : V \to \{T, F\}$ satisfying C." The pair (V, C) forms an* instance *of* SAT.

The weak system on which relations are based is called RCA_0, standing for Recursive Comprehension Axiom. It consists of basic Peano axioms together with a comprehension scheme restricted to Δ_1^0 formulas and an the induction restricted to Σ_1^0 formulas.

Theorem 2 (Simpson [10]). $RCA_0 \vdash WKL_0 \leftrightarrow SAT$

RWKL, a weakening of WKL_0, has been recently introduced by Flood in [5]. Given an infinite binary tree, the principle does not assert the existence of a path, but rather of an infinite subset of a path in the tree. Initially called RKL, it has been renamed to RWKL in [2] to give a consistent R prefix to Ramseyan principles. This principle has been shown to be strictly weaker than SRT_2^2 and WKL_0 by Flood, and stricly stronger than DNR by Bienvenu & al. in [2]. By analogy with RWKL, we formulate Ramsey-type versions of satisfiability problems.

Definition 3. *Let C be a set of Boolean formulas over an infinite set of variables V. A set H is* homogeneous *for C if there is a $c \in \{T, F\}$ such that every conjunction of a finite set of formulas in C is satisfiable by a truth assignment v such that $(\forall a \in H)(v(a) = c)$.*

Definition 4. LRSAT *is the statement " Let C be a satisfiable set of Boolean formulas over an infinite set of variables V For every infinite set $L \subseteq V$ there exists an infinite set $H \subseteq L$ homogeneous for C." The corresponding* instance *of* LRSAT *is the tuple (V, C, L).* RSAT *is obtained by restricting* LRSAT *to $L = V$. Then an* instance *of* RSAT *is an ordered pair (V, C).*

The equivalence between WKL_0 and SAT over RCA_0 extends to their Ramseyan version. The proof is relatively easy and directly adaptable from proof of Theorem 2.

Theorem 5 (Bienvenu & al. [2]). $RCA_0 \vdash RWKL \leftrightarrow RSAT \leftrightarrow LRSAT$

1.1   Definitions and Notations

Some classes of Boolean formulas – bijunctive, affine, horn, ... – have been extensively studied in Complexity Theory, leading to the well-known dichotomy theorem due to Schaefer. We give a precise definition of those classes in order to state our dichotomy theorems.

Definition 6. *A* literal *is either a Boolean variable (positive literal), or its negation (negative literal). A* clause *is a disjunction of literals. A clause is* horn *if it has at most one positive literal,* co-horn *if it has at most one negative literal and* bijunctive *if it has at most 2 literals. If we number Boolean variables, we can associate to each Boolean formula φ with Boolean variables $x_1, \ldots, x_n$ a relation $[\varphi] \subseteq \{F, T\}^n$ such that $a \in [\varphi]$ iff $\varphi(a)$. If S is a set of relations, an* S-formula *over a set of variables V is a formula of the form $R(y_1, \ldots, y_n)$ for some $R \in S$ and $y_1, \ldots, y_n \in V$.*

Example 7. Let $S = \{\rightarrow\}$. $(x \rightarrow y)$ is an S-formula but $(x \rightarrow \neg y)$ is not. Neither is $(x \rightarrow y) \wedge (y \rightarrow z)$. The formula $(x \rightarrow y)$ is equivalent to the horn clause $(\neg x \vee y)$ where the literals are $\neg x$ and y.

Definition 8. *A formula φ is i-valid for $i = 0, 1$ if $\varphi(i, \ldots, i)$ is true. It is* horn *(resp.* co-horn, bijunctive*) if it is a conjunction of horn (resp. co-horn, bijunctive) clauses. A formula is* affine *if it is a conjunction of formulas of the form $x_1 \oplus \ldots \oplus x_n = i$ for $i \in \{0, 1\}$ where $\oplus$ is the exclusive or.*

A relation $R \subseteq \{0, 1\}^n$ is *bijunctive* (resp. *horn, co-horn, affine, i-valid*) if there is bijunctive (resp. horn, co-horn, affine, i-valid) formula φ such that $R = [\varphi]$. A relation R is *i-default* for $i = 0, 1$ if for every finite set $I \subseteq \mathbb{N}$, if $r \in R$ with $r(k) = i$ for every $k \in I$ then s, defined by $s(k) = 1 - i$ for every $k \in I$ and $s(k) = i$ otherwise, is also in R. In particular every i-default relation is i-valid, as witnessed by taking $I = \emptyset$. We denote by $\mathsf{ISAT}(S)$ the class of satisfiable conjunctions of S-formulas.

1.2 Dichotomies

Theorem 9 (Schaefer's dichotomy [9]). *Let S be a finite set of Boolean relations. If S satisfies one of the conditions $(a) - (f)$ below, then $\mathsf{ISAT}(S)$ is polynomial-time decidable. Otherwise, $\mathsf{ISAT}(S)$ is log-complete in NP.*

- *(a) Every relation in S is 0-valid.*
- *(b) Every relation in S is 1-valid.*
- *(c) Every relation in S is horn*
- *(d) Every relation in S is co-horn*
- *(e) Every relation in S is affine.*
- *(f) Every relation in S is bijunctive.*

In the remainder of this paper, S will be a – possibly infinite – class of Boolean relations. Note that there is no effectiveness requirement on S.

Definition 10. *$\mathsf{SAT}(S)$ is the following statement: for every set C of S-formulas over an infinite set of variables V such that every finite set $C_0 \subseteq C$ is satisfiable there is an infinite assignment $v : V \rightarrow \{\mathrm{T}, \mathrm{F}\}$ satisfying C.*

We will prove the following dichotomy theorem based on Schaefer's theorem.

Theorem 11. *If S satisfies one of the conditions $(a) - (d)$ below, then $\mathsf{SAT}(S)$ is provable over RCA_0. Otherwise $\mathsf{SAT}(S)$ is equivalent to WKL_0 over RCA_0.*

- *(a) Every relation in S is 0-valid.*
- *(b) Every relation in S is 1-valid.*
- *(c) If $R \in S$ is not 0-default then $R = [x]$.*
- *(d) If $R \in S$ is not 1-default then $R = [\neg x]$.*

$\mathsf{SAT}(S)$ principles are not fully satisfactory as these are not robust notions: if we define $\mathsf{SAT}(S)$ in terms of satisfiable sets of *conjunctions* of S-formulas, this yields a different dichotomy theorems. In particular, $\mathsf{RCA}_0 \vdash \mathsf{SAT}([x], [\neg y])$ whereas $\mathsf{RCA}_0 \vdash \mathsf{SAT}([x \wedge \neg y]) \longleftrightarrow \mathsf{WKL}_0$. Ramseyan versions of satisfaction problems have better properties.

Definition 12. RSAT(S) *is the following statement: for every satisfiable set C of S-formulas over an infinite set of variables V, there is an infinite set $H \subseteq V$ homogeneous for C.*

Usual reductions between satisfiability problems involve fresh variable introductions. This is why it is natural to define a *localized* version of those principles, i.e. where the homogeneous set has to lie within a pre-specified set.

Definition 13. LRSAT(S) *is the following statement: for every satisfiable set C of S-formulas over an infinite set of variables V and every infinite set $X \subseteq V$, there is an infinite set $H \subseteq X$ homogeneous for C.*

In particular, we define LRSAT(0-valid) (resp. LRSAT(1-valid), LRSAT(Horn), LRSAT(CoHorn), LRSAT(Bijunctive) or LRSAT(Affine)) to denote LRSAT(S) where S is the set of all 0-valid (resp. 1-valid, horn, co-horn, bijunctive or affine) relations. We will prove the following dichotomy theorem.

Theorem 14. *Either* $\mathsf{RCA_0} \vdash$ LRSAT(S) *or* LRSAT(S) *is equivalent to one of the following principles over* $\mathsf{RCA_0}$:

1. LRSAT   3. LRSAT(*Affine*)
2. LRSAT($[x \neq y]$)   4. LRSAT(*Bijunctive*)

As we will see in Theorem 37, each of those principles are equivalent to their non localized version. As well, LRSAT($[x \neq y]$) coincides with an already existing principle about bipartite graphs called $\mathsf{RCOLOR_2}$ and LRSAT is equivalent to RWKL over $\mathsf{RCA_0}$. Hence LRSAT(S) is either provable over $\mathsf{RCA_0}$, or equivalent to one of $\mathsf{RCOLOR_2}$, RSAT(Affine), RSAT(Bijunctive) and RWKL over $\mathsf{RCA_0}$.

2 Schaefer's Dichotomy Theorem

Definition 15. *Let S be a class of Boolean relations and V be a set of variables. Let φ be an S-formula over V. We denote by* $\mathsf{Var}(\varphi)$ *the set variables occurring in φ. An assignment for φ is a function $v : \mathsf{Var}(\varphi) \to \{\mathrm{T}, \mathrm{F}\}$. An assignment can be naturally extended to a function over formulas by the natural interpretation rules for logical connectives. Then an assignment v satisfies φ if $v(\varphi) = \mathrm{T}$. The set of assignments of φ is written* $\mathsf{Assign}(\varphi)$. *Variable substitution is defined in the usual way and is written $\varphi[y/x]$, meaning that all occurrences of x in φ are replaced by y. We will also write $\varphi[y/X]$ where X is a set of variables to denote substitution of all occurrences of a variable of X in φ by y. A constant is either 0 or 1.*

Definition 16. *Let S be a class of relations over Booleans. The class of existentially quantified S-formulas with constants – i.e. of the form $(\exists x)\varphi[x, y, \mathrm{T}, \mathrm{F}]$ with $\varphi \in S$ – is denoted by* $\mathsf{Gen}(S)$. *We also define* $\mathsf{Rep}(S) = \{[R] : R \in \mathsf{Gen}(S)\}$, *ie. the relations represented by existentially quantified S-formula with constants. By abuse of notation, we may use $\mathsf{Rep}(R)$ when R is a relation to denote $\mathsf{Rep}(\{R\})$. We can also define similar relations without constants, denoted by* Gen_{NC} *and* Rep_{NC}.

Lemma 17 (Schaefer in [9, 4.3]). *At least one of the following holds:*

   (a) Every relation in S is 0-valid.
   (b) Every relation in S is 1-valid.
   (c) $[x]$ *and* $[\neg x]$ *are contained in* $Rep_{NC}(S)$.
   (d) $[x \neq y] \in Rep_{NC}(S)$.

One easily sees that if every relation in S is 0-valid (resp. 1-valid) then $RCA_0 \vdash$ SAT(S) as the assignment always equal to F (resp. T) is a valid assignment and is computable. We will now see that problems parameterized by relations either 0-default or $[x]$ (resp. 1-default or $[\neg x]$) are also solvable.

Lemma 18. *If the only relation in S which is not 0-default is* $[x]$ *or the only relation which is not 1-default is* $[\neg x]$ *then* $RCA_0 \vdash$ SAT(S).

The strategy for solving such an instance (V, C) of SAT(S) consists in defining an assignment which given a variable x will give it the default value F unless it finds the clause $(x \vee x) \in C$.

Lemma 19. *If* $[x \neq y] \in Rep_{NC}(S)$ *then* $RCA_0 \vdash WKL_0 \longleftrightarrow$ SAT(S).

Lemma 19 holds because SAT($[x \neq y]$) can be seen as a reformulation of $COLOR_2$ which is equivalent to WKL_0 over RCA_0 [6].

Theorem 11 is proven by a case analysis using Lemma 17, by noticing that when we are not in cases already handled by Lemma 18 and Lemma 19, we can find n-ary formulas encoding $[x]$ and $[\neg x]$ with $n \geq 2$. Thus diagonalizing against some values becomes a Σ_1^0 event.

3 Ramsey-Type Schaefer's Dichotomy Theorem

Proof of Theorem 14 can be split into four steps, each of them being dichotomies themselves. The first one, Theorem 22, states the existence of a gap between provability in RCA_0 and implying $RCOLOR_2$ over RCA_0. Then we focus successively on two classes of boolean formulas: bijunctive formulas (Theorem 29) and affine formulas (Theorem 33) whose corresponding principles happen to be either a consequence of $RCOLOR_2$ or equivalent to the full class of bijunctive (resp. affine) formulas. Remaining cases are handled by Theorem 34. We first state a trivial relation between a satisfaction principle and its Ramseyan version.

Lemma 20. $RCA_0 \vdash$ SAT(S) $\rightarrow$ LRSAT(S)

Lemma 21. *Let T be a c.e. set of Boolean relations such that* $[x \neq y] \in Rep_{NC}(T)$. *If* $S \subseteq Rep_{NC}(T \cup \{[x], [\neg x]\})$ *then* $RCA_0 \vdash$ LRSAT(T) $\rightarrow$ LRSAT(S).

3.1 From Provability to LRSAT($[x \neq y]$)

Our first dichotomy for Ramseyan principles is between RCA_0 and LRSAT($[x \neq y]$).

Theorem 22. *If S satisfies one of the conditions (a)-(d) below then $\mathrm{RCA}_0 \vdash \mathrm{LRSAT}(S)$. Otherwise $\mathrm{RCA}_0 \vdash \mathrm{LRSAT}(S) \to \mathrm{LRSAT}([x \neq y])$.*
  (a) Every relation in S is 0-valid.   *(c) Every relation in S is horn.*
  (b) Every relation in S is 1-valid.   *(d) Every relation in S is co-horn.*

Lemma 23 (Schaefer in [9, 3.2.1]). *If S contains some relation which is not horn and some relation which is not co-horn, then $[x \neq y] \in Rep(S)$.*

Lemma 24. *At least one of the following holds:*
  (a) Every relation in S is 0-valid.   *(d) Every relation in S is co-horn.*
  (b) Every relation in S is 1-valid.   *(e) $[x \neq y] \in Rep_{NC}(S)$.*
  (c) Every relation in S is horn.

Proof. Assume none of cases (a), (b) and (e) holds. Then by Lemma 17, $[x]$ and $[\neg x]$ are contained in $Rep_{NC}(S)$, hence $Rep_{NC}(S) = Rep(S)$. So by Lemma 23, either every relation in S is horn, or every relation in S is co-horn.   □

It is easy to see that LRSAT(0-valid) and LRSAT(1-valid) both hold over RCA_0. We will now prove that so do LRSAT(Horn) and LRSAT(CoHorn), but first we must introduce the powerful tool of *closure under functions*.

Definition 25. *We say that a relation $R \subseteq \{0,1\}^n$ is closed or invariant under an m-ary function f and that f is a polymorphism of R if for every m-tuple $\langle v_1, \dots, v_m \rangle$ of vectors of R, $f(v_1, \dots, v_m) \in R$ where f is the coordinate-wise application of the function f.*

We denote the set of all polymorphisms of R by $\mathrm{Pol}(R)$, and for a set Γ of Boolean relations we define $\mathrm{Pol}(\Gamma) = \{f : f \in \mathrm{Pol}(R) \text{ for every } R \in \Gamma\}$. Similarly for a set B of Boolean functions, $\mathrm{Inv}(B) = \{R : B \subseteq \mathrm{Pol}(R)\}$ is the set of *invariants* of B. For any set S of Boolean relations, $\mathrm{Pol}(R)$ is in Post's lattice.

Definition 26. *The conjunction function* $\mathrm{conj} : \{0,1\}^2 \to \{0,1\}$ *is defined by* $\mathrm{conj}(a,b) = a \wedge b$, *the disjunction function* $\mathrm{disj} : \{0,1\}^2 \to \{0,1\}$ *by* $\mathrm{disj}(a,b) = a \vee b$, *the affine function* $\mathrm{aff} : \{0,1\}^3 \to \{0,1\}$ *by* $\mathrm{aff}(a,b,c) = a \oplus b \oplus c = 1$ *and the majority function* $\mathrm{maj} : \{0,1\}^3 \to \{0,1\}$ *by* $\mathrm{maj}(a,b,c) = (a \wedge b) \vee (a \wedge c) \vee (b \wedge c)$.

The following theorem due to Schaefer characterizes relations in terms of closure under some functions. The proof involves finite objects and hence can be easily proven to hold over RCA_0.

Theorem 27 (Schaefer [9]). *A relation is*

  1. horn iff it is closed under conjunction function
  2. co-horn iff it is closed under disjunction function
  3. affine iff it is closed under affine function
  4. bijunctive iff it is closed under majority function

In other words, using Post's lattice, a relation R is horn iff $E_2 \subseteq \mathrm{Pol}(R)$, co-horn iff $V_2 \subseteq \mathrm{Pol}(R)$, affine iff $L_2 \subseteq \mathrm{Pol}(R)$ and bijunctive iff $D_2 \subseteq \mathrm{Pol}(R)$.

Theorem 27 is powerful because it does not only imply the closure of valid assignments under some functions. As we will see in Theorem 37, this can be interpreted as "the localized version of the principles parametrized by one of classes 1-4 is not stronger than their corresponding non-localized versions". The closure of valid assignments under some functions enables us to prove Theorem 28 below.

Theorem 28. *If every relation in S is horn (resp. co-horn) then* $\mathsf{RSAT} \vdash \mathsf{LRSAT}(S)$.

Proof. We will prove it over $\mathsf{RCA_0}$ for the horn case. The proof for co-horn relations is similar. Let (V, C, L) be an instance of $\mathsf{LRSAT}(\mathsf{Horn})$ and $F \subseteq L$ be the collection of variables $x \in L$ such that there is a finite $C_{fin} \subseteq C$ for which every valid assignment v for C_{fin} satisfies $v(x) = \mathsf{T}$.

Case 1: F is infinite. Because F is Σ_1^0, we can take a infinite Δ_1^0 subset of F as homogeneous set for C with color T.

Case 2: F is finite. We take $H = L \smallsetminus F$ as infinite set homogeneous for C with color F. If H is not homogeneous for C, then there exists a finite $C_{fin} \subseteq C$ witnessing it. Let $H_{fin} = \mathsf{Var}(C_{fin}) \cap H$. For every valid assignment v for C_{fin}, there is an $x \in H_{fin}$ such that $v(x) = \mathsf{T}$. By definition of H, for each $x \in H$ there is a valid assignment v_x such that $v_x(x) = \mathsf{F}$. By Theorem 27, the class valid assignments of a finite horn formula is closed under conjunction. So $v = \bigwedge_{x \in H_{fin}} v_x$ is a valid assignment for C_{fin} such that $v(x) = \mathsf{F}$ for each $x \in H_{fin}$. Contradiction.   □

Proof (of Theorem 22). If every relation in S is 0-valid (resp. 1-valid) then $\mathsf{LRSAT}(S)$ holds obviously over $\mathsf{RCA_0}$. If every relation in S is horn (resp. co-horn) then by Theorem 28, $\mathsf{LRSAT}(S)$ holds also over $\mathsf{RCA_0}$. By Lemma 24, it remains the case where $[x \neq y] \in Rep_{NC}(S)$. By Lemma 21, $\mathsf{RCA_0} \vdash \mathsf{LRSAT}(S) \to \mathsf{LRSAT}([x \neq y])$.   □

3.2 Bijunctive Satisfiability

Our second dichotomy theorem concerns bijunctive relations. Either the related principle is a consequence of $\mathsf{LRSAT}([x \neq y])$ over $\mathsf{RCA_0}$, or it has full strength of $\mathsf{LRSAT}(\mathsf{Bijunctive})$. In the remaining of this subsection, we will assume that S contains only bijunctive relations and $[x \neq y] \in Rep_{NC}(S)$. In other words we suppose that $D_2 \subseteq \mathsf{Pol}(S) \subseteq D$.

Theorem 29. *If S contains only affine relations then* $\mathsf{RCA_0} \vdash \mathsf{LRSAT}([x \neq y]) \to \mathsf{LRSAT}(S)$. *Otherwise* $\mathsf{RCA_0} \vdash \mathsf{LRSAT}(S) \longleftrightarrow \mathsf{LRSAT}(\mathit{Bijunctive})$.

Definition 30. *For any set S of relations, the* co-clone *of S is the closure of S by existential quantification, equality and conjunction. We denote it by $\langle S \rangle$.*

Remark that in general, $Rep_{NC}(S)$ may be different from $\langle S \rangle$ if $[x = y] \notin Rep_{NC}(S)$. However in our case, we assume that $[x \neq y] \in Rep_{NC}(S)$, hence $[x = y] \in Rep_{NC}(S)$ and $Rep_{NC}(S) = \langle S \rangle$. The following property will happen to be very useful for proving that a relation $R \in Rep_{NC}(S)$.

Lemma 31 (Folklore). $\mathsf{Inv}(\mathsf{Pol}(S)) = \langle S \rangle$

Lemma 32. *One of the following holds:*

   (a) $Rep_{NC}(S)$ *contains all bijunctive relations.*
   (b) $S \subseteq Rep_{NC}(\{[x], [x \neq y]\})$.

Proof. By hypothesis, $D_2 \subseteq \text{Pol}(S) \subseteq D$. Either $D_1 \subseteq \text{Pol}(S)$ – meaning that every relation in S is affine – in which case $S \subseteq \text{Inv}(D_1) = \text{Rep}_{NC}(\{[x], [x \neq y]\})$. Or $\text{Pol}(S) = D_2$. Then $\text{Rep}_{NC}(S) = \langle S \rangle = \text{Inv}(\text{Pol}(S)) = \text{Inv}(D_2)$ which is the set of all bijunctive relations.   □

Proof (of Theorem 29). By Lemma 32, either $\text{Rep}_{NC}(S)$ contains all bijunctive relations or $S \subseteq \text{Rep}_{NC}(\{[x], [x \neq y]\})$. In the latter case, by Lemma 21 $\text{LRSAT}([x \neq y])$ implies $\text{LRSAT}(S)$ over RCA_0. In the former case, there exists a finite basis $S_0 \subseteq S$ such that $\text{Rep}_{NC}(S_0)$ contains all bijunctive relations. In particular S_0 is a c.e. set, so $\text{RCA}_0 \vdash \text{LRSAT}(S_0) \rightarrow \text{LRSAT}(\text{Bijunctive})$. Any instance of $\text{LRSAT}(S_0)$ being an instance of $\text{LRSAT}(S)$, $\text{RCA}_0 \vdash \text{LRSAT}(S) \rightarrow \text{LRSAT}(\text{Bijunctive})$. The reverse implication follows directly from the assumption that every relation in S is bijunctive. So $\text{RCA}_0 \vdash \text{LRSAT}(S) \leftrightarrow \text{LRSAT}(\text{Bijunctive})$.   □

3.3 Affine Satisfiability

We now suppose that $L_2 \subset \text{Pol}(S) \subsetneq D$, i.e. S contains only affine relations, $[x \neq y] \in \text{Rep}_{NC}(S)$ and S contains a relation which is not bijunctive.

Theorem 33. $\text{RCA}_0 \vdash \text{LRSAT}(S) \leftrightarrow \text{LRSAT}(\textit{Affine})$

Proof. By assumption, every relation in S is affine. Hence $\text{RCA}_0 \vdash \text{LRSAT}(\text{Affine}) \rightarrow \text{LRSAT}(S)$. As $L_2 \subseteq \text{Pol}(S) \subsetneq D$, $\text{Pol}(S)$ is either L_3 or L_2. In particular, $\text{Pol}(S \cup \{[x], [\neg x]\}) = L_2$. Considering the corresponding invariants, $\text{Inv}(L_2) \subseteq \text{Inv}(\text{Pol}(S \cup \{[x], [\neg x]\})) = \langle S \cup \{[x], [\neg x]\} \rangle = \text{Rep}_{NC}(S \cup \{[x], [\neg x]\})$. $\text{Inv}(L_2)$ being the set of affine relations, by Lemma 21, $\text{RCA}_0 \vdash \text{LRSAT}(S) \rightarrow \text{LRSAT}(\text{Affine})$.   □

3.4 Remaining Cases

Based on Post's lattice, the only remaining cases are $\text{Pol}(S) = N_2$ or $\text{Pol}(S) = I_2$.

Theorem 34. *If* $\text{Pol}(S) \subseteq N_2$ *then* $\text{RCA}_0 \vdash \text{LRSAT}(S) \leftrightarrow \text{LRSAT}$.

Proof. The direction $\text{RCA}_0 \vdash \text{LRSAT} \rightarrow \text{LRSAT}(S)$ is obvious. We will prove the converse. Because $\text{Pol}(S) \subseteq N_2$, $\text{Pol}(S \cup \{[x]\}) = I_2$. $\text{Rep}_{NC}(S \cup \{[x]\}) = \langle S \cup \{[x]\} \rangle = \text{Inv}(\text{Pol}(S \cup \{[x]\})) \supseteq \text{Inv}(I_2)$. But $\text{Inv}(I_2)$ is the set of all Boolean relations. As $\text{Inv}(I_2)$ has a finite basis, there exists a finite $S_0 \subseteq S$ such that $\text{Rep}_{NC}(S_0 \cup \{[x]\})$ contains all Boolean relations. By Lemma 21, $\text{RCA}_0 \vdash \text{LRSAT}(S_0) \rightarrow \text{LRSAT}$. Hence $\text{RCA}_0 \vdash \text{LRSAT}(S) \leftrightarrow \text{LRSAT}$.   □

Proof (of Theorem 14). By case analysis over $\text{Pol}(S)$. If I_1, I_0, V_2 and E_2 are included in $\text{Pol}(S)$ then by Theorem 22, $\text{RCA}_0 \vdash \text{LRSAT}(S)$. If $D_1 \subseteq \text{Pol}(S) \subseteq D$ then $\text{RCA}_0 \vdash \text{LRSAT}(S) \leftrightarrow \text{LRSAT}([x \neq y])$ by Theorem 29. By the same theorem, if $\text{Pol}(S) = D_2$ then $\text{RCA}_0 \vdash \text{LRSAT}(S) \leftrightarrow \text{LRSAT}(\text{Bijunctive})$. If $L_2 \subseteq \text{Pol}(S) \subseteq L_3$ then by Theorem 33, $\text{RCA}_0 \vdash \text{LRSAT}(S) \leftrightarrow \text{LRSAT}(\text{Affine})$. Otherwise, $I_2 \subseteq \text{Pol}(S) \subseteq N_2$ in which case $\text{RCA}_0 \vdash \text{LRSAT}(S) \leftrightarrow \text{LRSAT}$.   □

In fact, LRSAT($[x \neq y]$) coincides with an already existing principle about bipartite graphs. For $k \in \mathbb{N}$, we say that a graph $G = (V, E)$ is *k-colorable* if there is a function $f : V \rightarrow k$ such that $(\forall(x, y) \in E)(f(x) \neq f(y))$, and we say that a graph is *finitely k-colorable* if every finite induced subgraph is k-colorable.

Definition 35. *Let $G = (V, E)$ be a graph. A set $H \subseteq V$ is homogeneous for G if every finite $V_0 \subseteq V$ induces a subgraph that is k-colorable by a coloring that colors every $v \in V_0 \cap H$ color 0. LRCOLOR$_k$ is the following statement: for every infinite, finitely k-colorable graph $G = (V, E)$ and every infinite $L \subseteq V$ there is an infinite $H \subseteq L$ that is homogeneous for G. RCOLOR$_k$ is the restriction of LRCOLOR$_k$ with $L = V$. An instance of LRCOLOR$_k$ is a pair (G, L). For RCOLOR$_k$, it is simply the graph G.*

Theorem 36. RCA$_0$ ⊢ RCOLOR$_2$ ↔ LRSAT($[x \neq y]$)

4 The Strength of Satisfiability

Localized principles are relatively easy to manipulate as they can express relations defined using existential quantifier by restricting the localized set L to the variables not captured by any quantifier. However we will see that when the set of relations has some good closure properties, the unlocalized version of the principle is as expressive as its localized one.

Theorem 37. *Let S be a c.e. co-clone.* RCA$_0$ ⊢ RSAT(S) ↔ LRSAT(S)

Noticing that affine (resp. bijunctive) relations form a co-clone, we immediately deduce the following corollary.

Corollary 38. RSAT(*Affine*) *and* RSAT(*Bijunctive*) *are equivalent to their local version over* RCA$_0$.

A useful principle below WKL$_0$ for studying the strength of a statement is the notion of *diagonally non-computable function*.

Definition 39. *A total function f is diagonally non-computable if $(\forall e)f(e) \neq \Phi_e(e)$. DNR is the corresponding principle, i.e. for every X, there exists a function d.n.c. relative to X.*

DNR is known to coincide with the restriction of RWKL to trees of positive measure ([5,2]). On the other side, there exists an ω-model of DNR which is not a model of RCOLOR$_2$ ([2]). We will now prove that we can compute a diagonally non-computable function from any infinite set homogeneous for a particular set of affine formulas. As RSAT implies LRSAT(Affine) over RCA$_0$, it gives another proof of RCA$_0$ ⊢ RWKL → DNR.

Theorem 40. *There exists a computable set C of affines formulas over a computable set V of variables such that every infinite set homogeneous for C computes a diagonally non-computable function.*

Corollary 41. RCA$_0$ ⊢ RSAT(*Affine*) → DNR.

5 Conclusions

Satisfaction principles happen to collapse in the case of a full assignment existence statement. The definition is not robust and the conditions of the corresponding dichotomy theorem evolve if we make the slight modification of allowing conjunctions in our definition of formulas.

However, the proposed Ramseyan version leads to a much more robust dichotomy theorem with four main subsystems. The conditions of "tractability" – here provability over RCA_0 – differ from those of Schaefer dichotomy theorem but the considered classes of relations remain the same. We obtain the surprising result that infinite versions of Horn and co-Horn satisfaction problems are provable over RCA_0 and strictly weaker than bijunctive and affine corresponding principles, whereas the complexity classification of [1] has shown that Horn satisfiability was P-complete under AC^0 reduction, hence at least as strong as Bijunctive satisfiability which is NL-complete.

Question 42. Does $RCOLOR_2$ imply DNR over RCA_0 ? Does it imply RWKL ?

References

1. Allender, E., Bauland, M., Immerman, N., Schnoor, H., Vollmer, H.: The complexity of satisfiability problems: Refining schaefer's theorem. In: Jedrzejowicz, J., Szepietowski, A. (eds.) MFCS 2005. LNCS, vol. 3618, pp. 71–82. Springer, Heidelberg (2005)
2. Bienvenu, L., Patey, L., Shafer, P.: A Ramsey-Type König's lemma and its variants (in preparation)
3. Bulatov, A.A.: A dichotomy theorem for constraints on a three-element set. In: Proceedings of the 43rd Annual IEEE Symposium on Foundations of Computer Science, pp. 649–658. IEEE (2002)
4. Creignou, N., Hermann, M.: Complexity of generalized satisfiability counting problems. Information and Computation 125(1), 1–12 (1996)
5. Flood, S.: Reverse mathematics and a Ramsey-type König's Lemma. Journal of Symbolic Logic 77(4), 1272–1280 (2012)
6. Hirst, J.: Marriage theorems and reverse mathematics. Logic and Computation 106 (1990)
7. Khanna, S., Sudan, M.: The optimization complexity of constraint satisfaction problems. In: Electonic Colloquium on Computational Complexity, Citeseer (1996)
8. Marek, V.W., Remmel, J.B.: The complexity of recursive constraint satisfaction problems. Annals of Pure and Applied Logic 161(3), 447–457 (2009)
9. Schaefer, T.J.: The complexity of satisfiability problems. In: Proceedings of the Tenth Annual ACM Symposium on Theory of Computing, pp. 216–226 (1978)
10. Simpson, S.G.: Subsystems of second order arithmetic, vol. 1. Cambridge University Press (2009)

An Early Completion Algorithm: Thue's 1914 Paper on the Transformation of Symbol Sequences

James F. Power

Department of Computer Science, National University of Ireland,
Maynooth, Co. Kildare, Ireland
jpower@cs.nuim.ie

Abstract. References to Thue's 1914 paper on string transformation systems are based mainly on a small section of that work defining Thue systems. A closer study of the remaining parts of that paper highlight a number of important themes in the history of computing: the transition from algebra to formal language theory, the analysis of the "computational power" (in a pre-1936 sense) of rules, and the development of algorithms to generate rule-sets.

Of the many current models of computation, one of the oldest is the *Thue system*, first specified by Axel Thue 100 years ago [1]. A Thue system is typically presented as a sequence of string-pairs over some fixed alphabet:

$$A_1, \quad A_2, \quad A_3, \quad ..., \quad A_n$$
$$B_1, \quad B_2, \quad B_3, \quad ..., \quad B_n,$$

Any other two strings P and Q over the same alphabet are said to be *similar* if it is possible to transform P into Q by replacing a substring matching some A_i with the corresponding string B_i (or vice versa, replacing some B_i with A_i). Two strings are said to be *equivalent* if we can form a finite sequence of strings, each similar to the former, taking us from P into Q.

Emil Post showed how this could be recast as a special form of one of his canonical systems and then to the decision problem for Turing machines [2]. At the time, it was important as one of the first undecidable problems outside of the original set from 1936, and Thue systems, with their close resemblance to unrestricted grammars, have since been established as one of the classical models of computation [3,4].

However, only the first two pages of Thue's paper are directly relevant to Post's proof, and the remainder of the paper seems to have been rarely explored. In what follows we review some of the remaining contributions of the paper, and to advocate its relevance for the history of computing.

Background to Thue's 1914 Paper

Axel Thue (1863-1922) was a Norwegian mathematician who published a range of papers, 35 of which are collected in his *Selected Mathematical Papers* [5].

A. Beckmann, E. Csuhaj-Varjú, and K. Meer (Eds.): CiE 2014, LNCS 8493, pp. 343–346, 2014.

Most of these relate to algebra and Diophantine approximations (he also worked in geometry and mechanics), and a recent conference was dedicated to his contributions in this area[1]. However, Axel Thue also published four papers directly relating to the theory of words and languages.

Two of these, published in 1906 and 1912, dealt with patterns in infinite strings [6,7] (Berstel provides a translation and discussion [8]). They are known for being an early contribution to the field of combinatorics (though not the earliest [9]) and, in particular, for the Thue-Morse sequence. This sequence can be specified by giving a morphism μ defining a mapping over strings (applied like the rules of an L-system) $\mu(a) = ab$, $\mu(b) = ba$. Thus, for example, starting with the string a we can produce the strings: a, ab, $abba$, $abbabaab$, $abbabaabbaababba$, ...

These strings have some interesting properties: in particular they are all *overlap-free*. Two strings have an overlap if they are of the form CU and UD, with the common substring U forming the overlap. A special case is where a string overlaps with itself, and a string is overlap-free if it does not contain any substring that overlaps with itself. Thue proves that the morphism μ preserves this property: it will always map overlap-free words to overlap-free words.

Thue's other two "language theory" papers from 1910 and 1914 discuss the more general problem of transformations [10,1]. Thue's 1910 paper deals with transformations between trees, and is thus a more direct predecessor of his 1914 paper. It been discussed by Steinby and Thomas [11].

The Importance of Critical Pairs

The 1914 paper, whose title translates roughly as *Problems concerning the transformation of symbol sequences according to given rules* specifically articulates the central problem in algorithmic terms:

> *Problem I:* For any arbitrary given sequences A and B, to find a method, where one can always calculate in a predictable number of operations, whether or not two arbitrary given symbol sequences are equivalent in respect of the sequences A and B.

Thue observes that this task of solving this problem is "extensive and of the utmost difficulty" and notes that he must settle for dealing with some special cases of the problem. Having posed the general problem in §II of his paper, Thue then presents an early example of a proof of (what we would now call) *termination* and *local confluence* for the special case where the rules are non-overlapping and non-increasing in size.

When reducing some string P, we must find some occurrence of A_i and replace it with B_i. A difficulty arises if there is an overlap: some substring CUD in P, such that A_i matches both CU and UD, and thus choosing one option will eliminate our ability to later choose the other. In the modern setting of term rewriting, CU and UD are known as a *critical pair*, and the problem has been well-studied in the literature [12], starting at least from Newman [13].

---

[1] *Thue 150*, held in Bordeaux, France from Sept 30 - Oct 4, 2013.

Thus, having studied overlap-free strings in his previous papers, Thue's focus in 1914 is the converse, and the overlap situation of strings CU and UD is the focus of study for most of the paper.

Completion in the Context of a Monoid Presentation

Thue deals with the special case where a language is defined by specifying some identity string, R. This is not the usual case in language theory but is not an unusual approach when presenting an algebraic group. In Thue's case he is presenting a *monoid*: a set with an associative binary operator and an identity element (but no inverse function).

So, given a monoid, represented by specifying the identity string, the word problem here simply involves transforming some string P to some string Q by repeated insertions and deletions of R. Thue calls this relation "equivalence with respect to R", writes it as $P = Q$, and formulates:

> *Problem II:* Given an arbitrary sequence R, to find a method where one can always decide in a finite number of investigations whether or not two arbitrary given sequences are equivalent with respect to R.

As before, a difficulty arises when two overlapping instances of R occur as substrings of P. If we represent these as CU and UD as above, then we have $R \equiv CU \equiv UD$. But in this case $C = CR \equiv C(UD) \equiv (CU)D \equiv RD = D$. This tells us that $C = D$ (modulo R). The importance of this equation is that if we choose to delete either CU and UD from the string containing CUD we are left with either C or D, but adding the equation $C = D$ restores the confluence of our derivation.

Moreover, since both C and D are constructed from R by removing the common substring U they have the same length and contain the same symbols. In this case, as Thue notes, it is relatively easy to derive an algorithm for solving the word problem, and Thue describes one in §V of his paper.

Given this solution for the special case of Problem I, Thue now can outline his *completion algorithm* to solve Problem II:

1. Start with the given identity word R.
2. Form equations $C = D$ based on the remainder from the overlaps within R. For all these equations C and D will have the same length, and this will be less than the length of R.
3. Form a new set of identity strings $R', R'', \ldots$ by applying the equivalences from step 2 in R. These new identity strings will all have the same length as the original R.
4. Iterate steps 2 and 3 until we reach the fixed point. We know the process terminates, since each new identity string we create can only be a permutation of the original identity string (and there are only finitely many of these).

Thue's algorithm lacks typical features of modern completion algorithms such as the Knuth-Bendix algorithm: in particular there is no need for a complex unification process when we are dealing with concrete strings. However, it certainly contains many of the "basic features" of the algorithm as described by Buchberger [12], and could be considered as an embryonic version of it.

An Early Computational Flavour

Throughout Thue's paper he distinguishes between the case where two strings are equal (modulo some R), and when two strings are *provably* equal with respect to some given set of rules. He also investigates special cases where these two relations coincide, and where he can formulate (what we would now call) an algorithm to solve the word problem.

Thue's perspective is vital from a computational point of view and is neatly summarised by Matiyasevich and Sénizergues [14]:

> "put[ting] more attention to the process of transformation of words rather than to its result [...] is typical to computer science but has no counterpart in, say, algebra"

Thue was writing as a mathematician, and well before the identification of computer science as a discipline, but in his 1914 paper we can recognise much of the computational DNA that would allow algebra to evolve into language theory.

References

1. Thue, A.: Probleme über Veränderungen von Zeichenreihen nach gegebenen Regeln. Christiana Videnskabs-Selskabs Skrifter, I. Math.-naturv. Klasse 10 (1914)
2. Post, E.L.: Recursive unsolvability of a problem of Thue. Journal of Symbolic Logic 12(1), 1–11 (1947)
3. Rozenberg, G., Salomaa, A. (eds.): Handbook of Formal Languages. Springer (1997)
4. Book, R.V., Otto, F.: String-rewriting Systems. Springer (1993)
5. Nagell, T., Selberg, A., Selberg, S., Thalberg, K. (eds.): Selected Mathematical Papers of Axel Thue. Universitetsforlaget, Oslo (1977)
6. Thue, A.: Über unendliche Zeichenreihen. Christiana Videnskabs-Selskabs Skrifter. I. Math. -Naturv. Klasse 7 (1906)
7. Thue, A.: Über die gegenseitige Lage gleicher Teile gewisser Zeichenreihen. Christiana Videnskabs-Selskabs Skrifter, I. Math. -Naturv. Klasse 1 (1912)
8. Berstel, J.: Axel Thue's papers on repetitions in words: a translation. Publications du LaCIM, Université du Québec à Montréal (1995)
9. Marcus, S.: Words and languages everywhere. In: Martin-Vide, C., Mitrana, V., Paun, G. (eds.) Formal Languages and Applications, pp. 11–53. Springer (2004)
10. Thue, A.: Die Lösung eines Spezialfalles eines generellen logischen Problems. Christiana Videnskabs-Selskabs Skrifter, I. Math.-naturv. Klasse 8 (1910)
11. Steinby, M., Thomas, W.: Trees and term rewriting in 1910: On a paper by Axel Thue. EATCS Bull. 72, 256–269 (2000)
12. Buchberger, B.: History and basic features of the critical-pair/completion procedure. Journal of Symbolic Computation 3(1-2), 3–38 (1987)
13. Newman, M.: On theories with a combinatorial definition of "equivalence". Annals of Mathematics 43(2), 223–243 (1942)
14. Matiyasevicha, Y., Sénizergue, G.: Decision problems for semi-Thue systems with a few rules. Theoretical Computer Science (330), 145–169 (2005)

Metric-Driven Grammars and Morphogenesis (Extended Abstract)

Przemyslaw Prusinkiewicz, Brendan Lane, and Adam Runions

Department of Computer Science, University of Calgary
2500 University Dr. N.W., Calgary, AB T2N 1N4, Canada
{pwp,laneb,runionsa}@cpsc.ucalgary.ca
http://algorithmicbotany.org

Abstract. Expansion of space, rather than the progress of time, drives many developmental processes in plants. Metric-driven grammars provide a formal method for specifying and simulating such processes. We illustrate their operation using cell division patterns, phyllotactic patterns, and several aspects of leaf development.

Keywords: natural computing, computational modeling of plant development, growth and form, L-system, cell complex.

Mathematical studies relating the growth and form of organisms were pioneered at the beginning of the XX century by d'Arcy Wentworth Thompson [24]. Among other concepts, he proposed a "theory of transformations" to describe how the forms of related species can be continuously mapped into each other. He also suggested that similar mappings could be used to describe gradual changes of form due to growth. These ideas have been followed and elaborated over time, leading to the characterization of growth in terms of growth tensor fields [7], which are widely used today [3]. Continuous transformations do not capture, however, the emergence and differentiation of new components of organisms, such as cells and organs. A mathematical description of this aspect of development was pioneered by Aristid Lindenmayer, who in 1968 introduced L-systems as a formalism for modeling the development of structures composed of a changing number of discrete components. L-systems were initially defined in terms of cellular automata [8], but soon afterwards were re-defined more elegantly in terms of formal grammars [9]. In this form they are known and used today. A distinctive feature of L-systems is their parallel operation, which lets us view derivation steps as advancing time by some interval. Correspondingly, consecutive words generated by an L-system can represent a sequence of developmental stages of an organism.

According to their original definition, L-systems describe developing structures at the level of topology, i.e., the adjacency relations between the structure components. L-systems are particularly well suited to model linear (filamentous) and branching structures, although extensions to discretized surfaces (maps) and volumes have also been considered [11,12]. Geometric representations, when needed, are introduced by the draftsperson illustrating the models, or calculated

A. Beckmann, E. Csuhaj-Varjú, and K. Meer (Eds.): CiE 2014, LNCS 8493, pp. 347–351, 2014.
© Springer International Publishing Switzerland 2014

algorithmically as a graphical interpretation of the generated structures [17]. This focus on topology has two implications. First, time is the only independent variable that can drive simulations. Second, geometric factors, such as size and shape, have no direct impact on the progress of the simulations (this limitation was partially addressed in extensions of L-systems aimed at the animation of plant development in continuous time [14] and the simulation of interaction between plants and their environment [13,15]). In many developmental processes, however, geometry plays a fundamental morphogenetic role [18]. For example, according to the Errera rule [1,6], the shortest wall passing through the centroid of the cell determines the most likely orientation of cell division in the absence of specific polarizing factors. Furthermore, the expansion of space may have a more direct impact on the progress of morphogenesis than the progress of time. For instance, according to the conceptual model of phyllotaxis by Snow and Snow [23] and its numerous computational implementations (e.g. [4,21,22]), new primordia (precursors of organs such as leaves and flowers) emerge in the growing plant apices when and where there is enough space for them. The plastochron, or the time interval between the appearance of consecutive primordia [5], is not an independent variable, but a result of the changing spatial relations in the plant.

Often it is not known whether an observed morphogenetic process is best described as being driven by the progress of time, the expansion of space, or some combination of both factors. Construction of models exploring alternative hypotheses is then an important part of discovery. To provide a methodology and a formal basis for this exploration, we employ metric-driven grammars as a complement of time-driven L-systems.

A metric-driven grammar operates on a cell complex. A justification for the use of cell complexes as models of biological structures, and examples of L-systems operating on 1-dimensional cell complexes, are presented in [16]. A metric of the cell complex specifies the distances between different elements of the structure. These distances change over time as a result of growth. Functions of distances measured within cells and/or their neighborhood control the application of productions, which locally modify the topology of the complex.

An example of the operation of a metric-driven grammar is shown in Figure 1. The production replaces a line segment that exceeds a predefined threshold length with a simple branching structure (compare the first and the second row in Figure 1). The structures are embedded in surfaces with different growth distributions. In the case of uniform growth (left column), all segments reach the threshold length and produce the successor structure simultaneously. The derivation sequence is then indistinguishable from that generated by an L-system: productions are applied in parallel. In contrast, in the case of non-uniform growth (middle and right columns), faster growing segments reach the threshold length before those in the slower growing parts. Productions are applied asynchronously, yielding patterns that depend on the distribution of growth.

A fertile area in which metric-driven grammars provide useful insights is leaf development. There, growing distances appear to trigger the emergence of

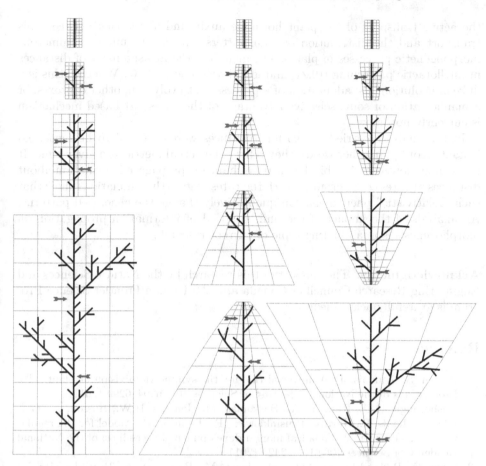

Fig. 1. Selected developmental stages of three branching structures simulated using the same metric-driven grammar. The grammar operates in a space that expands uniformly (left column), grows faster at the bottom than at the top (middle column) and grows faster at the top than at the bottom (right column). Arrows indicate positions of the branching points resulting from the first production application.

serrations [2], lobes [16], leaflets, veins [20], and trichomes. Model exploration suggests that the observed diversity of leaf forms and patterns may result from the variation of a small number of metric-related parameters of development. Further examples of patterning that is likely metric-driven include the initiation of flowers in compound inflorescences and the arrangement of organs within individual flowers.

From a biological perspective, an important question is how distances are measured. The measurement of small distances (on the order of millimeters and less) can be accomplished by diffusion and decay: the concentration of a diffusing substance decreases away from the source, and crosses a threshold value at some distance from it (c.f. [10]). Nevertheless, a different mechanism, based on the active transport of the plant hormone auxin and a feedback between this transport and the distribution of transporters, appears to underlie numerous morphogenetic processes in plants [18], including the measurement of distances in phyllotactic patterning [19,21] and leaf development [2,16]. Whether this is a fluke of evolution, the adaptation of a process that evolved in other contexts, or a manifestation of some selective advantage of the transport-based mechanism is currently not known.

In the analyses carried out so far, distances were assumed to be measured instantaneously; in other words, they reflect the actual metric at a given time. It is possible, however, that biochemical mechanisms propagate information about distances at rates commensurate with the rates of growth. Simulations show that such "relativistic" phenomena can qualitatively change the generated patterns. An analysis of the impact of the limited speed of information propagation on morphogenesis is a fascinating topic of current research.

Acknowledgments. The support of this research by the Natural Sciences and Engineering Research Council of Canada and the Human Frontier Science Program is gratefully acknowledged.

References

1. Besson, S., Dumais, J.: A universal rule for the symmetric division of plant cells. Proceedings of the National Academy of Sciences 108, 6294–6299 (2011)
2. Bilsborough, G.D., Runions, A., Barkoulas, M., Jenkins, H.W., Hasson, A., Galinha, C., Laufs, P., Hay, A., Prusinkiewicz, P., Tsiantis, M.: Model for the regulation of Arabidopsis thaliana leaf margin development. Proceedings of the National Academy of Sciences 108, 3424–3429 (2011)
3. Coen, E., Rolland-Lagan, A.-G., Matthews, M., Bangham, A., Prusinkiewicz, P.: The genetics of geometry. Proceedings of the National Academy of Sciences 101, 4728–4735 (2004)
4. Douady, S., Couder, Y.: Phyllotaxis as a dynamical self organizing process. Parts I–III. Journal of Theoretical Biology 178, 255–312 (1996)
5. Erickson, R.O., Michelini, F.J.: The plastochron index. American Journal of Botany 44(4), 297–305 (1957)
6. Errera, L.: Sur une condition fondamentale d'équilibre des cellules vivantes. Comptes Rendus Hebdomadaires des Séances de l'Académie des Sciences 103, 822–824 (1886)
7. Hejnowicz, Z., Romberger, J.A.: Growth tensor of plant organs. Journal of Theoretical Biology 110, 93–114 (1984)
8. Lindenmayer, A.: Mathematical models for cellular interaction in development, Parts I and II. Journal of Theoretical Biology 18, 280–315 (1968)
9. Lindenmayer, A.: Developmental systems without cellular interaction, their languages and grammars. Journal of Theoretical Biology 30, 455–484 (1971)
10. Lindenmayer, A.: Adding continuous components to L-systems. In: Rozenberg, G., Salomaa, A. (eds.) L Systems. LNCS, vol. 15, pp. 53–68. Springer, Berlin (1974)
11. Lindenmayer, A.: Models for plant tissue development with cell division orientation regulated by preprophase bands of microtubules. Differentiation 26, 1–10 (1984)

12. Lindenmayer, A., Rozenberg, G.: Parallel generation of maps: Developmental systems for cell layers. In: Claus, V., Ehrig, H., Rozenberg, G. (eds.) Graph Grammars 1978. LNCS, vol. 73, pp. 301–316. Springer, Berlin (1979)

13. Měch, R., Prusinkiewicz, P.: Visual models of plants interacting with their environment. In: Proceedings of SIGGRAPH 1996, New Orleans, Louisiana, August 4-9, pp. 397–410. ACM SIGGRAPH, New York (1996)

14. Prusinkiewicz, P., Hammel, M., Mjolsness, E.: Animation of plant development. In: Proceedings of SIGGRAPH 1993, Anaheim, California, August 1-6, pp. 351–360. ACM SIGGRAPH, New York (1993)

15. Prusinkiewicz, P., James, M., Měch, R.: Synthetic topiary. In: Proceedings of SIGGRAPH 1994, Orlando, Florida, July 24-29, pp. 351–358. ACM SIGGRAPH, New York (1994)

16. Prusinkiewicz, P., Lane, B.: Modeling morphogenesis in multicellular structures with cell complexes and L-systems. In: Capasso, V., Gromov, M., Harel-Bellan, A., Morozova, N., Pritchard, L. (eds.) Pattern Formation in Morphogenesis, pp. 137–151. Springer, Berlin (2012)

17. Prusinkiewicz, P., Lindenmayer, A.: The Algorithmic Beauty of Plants. In: Hanan, J.S., Fracchia, F.D., Fowler, D.R., de Boer, M.J.M., Mercer, L. (eds.). Springer, New York (1990)

18. Prusinkiewicz, P., Runions, A.: Computational models of plant development and form. New Phytologist 193(3), 549–569 (2012)

19. Reinhardt, D., Pesce, E.R., Stieger, P., Mandel, T., Baltensperger, K., Bennett, M., Traas, J., Friml, J., Kuhlemeier, C.: Regulation of phyllotaxis by polar auxin transport. Nature 426, 255–260 (2003)

20. Runions, A., Fuhrer, M., Lane, B., Federl, P., Rolland-Lagan, A.-G., Prusinkiewicz, P.: Modeling and visualization of leaf venation patterns. ACM Transactions on Graphics 24, 702–711 (2005)

21. Smith, R.S., Guyomarc'h, S., Mandel, T., Reinhardt, D., Kuhlemeier, C., Prusinkiewicz, P.: A plausible model of phyllotaxis. Proceedings of the National Academy of Sciences 103, 1301–1306 (2006)

22. Smith, R.S., Kuhlemeier, C., Prusinkiewicz, P.: Inhibition fields for phyllotactic pattern formation: A simulation study. Canadian Journal of Botany 84, 1635–1649 (2006)

23. Snow, M., Snow, R.: Experiments on phyllotaxis. I. The effect of isolating a primordium. Philosophical Transactions of the Royal Society of London B 221, 1–43 (1932)

24. Thompson, D.: On Growth and Form, 2nd edn. University Press, Cambridge (1942)

Hyperprojective Hierarchy of qcb$_0$-Spaces

Matthias Schröder[1,*] and Victor Selivanov[2,**]

[1] Kurt Gödel Research Center, University of Vienna
Austria
[2] A.P. Ershov Institute of Informatics Systems SB RAS
Novosibirsk, Russia

Abstract. We extend the Luzin hierarchy of qcb$_0$-spaces introduced in [ScS13] to all countable ordinals, obtaining in this way the hyperprojective hierarchy of qcb$_0$-spaces. We generalize all main results of [ScS13] to this larger hierarchy. In particular, we extend the Kleene-Kreisel continuous functionals of finite types to the continuous functionals of countable types and relate them to the new hierarchy. We show that the category of hyperprojective qcb$_0$-spaces has much better closure properties than the category of projective qcb$_0$-space. As a result, there are natural examples of spaces that are hyperprojective but not projective.

Keywords: Hyperprojective hierarchy, qcb$_0$-space, continuous functionals of countable types, cartesian closed category.

1 Introduction

A basic notion of Computable Analysis [We00] is the notion of an *admissible representation* of a topological space X. This is a partial continuous surjection δ from the Baire space $\mathcal{N}$ onto X satisfying a certain universality property (see Subsection 2.3 for some more details). Such a representation of X usually induces a reasonable computability theory on X, and the class of admissibly represented spaces is wide enough to include most spaces of interest for Analysis or Numerical Mathematics. As shown by the first author [Sch03], this class coincides with the class of the so-called qcb$_0$-spaces, i.e. T_0-spaces which are quotients of countably based spaces, and it forms a cartesian closed category (with the continuous functions as morphisms). Thus, among qcb$_0$-spaces one meets many important function spaces including the continuous functionals of finite types [Kl59, Kr59] interesting for several branches of logic and computability theory.

Along with the mentioned nice properties of qcb$_0$-spaces, this class seems to be too broad to admit a deep understanding. Hence, it makes sense to search for natural subclasses of this class which still include "practically" important spaces but are (hopefully) easier to study. Interesting examples of such subclasses are

* Supported by the FWF project "Definability and computability".
** Supported by the DFG Mercator professorship at the University of Würzburg, by the RFBR-FWF project "Definability and computability", and by the RFBR project 13-01-00015.

A. Beckmann, E. Csuhaj-Varjú, and K. Meer (Eds.): CiE 2014, LNCS 8493, pp. 352–361, 2014.
© Springer International Publishing Switzerland 2014

obtained if we consider, for each level Γ of the classical Borel or Luzin (projective) hierarchies of Descriptive Set Theory [Ke95], the class of spaces which have an admissible representation of the complexity Γ (below we make this precise). A study of the resulting Borel and Luzin hierarchies of qcb$_0$-spaces was undertaken in [ScS13]. In particular, it was shown that the Luzin hierarchy of qcb$_0$-spaces is closely related to the Kleene-Kreisel continuous functionals of finite types, and that the category of projective qcb$_0$-spaces is cartesian closed.

However, the class of projective qcb$_0$-spaces is in a sense too restricted. In particular, it is not closed under some natural constructions (e.g., countable products and countable coproducts) and does not contain some spaces of interest for Computable Analysis.

In this paper we extend the Luzin hierarchy of qcb$_0$-spaces to all countable ordinals, obtaining in this way the hyperprojective hierarchy of qcb$_0$-spaces. We generalize all main results of [ScS13] concerning the Luzin hierarchy to this larger hierarchy. In particular, we extend the Kleene-Kreisel continuous functionals of finite types to the continuous functionals of countable types and relate them to the new hierarchy. We show that the category of hyperprojective qcb$_0$-spaces has much better closure properties than the category of projective qcb$_0$-space. As a result, there are natural examples of spaces that are hyperprojective but not projective.

After recalling some notions and known facts in the next section, we summarize some basic facts on the hyperprojective hierarchy of sets in Section 3. In Section 4 we study the hyperprojective hierarchy of qcb$_0$-spaces, in particular we show that the category of hyperprojective qcb$_0$-spaces is closed under countable limits and countable colimits. In Section 5 we introduce the continuous functionals of countable types and relate them to the hyperprojective hierarchy of qcb$_0$-spaces. In Section 6 we establish some properties of categories of hyperprojective qcb$_0$-spaces. Because of space limitations, we omit the proofs in this conference paper.

2 Notation and Preliminaries

2.1 Notation

We freely use the standard set-theoretic notation like $dom(f)$, $rng(f)$, $graph(f)$ for the domain, range and graph of a function f, respectively, $X \times Y$ for the Cartesian product, $X \oplus Y$ for the disjoint union of sets X and Y, Y^X for the set of functions $f : X \to Y$ (but in the case when X, Y are qcb$_0$-spaces we use the same notation to denote the set of continuous functions from X to Y), and $P(X)$ for the set of all subsets of X. For $A \subseteq X$, $\overline{A}$ denotes the complement $X \setminus A$ of A in X. We identify the set of natural numbers with the first infinite ordinal ω. The first uncountable ordinal is denoted by ω_1. The notation $f : X \to Y$ means that f is a (total) function from a set X to a set Y.

2.2   Topological Spaces

We assume the reader to be familiar with the basic notions of topology. The collection of all open subsets of a topological space X (i.e. the topology of X) is denoted by τ_X; for the underlying set of X we will write X in abuse of notation. We will usually abbreviate "topological space" to "space". Remember that a space is *zero-dimensional*, if it has a basis of clopen sets. A *basis* for the topology on X is a set $\mathcal{B}$ of open subsets of X such that for every $x \in X$ and open U containing x, there is $B \in \mathcal{B}$ satisfying $x \in B \subseteq U$. A space is *countably based*, if it has a countable basis. By a cb_0-space we mean a countably based T_0-space. The class of cb_0-spaces is denoted by $\mathsf{CB_0}$. We write $X \cong Y$, if X and Y are homeomorphic.

A space Y is called a *(continuous) retract* of a space X if there are continuous functions $s : Y \to X$ and $r : X \to Y$ such that composition rs coincides with the identity function id_Y on Y. Such a pair of functions (s, r) is called a *section-retraction* pair. Note that the section s is a homeomorphism between Y and the subspace $s(Y) = \{x \in X \mid sr(x) = x\}$ of X, and $s^{-1} = r|_{s(Y)}$.

Let ω be the space of non-negative integers with the discrete topology. Of course, the spaces $\omega \times \omega = \omega^2$, and $\omega \oplus \omega$ are homeomorphic to ω, the first homeomorphism is realized by the Cantor pairing function $\langle \cdot, \cdot \rangle$.

Let $\mathcal{N} = \omega^\omega$ be the set of all infinite sequences of natural numbers (i.e., of all functions $\xi \colon \omega \to \omega$). Let ω^* be the set of finite sequences of elements of ω, including the empty sequence. For $\sigma \in \omega^*$ and $\xi \in \mathcal{N}$, we write $\sigma \sqsubseteq \xi$ to denote that σ is an initial segment of the sequence ξ. By $\sigma\xi = \sigma \cdot \xi$ we denote the concatenation of σ and ξ, and by $\sigma \cdot \mathcal{N}$ the set of all extensions of σ in $\mathcal{N}$. For $x \in \mathcal{N}$, we can write $x = x(0)x(1) \ldots$ where $x(i) \in \omega$ for each $i < \omega$. For $x \in \mathcal{N}$ and $n < \omega$, let $x^{<n} = x(0) \ldots x(n-1)$ denote the initial segment of x of length n.

By endowing $\mathcal{N}$ with the product of the discrete topologies on ω, we obtain the so-called *Baire space*. The product topology coincides with the topology generated by the collection of sets of the form $\sigma \cdot \mathcal{N}$ for $\sigma \in \omega^*$. The Baire space is of primary importance for DST and CA. The importance stems from the fact that many countable objects are coded straightforwardly by elements of $\mathcal{N}$, and it has very specific topological properties. In particular, it is a perfect zero-dimensional space, and the spaces $\mathcal{N}^2, \mathcal{N}^\omega, \omega \times \mathcal{N} = \mathcal{N} \oplus \mathcal{N} \oplus \ldots$ (endowed with the product topology) are all homeomorphic to $\mathcal{N}$. Let $(x, y) \mapsto \langle x, y \rangle$ be a homeomorphism between $\mathcal{N}^2$ and $\mathcal{N}$. Let $(x_0, x_1, \ldots) \mapsto \langle x_0, x_1, \ldots \rangle$ be the homeomorphism between $\mathcal{N}^\omega$ and $\mathcal{N}$ defined by $\langle x_0, x_1, \ldots \rangle \langle m, n \rangle = x_m(n)$.

The space $P\omega$ is formed by the set of subsets of ω equipped with the Scott topology, the basic open sets of which are the sets $\{A \subseteq \omega \mid F \subseteq A\}$, where F ranges over the finite subsets of ω. It has the following well-known universality property:

Proposition 1. *A topological space X embeds into $P\omega$ iff X is a cb_0-space.*

Remember that a space X is *Polish*, if it is countably based and metrizable with a metric d such that (X, d) is a complete metric space. Important examples of Polish spaces are ω, $\mathcal{N}$, the space of reals $\mathbb{R}$ and its Cartesian powers $\mathbb{R}^n$ ($n <$ ω), the closed unit interval $[0, 1]$, the Hilbert cube $[0, 1]^\omega$ and the Hilbert space $\mathbb{R}^\omega$. Simple examples of non-Polish space are the *Sierpinski space* $\mathbb{S} = \{\bot, \top\}$, where the set $\{\top\}$ is open but not closed, and the space of rationals.

2.3   Admissible Representations and qcb$_0$-spaces

A *representation* of a space X is a surjection of a subspace of the Baire space $\mathcal{N}$ onto X. A representation δ of X is *admissible*, if it is continuous and any continuous function $\nu : Z \to X$ from a subspace Z of $\mathcal{N}$ to X is continuously reducible to δ, i.e. $\nu = \delta g$ for some continuous function $g : Z \to \mathcal{N}$. A topological space is *admissibly representable* if it has an admissible representation.

The notion of admissibility was introduced in [KW85] for representations of countably based spaces (in a different but equivalent formulation) and was extensively studied by many authors. In [Sch02, Sch03] the notion was extended to non-countably based spaces and a nice characterization of the admissibly represented spaces was achieved. Namely, the admissibly represented sequential topological spaces coincide with the qcb$_0$-spaces. Spaces which arise as topological quotients of countably based spaces are called *qcb-spaces*, and qcb-spaces that have the T_0-property are called *qcb$_0$-spaces*.

The category QCB of qcb-spaces as objects and continuous functions as morphisms is known to be cartesian closed (cf. [ELS04, Sch03]). The same is true for its full subcategory QCB$_0$ of qcb$_0$-spaces. The exponential Y^X to qcb-spaces X, Y has the set of continuous functions from X to Y as the underlying set, and its topology is the sequentialization of the compact-open topology on Y^X. By the *sequentialization* of a topology τ we mean the family of all sequentially open sets pertaining to this topology. (Remember that *sequentially open* sets are defined to be the complements of the sets that are closed under forming limits of converging sequences.) The sequentialization of τ is finer than or equal to τ. The topology of the QCB-product to X and Y, which we denote by $X \times Y$, is the sequentialization of the well-known Tychonoff topology on the cartesian product of the underlying sets of X and Y. So products and exponentials in QCB and in QCB$_0$ are formed in the same way as in its supercategory Seq of sequential topological spaces. The category QCB$_0$ is closed under many other constructions. We will discuss them in Section 4.

Given admissible representations δ_i for QCB$_0$-spaces X_i, there are canonical admissible representations for the binary product $X_1 \times X_2$, the countable product $\prod_i X_i$ and the function space $X_2^{X_1}$ (see [Sch03, We00]). We denote them by $[\delta_1 \times \delta_2]$, $[\delta_0, \delta_1, \dots]$ and $[\delta_1 \to \delta_2]$, respectively. The function space representation is constructed using the following well-known fact (see e.g. [Sch03]).

Proposition 2. *There is a partial continuous function $u :\subseteq \mathcal{N}^2 \to \mathcal{N}$ such that $dom(u) \in \mathbf{\Pi}_2^0(\mathcal{N}^2)$, for any partial continuous function g on $\mathcal{N}$ there is some $p \in \mathcal{N}$ such that $u_p := \lambda x.u(p, x)$ is an extension of g, and for any partial continuous function $G :\subseteq \mathcal{N} \times \mathcal{N} \to \mathcal{N}$ there is a total continuous function g on $\mathcal{N}$ such that $u(g(p), q) = F(p, q)$ for all $(p, q) \in dom(G)$.*

3 Hyperprojective Hierarchy of Sets

Here we recall some facts on hierarchies in arbitrary spaces, with the emphasis on the hyperprojective hierarchy in the Baire space. Additional information on the hyperprojective hierarchy may be found in [Ke83].

A *pointclass* on X is simply a collection $\Gamma(X)$ of subsets of X. A *family of pointclasses* [Se13] is a family $\Gamma = \{\Gamma(X)\}$ indexed by arbitrary topological spaces X such that each $\Gamma(X)$ is a pointclass on X and Γ is closed under continuous preimages, i.e. $f^{-1}(A) \in \Gamma(X)$ for every $A \in \Gamma(Y)$ and every continuous function $f\colon X \to Y$. A basic example of a family of pointclasses is given by the family $\mathcal{O} = \{\tau_X\}$ of the topologies of all the spaces X.

We will use some operations on families of pointclasses. First, the usual set-theoretic operations will be applied to the families of pointclasses pointwise: for example, the union $\bigcup_i \Gamma_i$ of the families of pointclasses $\Gamma_0, \Gamma_1, \ldots$ is defined by $(\bigcup_i \Gamma_i)(X) = \bigcup_i \Gamma_i(X)$.

Second, a large class of such operations is induced by the set-theoretic operations of L.V. Kantorovich and E.M. Livenson (see e.g. [Se13] for the general definition). Among them are the operation $\Gamma \mapsto \Gamma_\sigma$, where $\Gamma(X)_\sigma$ is the set of all countable unions of sets in $\Gamma(X)$, the operation $\Gamma \mapsto \Gamma_\delta$, where $\Gamma(X)_\delta$ is the set of all countable intersections of sets in $\Gamma(X)$, the operation $\Gamma \mapsto \Gamma_c$, where $\Gamma(X)_c$ is the set of all complements of sets in $\Gamma(X)$, the operation $\Gamma \mapsto \Gamma_d$, where $\Gamma(X)_d$ is the set of all differences of sets in $\Gamma(X)$, the operation $\Gamma \mapsto \Gamma_\exists$ defined by $\Gamma_\exists(X) := \{\exists^\mathcal{N}(A) \mid A \in \Gamma(\mathcal{N} \times X)\}$, where $\exists^\mathcal{N}(A) := \{x \in X \mid \exists p \in \mathcal{N}.(p,x) \in A\}$ is the projection of $A \subseteq \mathcal{N} \times X$ along the axis $\mathcal{N}$, and finally the operation $\Gamma \mapsto \Gamma_\forall$ defined by $\Gamma_\forall(X) := \{\forall^\mathcal{N}(A) \mid A \in \Gamma(\mathcal{N} \times X)\}$, where $\forall^\mathcal{N}(A) := \{x \in X \mid \forall p \in \mathcal{N}.(p,x) \in A\}$.

The operations on families of pointclasses enable to provide short uniform descriptions of the classical hierarchies in arbitrary spaces. E.g., the Borel hierarchy is the family of pointclasses $\{\Sigma_\alpha^0\}_{\alpha < \omega_1}$ defined by induction on α as follows [Se06, Br13]: $\Sigma_0^0(X) := \{\emptyset\}$, $\Sigma_1^0 := \mathcal{O}$, $\Sigma_2^0 := (\Sigma_1^0)_{d\sigma}$, and $\Sigma_\alpha^0(X) := (\bigcup_{\beta < \alpha} \Sigma_\beta^0(X))_{c\sigma}$ for $\alpha > 2$. The sequence $\{\Sigma_\alpha^0(X)\}_{\alpha < \omega_1}$ is called *the Borel hierarchy* in X. We also let $\Pi_\beta^0(X) := (\Sigma_\beta^0(X))_c$ and $\Delta_\alpha^0(X) := \Sigma_\alpha^0(X) \cap \Pi_\alpha^0(X)$. The classes $\Sigma_\alpha^0(X), \Pi_\alpha^0(X), \Delta_\alpha^0(X)$ are called the *levels* of the Borel hierarchy in X.

For this paper, the hyperprojective hierarchy is of main interest.

Definition 1. The hyperprojective hierarchy is the family $\{\Sigma_\alpha^1\}_{\alpha < \omega_1}$ of pointclasses defined by induction on α as follows: $\Sigma_0^1 = \Sigma_2^0$, $\Sigma_{\alpha+1}^1 = (\Sigma_\alpha^1)_{c\exists}$, $\Sigma_\lambda^1 = (\Sigma_{<\lambda}^1)_{\delta\exists}$, where $\alpha, \lambda < \omega_1$, λ is a limit ordinal, and $\Sigma_{<\lambda}^1(X) := \bigcup_{\alpha < \lambda} \Sigma_\alpha^1(X)$.

In this way, we obtain for any topological space X the ω_1-sequence $\{\Sigma_\alpha^1(X)\}$, which is called here *the hyperprojective hierarchy in X*. The pointclasses $\Sigma_\alpha^1(X)$, $\Pi_\alpha^1(X) := (\Sigma_\alpha^1(X))_c$ and $\Delta_\alpha^1(X) := \Sigma_\alpha^1(X) \cap \Pi_\alpha^1(X)$ are called *levels of the hyperprojective hierarchy in X*. The finite non-zero levels of the hyperprojective hierarchy coincide with the corresponding levels of the Luzin's projective hierarchy [Br13, ScS13]. The class of *hyperprojective sets* in X is defined as the union of all levels of the hyperprojective hierarchy in X.

Note that if X is Polish then, as it is well known, we can take $\Sigma_0^1 = \Sigma_1^0$ in the definition of the hyperprojective hierarchy and obtain the same non-zero levels as above. In this case our "hyperprojective hierarchy" is an initial segment of the hyperprojective hierarchy from [Ke83]. For non-Polish spaces our definition guarantees the "right" inclusions of the levels, as the first item of the next assertion states. The assertion collects also some other properties of the hyperprojective hierarchy which are proved just in the same way as for the classical projective hierarchy [Ke95].

Proposition 3. (1) *For any* $\alpha < \beta < \omega_1$, $\Sigma_\alpha^1 \cup \Pi_\alpha^1 \subseteq \Delta_\beta^1$.

(2) *For any limit countable ordinal* λ, $\Sigma_{<\lambda}^1 = \Pi_{<\lambda}^1$.

(3) *For any non-zero* $\alpha < \omega_1$, $\Sigma_\alpha^1 = (\Sigma_\alpha^1)_\sigma = (\Sigma_\alpha^1)_\delta = (\Sigma_\alpha^1)_\exists$. *In particular, the class* $\Sigma_\alpha^1(\mathcal{N})$ *is closed under countable unions, countable intersections, continuous images, and continuous preimages of functions with a* Π_2^0-*domain.*

(4) *For any non-zero* $\alpha < \omega_1$, $\Pi_\alpha^1 = (\Pi_\alpha^1)_\sigma = (\Pi_\alpha^1)_\delta = (\Pi_\alpha^1)_\forall$. *In particular, the class* $\Pi_\alpha^1(\mathcal{N})$ *is closed under countable unions and countable intersections, and continuous preimages of functions with a* Π_2^0-*domain.*

(5) *For any uncountable Polish space (and also for any uncountable quasi-Polish space [Br13]) X, the hyperprojective hierarchy in X does not collapse, i.e.* $\Sigma_\alpha^1(X) \not\subseteq \Pi_\alpha^1(X)$ *for each* $\alpha < \omega_1$.

4   Hyperprojective Hierarchy of qcb$_0$-spaces

Here we discuss the hyperprojective hierarchy of qcb$_0$-spaces. In particular we extend all results from [ScS13] concerning the Luzin's projective hierarchy of qcb$_0$-spaces.

For any representation δ of a space X, let $EQ(\delta) := \{(p,q) \in \mathcal{N}^2 \mid p,q \in dom(\delta) \wedge \delta(p) = \delta(q)\}$. Let Γ be a family of pointclasses. A topological space X is called Γ-*representable*, if X has an admissible representation δ with $EQ(\delta) \in \Gamma(\mathcal{N}^2)$. The class of all Γ-representable spaces is denoted $QCB_0(\Gamma)$. This notion from [ScS13] enables to transfer hierarchies of sets to the corresponding hierarchies of qcb$_0$-spaces. In particular, we arrive at the following definition.

Definition 2. The sequence $\{QCB_0(\Sigma_\alpha^1)\}_{\alpha < \omega_1}$ is called the *hyperprojective hierarchy* of qcb$_0$-spaces. By *levels* of this hierarchy we mean the classes $QCB_0(\Sigma_\alpha^1)$ as well as the classes $QCB_0(\Pi_\alpha^1)$ and $QCB_0(\Delta_\alpha^1)$.

The next assertion summarizes extensions of the corresponding results from [ScS13] about the Luzin hierarchy. They are proved just in the same way as in [ScS13].

Proposition 4. (1) *Let* $\Gamma \in \{\Sigma_\alpha^1, \Pi_\alpha^1 \mid 0 \leq \alpha < \omega_1\}$ *and let X be a Hausdorff space. Then X is Γ-representable, if X has an admissible representation δ with* $dom(\delta) \in \Gamma(\mathcal{N})$.

(2) *Let* $\Gamma \in \{\Sigma_\alpha^1, \Pi_\alpha^1 \mid 0 \leq \alpha < \omega_1\}$. *Then any continuous retract of a Γ-representable space is a Γ-representable space.*

(3) *The hyperprojective hierarchy of qcb$_0$-spaces does not collapse, more precisely,* $\mathsf{QCB}_0(\Sigma^1_\alpha) \not\subseteq \mathsf{QCB}_0(\Pi^1_\alpha)$ *for each* $\alpha < \omega_1$.

(4) *For any* $\Gamma \in \{\Sigma^1_\alpha, \Pi^1_\alpha \mid 1 \le \alpha < \omega_1\}$, *we have* $\mathsf{QCB}_0(\Gamma) \cap \mathsf{CB}_0 = \mathsf{CB}_0(\Gamma)$, *where* $\mathsf{CB}_0(\Gamma)$ *is the class of spaces homeomorphic to a* Γ-*subspace of* $P\omega$.

Now we establish some closure properties of the hyperprojective hierarchy of qcb$_0$-spaces. We begin with exponentiation in QCB_0. The next proposition extends and improves Theorem 7.1 in [ScS13]:

Proposition 5. *Let* $1 \le \alpha < \omega_1$, $X \in \mathsf{QCB}_0(\Sigma^1_\alpha)$ *and* $Y \in \mathsf{QCB}_0(\Pi^1_\alpha)$. *Then* $Y^X \in \mathsf{QCB}_0(\Pi^1_\alpha)$.

The next proposition provides some complexity bounds on products and co-products formed in QCB_0.

Proposition 6. (1) *Any non-zero level of the hyperprojective hierarchy of qcb$_0$-spaces is closed under countable* QCB_0-*products and coproducts.*

(2) *Let* λ *be a countable limit ordinal. Let* $\{X_k\}_{k<\omega}$ *be a sequence of qcb$_0$-spaces such that* $X_k \in \mathsf{QCB}_0(\Sigma^1_{<\lambda})$ *for all* k. *Then the* QCB_0-*product* $\prod_{k<\omega} X_k$ *is in* $\mathsf{QCB}_0((\Sigma^1_{<\lambda})_\delta)$ *and the co-product* $\bigoplus_{k<\omega} X_k$ *is in* $\mathsf{QCB}_0((\Sigma^1_{<\lambda})_\sigma)$.

For equalizers we have the following result.

Proposition 7. *Let* α *be a non-zero countable ordinal. Then* $\mathsf{QCB}_0(\Sigma^1_\alpha)$ *and* $\mathsf{QCB}_0(\Pi^1_\alpha)$ *are closed under forming equalizers.*

Now we turn our attention to co-equalizers. Co-equalizers in QCB_0 are constructed by first forming a co-equalizer in the category QCB and then, if the resulting space is non-T_0, identifying points with the same neighbourhoods. Non-T_0 qcb-spaces do not have an admissible representation, but some of them have a quotient representation. This motivates the following definition generalizing the one from above. For a given family Γ of pointclasses, we say that a topological space X is Γ-*quotient-representable*, if X has a quotient representation δ such that $EQ(\delta) \in \Gamma(\mathcal{N}^2)$. We denote the class of Γ-quotient-representable spaces by $\mathsf{QTE}(\Gamma)$ and the class of Γ-quotient-representable T_0-spaces by $\mathsf{QTE}_0(\Gamma)$. Since any admissible representation of a sequential space is a quotient representation, we have $\mathsf{QCB}_0(\Gamma) \subseteq \mathsf{QTE}_0(\Gamma)$.

We study the (non-uniform) descriptive complexity of the Kolmogorov operator $\mathcal{T}_0$. It maps any T_0-space to itself and sends a non-T_0-space X to the quotient space induced by the equivalence relation $\equiv_X$ given by the specification order of X, i.e., $x \equiv_X x'$ iff x and x' have the same open neighbourhoods.

Proposition 8. *Let* α *be a non-zero countable ordinal. Then* $X \in \mathsf{QTE}(\Sigma^1_\alpha)$ *implies* $\mathcal{T}_0(X) \in \mathsf{QCB}_0(\Sigma^1_{\alpha+2})$. *Moreover,* $\mathsf{QCB}_0(\Sigma^1_\alpha) \subseteq \mathsf{QTE}_0(\Sigma^1_\alpha) \subseteq \mathsf{QCB}_0(\Sigma^1_{\alpha+2})$.

Now we can formulate our result about forming co-equalizers.

Proposition 9. *Let* λ *be a countable limit ordinal. Then* $\mathsf{QCB}_0(\Sigma^1_{<\lambda})$ *is closed under forming co-equalizers in* QCB_0.

We do not know whether $\mathsf{QCB}_0(\Sigma^1_\alpha)$ is closed under forming co-equalizers in QCB_0.

5   Continuous Functionals of Countable Types

Here we extend all results in [ScS13] about the continuous functionals of finite types to the continuous functionals of countable types defined as follows:

Definition 3. Using the function space construction of QCB$_0$, we define the sequence of qcb$_0$-spaces $\{\mathbb{N}\langle\alpha\rangle\}_{\alpha<\omega_1}$ by induction on countable ordinals α as follows:

$$\mathbb{N}\langle 0\rangle := \omega, \ \mathbb{N}\langle\alpha+1\rangle := \omega^{\mathbb{N}\langle\alpha\rangle} \text{ and } \mathbb{N}\langle\lambda\rangle := \prod_{\alpha<\lambda}\mathbb{N}\langle\alpha\rangle,$$

where ω denotes the space of natural numbers endowed with the discrete topology, $\alpha, \lambda < \omega_1$ and λ is a limit ordinal. We call $\mathbb{N}\langle\alpha\rangle$ *the space of continuous functionals of type* α *over* ω.

Obviously, for $k < \omega$ the space $\mathbb{N}\langle k\rangle$ coincides with the space of Kleene-Kreisel continuous functionals of type k extensively studied in the literature [No80, No81, No99], and $\mathbb{N}\langle 1\rangle$ coincides with the Baire space $\mathcal{N}$. For any finite $k \geq 2$, the sequential topology on $\mathbb{N}\langle k\rangle$ is strictly finer than the corresponding compact-open topology [Hy79]. Furthermore it is neither zero-dimensional nor regular [Sch09].

Any of the introduced spaces has a natural canonical admissible representation $\delta_\alpha : D_\alpha \to \mathbb{N}\langle\alpha\rangle$ induced by the standard constructions mentioned in Section 2.3, starting with the admissible representation $\delta_0 : \mathcal{N} \to \omega$ defined by $\delta_0(p) = p(0)$.

The next proposition collects some basic properties of the continuous functionals of countable types. It follows from the fact that QCB$_0$ is cartesian closed and closed under countable limits and countable colimits.

Proposition 10. (1) *For all* $\alpha < \omega_1$, *the space* $\mathbb{N}\langle\alpha\rangle$ *is homeomorphic to* $\mathbb{N}\langle\alpha\rangle\times$
$\mathbb{N}\langle\alpha\rangle$.
(2) *For all* $\alpha < \beta < \omega_1$, *the spaces* $\mathbb{N}\langle\alpha\rangle$, $\omega\times\mathbb{N}\langle\beta\rangle$ *and* $\mathcal{N}\times\mathbb{N}\langle\beta\rangle$ *are continuous retracts of* $\mathbb{N}\langle\beta\rangle$.
(3) *For all* $1 \leq \alpha < \omega_1$, *the space* $(\mathbb{N}\langle\alpha\rangle)^\omega$ *is a continuous retract of* $\mathbb{N}\langle\alpha\rangle$.

The next result is an extension of Theorem 7.6 in [ScS13]. It can be proved in a similar way as in [ScS13] by using Proposition 10.

Theorem 1. *Let* α *be a non-zero countable ordinal and* B *a non-empty subset of* $\mathcal{N}$. *Then* $B \in \Sigma^1_\alpha(\mathcal{N})$ *iff there is a continuous function* $f : \mathbb{N}\langle\alpha\rangle \to \mathcal{N}$ *with* $rng(f) = B$.

Finally, we relate the continuous functionals of countable types to the hyperprojective hierarchy of qcb$_0$-spaces (extending Theorem 7.7 of [ScS13]). The next result provides the exact estimation of the spaces of continuous functionals of countable types in the hyperprojective hierarchy of qcb$_0$-spaces. On the other hand, the result provides "natural" witnesses for the non-collapse property of this hierarchy.

Theorem 2. *For any* $0 < \alpha < \omega_1$, $\mathbb{N}\langle\alpha+1\rangle \in \text{QCB}_0(\mathbf{\Pi}^1_\alpha)\setminus\text{QCB}_0(\mathbf{\Sigma}^1_\alpha)$. *For any countable limit ordinal* λ, $\mathbb{N}\langle\lambda\rangle \in \text{QCB}_0((\mathbf{\Pi}^1_{<\lambda})_\delta)\setminus\text{QCB}_0((\mathbf{\Sigma}^1_{<\lambda})_\sigma)$.

6   Categories of Hyperprojective qcb$_0$-spaces

We call any space in $\bigcup_{\alpha<\omega_1} \mathrm{QCB}_0(\Sigma_\alpha^1)$ a *hyperprojective qcb$_0$-space*. We denote the full subcategory of the category QCB_0 consisting of all hyperprojective qcb$_0$-spaces by $\mathrm{QCB}_0(\mathbf{HP})$. By Section 4, it has excellent closure properties.

Theorem 3. *The category* $\mathrm{QCB}_0(\mathbf{HP})$ *of hyperprojective qcb$_0$-spaces is cartesian closed, countably complete and countably co-complete. Countable limits, countable colimits and function spaces are formed as in* QCB_0.

The next result provides a characterization of $\mathrm{QCB}_0(\mathbf{HP})$ that avoids explicit mention of the hyperprojective hierarchy. We thank Matthew de Brecht for pointing out this fact to us.

Theorem 4. *Let* C *be a subcategory of* QCB_0 *which contains the Sierpinski space as an object, is closed under homeomorphism, and is closed under* QCB_0-*exponentials, countable* QCB_0-*limits and countable* QCB_0-*colimits. Then* C *contains* $\mathrm{QCB}_0(\mathbf{HP})$.

Regarding cartesian closed subcategories of QCB_0 that contain the discrete space of natural numbers, we have the following minimality result for $\mathrm{QCB}_0(\mathbf{HP})$, showing that any level of the hierarchy is needed.

Proposition 11. *There is no full cartesian closed subcategory* C *of* QCB_0 *such that* C *inherits countable products from* QCB_0, *contains the discrete space* ω *of natural numbers and is contained itself in* $\mathrm{QCB}_0(\Sigma_\alpha^1)$ *for some* $\alpha < \omega_1$.

7   Conclusion

We have defined the category $\mathrm{QCB}_0(\mathbf{HP})$ of hyperprojective qcb$_0$-spaces and shown that it has excellent closure properties. The latter exhibit $\mathrm{QCB}_0(\mathbf{HP})$ as a nice category in which to study many spaces of interest for Computable Analysis.

References

[Br13]   de Brecht, M.: Quasi-Polish spaces. Annals of Pure and Applied Logic 164, 356–381 (2013)

[En89]   Engelking, R.: General Topology. Heldermann, Berlin (1989)

[ELS04]   Escardó, M., Lawson, J., Simpson, A.: Comparing Cartesian closed Categories of Core Compactly Generated Spaces. Topology and its Applications 143, 105–145 (2004)

[GH80]   Giertz, G., Hoffmann, K.H., Keimel, K., Lawson, J.D., Mislove, M.W., Scott, D.S.: A compendium of Continuous Lattices. Springer, Berlin (1980)

[Hy79]   Hyland, J.M.L.: Filter spaces and continuous functionals. Annals of Mathematical Logic 16, 101–143 (1979)

[Ke83]   Kechris, A.S.: Suslin cardinals, k-Suslin sets and the scale property in the hyperprojective hierarchy. In: Kechris, A.S., Löwe, B., Steel, J.R. (eds.) The Cabal Seminar, v. 1: Games, Scales and Suslin Cardinals. Lecture Notes in Logic, vol. 31, pp. 314–332 (2008); (reprinted from Lecture Notes in Mathematica, No 1019. Springer, Berlin 1983)

[Ke95]   Kechris, A.S.: Classical Descriptive Set Theory. Springer, New York (1995)

[Kl59]   Kleene, S.C.: Countable functionals. In: Heyting, A. (ed.) Constructivity in Mathematics, pp. 87–100. North Holland, Amsterdam (1959)

[Kr59]   Kreisel, G.: Interpretation of analysis by means of constructive functionals of finite types. In: Heyting, A. (ed.) Constructivity in Mathematics, pp. 101–128. North Holland, Amsterdam (1959)

[KW85]   Kreitz, C., Weihrauch, K.: Theory of representations. Theoretical Computer Science 38, 35–53 (1985)

[No80]   Normann, D.: Recursion on the Countable Functionals. Lecture Notes in Mathematics, vol. 811. Springer, Heidelberg (1980)

[No81]   Normann, D.: Countable functionals and the projective hierarchy. Journal of Symbolic Logic 46(2), 209–215 (1981)

[No99]   Normann, D.: The continuous functionals. In: Griffor, E.R. (ed.) Handbook of Computability Theory, pp. 251–275. Elsevier, Amsterdam (1999)

[Sch02]   Schröder, M.: Extended admissibility. Theoretical Computer Science 284, 519–538 (2002)

[Sch03]   Schröder, M.: Admissible representations for continuous computations. PhD thesis, Fachbereich Informatik, FernUniversität Hagen (2003)

[Sch09]   Schröder, M.: The sequential topology on $\mathbb{N}^{\mathbb{N}^{\mathbb{N}}}$ is not regular. Mathematical Structures in Computer Science 19, 943–957 (2009)

[ScS13]   Schröder, M., Selivanov, V.: Some Hierarchies of qcb₀-Spaces. Mathematical Structures in Computer Science, arXiv:1304.1647 (to appear)

[Se06]   Selivanov, V.L.: Towards a descriptive set theory for domain-like structures. Theoretical Computer Science 365, 258–282 (2006)

[Se13]   Selivanov, V.L.: Total representations. Logical Methods in Computer Science 9(2), 1–30 (2013), doi:10.2168/LMCS-9(2:5)2013

[We00]   Weihrauch, K.: Computable Analysis. Springer, Berlin (2000)

Online Bin Packing:
Old Algorithms and New Results

Jiří Sgall

Computer Science Institute of Charles University,
Faculty of Mathematics and Physics, Praha, Czech Republic
sgall@iuuk.mff.cuni.cz

Abstract. In the bin packing problem we are given an instance consisting of a sequence of items with sizes between 0 and 1. The objective is to pack these items into the smallest possible number of bins of unit size. FIRSTFIT and BESTFIT algorithms are simple online algorithms introduced in early seventies, when it was also shown that their asymptotic approximation ratio is equal to 1.7. We present a simple proof of this bound and survey recent developments that lead to the proof that also the absolute approximation ratio of these algorithms is exactly 1.7. More precisely, if the optimum needs OPT bins, the algorithms use at most $\lfloor 1.7 \cdot \text{OPT} \rfloor$ bins and for each value of OPT, there are instances that actually need so many bins. We also discuss bounded-space bin packing, where the online algorithm is allowed to keep only a fixed number of bins open for future items. In this model, a variant of BESTFIT also has asymptotic approximation ratio 1.7, although it is possible that the bound is significantly smaller if also the offline solution is required to satisfy the bounded-space restriction.

1 Introduction

Johnson's thesis [13] on bin packing together with Graham's work on scheduling [11,12] belong to the early influential works that started and formed the whole area of approximation algorithms. The proof that the asymptotic approximation ratio of FIRSTFIT and BESTFIT bin packing is 1.7 given by Ullman [20] and subsequent works by Garey et al. and Johnson et al. [10,15] were among these first results on approximation algorithms.

We survey this area with emphasis on the new results, approaches and open problems.

1.1 The Algorithms

Bin packing is a classical combinatorial optimization problem in which we are given an instance consisting of a sequence of items with rational sizes between 0 and 1, and the goal is to pack these items into the smallest possible number of bins of unit size. Since bin packing is NP-hard, one particularly active branch of

A. Beckmann, E. Csuhaj-Varjú, and K. Meer (Eds.): CiE 2014, LNCS 8493, pp. 362–372, 2014.

research has concentrated on approximation algorithms that find near-optimal packings.

The main topic of this survey are BESTFIT, FIRSTFIT and related algorithms. BESTFIT algorithm packs each item into the most full bin where it fits, possibly opening a new bin if the item does not fit into any currently open bin. FIRSTFIT algorithm packs each item into the first bin where it fits, again opening a new bin only if the item does not fit into any currently open bin.

Both BESTFIT and FIRSTFIT as well as any algorithm that we will consider belong to a wide class of algorithms called ANYFIT algorithms. An algorithm is called an ANYFIT algorithm if it opens a new bin only if the item being packed does not fit into any previously open bin. Otherwise the item is allowed to be packed into any bin where it fits.

A relevant subclass are ALMOSTANYFIT algorithms (AAF algorithms for short). An algorithm is an AAF algorithm if it is an ANYFIT algorithm and additionally, if an item fits into more than one of the open bins, it is not packed into the smallest open bin (breaking ties arbitrarily).

The worst-case performance of an approximation algorithm A is measured by comparing it against the optimal packing. Let OPT denote the number of bins used in an optimal packing for instance I, and let A denote the number of bins used by algorithm A for the same instance. The algorithm A is an **absolute R-approximation** algorithm if for every instance I

$$A \leq R \cdot \text{OPT},$$

and it is an **asymptotic R-approximation** algorithm if for every instance I

$$A \leq R \cdot \text{OPT} + f(\text{OPT}),$$

where $f(n) \in o(n)$; typically, and in all cases relevant for us, $f(n)$ is a global constant.

As far as the worst-case ratios are concerned, FIRSTFIT can be considered as a special case of BESTFIT or any other ANYFIT algorithm: The items in any instance can be reordered so that they arrive in the order of bins in the FIRSTFIT packing. This changes neither FIRSTFIT, nor the optimal packing. Thus it is sufficient to analyze FIRSTFIT on such instances. On the other hand, on them any ANYFIT algorithm behaves exactly as FIRSTFIT, as there is always a single bin where the new item fits. Thus any lower bound for FIRSTFIT applies immediately to an arbitrary ANYFIT algorithms and any upper bound for FIRSTFIT is equivalent to a bound for this very restricted subset of instances and BESTFIT or another ANYFIT algorithm.

Both FIRSTFIT and BESTFIT are **online** algorithms. An online bin packing algorithm places each item without knowledge of the future items, that is based only on the size of this item and the already packed items, and cannot change the packing of the items in the future.

In the context of online algorithms, a very interesting concept is that of bounded-space algorithms. We distinguish the bins to be either open or closed and items are allowed to be packed only into open bins. When a new bin is started, we set the status of the bin to be open. At some point, the algorithm may decide to close some bin(s). This means that no future items can be packed there; once a bin is closed, it cannot be reopened. A bin packing algorithm is a **k-bounded-space** algorithm, if the number of open bins is at most k at all times. That is, if there are k open bins and an item is packed into a new bin, one of the open bins needs to be closed.

The natural bounded-space variant of BESTFIT is BBF_k, which packs an item into the most full open bin where it fits. If the item does not fit into any of them, a new bin is opened; if already k bins are open, the most full bin is closed. Note that BBF_k is not an ANYFIT algorithm. It may happen that an item fits in one of the already closed bins but not into any open one.

1.2   The Old Results

The upper bound on BESTFIT (and FIRSTFIT) was first shown by Ullman in 1971 [20]; he proved that for any instance, $BF, FF \leq 1.7 \cdot OPT + 3$, where BF, FF and OPT denote the number of bins used by BESTFIT, FIRSTFIT and the optimum, respectively. Still in seventies, the additive term was improved first in [10] to 2 and then in [9] to $BF \leq \lceil 1.7 \cdot OPT \rceil$; due to integrality of BF and OPT this is equivalent to $BF \leq 1.7 \cdot OPT + 0.9$. These bounds actually hold for any AAF algorithms, see [13,14].

For the lower bound, the early works give examples both for the asymptotic and absolute ratios. The example for the asymptotic bound gives $FF = 17k$ whenever $OPT = 10k + 1$, thus it shows that the asymptotic upper bound of 1.7 is tight, see [20,10,15]. For the absolute ratio, an example is given with $FF = 17$ and $OPT = 10$, i.e., an instance with approximation ratio exactly 1.7 [10,15], but no such example was known for large OPT. (Also an example with $FF = 34$ and $OPT = 20$ is claimed and it can be constructed in a similar manner, but it seems that this example has never been published.)

Johnson [13,15] has also analyzed the Best Fit Decreasing and First Fit Decreasing algorithms, which behave like BESTFIT and FIRSTFIT but receive the items on the input sorted from the largest one to the smallest, and proved that the asymptotic approximation ratio is equal to 11/9. Johnson's bound had an additive constant of 4; this was improved several times.

Turning to the bounded-space algorithms, a central result of Lee and Lee [16] designs k-bounded-space online bin packing algorithms whose asymptotic ratios come arbitrarily close to the magic harmonic number $h_\infty \approx 1.69103$, as the space bound k tends to infinity. They also show that every bounded-space online bin packing algorithm A satisfies $R^\infty(A) \geq h_\infty$

Csirik and Johnson [5] show that the k-space-bounded Best Fit algorithm BBF_k has the asymptotic worst case ratio 1.7 for any $k \geq 2$. (The case of $k = 1$ is trivial, 2-approximation is possible and no better algorithm exists.) Among all 2-space-bounded online algorithms in the bin packing literature, this online algorithm is the champion with respect to worst-case ratios.

We have mentioned only directly relevant work. Of course, there is much more work on bin packing, in particular there exist asymptotic approximation schemes for this problem, as well as many other algorithms. We refer to the surveys [3,4] or to the recent excellent book [21].

1.3   The New Results

The absolute approximation ratio of FIRSTFIT and BESTFIT was proven to be at most 1.75 by Simchi-Levy [19]. This started a renewed interest in these algorithm and their absolute ratio and the closely related question of decreasing the additive constant in the approximation ratio. A natural approach to improve absolute upper bounds is to study fixed small values of OPT and to exclude the possibility of a higher absolute ratio for them. Indeed, solving a few such cases necessarily improves upper bounds on the absolute ratio—but cannot give a tight result. Of course, this is still far from trivial: Even for a fixed OPT, each such problem seems to lead to a new and more extensive case analysis.

Such analysis lead to a sequence of improvements for FIRSTFIT: Xia and Tan [22] showed that $FF \leq 1.7 \cdot OPT + 0.7$ and that the absolute ratio of FIRSTFIT is at most $12/7 \approx 1.7143$; the second bound was given independently by Boyar et al. [1] and later improved to $101/59 \approx 1.7119$ by Németh [17].

In [7,8], Dósa and the author have shown that $FF, BF \leq 1.7 \cdot OPT$ and that this bound is the best possible for every single value of OPT. Thus we have a complete understanding of the worst-case behavior of these algorithm. In particular, we know that the asymptotic and absolute bounds coincide and no additive term is needed. These results are based on new insights that also make possible a simple proof and a slight generalization and strengthening of the asymptotic bound for AAF algorithms.

Turning to other online algorithms, these new techniques can be also applied to BBF_k and simplify the proof that BBF_k is asymptotic 1.7-approximation, even though BBF_k is not ANYFIT algorithm [18].

It is easy to observe that no online algorithm has an absolute ratio less than $5/3 \approx 1.666$. Recently, again using the new techniques, we (G. Dósa, R. van Stee and the author) have been able to find a matching algorithm (article in preparation).

For First Fit Decreasing, it was shown that the additive constant is exactly $2/3$ [6]. That is, $\frac{11}{9}OPT + \frac{2}{3}$ bins are sufficient for First Fit Decreasing, but this number of bins is actually also necessary for some instances for infinitely many values of OPT. Thus for First Fit Decreasing, the asymptotic and absolute approximation ratios are not equal. In fact, the results of [6] give the exact value of the worst case for every value of OPT. In light of this result, it is rather surprising that for BESTFIT and FIRSTFIT the asymptotic and absolute approximation ratios are equal and no additive term is needed.

1.4   This Survey

We focus on the results connected to the recent optimal analysis of FIRSTFIT and BESTFIT. We start in Section 2 by sketching the lower bound, as it gives a good intuition for the choices made in the remaining parts of the paper.

In the Section 3 we give a simple proof of the asymptotic upper bound of 1.7 for AAF algorithms, in fact, even for a slightly more generalized class of RAAF algorithms (Relaxed AAF algorithms). This proof uses the same insights as the optimal analysis of FIRSTFIT and BESTFIT, namely a combination of weight functions and amortized analysis. In addition, to demonstrate that the equality of the absolute and asymptotic bounds for FIRSTFIT and BESTFIT is by far not automatic, we show a RAAF packing with an absolute ratio strictly larger than 1.7.

Finally, in Section 4 we discuss a new measure for bounded space algorithm introduced in [2].

2   The Lower Bound

Suppose for a moment that the algorithm is not allowed to pack a bin with item sizes summing to exactly 1, while the optimum can. This model gives an unfair advantage to the optimum, but we hope that lower bound examples in this model can be later modified by changing the item sizes by very small amount to provide a correct lower bound. This is a bold assumption that is not quite true, however, this simplified model provides a very good high-level intuition in a number of different bin packing scenarios.

In the simplified model, the lower bound instance for $\text{OPT} = 10m$ is this: Start by OPT items of size $1/6$, continue by OPT items of size $1/3$, and conclude by OPT items of size $1/2$. The optimum uses $\text{OPT} = 10m$ bins by packing three items of different sizes $1/6 + 1/3 + 1/2 = 1$ in each bin. FIRSTFIT uses $2m$ bins with 5 items of size $1/6$, $5m$ bins with 2 items of size $1/3$ and $10m$ bins with a single item of size $1/2$. Total of $17m$ bins, giving a lower bound of 1.7.

To modify the instance to be valid in the real model, we modify the item sizes. First, for all items $1/2$ the size is changed into $1/2 + \varepsilon$ for a tiny $\varepsilon > 0$. The sizes of items $1/3$ is changed into carefully chosen $1/3 + \delta_i$ and $1/3 - \delta_i$, for δ_i tiny and exponentially decreasing to 0. This guarantees that the previous packing works if we arrange the sequence so that FIRSTFIT packs $1/3 + \delta_i$ and $1/3 - \delta_{i-1}$ together. Now, to allow the optimal packing, the remaining items have size $1/6 + \delta_i - \varepsilon$ and $1/6 - \delta_i - \varepsilon$. We need to order them so that the FIRSTFIT packing works as we wish. Here we come to a point where the simplified model fails. We modify the FIRSTFIT packing so that the first bin contains 6 items instead of 5 and the last bin of the first phase contains 4 items instead of 5. Of course, this needs to be carefully checked, but the main idea why the construction works is that the exponential decrease of δ_i guarantees that only the item with the largest δ_i in a bin is relevant for its final size. Also, we need a proof for other residue classes of OPT, but the case of $\text{OPT} = 10m$ is tightest and easily modified to other $\text{OPT} = 10m + i$. We obtain

Theorem 2.1. *For all values of* OPT, *there exists an instance* I *with such that* $FF = \lfloor 1.7 \cdot \text{OPT} \rfloor$.

Since FIRSTFIT is equivalent to an arbitrary ANYFIT algorithm on a subset of instances, the lower bound holds for a general ANYFIT algorithm.

3   The Simplified and Generalized Asymptotic Upper Bound

We now present the simple proof of the asymptotic ratio 1.7. This proof holds for a wide class of any-fit-type algorithms: Call an ANYFIT algorithm a RAAF algorithm, if it uses the bin with level at most $1/2$ only when the item does not fit into any previous bin. (It is easy to verify that there is always at most one such bin, see Lemma 3.1(i).) The RAAF condition say that the AAF condition holds whenever the smallest bin has level at most $1/2$; thus any AAF algorithm is a RAAF algorithm.

The asymptotic bound for AAF algorithms was proved in [13,14]. Theorem 3.3 improves the additive term and generalizes the bound to the slightly less restrictive RAAF condition (although it seems that the original proof also uses only the RAAF condition).

Let us fix an instance I with items $a_1, \ldots, a_n$ and denote the number of bins in the BESTFIT and optimal solutions by BF and OPT, respectively. We will often identify an item and its size. For a set of items A, let $s(A) = \sum_{a \in A} a$, i.e., the total size of items in A and also for a set of bins $\mathcal{A}$, let $s(\mathcal{A}) = \sum_{A \in \mathcal{A}} s(A)$. Furthermore, let $S = s(I)$ be the total size of all items of I. Obviously $S \leq$ OPT.

We classify the items by their sizes: items $a \leq 1/6$ are **small**, items $a \in (1/6, 1/2]$ are **medium**, and items $a > 1/2$ are **huge**.

The bins in the BF packing are ordered by the time they are opened (i.e., when the first item is packed into them). Expressions like "before", "after", "first bin", "last bin" refer to this ordering. At any time during the packing, the **level of a bin** is the total size of items currently packed in it, while by **size of a bin** we always mean its final level.

Lemma 3.1. *At any moment, in any* RAAF *packing the following holds:*

 (i) *The sum of levels of any two bins is greater than 1. In particular, there is at most one bin with level at most $1/2$.*

 (ii) *Any item a packed into a bin with level at most $1/2$ (i.e., a new bin or the single bin with level at most $1/2$ guaranteed by (i)) does not fit into any bin open at the time of its arrival, except for the bin where the item a is packed.*

 (iii) *If there are two bins B, B' with level at most $2/3$, in this order, then either B' contains a single item or the first item in B' is huge.*

Proof. **(i):** The first item in any bin does not fit in any open bin by the definition of RAAF (in fact, the ANYFIT condition is sufficient here), thus the sum of the levels of the two bins is greater than 1 already at the time when the second bin is opened.

(ii): If a is packed into a new bin, this follows by the RAAF (or ANYFIT) condition again. Otherwise Let $x \leq 1/2$ be the level of the bin where a is packed, just before a is packed there. By (i), there is at most one bin with level at most $1/2$, thus at the time of packing of a all the other bins have level strictly greater than x. By the definition of RAAF, a does not fit into any of these bins.

(iii): If B' contains two items and the first one is not huge, then by (ii) the first two items in B' do not fit into B. Thus they are larger than $1/3$ and the level of B' is greater than $2/3$.   □

The short proof of the asymptotic ratio 1.7 for RAAF algorithms uses the same weight function as the traditional analysis of BESTFIT and FIRSTFIT. (In some variants the weight of an item is capped to be at most 1, which makes almost no difference in the analysis.)

The weight function assigns a weight to each item, depending on its size. Intuitively, the weight measures how much space is needed to pack each item. For small items, it should be proportional to its size, but large items may generate empty space in some of the bins and their weight is accordingly larger. The overall idea is to show that the weight of each optimal bin is at most 1.7 while the average weight of each algorithm's bin is at least 1. Looking at the lower bound example with this outline in mind, it is easy to see that we should set $w(0) = 0$, $w(1/6) = 0.2$, $w(1/3) = 0.5$, and $w(1/2 + \varepsilon) \geq 1$. The weight function we define below fits these values by a piecewise linear function, with a discontinuity at $1/2$.

To use amortization, we split the weight of each item a into two parts, namely its bonus $\overline{w}(a)$ and its scaled size $\overline{\overline{w}}(a)$, defined as

$$
\overline{w}(a) = \begin{cases}
0 & \text{if } a \leq \frac{1}{6}, \\
\frac{3}{5}(a - \frac{1}{6}) & \text{if } a \in \left(\frac{1}{6}, \frac{1}{3}\right), \\
0.1 & \text{if } a \in \left[\frac{1}{3}, \frac{1}{2}\right], \\
0.4 & \text{if } a > \frac{1}{2}.
\end{cases}
$$

For every item a we define $\overline{\overline{w}}(a) = \frac{6}{5}a$ and its weight is $w(a) = \overline{\overline{w}}(a) + \overline{w}(a)$. For a set of items B, $w(B) = \sum_{a \in B} w(a)$ denotes the total weight, similarly for $\overline{w}$ and $\overline{\overline{w}}$.

It is easy to observe that the weight of any bin B, i.e., of any set with $s(B) \leq 1$, is at most 1.7: The scaled size of B is at most 1.2, so we only need to check that $\overline{w}(B) \leq 0.5$. If B contains no huge item, there are at most 5 items with non-zero $\overline{w}(a)$ and $\overline{w}(a) \leq 0.1$ for each of them. Otherwise the huge item has bonus 0.4; there are at most two other medium items with non-zero bonus and it is easy to check that their total bonus is at most 0.1. This implies that the weight of the whole instance is at most $1.7 \cdot \text{OPT}$.

The key part is to show that, on average, the weight of each BF-bin is at least 1. Lemma 3.2 together with Lemma 3.1 implies that for almost all bins with two or more items, its scaled size plus the bonus of the **following** such bin is at least 1.

Lemma 3.2. *Let B be a bin such that $s(B) \geq 2/3$ and let c, c' be two items that do not fit into B, i.e., $c, c' > 1 - s(B)$. Then $\overline{\overline{w}}(B) + \overline{w}(c) + \overline{w}(c') \geq 1$.*

Proof. If $s(B) \geq 5/6$, then $\overline{\overline{w}}(B) \geq 1$ and we are done. Otherwise let $x = 5/6 - s(B)$. We have $0 < x \leq 1/6$ and thus $c, c' > 1/6 + x$ implies $\overline{w}(c), \overline{w}(c') > \frac{3}{5}x$. We get $\overline{\overline{w}}(B) + \overline{w}(c) + \overline{w}(c') > \frac{6}{5}(\frac{5}{6} - x) + \frac{3}{5}x + \frac{3}{5}x = 1$.   □

Any RAAF-bin (i.e., a bin of the RAAF algorithm) D with a huge item has $\overline{w}(D) \geq 0.4$ and $\frac{6}{5}s(D) > 0.6$, thus $w(D) > 1$.

For the amortization, consider all RAAF-bins B with two or more items, size $s(B) \geq 2/3$, and no huge item. For any such bin except for the last one choose C as the next bin with the same properties. Since C has no huge item, its first two items c, c' have level at most $1/2$ and by Lemma 3.1(ii) they do not fit into B. Lemma 3.2 implies $\overline{\overline{w}}(B) + \overline{w}(C) \geq \overline{\overline{w}}(B) + \overline{w}(c) + \overline{w}(c') \geq 1$.

Summing all these inequalities (note that each bin is used at most once as B and at most once as C) and $w(D) > 1$ for the bins with huge items we get $w(I) \geq BF - 3$. The additive constant 3 comes from the fact that we are missing an inequality for at most three BF-bins: the last one from the amortization sequence, possibly one bin B with two or more items and $s(B) < 2/3$ (cf. Lemma 3.1(iii)) and possibly one bin B with a single item and $s(B) < 1/2$ (cf. Lemma 3.1(i)). Combining this with the previous bound on the total weight, we obtain $RAAF - 3 \leq w(I) \leq 1.7 \cdot \text{OPT}$, where RAAF denotes the number of bins of the RAAF algorithm, and the asymptotic bound follows.

This simple proof of the asymptotic ratio can be tightened so that the additive constant is smaller. We save one of the three bins by noticing that we do not need to do amortization for bins that are after the bin of size smaller than $2/3$. The remaining two bins have total size larger than 1, which brings the additive constant further down to 0.7.

Theorem 3.3. *For any* RAAF *algorithm and any instance of bin packing we have* $GAAF \leq \lfloor 1.7 \cdot \text{OPT} + 0.7 \rfloor$.

Proof. Any RAAF-bin D with a huge item has $\overline{w}(D) \geq 0.4$ and $\frac{6}{5}s(D) > 0.6$, thus $w(D) > 1$. Similarly, any RAAF-bin with two items larger than $1/3$ has $\overline{w}(D) \geq 0.2$ and $\frac{6}{5}s(D) > 0.8$, thus $w(D) > 1$.

For the amortization, consider all the RAAF-bins B, with (i) two or more items, (ii) no huge item, and (iii) no pair of items both larger than $1/3$. For any such bin except for the last one choose C as the next bin with the same properties. Since C has no huge item, its first two items c, c' have level at most $1/2$ and by the RAAF condition they do not fit into B. Since C has no pair of items larger than $1/3$, we have $c \leq 1/3$ or $c' \leq 1/3$ and thus $s(B) > 2/3$. Lemma 3.2 now implies $\overline{\overline{w}}(B) + \overline{w}(C) \geq \overline{\overline{w}}(B) + \overline{w}(c) + \overline{w}(c') \geq 1$.

Let $\overline{C}$ be the last bin used in the amortization, if it exists, and $\overline{D}$ be the single bin with $s(\overline{D}) \leq 1/2$, if it exists.

If $\overline{C}$ and $\overline{D}$ both exist, we have $s(\overline{C}) + s(\overline{D}) > 1$ by Lemma 3.1(i) and thus $\overline{\overline{w}}(\overline{C}) + w(\overline{D}) > 1.2$. Summing this, all the amortization inequalities (note that each bin is used at most once as B or $\overline{C}$ and at most once as C) and $w(D) > 1$ for the bins with huge items or two items larger than $1/3$ we get $w(I) > RAAF - 0.8$. Combining this with the previous bound $w(I) \leq 1.7 \cdot \text{OPT}$ on the total weight, we obtain $RAAF < w(I) + 0.8 \leq 1.7 \cdot \text{OPT} + 0.8$ and the theorem follows from the integrality of RAAF and OPT.

If $\overline{C}$ exists but $\overline{D}$ does not, we have $s(\overline{C}) > 1/2$ and thus $\overline{\overline{w}}(\overline{C}) > 0.6$. Summing this and again both all the amortization inequalities and $w(D) > 1$ for the bins with huge items or two items larger than 1/3 we get $w(I) \geq A - 0.4$ and the theorem follows again.

If $\overline{C}$ does not exist but $\overline{D}$ does, let C be an arbitrary bin other than $\overline{D}$ (if none exists, RAAF $= 1 =$ OPT and the theorem is trivial). We have $s(C) + s(\overline{D}) > 1$ and thus $w(\overline{C}) + w(\overline{D}) > 1.2$. Summing this and $w(D) > 1$ for all the remaining bins, we get $w(I) > \text{RAAF} - 0.8$ and the theorem follows as above.

Finally, if neither $\overline{C}$ nor $\overline{D}$ exists, we have $w(D) > 1$ for all the RAAF-bins, thus $w(I) > A$ and the theorem follows as well. $\square$

Next we give an example showing that RAAF algorithms do not have an absolute approximation ratio 1.7. In particular, we give an instance with OPT $=$ 7 and RAAF packing with 12 bins.

We first describe the RAAF packing; the input sequence contains items in the order of the bins, i.e., it starts by all the items from the first bin, then continues by items from the second bin, etc. The first bin contains 6 items of size 0.12, total of 0.72. The next three bins contain each 2 items of size 0.34; note that these do not fit into any previous bin. The fifth bin has items 0.52 and 0.01; the item 0.01 fits into the previous bins, but it is packed at a level larger than 0.5, so this satisfies the RAAF condition. The sixth bin contains a single item of size 0.48 and the remaining six bins contain each an item of size 0.53; again, these items do not fit into any previous bin.

OPT contains a bin with two items of sizes 0.52 and 0.48. The remaining 6 bins contain each three items of sizes 0.53, 0.34, and 0.12, total of 0.99; in addition one of them contains also the item 0.01. This packs all the items in the 7 bins and completes the example.

Thus removing the additive constant completely is impossible for RAAF algorithms, and thus it needs to use additional properties of the algorithm. We are able to do this for FIRSTFIT and BESTFIT, using additional ideas [7,8]. It remains an interesting open problem whether this is possible for all AAF algorithms.

4   Bounded Space Algorithms

The lower bound constructions in [16] is a variant of the FIRSTFIT lower bound which we have presented above. In its first approximation, giving a lower bound of $5/3 \approx 1.666$, we give the algorithm the same number of items of sizes $1/7 + \varepsilon$, $1/3 + \varepsilon$, and $1/2 + \varepsilon$. The optimal packing uses almost full bins with one item of each size. The full construction repeatedly prepends this sequence by another group of same-size items, the size is set to $1/t + \varepsilon$ for the smallest integer so that this item fits into the optimal bin (for a tiny $\varepsilon > 0$). On any such sequence, **any** k-bounded-space algorithm uses bins grouping items of the same size, with a constant number of exceptions. This shows that every bounded-space algorithm has asymptotic worst case ratio at least $h_\infty \approx 1.69103$; the number h_∞ is defined from the sequence of reciprocals $1/2, 1/3, 1/7, 1/43, \ldots$, which we described above.

The optimal packing above can be achieved only by an algorithm which keeps many bins open even if the whole sequence is known in advance. Or, from another viewpoint, we are allowed the optimal algorithm to reorder sequence. It is natural to ask how much inefficiency is caused by not knowing the future, i.e., by the fact that the algorithm is online. For this purpose, we restrict the offline algorithms so that they can also have only k bins open. For a given instance I, let $\text{OPT}_k = \text{OPT}_k(I)$ be the smallest possible number of bins used in a packing produced by a k-bounded-space bin packing algorithm that know the whole instance I before it starts serving it. We say that the algorithm A it is an **asymptotic R-approximation** algorithm **w.r.t. the k-bounded-space optimum** if for every instance I

$$A \leq R \cdot \text{OPT}_k + f(\text{OPT}_k),$$

where $f(n) \in o(n)$; again, typically $f(n)$ is a global constant. The smallest such R is called the k-**bounded-space ratio** of algorithm A in [2], where it was introduced.

On the lower-bound instances described above, BBF_k is an asymptotic 1-approximation w.r.t. the k-bounded-space optimum, so there is no inefficiency caused by the online environment. It is not hard to construct instances showing that the 2-bounded-space ratio of BBF_2 is at least 1.5, see [2]. However, the best upper bound on the 2-bounded-space ratio of BBF_2 or any other bounded-space algorithm is 1.7 and closing this gap is an interesting question. Levin and Epstein (private communication) have shown that the 2-bounded-space ratio of any online algorithm is at least 1.5, which leaves open the possibility that BBF_2 is the best possible algorithm. The case of $k > 2$ is widely open.

One of the difficulties of studying the k-bounded-space ratio is that it seems non-trivial to understand OPT_k. Unlike OPT, for which we have an asymptotic approximation scheme, OPT_k is hard to approximate. In [2], it is shown that the 2-bounded-space ratio of any polynomial algorithm is at least $5/4$. There exists a $(3/2 + \varepsilon)$-approximation algorithm for OPT_k, based on a partial enumeration of the solutions [2]. We are missing a deeper insight in the structure of the solutions and also nothing is known for $k > 3$.

Acknowledgements. Partially supported by the Center of Excellence – Inst. for Theor. Comp. Sci., Prague (project P202/12/G061 of GA ČR).

References

1. Boyar, J., Dósa, G., Epstein, L.: On the absolute approximation ratio for First Fit and related results. Discrete Appl. Math. 160, 1914–1923 (2012)
2. Chrobak, M., Sgall, J., Woeginger, G.J.: Two-bounded-space bin packing revisited. In: Demetrescu, C., Halldórsson, M.M. (eds.) ESA 2011. LNCS, vol. 6942, pp. 263–274. Springer, Heidelberg (2011)
3. Coffman, E.G., Garey, M.R., Johnson, D.S.: Approximation algorithms for bin packing: A survey. In: Hochbaum, D. (ed.) Approximation algorithms. PWS Publishing Company (1997)

4. Coffman, E.G., Csirik, J., Woeginger, G.J.: Approximate solutions to bin packing problems. In: Pardalos, P.M., Resende, M.G.C. (eds.) Handbook of Applied Optimization, pp. 607–615. Oxford University Press, New York (2002)
5. Csirik, J., Johnson, D.S.: Bounded space on-line bin packing: Best is better than first. Algorithmica 31, 115–138 (2001)
6. Dósa, G.: The tight bound of First Fit Decreasing bin-packing algorithm is $FFD(I) \leq 11/9\,OPT(I) + 6/9$. In: Chen, B., Paterson, M., Zhang, G. (eds.) ESCAPE 2007. LNCS, vol. 4614, pp. 1–11. Springer, Heidelberg (2007)
7. Dósa, G., Sgall, J.: First Fit bin packing: A tight analysis. In: Proc. of the 30th Ann. Symp. on Theor. Aspects of Comput. Sci (STACS). LIPIcs, vol. 3, pp. 538–549. Schloss Dagstuhl (2013)
8. Dósa, G., Sgall, J.: Optimal analysis of Best Fit bin packing (2013) (submitted)
9. Garey, M.R., Graham, R.L., Johnson, D.S., Yao, A.C.-C.: Resource constrained scheduling as generalized bin packing. J. Combin. Theory Ser. A 21, 257–298 (1976)
10. Garey, M.R., Graham, R.L., Ullman, J.D.: Worst-case analysis of memory allocation algorithms. In: Proc. 4th Symp. Theory of Computing (STOC), pp. 143–150. ACM (1973)
11. Graham, R.L.: Bounds for certain multiprocessing anomalies. Bell System Technical J. 45, 1563–1581 (1966)
12. Graham, R.L.: Bounds on multiprocessing timing anomalies. SIAM J. Appl. Math. 17, 263–269 (1969)
13. Johnson, D.S.: Near-optimal bin packing algorithms. PhD thesis, MIT, Cambridge, MA (1973)
14. Johnson, D.S.: Fast algorithms for bin packing. J. Comput. Syst. Sci. 8, 272–314 (1974)
15. Johnson, D.S., Demers, A., Ullman, J.D., Garey, M.R., Graham, R.L.: Worst-case performance bounds for simple one-dimensional packing algorithms. SIAM J. Comput. 3, 256–278 (1974)
16. Lee, C.C., Lee, D.T.: A simple on-line bin-packing algorithm. Journal of the ACM 32, 562–572 (1985)
17. Németh, Z.: A first fit algoritmus abszolút hibájáról, Eötvös Loránd Univ., Budapest, Hungary (2011) (in Hungarian)
18. Sgall, J.: A new analysis of Best Fit bin packing. In: Kranakis, E., Krizanc, D., Luccio, F. (eds.) FUN 2012. LNCS, vol. 7288, pp. 315–321. Springer, Heidelberg (2012)
19. Simchi-Levi, D.: New worst case results for the bin-packing problem. Naval Research Logistics 41, 579–585 (1994)
20. Ullman, J.D.: The performance of a memory allocation algorithm. Technical Report 100, Princeton Univ., Princeton, NJ (1971)
21. Williamson, D.P., Shmoys, D.B.: The Design of Approximation Algorithms. Cambridge University Press (2011)
22. Xia, B., Tan, Z.: Tighter bounds of the First Fit algorithm for the bin-packing problem. Discrete Appl. Math. 158, 1668–1675 (2010)

Pluralism Ignored: The Church-Turing Thesis and Philosophical Practice

G. Graham White

Electronic Engineering and Computer Science
Queen Mary, University of London
g.graham.white@gmail.com
http://www.eecs.qmul.ac.uk/~graham/

Abstract. The Church-Turing thesis is widely stated in terms of three equivalent models of computation (Turing machines, the lambda calculus, and rewrite systems), and it says that the intuitive notion of a computable function is what is defined by any one of these models. Despite this well-established equivalence, the philosophical literature concentrates almost exclusively on the Turing machine model. We argue that this has been to the detriment of the philosophy of computation, and specifically that it ignores two issues: firstly, equivalence in the Church-Turing sense is extensional equivalence, whereas many of the delicate issues in the philosophy of mind, and in theoretical computer science, are to do with fine-grained intensional equivalence of algorithms. Secondly, real computers are not in any meaningful sense Turing machines: they are nondeterministic, their memory may fail to be in a determinate state due to cache coherence issues, and the boundaries between inside and outside are ill-defined and permeable. We explore the philosophical significance of these issues and give some examples.

1 The Church-Turing Thesis

We start with a cognitive scientist's description of what has become known as the Church-Turing thesis:

> The same work that provided demonstrations of some in-principle limitations of formalisation provided demonstrations of formalisation's universality as well. Thus Alan Turing, Emil Post, and Alonzo Church independently developed distinct formalisms that are powerful enough to formally (that is, "mechanistically") generate all sequences of expressions capable of interpretation as proofs, and hence, can generate all provable theorems of logic. In Turing's work this took the form of showing that there exists a universal mechanism ... that can simulate any mechanism describable in its formalism. [1, p. 50]

This description is in terms of procedures that generate sequences of proofs. A large number of other treatments – for example [2] – talk about functions from the integers to the integers (the two versions can easily be shown to be

A. Beckmann, E. Csuhaj-Varjú, and K. Meer (Eds.): CiE 2014, LNCS 8493, pp. 373–382, 2014.
© Springer International Publishing Switzerland 2014

equivalent). We will, in what follows, use functions from the integers to the integers.

In those terms, the Church-Turing thesis describes the set of functions, from the integers to the integers, which are capable of being computed by *some* algorithm in *any* of the equivalent computational formalisms (Turing's machines, the lambda calculus as used by Church, or rewrite systems as used by Post). If we have an algorithm in any of these three formalisms, then we can easily translate it into an equivalent algorithm in the other formalisms. There is nothing special about this particular list of formalisms: most books on computability will probably define other variants (register machines, machines with numerous tapes, and so on) purely in order to make the proofs easier.

Note that this set is specified extensionally. Functions here are regarded extensionally, as sets of ordered pairs. The definition talks about algorithms, but it quantifies over *all* (suitable) algorithms, so there is no bias in favour of one algorithm or the other: similarly, it quantifies over all models of computation, since the definition is perfectly egalitarian as far as algorithms and models of computation go.

Since the three computational formalisms are provably equivalent, the definition is also perfectly egalitarian as far as computational formalisms are concerned. It is worth noting that equivalences like these are quite common in mathematics.[1] In such a situation, the various definitions could well each belong to a different area of mathematics, which means that mathematicians can, and do, prove theorems using whatever definition, and hence whatever technical means, are the most convenient or the most illuminating.

1.1   Philosophers and The Church-Turing Thesis

Under these circumstances, it is a little surprising to see that a great deal of the philosophical literature which quotes the Church-Turing thesis is devoted exclusively to the Turing model. Pylyshyn [1], for example, mentions Post twice (but in a historical survey), Church three times (twice in the same historical survey, once for his undecidability theorem). However, Turing is discussed seven times (many of these discussions being several pages in length), and the Turing machine eleven times. There is a discrepancy here, and this discrepancy is quite typical of the philosophical literature.

Why is this bad? If we neglect the egalitarian nature of the definition, we may well end up attributing merely accidental properties of Turing machine computation to the abstract concept of computable function. This is an easy mistake to make, since philosophers naturally want to have more intensional information than merely the concept of a computable function will give them.

---

[1] Here is an example (the technical details are not important, but the fact of equivalent definitions is). Riemann surfaces can be defined in algebraic geometry as one-dimensional algebraic curves, in complex analytic geometry as one-dimensional complex manifolds, in complex analysis as the graphs of many-valued complex analytic functions, and so on; see Donaldson [3].

But it is, I would argue, fallacious. I will give two examples of such reasoning, one by Searle and one by Fodor, and then I will go on to talk about real computers and about intensionality.

2 Searle

Searle, writing about machines and simulation, says

> There is no question that an artificially made machine could, in principle, think. Just as we can build an artificial heart, so there is no reason why we cannot build an artificial brain. The point, however, is that any such artificial machine would have to be able to duplicate, and not merely simulate, the causal powers of the original biological machine. An artificial heart does not merely simulate pumping, it actually pumps. And an artificial brain would have to do something more than simulate consciousness, it would have to be able to *produce* consciousness. [4, p. 56]

And, at numerous places in his *oeuvre*, Searle emphasises that "formal symbol manipulation" is not sufficient for understanding (see [5–7] for summaries). "Formal symbol manipulation" is usually described, by Searle, in terms of zeroes and ones.

Now zeroes and ones are the way that Turing machines typically encode numbers. But Church's formalism uses the lambda calculus, and encodes numbers using an iterative formalism known as the Church numerals. Whereas it seems quite easy to argue that Turing machines simulate counting, rather than actually counting, I shall argue that a machine doing symbol manipulation using the lambda calculus actually *counts*.

2.1 The Church Numerals

The lambda calculus is based on the idea of function application: in the untyped lambda calculus, there is only one sort of object, and these objects can equally well be functions and the arguments of functions. We call these objects *terms*.

Suppose we have a term f: we can apply it to an argument a simply by juxtaposition, fa. Conversely, if we have an expression $\Phi(x)$ involving the free variable x, then the lambda-abstraction $\lambda x.\Phi(x)$ is the lambda-term which, when applied to a, yields $\Phi(a)$: i.e.

$$(\lambda x.\Phi(x))\, a \quad = \quad \Phi(a) \tag{1}$$

Consider now the expression $\mathsf{ap} = \lambda f.\lambda x.fx$; for any g and a,

$$
\begin{aligned}
\mathsf{ap}\, ga &= (\lambda f.\lambda x.fx)ga \\
&= (\lambda x.gx)a & \text{by(1)} \\
&= ga & \text{by(1)}
\end{aligned}
$$

so ap simply applies terms to each other. Now consider $\mathsf{cmp} = \lambda f.\lambda g.\lambda x.f(gx)$; a similar computation shows that $\mathsf{cmp}\,hk$, when applied to a, yields $h(ka)$, that is, cmp does function composition. So, given a natural number n, we can define by induction a λ-term Φ_n:

$$\Phi_n = \begin{cases} \lambda f\lambda x.x & n = 0 \\ \lambda f\lambda x.fx & n = 1 \\ \lambda f\lambda x.\mathsf{cmp}(\Phi_{n-1}f)fx & n > 1 \end{cases}$$

Φ_n is an expression which, when applied to the two arguments g and a, yields $\overbrace{g(\cdots(g\,a)\cdots)}^{n}$: the term $\Phi_n g$ is the nth iteration of g, $\overbrace{\mathsf{cmp}\,g(\mathsf{cmp}\,g(\cdots))}^{n}$.

Thus, the terms Φ_n have a double life. Concretely, when applied to a term f, they yield the nth iteration of f: but abstractly, before it is applied to anything, Φ_n represents the operation of n-fold iteration. These terms Φ_n can thus be used to represent the natural numbers: the usual arithmetic operations can be defined on them purely by the operations of the lambda calculus [8].

I claim the following:

1. Application of symbolic expressions by a machine actually is application. That is, the machine transforms an argument by applying a function to it, and the steps in the computation are (apart from being represented in electronic form rather than on paper, and with rather different symbols) the same as a human calculator would use.
2. The application of a church numeral to a pair of arguments actually involves iteration, rather than simulating iteration: this is because, when we unwind the application in the way we have shown above, we actually do apply the first argument n times to the second one. So, iteration is iteration whether it happens inside or outside of the computer.

Suppose, as a working hypothesis, that a grasp of numbers could be based on the concept of iteration of an operation – let us say, a mental operation – there is no reason why a computer, using the Church numerals, would not have the same access to numbers as we do when we calculate with them. So the computer here would be actually calculating rather than simulating a calculation.

One would be foolish here to argue that a computer which used the Church numerals actually *understood* numbers thereby. But, I claim, these considerations show that Searle's argument that computers simulate, rather than perform, breaks down if we think of numbers as being defined by iteration and if the computer uses the Church numerals for calculation.

2.2   Intensionality

What this argument shows is that there are important differences between computations: some computations actually implement numbers as iterators and some do not. Different devices can both implement the Church-Turing definition of computation, and one of them may implement numbers as iterators and one of them may not. This is not paradoxical: it simply shows that property of implementing numbers as iterators is an *intensional* property of computations.

3   Fodor

Fodor, in [9], distinguishes between "local" and "global" properties of mental representations; local properties are "constituted solely by what parts a representation has and how those parts are arranged" [9, p. 20]. Syntactic properties are local properties of representations, as is logical form [9, pp. 20f]. Mental processes are computational, and, more than that, *syntactically* computational: "a mental process, qua computation, is a formal operation on syntactically structured representations" [9, p. 11]. And thus "[m]ental properties are ipso facto insensitive to context dependent properties of mental representations" [9, p. 25] Fodor calls such properties *intrinsic* [9, p. 26].

This line of argument seems to lead to problems, however. There seem to be context dependent properties of mental representations which play a role in mental processes: Fodor gives the example of simplicity [9, p. 25ff] (deciding between rival explanations is a mental process, people, when doing so, quite often decide to go for the simplest explanation of a phenomenon, but simplicity is, as examples show, a context dependent property).

So, Fodor then weakens the requirement for mental processes to be sensitive only to intrinsic semantic properties, and considers, instead, what he calls the Minimal Computational Theory of Mind: "[t]he role of a mental representation in cognitive processes supervenes on some syntactic facts or other" [9, p. 29]. But now there are further problems, because some syntactic facts are relational (i.e. they have to do with the relation of a particular syntactic item to other items): and "these [relational] facts are not ipso facto accessible to computations for with the representation provides a domain" [9, p. 30].

Now in the language of local and global, the problem seems to be this. If we look at the local domain suitable for computations with a particular mental representation, then that local domain is very small: it is the representation and its parts. Fodor allows us to go beyond that small domain, but that leads to a drastic expansion of the domain: it will consist of the relations between the given mental representation and all other mental representations. It is, as Fodor says, "global". So what Fodor needs is a notion of computation which has a wider notion of locality than simply a single expression. We have, then, to show how to define a domain intermediate between local and global: large enough for the computations one needs, but small enough to be still computable. We have suggested a *logical* approach in [10], but we are here investigating models of computation, so we will consider it from that direction.

3.1   Locality in Models of Computation

Fodor talks of mental representations in terms of whole and parts. It is not clear how to derive this merely from the idea of a Turing machine: generally, the items in the cells of a Turing machine tape are *bits*, and there is no reason why the bits corresponding to a particular syntactic item should even be contiguous, so we get no help there. We might think of allowing larger items in a cell, but cells have to be of bounded size, whereas syntactic structures can be arbitrarily large. We can represent structured items in computer memory, but it usually depends on storing, instead of a constituent, the address of the constituent in memory: memory addresses used in this way are known as *pointers* [11, ch. 3]. We need pointers, because we cannot generally preserve contiguity in memory (for example, we may want to replace a constituent by a constituent of a larger size, without recopying all of the structured item in question). Memory is, in any case, one-dimensional, whereas many structured objects are higher dimensional, so there is not much hope of reconstructing structured objects from contiguity relations in memory.

But there *is* a model of computation which treats structured items perspicuously, namely the Church model based on the λ-calculus. One might think that this model is of theoretical interest only, but, in fact, there is a very elegant abstract machine for evaluating lambda expressions: Landin's SECD machine [12–15]. It is not only of theoretical interest: it has influenced the implementation and design of very many functional programming languages.

The SECD machine is a machine with four registers, S, E, C, and D, each of which contains *pointers*, i.e. memory addresses: these addresses are generally the heads of lists. These lists are initially empty, and change during evaluation of the expression. E (the environment) contains bindings of variables to values: this will change as new variables are defined or as variables go out of scope. C (the code) is a list of expressions to be evaluated: it contains λ-expressions in reverse Polish notation, so that the application FG becomes the list $(G :: F :: \mathsf{app})$. If we want to evaluate this, then first G, then F, have to be evaluated, and then the value of F applied to the value of G. The values have to be saved, of course: we use the list S, or the *stack*. F and G may be complex terms: if so, then, in order to evaluate them, we have to perform subsidiary computations, and these computations may change the state of the machine. So, when we start such a subsidiary calculation, we save the state of the machine in the list D, so that we can later restore it.

What is important here is not the details, but two important facts. Firstly, the four lists contain all the information necessary in order to evaluate the λ-expression that we started with; everything that ends up in one of the lists ends up there because there was a good reason to put it there. So they are not simply random collections of values, but rather lists of *relevant* values. Secondly, the general state of the computation is not the evaluation of a single expression, but the evaluation of that expression given the contents of the lists: that is, the evaluation of an expression *in a given context*. And if we want to know what evaluation of λ-expressions amounts to, we want to know not just how the evaluation starts – when we have only an expression – but how it goes on, when we have the lists populated with items which will be necessary in the course of the calculation. So the idea of context can be regarded as a less problematic version of the idea of locality, which, as we have seen, depended on contiguity in an unsustainable way.

3.2 Expressions in Context

This idea – of evaluating an expression in context – gives us the wider notion of locality which Fodor needs. It is also tremendously pervasive: almost all modern programming languages have such a concept. It is what makes them easy to use, and which, in particular, makes it possible to write programs that can be read and understood by humans as well as by computers.

When we read a novel, we have to remember, at any point, the back story: who the characters are and how they came to be in the situations that they are in. Without that back story the descriptions of the actions being performed at a particular moment would be incomprehensible. Similarly, when we evaluate something in a program, there is a back story: this story tells us how we came to be evaluating this expression – and, in particular, it tells us what the variables stand for – and it also tells us what happens to the results of the evaluation (whether they come from some subsidiary computation, for example), and thus it tells us what we are to do when we finish the computation. In other words, it tells us what is called the *context* of the calculation. And, as well as helping to design programming languages, the idea of a context shows us how we might arrive at a notion of locality which is subtle enough to be used in the philosophy of mind.

The phenomenologists tell us that, when we are directly aware of some salient item, we also implicitly know a great deal of other things which put our awareness of the salient item into context. Thinking that all you need to know about computation is to know about Turing machines is to forget about the context in which any computation is performed. It was the great achievement of the modern theorists of programming languages that, starting from the bare idea of calculation given by a Turing machine, they systematically constructed such a context, such a back story, which would render our computations comprehensible to us [16].

4 Real Computers

We have seen, then, that there is a great deal of the formal and intellectual content of the idea of computation is not contained in the idea of a Turing machine. There is still, however, the temptation to think that, in any event, Turing machines might constitute a foundation for computing: that, if we took the hardware of a real computer, such as the laptop which I am typing this on, it could be formally described as a Turing machine, and that any computation written in a programming language on it, contexts and all, could be thought of as running on the Turing machine using particular disciplines about where the contexts lay on the tape and how they were handled.

Unfortunately, this is not true. Here are just two of the problems (there are more). Firstly, modern computers are in practice quite badly nondeterministic.

Hard drives, for example, respond to a request for data after an unpredictable interval. Part of the delay is due to the time the head takes to move to the required position, and this time, as well as depending on how far the head has to move, is affected by air turbulence inside the disk enclosure: this means that hard disks can be used as a source of cryptographic randomness [17]. This effect may seem very small, but one should notice that the CPU of a modern computer works on a timescale of nanoseconds, whereas hard disk seek times are about a millisecond or longer: this is about a million times longer than durations of events in the CPU. Consequently, even small variations in hard disk seek times can potentially affect the CPU.

"Potentially" is very often "really". Modern computers are multithreaded: they run many computations at once, and these computations compete with each other for resources (disk access, CPU time, input and output). So if one process is delayed by even a small amount of time, the competitive environment in which it is running may amplify that small time into a humanly noticeable time. Consequently, we have nondeterminism. We similarly get nondeterminism because computers are generally connected to the internet, and the internet has chaotic behaviour: so any process which reads from or writes to the internet can be delayed by arbitrary and unpredictable amounts. This internet access might not be anything visible: it might be, for example, checking for email, or setting the computer clock from a timeserver.

The second problem is this. As we have seen, modern computers are genuinely concurrent, and they consist several independent processors which can be regarded as computers in their own right. These processors all have access to the same memory, and, even worse, they access it through a cache. Under these circumstances different processors may simultaneously update their cached versions of the same data item in different ways: we need to be able to deal with this possibility.

The problem is called *cache coherence*, and is to do with what you get when you read some replica of a data item (you would like to get the most recently written version of it, whatever processor wrote it, but this is impossible in general) [18, p. 466]. Similar properties (for example, that each processor sees updates to memory (by any of the processors) in the same order) are similarly impossible to achieve in general.

What turns out to be achievable, and which still allows comprehensible programs to be written, is what is called *eventual consistency*: that is,

> An update executes at some replica, without synchronisation; later, it is sent to the other replicas. All updates eventually take effect at all replicas, asynchronously and possibly in different orders. Concurrent updates may conflict; conflict arbitration may require a consensus and a roll-back [19].

In the multicore context, this would mean that processors write values to their caches. These values eventually propagate to RAM and to other caches, but not immediately. If two processors update the same item at their replicas in inconsistent ways, then they may have to retrace their steps (what is called a *rollback*).

Here we have left the world of Turing machines: in a Turing machine, there is a single tape, cells on the tape are instantly and deterministically written to whenever the machine says so, and they are instantly and deterministically read from whenever the machine says so. A Turing machine has no need of rollback, and no mechanism for rollback. We have a machine – i.e. a CPU – connected to a memory, i.e. RAM: but there the similarity to a Turing machine ends.

5 The Intensional Disciplines

We have seen, then, that modern computers differ from Turing machines in ways which are invisible extensionally (up to Turing equivalence, modern computers can still perform all and only the computations that Turing machines can), but which differ intensionally, in ways which make a great difference to, for example, how you program them.

It is, then, not surprising that there is a substantial amount of theory about what these intensional differences are, and about how they affect the ways that one can program such machines. This theory uses what we might call the *intensional disciplines*: game theory [20], dependent type theory, category theory [21]. It is, although technical, very illuminating.

It is also generally unknown to philosophers. Searle, for example, writes

> there is little theoretical agreement among practitioners on such absolutely fundamental questions as, What exactly is a digital computer? What exactly is a symbol? What exactly is an algorithm? What exactly is a computational process? Under what physical conditions exactly are two systems implementing the same program? [22, Ch. 9 §III]

These are, strikingly, many of the questions which have proved to be important in modern theoretical computer science. That is not surprising: they are good questions, Searle has very good instincts, and he knows a good question when he sees one. But in order to answer these questions, it is necessary to gain some acquaintance with the intensional world and with the methods that one needs in order to work in it.

Most of these problems are, it must be admitted, quite hard. Abramsky [23] admits that we do not know "what are the fundamental structures of concurrency". To that extent, Searle is right that many of these fundamental questions remain unanswered, but it is not through want of trying, and it is not that a great deal of progress has nevertheless been made.

References

1. Pylyshyn, Z.W.: Computation and Cognition: Toward a Foundation for Cognitive Science. MIT Press (1986)
2. Copeland, B.J.: The church-turing thesis. In: Zalta, E.N. (ed.) The Stanford Encyclopedia of Philosophy, Fall 2008 edn. (2008)
3. Donaldson, S.: Riemann Surfaces. Oxford (2011)

4. Searle, J.R.: Twenty-one years in the Chinese room. In: [24], pp. 51–69
5. Copeland, B.J.: The Chinese room from a logical point of view: New essays on Searle and artificial intelligence. In: [24], pp. 109–122
6. Haugeland, J.: Syntax, semantics, physics. In: [24], pp. 379–392
7. Bishop, M.: Dancing with picies: Strong artificial intelligence and panpsychism. In: [24], pp. 360–378.
8. Wikipedia: Church encoding — wikipedia, the free encyclopedia (2014), `http://en.wikipedia.org/w/index.php?title=Church_encoding&oldid=590011645` (accessed January 10, 2014)
9. Fodor, J.A.: The Mind Doesn't Work That Way: The Scope and Limits of Computational Psychology. Representation and Mind. MIT Press, Cambridge (2000)
10. White, G.G.: Causality, modality and explanation. Notre Dame Journal of Formal Logic 49(3), 313–343 (2008)
11. Watt, D.A.: Programming Language Concepts and Paradigms. Prentice Hall (1990)
12. Landin, P.J.: The mechanical evaluation of expressions. Computer Journal 6(4), 308–320 (1964)
13. Landin, P.J.: The next 700 programming languages. Communications of the ACM 9(3), 157–166 (1966)
14. Danvy, O.: A rational deconstruction of Landin's SECD machine. Technical Report RS-04-30, BRICS (2004)
15. Wikipedia: SECD machine — wikipedia, the free encyclopedia (2013), `http://en.wikipedia.org/w/index.php?title=SECD_machine&oldid=586114473` (accessed January 8, 2014)
16. White, G.G.: The philosophy of programming languages. In: Floridi, L. (ed.) The Blackwell Guide to the Philosophy of Computing and Information, pp. 237–247. Blackwell (2004)
17. Davis, D., Ihaka, R., Fenstermacher, P.: Cryptographic randomness from air turbulence in disk drives. In: Desmedt, Y.G. (ed.) CRYPTO 1994. LNCS, vol. 839, pp. 114–120. Springer, Heidelberg (1994), at `http://world.std.com/~dtd/`
18. Hennessy, J.L., Patterson, D.A.: Computer Organisation and Design: The Hardware-Software Interface, 5th edn. Morgan Kaufmann (2014)
19. Shapiro, M., Preguiça, N., Baquero, C., Zawirski, M.: Conflict-free replicated data types. In: Défago, X., Petit, F., Villain, V. (eds.) SSS 2011. LNCS, vol. 6976, pp. 386–400. Springer, Heidelberg (2011)
20. Abramsky, S.: Algorithmic game semantics: A tutorial introduction. In: Schwichtenberg, H., Steinbrüggen, R. (eds.) Proceedings of the NATO Advanced Study Institute, pp. 21–47. Kluwer Academic Publishers, Marktoberdorf (2001)
21. Crole, R.: Categories for Types. Cambridge University Press (1993)
22. Searle, J.R.: The Rediscovery of the Mind. MIT Press (1994)
23. Abramsky, S.: What are the fundamental structures of concurrency? We still don't know? Electronic Notes in Theoretical Computer Science 162, 37–41 (2006)
24. Preston, J., Bishop, M. (eds.): Views into the Chinese Room: New Essays on Searle and Artificial Intelligence. Oxford University Press (2002)

The FPGA-Based High-Performance Computer RIVYERA for Applications in Bioinformatics

Lars Wienbrandt

Department of Computer Science
Christian-Albrechts-University of Kiel, Germany
lwi@informatik.uni-kiel.de

Abstract. Bioinformatics specifies a wide field of applications with generally long runtimes or huge amounts of data to be processed – or even both. Typically, large computing clusters or special computing platforms are harnessed to solve problems in this field in reasonable time. One such platform is represented by the FPGA-based high-performance computer RIVYERA, which was intentionally developed for problems in cryptanalysis. On the basis of three easy examples taken from our current research field, we show how RIVYERA can be applied to different kinds of problems regarding bioinformatics. RIVYERA is able to significantly speed up the process of exact sequence alignment using the Smith-Waterman [1] algorithm, querying protein sequence databases using BLASTp [2], and running genome-wide association studies (GWAS) using iLOCi [3] or similar methods based on contingency tables. Likewise, energy savings with RIVYERA are in the same order as runtime reductions compared to standard PCs or computing clusters.

1 Introduction

Applications and algorithms in bioinformatics have to deal with the ever-growing amount of biological data stored in large sequence databases. NCBI's Genbank database [4] or the UniprotKB/TrEMBL database [5] are prominent examples for an exponentially growing amount of data. Likewise, runtimes of algorithms with quadratic or higher complexity easily become unreasonable if applied on large datasets with standard hardware. To keep up with this rising demand on computational power in bioinformatics, the focus is set to parallel processing. Commonly, these problems are addressed by standard computing clusters. However, a linear increment in computing nodes may only provide a linear speedup for processing time, but costs for acquisition, energy, and maintenance grow linearly as well.

These problems can be addressed by moving away from standard architectures, e.g. with graphics processing units (GPUs). Unfortunately, GPUs, as CPUs, have to provide on-die resources for a large fixed instruction set consuming energy even if not required. Hence, no significant energy reductions are expected. In contrast, FPGAs are configurable to exactly meet the requirements of the application. Thus, no resources are spent for obsolete instructions and,

A. Beckmann, E. Csuhaj-Varjú, and K. Meer (Eds.): CiE 2014, LNCS 8493, pp. 383–392, 2014.
© Springer International Publishing Switzerland 2014

therefore, energy is only consumed on the chip for application essential processing. Fully utilizing all available FPGA resources results in very compact processing units with a fine-grained on-chip parallelism significantly reducing energy requirements and processing time. With a massively parallel utilization of FPGAs, the performance of a whole computing cluster can be fit in a single computer system, maintaining flexibility due to reconfigurability. The RIVYERA architecture [6] is an example for such a system, providing the resources of e.g. 128 Xilinx Spartan6-LX150 FPGAs.

In three examples, we present how this architecture can improve major problems in bioinformatics. Optimal sequence alignment using the Smith-Waterman algorithm [1], protein database searches with BLASTp [2,7], and iLOCi, an application for detecting epistatis in genome-wide association studies (GWAS).

2   The RIVYERA Architecture

In 2008 the computing platform RIVYERA [8], originally developed for cryptanalysis, was introduced for problems related to bioinformatics. Two specific models of RIVYERA were developed, the RIVYERA S3-5000 and the successor RIVYERA S6-LX150 [6]. This paper focuses on the newer model RIVYERA S6-LX150 equipped with 128 FPGAs of type Xilinx Spartan6-LX150.

The basic structure of the RIVYERA architecture consists of two elements, the FPGA computer and a server grade mainboard with standard PC components. The FPGA computer consists of up to 16 FPGA modules with 8 FPGAs each. Each FPGA is connected to a local memory of 256 MB DDR3-RAM. Upgrades may allow up to 16 FPGAs on each module or larger memory.

The configuration of the mainboard is variable as well. Here, the RIVYERA S6-LX150 is equipped with two Intel Xeon E5-2620 CPUs (6 cores @ 2 GHz each) with 128 GB of RAM running a Linux OS.

The bus system implemented on the RIVYERA FPGA computer is organized as a systolic chain, i.e. each FPGA on an FPGA module is directly connected to its neighbors forming a ring. A communication controller in this ring, provides the interconnection of each module to its neighboring modules. The uplink to the mainboard, further referred to as *host*, is realized via the communication controller of the first FPGA module. A picture of RIVYERA S6-LX150 is shown in Fig. 1.

3   Exact Sequence Alignment with Smith-Waterman

3.1   Smith-Waterman Algorithm

The Smith-Waterman algorithm [1] calculates and evaluates biological sequence alignments of two sequences. Alignments are required for instance to measure similarity of DNA sequences or to find similar occurrences of a short sequence in a longer one. Hence, for convenience, one sequence is referred to as *query q* and the other as *subject* or *database s*.

Fig. 1. The RIVYERA S6-LX150 system

The score of alignments is calculated by a simple *scoring function*, generally counting a positive score for matching characters and a negative score for mismatches. Additionally, alignments may contain insertions or deletions (shortly referred to as *gaps*) which are counted negative as well, but different for a gap opening and extension (*affine* gap penalty). Smith-Waterman alignments are *locally optimal*, i.e. there is no alignment of the two input sequences or of any of its subsequences that results in a higher score. Here, we focus on nucleotide (DNA) sequence alignment.

In the first step an alignment matrix $H_{n \times m}$ is calculated (n and m denote the lengths of the query and subject sequence respectively, g the gap penalty, S the scoring function, q_i and s_i the query and subject symbol at position i):

$$H_{i,j} = \max \begin{cases} H_{i-1,j-1} + S(q_i, s_j) & \text{match/mismatch} \\ H_{i-1,j} + g & \text{insertion opening/extension} \\ H_{i,j-1} + g & \text{deletion opening/extension} \\ 0 & \text{do not allow negative values} \end{cases} \quad (1)$$

Afterwards, in order to generate the final alignment, a backtracking step is performed. In brief, the backtracking starts at matrix cell $H_{i,j}$ with the highest value and follows the path through the alignment matrix that reflects the chain of matrix cells whose values were taken for the maximum calculation in (1). For each chosen direction (*up*, *left*, or *up-left*) the corresponding character or gap is inserted into the final alignment. The backtracking stops if a cell with $H_{x,y} = 0$ is encountered.

3.2   Implementation of Smith-Waterman

Time and memory complexity for the calculation of the alignment matrix are clearly of $\mathcal{O}(n \times m)$. For fine-grained parallel processing, if m processing elements are utilized concurrently, the runtime complexity can be reduced to $\mathcal{O}(n)$, i.e. linear to the length of the database sequence. Additionally, memory complexity is reduced to $\mathcal{O}(1)$ for each processing element since it is not necessary to store the alignment matrix due to the common usage of the algorithm to perform a huge number of alignments on a large dataset just to calculate the alignment score. For the candidates requiring the particular alignment (usually those with the best scores), the process is repeated on a small section of the

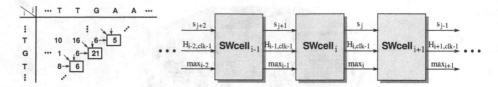

Fig. 2. Smith-Waterman chain structure and example for the calculation of an alignment matrix on an FPGA

original database sequence corresponding to the location where the main process has calculated the maximum score, now including storage of the alignment matrix and backtracking.

Hence, the FPGA implementation is realized by the following parallelization scheme. For every nucleotide in the query sequence a processing element *SWcell* is implemented on the FPGA. It calculates the values in the row of the alignment matrix corresponding to its assigned character, i.e. a direct implementation of (1), whereby i, the index of the row, is fixed for each cell. Each cell requires access to three neighboring values from the alignment matrix to calculate a new cell value. Therefore, all processing elements are connected in a chain such that each element has access to the cell value of its predecessor. Now, the database sequence can be streamed character by character with every clock cycle through the chain of processing elements. The three required values are accessed in the following way. $H_{i-1,j}$ (up) is the cell value from the previous clock cycle in the previous row (and therefore previous neighbor). $H_{i,j-1}$ (left) is the cell value from the previous clock cycle in the same row (and therefore the same cell). $H_{i-1,j-1}$ (up-left) is the cell value from two previous clock cycles in the previous row (previous neighbor again). The processing is now accomplished in anti-diagonals of the alignment matrix. Figure 2 shows a calculation step of the alignment matrix and a part of the chain structure.

3.3   Performance Evaluation

We have tested the Smith-Waterman implementation with a query length of $m = 100$ and variable database size. This results in five chains on one FPGA of RIVYERA S6-LX150, i.e. 1,024 queries may be processed concurrently on the whole RIVYERA. A test set of 1 million Illumina 100 bp paired reads (plus reverse complements) aligned against the human genome (*hg19*, $\sim$ 3 Mbp) requires only about 29h. This leads to a speed of 6,020 GCUPS (billion cell updates per second) which outperforms standard PC architectures by far. Compared to a commercial software solution for PCs and clusters provided by CLCbio [9], this is a speedup of up to 134 compared to two Xeon CPUs with four cores, or 463 compared to a dual-core standard PC (see Table 1).

Table 1. Smith-Waterman performance for DNA sequence alignment. 1 million 100 bp reads (plus reverse complements) are aligned against the human genome.

Architecture	Energy (W)	Energy (kWh)	Time	Speed (GCUPS)
RIVYERA S6-LX150	780	22.6	29 h 01 m	6,020
CLCbio 2x Xeon X3210 @ 2.13 GHz (2x4 cores)	210	408.3	~ 162 d	45
CLCbio Core2Duo @ 2.17 GHz (2 cores)	100	673.1	~ 560 d	13

4 Sequence Database Searches with BLASTp

4.1 BLAST Algorithm

To find sequence similarities in biological databases, a sequence alignment of the query sequence to all stored database sequences has to be performed. Unfortunately, optimal solutions as the Smith-Waterman algorithm (see Sect. 3) are unfeasible for this task. Thus, the Basic Local Alignment Search Tool (BLAST) [7,2] has been developed to perform a heuristic search with significantly reduced runtime.

BLAST is organized in several steps. In the first step, the query sequence is preprocessed to identify its *neighborhood*, i.e. a list of k-mers which exceed a certain threshold score when directly compared to k-mers of the query sequence. For the comparison a scoring matrix such as BLOSUM62 is taken. For BLASTp the value for k is fixed to $k = 3$.

In the second step, *hits* are located by searching for exact matches of words from the neighborhood in the database sequences. The hits are tested pairwise if both hits of a pair hold the same distance to each other in the query sequence and in the subject sequence. The pair is then referred to as *two-hit*. The equation $k \leq q_1 - q_0 = s_1 - s_0 < A$ shows the condition for a two-hit whereby s_0 and s_1 state the location of two hits in the subject and q_0 and q_1 their locations in the query, respectively. Overlapping hits and long distances are omitted by parameters k and A.

Each two-hit is further examined by an *ungapped extension* process. Both hits of a hit pair are extended by calculating a similarity score and then taking residue by residue to enlarge the aligned part. The extension is performed in both directions (backwards first) and stops if the score declines a certain cut-off distance below the so far calculated maximum (*X-drop* mechanism). The result, further referred to as *high-scoring pair* (HSP), is the pair of the two positions where the score is maximal. Figure 3 shows an example.

In the last step of the BLAST algorithm, the *gapped extension*, HSPs are simply analyzed with a modified version of the Needleman-Wunsch algorithm [10], which is very similar to the Smith-Waterman algorithm (see Sect. 3).

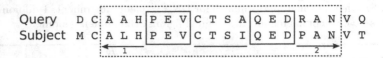

Fig. 3. Example for the ungapped extension of a two-hit in the NCBI BLAST implementation. The solid rectangles mark the hit pair, the dashed an extension.

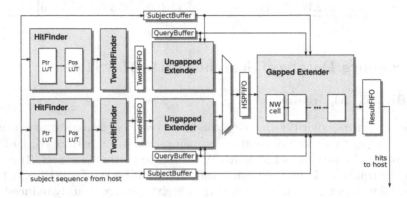

Fig. 4. Structure of two BLASTp hardware pipelines sharing one GappedExtender component

4.2   Application Structure and Implementation of BLASTp

The existing approach for protein database searches (BLASTp) implemented on RIVYERA S3-5000 [11] is based on the ideas of single FPGA solutions [12,13,14], modified and adapted to work with multiple FPGAs. Now, this approach has been adapted for the RIVYERA S6-LX150 architecture. Figure 4 shows an overview of the implementation.

The first of the previously described steps, generating the neighborhood, is performed as a preprocessing step on the host system. The others are implemented on each FPGA as a long pipeline consisting of four main components. The first component, the *HitFinder*, searches for occurences of k-mers of the subject sequence in the neighborhood, which is a simple look-up in a hashtable, organized in two separate tables. Afterwards, all possible pairs of hits are filtered with the two-hit condition by the *TwoHitFinder* component. This can be done in linear runtime with respect to the number of hits, basically by using a storage array for hit positions of a size corresponding to the query length.

The *UngappedExtender* analyzes all two-hits stored in a FIFO buffer. It extends a two-hit by a pair of residues in each clock cycle, calculating the new score and comparing it with the so far calculated maximum. If the X-drop condition holds, the extension stops and the maximum positions are buffered as HSP if the score exceeds a predefined threshold.

The *GappedExtender* component is applied as an effective prefilter before a HSP is reported to the host system for postprocessing. It performs a modified

Needleman-Wunsch alignment with a banded matrix and the HSP at its center. The width of the matrix band is set to $\omega = 64$ and corresponds to the number of generated processing elements ($NWcells$, similar to $SWcells$ in Sect. 3). The process uses the X-drop mechanism again to reduce runtime. All HSPs passing the gapped extension filter are buffered in the attached DRAM before the host system calculates the exact alignment including backtracking. Since the GappedExtender is utilized very infrequently but requires a lot of FPGA resources, each two processing pipelines share one GappedExtender component.

4.3   Performance Evaluation

In RIVYERA S6-LX150 two BLASTp pipelines are implemented on each FPGA. Hence, 256 queries are processed concurrently on a fully equipped system with 128 FPGAs. For evaluation, we have compared to a PC system equipped with an Intel Core i7-950 CPU (4 cores / 8 threads @ 3.07 GHz) and 12 GB RAM running NCBI BLASTp v2.2.25+ [15] on 8 threads. Three different query sets with proteoms of human, mouse and rat have been tested, each set reduced to exactly 10,000 queries. As reference, we have taken the first part of the NCBI *RefSeq* BLAST database, release 50, containing $2,996,372$ sequences (≈ 1 billion residues) [4].

The runtimes for the three datasets as well as the energy consumption are listed in Table 2. It shows that RIVYERA outperforms the PC system by a factor of about 19. The power consumption of RIVYERA is measured with 780 W while for the PC system only the TDP with 130 W is considered without peripherals etc. This alone results in energy savings of about 70%.

5   Genome-Wide Association Studies with iLOCi

5.1   iLOCi Algorithm

For detecting epistasis, i.e. gene-gene interactions, in genome-wide association studies (GWAS) the powerful method iLOCi [3] outperforms other available tools, such as MDR [16] or BOOST [17], in terms of accuracy and speed. iLOCi computes a simple statistical test for every possible pair of SNPs, i.e. $n(n-1)/2$

Table 2. BLASTp runtimes (in seconds) and energy consumption of three query sets and part one of the NCBI *RefSeq* database. The Xeon reference system runs NCBI BLASTp v. 2.2.25+.

Query set	#queries	RIVYERA S6-LX150 (780 W)		Intel Core i7-950 (8 threads, 130 W)		Speedup
		time	energy	time	energy	
Human	10,000	55 m 40 s	723.7 Wh	17 h 02 m 25 s	2,215.2 Wh	18.4
Mouse	10,000	53 m 33 s	696.2 Wh	15 h 43 m 55 s	2,045.2 Wh	17.6
Rat	10,000	55 m 47 s	725.2 Wh	17 h 33 m 03 s	2,281.6 Wh	18.9

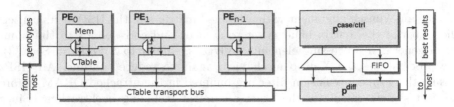

Fig. 5. Overview of the FPGA implementation of iLOCi

tests for n SNPs. This process consists of two simple steps. First, two contingency tables are created containing the counts of all combinations of genotypes for case and control samples respectively. There are three possible genotypes: homozygous wild, heterozygous and homozygous variant types. This results in nine possible combinations for a SNP pair for each type of samples, i.e. either case or control.

The second step is the calculation of a statistical test value p^{diff} based on the previously created contingency tables. Let n_{ij}^{case} denote the genotype counts for the combination i and j of the current SNP pair in all case samples ($i, j \in \{0, 1, 2\}$). p_{ij}^{case} then denotes the relative probability respective to n_{ij}^{case}. The same applies for n_{ij}^{ctrl} and p_{ij}^{ctrl} analogue to the control samples.

$$p^{\text{diff}} = |p^{\text{ctrl}} - p^{\text{case}}| \tag{2}$$

$$p^{\text{case}} = \frac{p_{00}^{\text{case}} - p_{02}^{\text{case}} - p_{20}^{\text{case}} + p_{22}^{\text{case}}}{\sqrt{(p_{0\bullet}^{\text{case}} + p_{2\bullet}^{\text{case}})(p_{\bullet0}^{\text{case}} + p_{\bullet2}^{\text{case}})}} \tag{3}$$

$$= \frac{n_{00}^{\text{case}} - n_{02}^{\text{case}} - n_{20}^{\text{case}} + n_{22}^{\text{case}}}{\sqrt{\left(\sum n_{0j}^{\text{case}} + \sum n_{2j}^{\text{case}}\right)\left(\sum n_{i0}^{\text{case}} + \sum n_{i2}^{\text{case}}\right)}} \tag{4}$$

iLOCi saves the n best results (e.g. $n = 1000$) and presents the corresponding SNP pairs in a sorted list.

5.2   Implementation of iLOCi

The goal for the implementation of iLOCi is to concurrently calculate as many statistical tests as possible. At first, the host distributes the genotype data of all SNPs to the available FPGAs, ordered by cases and controls. The creation of the contingency tables on the FPGA is organized in a systolic chain of processing elements (PEs), each capable of storing the genotype data for one SNP and the necessary counters for one contingency table. The genotype data is loaded from the FPGA attached DRAM and streamed genotype by genotype through the chain of PEs, whereby each PE stores data for another SNP in its local RAM. This way n PEs are able to create n contingency tables in parallel. For RIVYERA S6-LX150 100 PEs fit on each FPGA.

Since the input data is ordered by cases and controls, the calculation of (2) is divided into two subsequent calculations of (4). At first, p^{case} is calculated from all PEs and stored in a FIFO. p^{ctrl} is calculated afterwards. The calculation of p^{diff} directly follows. The unit calculating (4) is required only once for all PEs. It is organized in a large pipeline to ensure a new result in every clock cycle. This way, the FPGA resources are optimally utilized.

Each FPGA stores its own n best results such that in a short postprocessing phase the host system can generate a list of the total n best results. A schematic of the implementation structure is shown in Fig. 5.

5.3 Performance Evaluation

According to the authors of iLOCi, the analysis of a WTCCC dataset [18] with about 500,000 SNPs and 5,000 samples takes about 19 hours on a MacPro workstation with two Intel Xeon quad-core CPUs [3]. For a sample dataset of the same size, the RIVYERA S6-LX150 system takes only 4 minutes which leads to a speedup of more than 285. We also tested iLOCi on a system with a GeForce GTX480 GPU and the example test set provided by the software package. Due to a surprisingly poor performance, we extrapolated the runtime to match the WTCCC dataset. All runtimes including energy consumption are listed in Table 3.

Table 3. iLOCi performance for analysis of a dataset with 500,000 SNPs and 5,000 samples. GPU results are interpolated.

Architecture		Energy	Time	Speed (M tests/s)
RIVYERA S6-LX150	780 W	0.05 kWh	$4m$	520.833
2x Intel Xeon quad-core	~260 W	4.94 kWh	$19h$	1.827
nVidia GeForce GTX480	~250 W	42.00 kWh	~$7d$	~0.200

6 Summary

Three examples of different areas in bioinformatics show the capabilities of high-performance FPGA technology. The RIVYERA S6-LX150 system with 128 Xilinx Spartan6-LX150 FPGAs outperforms standard computer architectures by far in terms of runtime and energy consumption. The speedup for Smith-Waterman alignments is more than 134 compared to a commercial solution on a dual quad-core CPU, BLASTp is outperformed on a quad-core system by a factor of 19, and the speed for GWAS with iLOCi is increased even 285-fold compared to two quad-core CPUs. With this increment of speed, problems with intractable sizes of data become feasible again on one single system.

When compared to CPU cluster systems of a size large enough to handle the problems in the same time as RIVYERA, energy concsumption becomes important. Due to the optimal utilization of electrical power in FPGAs, energy savings are in the same range as the speedup, i.e. about 70% for BLASTp and between 95% and 99% for Smith-Waterman and iLOCi.

References

1. Smith, T.F., Waterman, M.S.: Identification of common molecular subsequences. Journal of Molecular Biology 147, 195–197 (1981)
2. Altschul, S.F., Madden, T.L., Schäffer, A.A., Zhang, J., Zhang, Z., Miller, W., Lipman, D.J.: Gapped BLAST and PSI-BLAST: a new generation of protein database search programs. Nucleic Acids Research 25, 3389–3402 (1997)
3. Piriyapongsa, J., Ngamphiw, C., Intarapanich, A., Kulawonganunchai, S., Assawamakin, A., Bootchai, C., Shaw, P.J., Tongsima, S.: iLOCi: a SNP interaction prioritization technique for detecting epistasis in genome-wide association studies. BMC Genomics 13(suppl. 7), S2+ (2012)
4. NCBI: GenBank database, http://www.ncbi.nlm.nih.gov/genbank/
5. EMBL EBI: UniProt Knowledgebase, http://www.ebi.ac.uk/uniprot/
6. SciEngines GmbH, http://www.sciengines.com
7. Altschul, S.F., Gish, W., Miller, W., Myers, E.W., Lipman, D.J.: Basic Local Alignment Search Tool. Journal. Journal of Molecular Biology 215(3), 403–410 (1990)
8. Pfeiffer, G., Baumgart, S., Schröder, J., Schimmler, M.: A Massively Parallel Architecture for Bioinformatics. In: Allen, G., Nabrzyski, J., Seidel, E., van Albada, G.D., Dongarra, J., Sloot, P.M.A. (eds.) ICCS 2009, Part I. LNCS, vol. 5544, pp. 994–1003. Springer, Heidelberg (2009)
9. CLCbio: High-Speed Smith-Waterman (2012), http://www.clcbio.com/index.php?id=1254
10. Needleman, S.B., Wunsch, C.D.: A general method applicable to the search for similarities in the amino acid sequence of two proteins. Journal of Molecular Biology 48(3), 443–453 (1970)
11. Wienbrandt, L., Siebert, D., Schimmler, M.: Improvement of BLASTp on the FPGA-based High-Performance Computer RIVYERA. In: Bleris, L., Măndoiu, I., Schwartz, R., Wang, J. (eds.) ISBRA 2012. LNCS, vol. 7292, pp. 275–286. Springer, Heidelberg (2012)
12. Mahram, A., Herbordt, M.C.: Fast and Accurate BLASTP: Acceleration with Multiphase FPGA-Based Prefiltering. In: Proceedings of ICS 2010, pp. 73–28 (2010)
13. Kasap, S., Benkrid, K., Liu, Y.: Design and Implementation of an FPGA-based Core for Gapped BLAST Sequence Alignment with the Two-Hit Method. Engineering Letters 16, 443–452 (2008)
14. Jacob, A., Lancaster, J., Buhler, J., Harris, B., Chamberlain, R.D.: Mercury BLASTp: Accelerating Protein Sequence Alignment. ACM Transactions on Reconfigurable Technology and Systems 1, 9:1–9:44 (2008)
15. NCBI: NCBI BLAST, http://blast.ncbi.nlm.nih.gov/Blast.cgi
16. Ritchie, M.D., Hahn, L.W., Roodi, N., Bailey, L.R., Dupont, W.D., Parl, F.F., Moore, J.H.: Multifactor-Dimensionality Reduction Reveals High-Order Interactions among Estrogen-Metabolism Genes in Sporadic Breast Cancer. Am. J. Hum. Genet. 69(1), 138–147 (2001)
17. Wan, X., Yang, C., Yang, Q., Xue, H., Fan, X., Tang, N.L., Yu, W.: BOOST: A Fast Approach to Detecting Gene-Gene Interactions in Genome-wide Case-Control Studies. Am. J. Hum. Genet. 87(3), 325–340 (2010)
18. The Wellcome Trust Case Control Consortium: Genome-wide association study of 14,000 cases of seven common diseases and 3,000 shared controls. Nature 447(7145), 661–78 (2007)

Exploiting Membrane Features to Compute

Claudio Zandron

Dipartimento di Informatica, Sistemistica e Comunicazione
Università degli Studi di Milano-Bicocca
Viale Sarca 336/14, 20126 Milano, Italy
zandron@disco.unimib.it

Abstract. P systems are a computational model inspired by the functioning of the cell and based upon the notion of cellular membrane. We show how different features of P systems with active membranes, a variant of the basic model where membranes can be multiplied by division, can be used to approach various problems in computation theory.

1 Introduction

Membrane systems (also known as *P systems*) have been introduced in [11] as a parallel, nondeterministic, synchronous and distributed model of computation inspired by the structure and functioning of living cells. The basic model consists of a hierarchical structure composed by several membranes, embedded into a main membrane called the *skin*. Membranes defines *regions* that contain multisets of *objects* (represented by symbols of an alphabet) and *evolution rules*.

Using these rules, the objects may evolve and/or move from a region to a neighboring one. Usually, the rules are applied in a maximally parallel and nondeterministic way: all objects which can evolve in a computation step must evolve; if different sets of rules can be applied in a computation step (in a maximal parallel way), then one of them is nondeterministically chosen.

A *computation* starts from an initial configuration of the system and terminates when no evolution rule can be applied. The result of a computation is the multiset of objects contained into an *output membrane*, or emitted from the skin of the system. For a systematic introduction to P systems we refer the reader to [13], [14], whereas some recent information can be found in [25].

The variant of P systems with active membranes has been introduced in [12] to consider the possibility to communicate objects according to electrical charges associated with the membranes, and the possibility to increase the number of membranes by division of existing ones. Such features can be used in many different ways: as an example, one can construct an exponential workspace in linear time to attack computationally hard problems.

In this paper we survey some results concerning P systems with active membranes, to show how different membrane features can be exploited in different ways to approach various classical problems in the theory of computation.

A. Beckmann, E. Csuhaj-Varjú, and K. Meer (Eds.): CiE 2014, LNCS 8493, pp. 393–402, 2014.

2   Definitions

In this section we recall the basic definition of P systems with active membranes.

Definition 1. *A P system with active membranes of initial degree $d \geq 1$ is a tuple $\Pi = (\Gamma, \Lambda, \mu, w_1, \ldots, w_d, R)$, where:*

- Γ *is an alphabet, a finite non-empty set of symbols, usually called* objects;
- Λ *is a finite set of labels for the membranes;*
- μ *is a membrane structure (i.e., a rooted unordered tree) consisting of d membranes enumerated by $1, \ldots, d$; each membrane is labeled by an element of Λ, not necessarily in a one-to-one way, and possesses an electrical charge (or polarization), that can be neutral (0), positive (+) or negative (−).*
- $w_1, \ldots, w_d$ *are strings over Γ, describing the initial multisets of objects placed in the d regions of μ;*
- *R is a finite set of rules.*

The rules are of the following kinds:

- *Object evolution rules, of the form $[a \rightarrow w]_h^\alpha$*
  They can be applied if the membrane h has charge α and contains an occurrence of the object a; the object a is rewritten into the multiset w.
- *Send-in communication rules, of the form $a\,[\,]_h^\alpha \rightarrow [b]_h^\beta$*
  They can be applied to a membrane labeled by h, having charge α and if the external region contains an occurrence of the object a; the object a is sent into h becoming b and, simultaneously, the charge of h is changed to β.
- *Send-out communication rules, of the form $[a]_h^\alpha \rightarrow [\,]_h^\beta\,b$*
  They can be applied to a membrane labeled by h, having charge α and containing an occurrence of a; the object a is sent out from h to the outside region becoming b. Simultaneously, the charge of h is changed to β.
- *Dissolution rules, of the form $[a]_h^\alpha \rightarrow b$*
  They can be applied to a membrane labeled by h, having charge α and containing an occurrence of the object a; the membrane h is dissolved and its contents are left in the surrounding region unaltered, except that an occurrence of a becomes b.
- *Elementary division rules, of the form $[a]_h^\alpha \rightarrow [b]_h^\beta\,[c]_h^\gamma$*
  They can be applied to a membrane labeled by h, having charge α, containing an occurrence of the object a but having no other membrane inside (an *elementary membrane*); the membrane is divided into two membranes having label h and charge β and γ; the object a is replaced, respectively, by b and c while the other objects in the initial multiset are copied to both membranes.
- *Nonelementary division rules, of the form*

$$[[\,]_{h_1}^+ \cdots [\,]_{h_k}^+ [\,]_{h_{k+1}}^- \cdots [\,]_{h_n}^-]_h^\alpha \rightarrow [[\,]_{h_1}^\delta \cdots [\,]_{h_k}^\delta]_h^\beta [[\,]_{h_{k+1}}^\epsilon \cdots [\,]_{h_n}^\epsilon]_h^\gamma$$

They can be applied to a membrane labeled by h, having charge α, containing the positively charged membranes $h_1, \ldots, h_k$, the negatively charged membranes $h_{k+1}, \ldots, h_n$, and possibly some neutral membranes. The membrane h is divided into two copies having charge β and γ, respectively; the positively charged membranes $h_1, \ldots, h_k$ are placed inside the former membrane, their charge set to δ, while the negative ones are placed inside the latter membrane, their charges set to ϵ. Neutral membranes inside h are duplicated and placed inside both copies.

Each instantaneous configuration of a P system with active membranes is described by the current membrane structure, including the electrical charges, together with the multisets located in the corresponding regions. A computation step changes the current configuration according to the following set of principles:

- Each object and membrane can be subject to at most one rule per step, except for object evolution rules (inside each membrane any number of evolution rules can be applied simultaneously).
- The application of rules is *maximally parallel*: each object appearing on the left-hand side of evolution, communication, dissolution or elementary division rules must be subject to exactly one of them (unless the current charge of the membrane prohibits it). The same reasoning applies to each membrane that can be involved to communication, dissolution, elementary or nonelementary division rules. In other words, all possible rules that can be applied must be applied at each computation step; the only objects and membranes that do not evolve are those associated with no rule, or only to rules that are not applicable due to the electrical charges.
- When several conflicting rules can be applied at the same time, a nondeterministic choice is performed; this implies that, in general, multiple possible configurations can be reached after a computation step (e.g. consider two rules $a \rightarrow b$ and $a \rightarrow c$ in a region h; if an object a is present in that region, then it can nondeterministically produce either b or c, by using respectively the first or the second rule).
- While all the chosen rules are considered to be applied simultaneously during each computation step, they are logically applied in a bottom-up fashion: first, all evolution rules are applied to the elementary membranes, then all communication, dissolution and division rules; then the application proceeds towards the root of the membrane structure. In other words, each membrane evolves only after its internal configuration has been updated.
- The outermost membrane cannot be divided or dissolved, and any object sent out from it cannot re-enter the system again.

A *halting computation* of the P system Π is a finite sequence of configurations $\mathcal{C} = (\mathcal{C}_0, \ldots, \mathcal{C}_k)$, where $\mathcal{C}_0$ is the initial configuration, every $\mathcal{C}_{i+1}$ is reachable by $\mathcal{C}_i$ via a single computation step, and no rules can be applied anymore in $\mathcal{C}_k$. The result of a halting computation is the multiset of objects emitted from the skin during the whole computation. A *non-halting* computation $\mathcal{C} = (\mathcal{C}_i : i \in \mathbb{N})$ consists of infinitely many configurations, again starting from the initial one and generated by successive computation steps, where the applicable rules are never exhausted. A non-halting computation produces no output.

P systems can also be used as *recognizers* (see, e.g. [3]) by employing two distinguished objects yes and no; exactly one of these must be sent out from the outermost membrane during each computation, in order to signal acceptance or rejection respectively; we also assume that all computations are halting. If all computations starting from the same initial configuration are accepting, or all are rejecting, the P system is said to be *confluent*. If this is not necessarily the case, then we have a *non–confluent* P system, and the overall result is established as for nondeterministic Turing machines: it is acceptance iff an accepting computation exists. All P systems considered in this paper are confluent.

In order to solve decision problems (i.e., decide languages), we use *families* of recognizer P systems $\Pi = \{\Pi_x : x \in \Sigma^\star\}$. Each input x is associated with a P system Π_x that decides the membership of x in the language $L \subseteq \Sigma^\star$ by accepting or rejecting. The mapping $x \mapsto \Pi_x$ must be efficiently computable for each input length [10].

Definition 2. *A family of P systems $\Pi = \{\Pi_x : x \in \Sigma^\star\}$ is said to be (polynomial-time) uniform if the mapping $x \mapsto \Pi_x$ can be computed by two deterministic polynomial-time Turing machines F (for "family") and E (for "encoding") as follows:*

- *The machine F, taking as input the length n of x in unary notation, constructs a P system Π_n, which is common for all inputs of length n, with a distinguished input membrane.*
- *The machine E, on input x, outputs a multiset w_x (an encoding of x).*
- *Finally, Π_x is simply Π_n with w_x added to the multiset placed inside its input membrane.*

Definition 3. *If the mapping $x \mapsto \Pi_x$ is computed by a single polynomial-time Turing machine, the family Π is said to be semi-uniform. In this case, inputs of the same size may be associated with P systems having possibly different membrane structures and rules.*

Any explicit encoding of Π_x is allowed as output of the construction, as long as the number of membranes and objects represented by it does not exceed the length of the whole description, and the rules are listed one by one. This restriction is enforced to mimic a (hypothetical) realistic process of construction of the P system, where membranes and objects are placed in a constant amount during each construction step, and require actual physical space proportional to their number. Moreover, notice that uniformity condition can also be restricted to be computed in classes below **P**, such as log–space Turing machines. We refer the reader to [10] for further details on the encoding of P systems.

Finally, we describe how time and space complexity for families of recognizer P systems are measured.

Definition 4. *A uniform or semi–uniform family of P systems $\Pi = \{\Pi_x : x \in \Sigma^\star\}$ is said to decide the language $L \subseteq \Sigma^\star$ (in symbols $L(\Pi) = L$) in time $f : \mathbb{N} \to \mathbb{N}$ iff, for each $x \in \Sigma^\star$,*

- the system Π_x accepts if $x \in L$, and rejects if $x \notin L$;
- each computation of Π_x halts within $f(|x|)$ computation steps.

Definition 5. *Let C be a configuration of a P system Π. The size $|C|$ of C is defined as the sum of the number of membranes in the current membrane structure and the total number of objects they contain. If $C = (C_0, \ldots, C_k)$ is a halting computation of Π, then the* space required by C *is defined as*

$$|\mathcal{C}| = \max\{|C_0|, \ldots, |C_k|\}$$

or, in the case of a non-halting computation $\mathcal{C} = (C_i : i \in \mathbb{N})$,

$$|\mathcal{C}| = \sup\{|C_i| : i \in \mathbb{N}\}.$$

Non-halting computations might require an infinite amount of space (in symbols $|\mathcal{C}| = \infty$). The space required by Π *itself is then*

$$|\Pi| = \sup\{|\mathcal{C}| : \mathcal{C} \text{ is a computation of } \Pi\}.$$

Notice that $|\Pi| = \infty$ occurs if either Π has a non-halting computation requiring infinite space, or Π has an infinite set of halting computations, such that for each bound $b \in \mathbb{N}$ there exists a computation requiring space larger than b.

3   Basic Results for P Systems

First investigations of P systems concentrated on their computational power, comparing them with other classic computation models like automata and Turing machines. Let us denote by $NOP_k(\delta)$ (resp. $NOP_k(n\delta)$) the family of natural numbers generated by P systems having k membranes and using (resp. not using) the dissolving membrane action. The following results are known ([13]):

Theorem 1. $NOP_*(n\delta) = NOP_1(n\delta) = NCF$
$NCF = NOP_*(n\delta) \subset (NE0L \subseteq)NOP_2(\delta)$
$NOP_*(\delta)(\subseteq ET0L) \subset NCS$

The theorem shows that by using a single membrane we can only generate the length sets of context–free languages, and the power cannot be extended by using more membranes. However, the membrane dissolving action used in a system with at least two membranes improves the computational power, even if universality cannot be obtained in this way.

If we consider the variant (called Rewriting P systems) where objects are structured in strings, then this fact is even more evident. Let us denote by RP_k the family of languages generated by Rewriting P systems using k membranes and context–free rewriting rules. The following theorem from [13] shows that using a single membrane only context–free languages can be obtained, but a structure with four membranes allow to obtain a strictly more powerful class.

Theorem 2. $RP_1(CF) = CF \subset RP_4(CF)$

Thus, it is evident from these results that the power of such systems can be improved (as expected) by exploiting membranes to define regions to keep separated specific subsets of rules and objects.

By making use of generic communication rules allowed by electrical charges, standard normal forms can be defined as showed in [23]:

Theorem 3. *Each language generated by a (rewriting) P system which make use of electrical charges to communicate objects, can be generated by a system of the same type with exactly three rules in each region.*

If we allow to set priority relations among rules to define the order in which they should be applied in case of a conflict, then the following result holds:

Theorem 4. *Each recursively enumerable language can be generated by a (rewriting) P system having exactly two different rules in each region.*

4   Using Membrane Division to Attack Computationally Hard Problems

As mentioned above, P systems with active membrane allow to create new membranes during the computation by division of existing membranes. In this way we can obtain a trade off between time and space resources that allows to solve NP–complete (or even harder) problems in polynomial time and exponential space (see, e.g., [12], [6], [7], [22], [24]).

Theorem 5. *The SAT problem can be solved in linear time (with respect to the number of variables and the number of clauses) by a confluent P-system with active membranes using elementary membrane division only.*

Proof. (sketch) Consider a boolean expression Φ in conjunctive normal form, with m clauses and n variables. We can build a P-system $\Pi = (\Gamma, \Lambda, \mu, w_1, w_2, R)$ having initial objects $a_1, a_2, \ldots, a_n$ in region 2 and such that R is defined to contain a polynomial number of rules (with respect to the size of the input formula) that operate as it follow.

By using the variables a_i and elementary membrane division rules, in $O(n)$ steps we generate 2^n copies of membrane 2, containing all possible truth assignments of the n variables of Φ.

Then, in $O(m)$ steps we verify if there is at least one membrane containing a truth assignment that satisfies all the m clauses of Φ. In this case, an object *yes* is sent out from the skin membrane; otherwise, an object *no* is sent out.   □

Let us denote by $\mathbf{PMC}_{\mathcal{NAM}}$, $\mathbf{PMC}_{\mathcal{EAM}}$, and $\mathbf{PMC}_{\mathcal{AM}}$ the class of problems solved by P systems with active membranes without membrane division, with division for elementary membranes only, and for both elementary and non–elementary membranes, respectively. An immediate consequence of the previous theorem is the following:

Theorem 6. NP $\subseteq$ PMC$_{\mathcal{EAM}}$

From this result and from the closure properties for **PMC$_{\mathcal{EAM}}$** it also follows:

Theorem 7. coNP $\subseteq$ PMC$_{\mathcal{EAM}}$

In [16] a stronger result was proved: the complexity class **PP** (Probabilistic Polynomial time: the class of decision problems solvable by a probabilistic Turing machine in polynomial time, with an error probability of less than $1/2$ for all instances) is also included in **PMC$_{\mathcal{EAM}}$**. The result is obtained by solving the **PP**–complete problem SQRT–3SAT (given a Boolean formula of n variables in 3CNF, do at least $\sqrt{(2^n)}$ among the 2^n possible truth assignments satisfy it?).

Of course, if we also allow the use of division for non-elementary membranes, harder problems can be solved. In particular, in [2], [21] and [19] it has been proved the following:

Theorem 8. PSPACE $\subseteq$ PMC$_{\mathcal{AM}}$ $\subseteq$ EXPTIME

One could conjecture that computationally hard problems could also be solved by means of P systems without the use of membrane division: in fact, by applying in a maximally parallel way the rewriting rules one can obtain an exponential number of objects in polynomial time. Nonetheless, if the considered P system is confluent, then all objects of the same type must be used in the same way inside a region. If we do not allow membrane division, then the languages accepted in polynomial time by (confluent) membrane systems can also be accepted by a deterministic Turing machine in polynomial time, as showed in [22]. In [5] a characterization of **P** was given in terms of P systems with active membranes, without membrane division:

Theorem 9. P $=$ PMC$_{\mathcal{NAM}}$

Another important (and somehow surprising) characterization of **P** was given in [4], where it was shown that membrane division alone does not suffice to speed-up computations. Let us denote by **PMC$_{\mathcal{AM}}(n\delta, nPol)$** the class of languages recognized in polynomial time by P–systems with active membranes, without dissolving membrane action nor polarization on membranes. Then:

Theorem 10. P $=$ PMC$_{\mathcal{AM}}(n\delta, nPol)$

5 Space Complexity of P–Systems and Polarization of Membranes

In order to clarify relations between the amount of time and space needed to solve various classes of problems, in [15] a definition of space complexity for P systems has been introduced. In [17] and [18] it has been shown, respectively, that the **PSPACE**–complete problem Quantified–3SAT can be solved by P–systems with active membranes using a polynomial amount of space, and that such P systems can be simulated by Turing machines with only a polynomial increase in space requirements, thus giving a precise characterization of the class **PSPACE** in terms of space complexity classes for membrane systems. A similar result to characterize the complexity class **EXPSPACE** can be obtained by considering exponential space P systems, as showed in [1].

The opposite direction was also considered by investigating classes of problems solved by P systems which make use of logarithmic space. We recall here one result obtained in this framework, as it clearly shows how the same membrane feature can be used in different ways.

In order to consider sublinear space, we first need to define a meaningful notion of sublinear space for P systems, inspired by sublinear space definition for Turing machines: we consider two distinct alphabets, an *INPUT* alphabet and a *WORK* alphabet, in the definition of a P systems. The input objects cannot be rewritten and do not contribute to the size of the configuration of a P system. The size of a configuration is defined as the sum of the number of membranes in the current membrane structure and the total number of working objects they contain. The space complexity of a P system is defined, as in section 2, as the maximum size among all configurations. Moreover, we need to define a uniformity condition for the families of P systems that is weaker than the usual **P** uniformity, to avoid the possibility to solve a problem directly by using the Turing machine that build the P systems we use to compute. We consider **DLOGTIME**-uniformity, defined on the basis of **DLOGTIME** Turing machines [9]. We refer the reader to [20] for formal definitions.

The efficient simulation of logarithmic space Turing machines (or other equivalent models) by employing standard techniques used in the papers previously cited seems not to work because of two main problems: we either need to use a polynomial number of working objects (thus violating the logarithmic space condition) or to use a polynomial number of rewriting rules (thus violating the uniformity condition). Nonetheless, it has been showed in [20] that such a simulation can be efficiently done by using membrane polarization both to communicate objects through membranes as well as to store some information:

Theorem 11. *Each log–space deterministic Turing machine M can be simulated by a **DLOGTIME**-uniform family Π of confluent recognizer P systems with active membranes in logarithmic space.*

Proof. (sketch) Consider a Turing machine M working in a logarithmic space. The P system Π_n that simulates M on input of length n is composed of:

- A skin membrane containing a *state object* object $q_{i,w}$ to indicate that M is currently in state q and its tape heads are on the i-th and w-th symbols of the input and work tape, respectively.
- $O(\log(n))$ nested membranes (INPUT tape membranes) containing, in the innermost one, the input symbols of M, and $O(\log(n))$ membranes to store the work tape of M (WORK tape membranes).
- Two sets of membranes, which size depends on the dimensions of the input and the working alphabets of M (SYMBOL membranes).

To simulate a computation step of M, the state object enters the INPUT membranes, storing the bits corresponding to the actual position of the IN-PUT head of M in the polarizations of the INPUT membranes. Only one object (corresponding to the INPUT symbol actually read) can travel to the outermost membrane. Then, the state object identifies the symbol actually under the WORK head (using WORK tape membranes) and proceed to simulate the transition of M using the SYMBOLS membranes.

Each P system Π_x (simulating each $M(x)$ such that $|x| = n$) only requires $O(\log |x|)$ membranes and objects besides the input objects and the family Π is **DLOGTIME**-uniform. The time required by the simulation is $O(n \cdot t(n))$, where $t(n)$ is the maximum number of steps performed by M on inputs of length n. $\square$

An immediate corollary of Theorem 11 is that the class **L** (the class of problems solved by log–space Turing machines) is contained in the class of problems solved by **DLOGTIME**-uniform, log–space P systems with active membranes.

6 Conclusions

We survey some results concerning P systems with active membranes, to show that the various features associated to membranes can be exploited in different ways to approach various classical problems in the theory of computation.

Of course, further and more powerful features can be considered. As an example, we recall here the so-called UREM P–systems ([8]), where rules are assigned to membranes (not to the regions as usually done in membrane computing) and the concept of polarization is extended so that every membrane carries an integer value representing an energy value that can be changed during a computation. Turing–completeness without the use of other features can be obtained in this way, as well as a definition of quantum-like membrane systems.

Acknowledgements. This work was partially supported by Università degli Studi di Milano-Bicocca, Fondo di Ateneo per la Ricerca (FAR) 2013.

References

1. Alhazov, A., Leporati, A., Mauri, G., Porreca, A.E., Zandron, C.: The computational power of exponential-space P systems with active membranes. In: Martínez-del-Amor, et al. (eds). Proc. of BWMC 2010, vol. I, pp. 35–60. Fénix Editora (2012)
2. Alhazov, A., Martin-Vide, C., Pan, L.: Solving a PSPACE–complete problem by P–systems with restricted active membranes. Fund. Inf. 58(2), 67–77 (2003)
3. Csuhaj-Varju, E., Oswald, M., Vaszil, G.: P automata, Handbook of Membrane Computing. In: Paun, G., et al. (eds.), pp. 144–167. Oxford University Press (2010)
4. Gutiérrez-Naranjo, M.A., Jesús Pérez-Jímenez, M., Riscos-Núñez, A., Romero-Campero, F.J.: P–systems with active membranes, without polarizations and without dissolution: a characterization of P. In: Calude, C.S., Dinneen, M.J., Păun, G., Jesús Pérez-Jímenez, M., Rozenberg, G. (eds.) UC 2005. LNCS, vol. 3699, pp. 105–116. Springer, Heidelberg (2005)

5. Gutierrez-Naranjo, M., Perez-Jimenez, M.J., Riscos-Nunez, A., Romero-Campero, F.J.: Characterizing tractability by cell-like membrane systems. In: Subramanian, K.G., et al. (eds.) Formal Models, Languages and Applications, Ser. Mach. Percept. Artif. Intell., vol. 66, pp. 137–154. World Scientific (2006)
6. Krishna, S.N., Rama, R.: A variant of P-systems with active membranes: Solving NP-complete problems. Rom. J. of Inf. Sci. and Tech. 2, 4 (1999)
7. Leporati, A., Zandron, C., Gutierrez-Naranjo, M.A.: P systems with input in binary form. Int. J. of Found. of Comp. Sci. 17(1), 127–146 (2006)
8. Leporati, A., Mauri, G., Zandron, C.: Quantum Sequential P Systems with Unit Rules and Energy Assigned to Membranes. In: Freund, R., Păun, G., Rozenberg, G., Salomaa, A. (eds.) WMC 2005. LNCS, vol. 3850, pp. 310–325. Springer, Heidelberg (2006)
9. Mix Barrington, D.A., Immerman, N., Straubing, H.: On uniformity within NC^1. Journal of Computer and System Sciences 41(3), 274–306 (1990)
10. Murphy, N., Woods, D.: The computational power of membrane systems under tight uniformity conditions. Natural Computing 10(1), 613–632 (2011)
11. Păun, G.: Computing with membranes. J. of Computer and System Sciences 61(1), 108–143 (2000)
12. Păun, G.: P systems with active membranes: Attacking NP-complete problems. J. of Automata, Languages and Combinatorics 6(1), 75–90 (2001)
13. Păunm, G.: Membrane Computing. An Introduction. Springer, Berlin (2002)
14. Păun, G., Rozenberg, G., Salomaa, A. (eds.): Handbook of Membrane Computing. Oxford University Press (2010)
15. Porreca, A.E., Leporati, A., Mauri, G., Zandron, C.: Introducing a space complexity measure for P systems. Int. J. of Comp., Comm. & Control 4(3), 301–310 (2009)
16. Porreca, A.E., Leporati, A., Mauri, G., Zandron, C.: P systems with Elementary Active Membranes: Beyond NP and coNP. In: Gheorghe, M., Hinze, T., Păun, G., Rozenberg, G., Salomaa, A. (eds.) CMC 2010. LNCS, vol. 6501, pp. 338–347. Springer, Heidelberg (2010)
17. Porreca, A.E., Leporati, A., Mauri, G., Zandron, C.: P Systems with Active Membranes: Trading Time for Space. Natural Computing 10(1), 167–182 (2011)
18. Porreca, A.E., Leporati, A., Mauri, G., Zandron, C.: P systems with active membranes working in polynomial space. Int. J. Found. Comp. Sc. 22(1), 65–73 (2011)
19. Porreca, A.E., Mauri, G., Zandron, C.: Complexity classes for membrane systems. RAIRO-Theor. Inform. and Applic. 40(2), 141–162 (2006)
20. Porreca, A.E., Leporati, A., Mauri, G., Zandron, C.: Sublinear-Space P Systems with Active Membranes. In: Csuhaj-Varjú, E., Gheorghe, M., Rozenberg, G., Salomaa, A., Vaszil, G. (eds.) CMC 2012. LNCS, vol. 7762, pp. 342–357. Springer, Heidelberg (2013)
21. Sosik, P.: The computational power of cell division in P systems: Beating down parallel computers? Natural Computing 2(3), 287–298 (2003)
22. Zandron, C., Ferretti, C., Mauri, G.: Solving NP-complete problems using P systems with active membranes. In: Antoniou, I., Calude, C.S., Dinneen, M.J. (eds.) Unconventional Models of Computation, pp. 289–301. Springer, London (2000)
23. Zandron, C., Ferretti, C., Mauri, G.: Two Normal Forms for Rewriting P systems, in Machines. In: Margenstern, M., Rogozhin, Y. (eds.) MCU 2001. LNCS, vol. 2055, pp. 153–164. Springer, Heidelberg (2001)
24. Zandron, C., Leporati, A., Ferretti, C., Mauri, G., Pérez-Jiménez, M.J.: On the Computational Efficiency of Polarizationless Recognizer P Systems with Strong Division and Dissolution. Fundamenta Informaticae 87(1), 79–91 (2008)
25. The P systems Web page, http://ppage.psystems.eu/

Short Lists with Short Programs in Short Time – A Short Proof

Marius Zimand*

Towson University, Baltimore, MD

Abstract. Bauwens, Mahklin, Vereshchagin and Zimand [1] and Teutsch [6] have shown that given a string x it is possible to construct in polynomial time a list containing a short description of it. We simplify their technique and present a shorter proof of this result, which also achieves better values for the main parameters.

1 Introduction

Given that the Kolmogorov complexity is not computable, it is natural to ask if given a string x it is possible to construct a short list containing a minimal (+ small overhead) description of x. Bauwens, Mahklin, Vereshchagin and Zimand [1] and Teutsch [6] show that, surprisingly, the answer is YES. Even more, in fact the short list can be computed in polynomial time. More precisely, [1] showed that one can effectively compute lists of quadratic size guaranteed to contain a description of x whose size is additively $O(1)$ from a minimal one (it is also shown that it is impossible to have such lists shorter than quadratic), and that one can compute in polynomial-time lists guaranteed to contain a description that is additively $O(\log n)$ from minimal. Finally, [6] improved the latter result by reducing $O(\log n)$ to $O(1)$.

Theorem 1 ([6]). *For every standard machine U there is a constant c and a polynomial-time algorithm f such that for every x, $f(x)$ outputs a list of programs that contains a c-short program for x.*

Let us explain the formal terms. Given a Turing machine U, a c-short program for x is a string p such that $U(p) = x$ and the length of p is bounded by $c+$ (length of a shortest program for x). A machine U is *optimal* if $C_U(x) \leq C_V(x) + O(1)$ for all machines V and all strings x (where C_U (C_V) is the Kolmogorov complexity induced by U (respectively, by V) and the constant $O(1)$ may depend on V). An optimal machine U is *standard* if for every machine V there is an efficient translator from any machine V to U, i.e., a polynomial-time computable function t such that for all p, y, $U(t(p)) = V(p)$ and $|t(p)| = |p| + O(1)$.

Both [1] and [6] prove their results regarding polynomial-time computable lists as corollaries of somewhat more general theorems. We present in this note a direct proof of Theorem 1, which is simpler and shorter than the one in [6].

---

* This work has been supported by NSF grant CCF 1016158.

A. Beckmann, E. Csuhaj-Varjú, and K. Meer (Eds.): CiE 2014, LNCS 8493, pp. 403–408, 2014.
© Springer International Publishing Switzerland 2014

We emphasize that there is no technical innovation in the proof that we present below. We use the same general approach and the same ingredients as in [1] and [6], but, because we go straight to the target, we can take some shortcuts that render the proof simpler. The proof given here also produces a smaller size of the list and a smaller value for the constant in the theorem.

Proof overview. Essentially we want to compress a string x in polynomial time to a (close to) minimal posible succinct description p, such that decompression (i.e., reconstructing back x from p) is computable (not necessarily in poynomial time). This is of course impossible in absolute terms, but here we compress in a weaker sense, because we obtain from x not a single compressed string, but a list guaranteed to contain a (close to) optimally succinct description of x. It is natural to think to use seeded extractors, because an extractor's output is close to being optimally compressed in the Shannon entropy sense. The problem is that we need an extractor with logarithmic seed (because we want a list of polynomial size) and no entropy loss (because we want to decompress). Unfortunately, such extractors have not yet been shown to exist. The key observation from [1], also used in [6], is that in fact a disperser is good enough, and then one can use the disperser from [5], which has the needed parameters. Now, why are dispersers sufficient? The answer, inspired by [4], stems from the idea from [1] to use for this kind of compression graphs that allow on-line matching. These are unbalanced bipartite graphs, which, in their simplest form, have LEFT $= \{0,1\}^n$, RIGHT $= \{0,1\}^{k+\text{small overhead}}$, and left degree $= \text{poly}(n)$, and which permit on-line matching up to size $K = 2^k$. This means that any set A of K left nodes, each one requesting to be matched to some adjacent right node, can be satisfied in the on-line manner(i.e., the requests arrive one by one and each request is satisfied before seeing the next one; in our proof we will allow a small number of requests to be discarded, but this should also happen before the next request arrives). The correspondence to our problem is roughly that strings in LEFT are the strings that we want to compress, and the strings in RIGHT are their compressed forms. We need on-line matching because we are going to enumerate left strings as they are produced by the universal machine and each time a string is enumerated we want to find it a match, i.e., to compress it. In order for a graph to allow matching, it needs to have good expansion properties. It turns out that it is enough if left subsets of a given size $K/O(1)$ expand to size K, and a disperser has this property. When we decompress, given the right node (the compressed string), we run the matching algorithm and see which left node has been matched to it. For this the decompressor needs to have n to be able to construct the graph, and this produces the $O(\log n)$ overhead. Thus this approach is good enough to obtain the result with $O(\log n)$-short programs from [1]. To reduce $O(\log n)$ to $O(1)$, we need the new ideas from [6]. The point is that this time we want LEFT to have strings not of a single length n, but of all lengths $n \geq k$ (because we can no longer afford to give n to the decompressor). In fact, it is not hard to see, that it is enough to restrict to lengths $k \leq n \leq 2^k$. This time we need expansion for all sets of size $\leq K$ (not just equal to a fixed $K/O(1)$, because we need each subset (of the match-requesting set A) of strings of a given length to expand. For this, the unbalanced lossless expander from [3] is good, except for one problem: The size of RIGHT in this expander is poly(K) and not the desired $O(K)$. This problem is fixed by compressing using again the disperser from [5] to a set of size $K \cdot \text{poly}(k)$, and, finally, using a simple trick, to size $O(K)$, which implies the $O(1)$ overhead we aim for.

2 Combinatorial Tools

We use bipartite graphs $G = (L, R, E \subseteq L \times R)$. We denote LEFT($G$) = L, RIGHT(G) = R. For integers n, m, k, d we denote $N = 2^n, M = 2^m, K = 2^k, D = 2^d$. We denote $[n] = \{1, 2, \ldots, n\}$. A bipartite graph G is explicit if there exists a polynomial-time algorithm that given $x \in$ LEFT(G) and i, outputs the i-th neighbor of x (in case i exceeds the number of neighbors, it will output a special symbol).

The main tools are the following expander and disperser graphs.

Definition 1. *A bipartite graph G is a (K, K')-expander if every subset of left nodes having size K, has at least K' right neighbors.*

Theorem 2 (Guruswami, Umans, Vadhan [3]). *For every constant $\alpha > 0$, every n, every $k \leq n$, and $\epsilon > 0$, there exists an explicit $(K', (1-\epsilon)DK')$ expander for every $K' \leq K$, in which every left node has degree $D = O((nk/\epsilon)^{1+1/\alpha})$, $L = [N], R = [M], M \leq D^2 \cdot K^{1+\alpha}$.*

Definition 2. *A bipartite graph $G = (L, R, E)$ is a (K, δ)-disperser, if every subset $B \subseteq L$ with $|B| \geq K$ has at least $(1 - \delta)|R|$ distinct neighbors.*

Theorem 3 (Ta-Shma, Umans, Zuckerman [5]). *For every K, n and constant δ, there exists explicit (K, δ)-dispersers $G = (L = \{0, 1\}^n, R = \{0, 1\}^m, E \subseteq L \times R)$ in which every node in L has degree $D = n2^{O((\log \log n)^2)}$ and $|R| = \frac{\alpha KD}{n^3}$, for some constant α.[1]*

3 The Proof

The key combinatorial object that we use is provided in the following lemma.

Lemma 1. *For every constant c and every sufficiently large k, there exists an explicit bipartite graph H_k with the following properties:*

1. *LEFT(H_k) = $\{0, 1\}^k \cup \{0, 1\}^{k+1} \cup \ldots \cup \{0, 1\}^{2^k}$, RIGHT($H_k$) = $\{0, 1\}^{k+1}$,*
2. *Each left node x has degree poly($|x|$),*
3. *H_k is a $(K/c^2, K)$-expander.*

We defer the proof of this lemma for later.

---

[1] [5] only indicates that $D = \text{poly}(n)$. The value $D = n2^{O((\log \log n)^2)}$ is obtained by reworking the proof in Lemma 6.4 [5] using the extractor with constant entropy loss from Theorem 4.21 in [3].

We show how the lemma implies Theorem 1. We start with the following lemma about on-line matching (recall that this means that one receives a sequence of requests to match left nodes with one of their adjacent right nodes and each request must be satisfied, or discarded, before seeing the next one).

Lemma 2. *If K on-line matching requests are made in a $(K/c^2, K)$-expander all but less than K/c^2 can be satisfied.*

Proof. Suppose there are K requests for matching left nodes and we attempt to satisfy them in the obvious greedy manner. Suppose that K/c^2 requests cannot be satisfied (because all their neighbors have been used to match previous requests). The K/c^2 left nodes that are not satisfied have K right neighbors and all of them have satisfied matching requests. This would imply that all the K requests have been satisfied, contradiction.

Proof of Theorem 1

We define the following machine V ("the decompressor") that reconstructs x from p. Essentially, the machine V on an input p is looking for a string x that is matched to p during an on-line matching process (this is handled in case (3) below). But first we handle two easy special situations: case (1) when x does not have a description that is shorter than its length, and case (2) when x has a description that is shorter than $\log(|x|)$.

Description of V:

(1) On inputs of the form $00p$, V outputs p.

(2) On inputs of the form $01p$, V simulates $U(p)$ and if $U(p) = x$ and $|x| > 2^{|p|}$, outputs x.

(3) On inputs of the form $1p$, V works as follows:
Let $k = |p| - 1$. Enumerate the elements of the set $\{x \mid \exists q$ of length $k, U(q) = x\}$. When an element x is enumerated and $|x|$ is between k and 2^k, pass x to the online matching algorithm for the graph H_k given by Lemma 1. If x is matched to p, then $V(p)$ outputs x and halts; otherwise the enumeration continues.

Observe that during computations of the form (3), at most K matching requests are made and therefore, by the property of H_k, there are fewer than K/c^2 rejections. It follows that if v is a rejected node then $C_U(v) \leq k - 2\log c + \log c + 2\log\log c + O(1) < k$, for c a large enough constant. Indeed a rejected string can be described by its index in the set of rejected strings written on exactly $k - 2\log c$ bits, and c (which is needed in order to reconstruct k and next enumerate the set of rejected strings). The additional $2\log\log c$ term is required for concatenating the index and c. It follows that if x is a string such that $C_U(x) = k$ and $k \in \{\log|x|, \ldots, |x|\}$, then there exists p of length $k+1$ such that $V(1p) = x$. Moreover, p is one of the right neighbors of x in H_k.

Now, for each x, let $list(x)$ be the list containing the following strings: $00x$, all strings of length $< \log|x|$ prefixed with 01, and all the neighbors of x in H_k prefixed with a 1, for $k = |x|, |x| - 1, \ldots, \log(|x|)$. Note that for every x, $list(x)$ can be computed in polynomial time, and there exists $v \in list(x)$, $|v| \leq$

$C_U(x) + O(1)$ such that $C_V(v) = x$. Finally, using the "translator" t from V programs to U programs, take $f(x) = \{t(v) \mid v \in list(x)\}$. Since t is computable in polynomial time, $U(t(v)) = V(v)$ and $|t(v)| = |v| + O(1)$, we are done.   □

It remains to prove Lemma 1. We use two types of graphs given in the following two lemmas.

Lemma 3. *For every n, and $k \leq n$, there exists an explicit bipartite graph $GUV_{n,k}$ with each left node having degree $D = \lambda(nk)^2$ (for some fixed constant λ), $\mathrm{LEFT}(GUV_{n,k}) = \{0,1\}^n$, $\mathrm{RIGHT}(GUV_{n,k}) = [M]$ with $M \leq D^2 K^2$, which is a $(K', (1/2)DK')$-expander for every $K' \leq K$.*

Proof. This is the Guruswami, Umans, Vadhan expander with parameters $\alpha = 1, \epsilon = 1/2$.

Lemma 4. *For every k, there exists a bipartite graph F_k with each left node having degree $D = O(k^3)$, $\mathrm{LEFT}(F_k) = \{0,1\}^{8k}$, $\mathrm{RIGHT}(F_k) = \{0,1\}^{k+1}$, which is a (K, K)-expander.*

Proof. Consider the Ta-Shma, Umans, Zuckerman $(K, 1/2)$-disperser G, with $\mathrm{LEFT}(G) = \{0,1\}^{8k}$, $\mathrm{RIGHT}(G) = \{0,1\}^m$, left degree $D = O(k2^{O((\log \log k)^2)})$ and $|\mathrm{RIGHT}(G)| = \frac{\alpha KD}{(8k)^3}$.

To increase the size of the right set to be at least $2K$, we make RIGHT consist of $2\lceil \frac{(8k)^3}{\alpha D} \rceil$ copies of $\mathrm{RIGHT}(G)$ connected to $\mathrm{LEFT}(G)$ in the same way as the original nodes. Thus each right node is labelled by a string of length $\geq k+1$ and the left degree is $O(k^3)$.

By merging the nodes whose labels have the same prefix of length $k + 1$, we obtain the graph F_k, which as desired has $\mathrm{RIGHT}(F_k) = \{0,1\}^{k+1}$ and is a $(K, 1/2)$-disperser (because the merge operation can only improve the dispersion property).

Thus, every left subset of size K has at least $(1/2) \cdot 2K$ right neighbors, i.e., F_k is a (K, K)-expander.

We are now prepared to prove Lemma 1.

Proof of Lemma 1

Let us fix c and a sufficiently large k.

We first construct the graph G_k as the union $GUV_{k,k} \cup GUV_{k+1,k} \cup \ldots \cup GUV_{2^k,k}$.

Note that $\mathrm{LEFT}(G_k)$ consists of all strings having length between k and 2^k. For $\mathrm{RIGHT}(G_k)$, we shift the numerical labels of the right nodes in each set in the obvious way before taking the union, so that the sets that we union are pairwise disjoint. We have

$$|\mathrm{RIGHT}(G_k)| \leq \sum_{n=k}^{2^k} \lambda^2(nk)^4 K^2 = \lambda^2 k^4 K^2 \sum_{n=k}^{2^k} n^4 \leq \lambda^2 k^4 \cdot K^7 < K^8,$$

for k sufficiently large. By padding each right node in G_k with $100\ldots0$, we label each right node by a string of length $8k$.

Note that, provided k is sufficiently large, G_k is a $(K/c^2, K)$-expander. Indeed take $B \subseteq \text{LEFT}(G_k)$, $|B| \leq K/c^2$. B has strings of different lengths. If we partition B into subsets of strings corresponding to the different lengths, each subset with strings of length say n expands according to $GUV_{n,k}$ by a factor of $(1/2)\lambda(nk)^2 \geq c^2$ (if k is large enough). Since different subsets of the partition map into disjoint right subsets, the above assertion follows.

The degree of every left node x in G_k is bounded by $\text{poly}(|x|)$ because the edges originating in x are those from the graph $GUV_{|x|,k}$. So G_k is almost what we need except that the right nodes have length $8k$ instead of $k+1$. We fix this issue by compressing strings of length $8k$ to length $k+1$ using the graph F_k from Lemma 4.

More precisely, we build the graph H_k by taking the product of the above graph G_k with the graph F_k. Thus $\text{LEFT}(H_k) = \text{LEFT}(G_k)$, $\text{RIGHT}(H_k) = \text{RIGHT}(F_k)$ and (x, y) is an edge in H_k if there exists $z \in \text{RIGHT}(G_k) \subseteq \text{LEFT}(F_k)$ such that (x, z) is an edge in G_k and (z, y) is an edge in F_k.

As desired, $\text{LEFT}(H_k)$ consists of all strings x having length between k and 2^k, $\text{RIGHT}(H_k) = \{0,1\}^{k+1}$, the degree of every left node x is bounded by $\text{poly}(|x|)\text{poly}(k) = \text{poly}(|x|)$ and H_k is a $(K/c^2, K)$-expander, because each left subset of size K/c^2 expands to size at least K in G_k and then it keeps its size at least K when passing through F_k.   □

Note. The above construction yields in Theorem 1 a list of size $O(n^8)$. If in Lemma 3 we take a small α (instead of $\alpha = 1$), we obtain list size $n^{6+\delta}$, for arbitrarily small positive constant δ. Bauwens and Zimand [2] have recently found a randomized algorithm that with high probability constructs in polynomial time a list of size n guaranteed to contain a c-short programs for $c = O(\log n)$.

Acknowledgements. We are grateful to Alexander Shen for his comments and for signalling an error in an earlier version. We thank Jason Teutsch for useful conversations that lead to a more precise estimation of the list size in Theorem 1.

References

1. Bauwens, B., Makhlin, A., Vereshchagin, N., Zimand, M.: Short lists with short programs in short time. In: Proceedings of 28th IEEE Conference on Computational Complexity, Stanford, California, USA (2013)
2. Bauwens, B., Zimand, M.: Linear list-approximation for short programs (or the power of a few random bits). CoRR, abs/1311.7278 (2013)
3. Guruswami, V., Umans, C., Vadhan, S.P.: Unbalanced expanders and randomness extractors from Parvaresh–Vardy codes. J. ACM 56(4) (2009)
4. Musatov, D., Romashchenko, A.E., Shen, A.: Variations on Muchnik's conditional complexity theorem. Theory Comput. Syst. 49(2), 227–245 (2011)
5. Ta-Shma, A., Umans, C., Zuckerman, D.: Lossless condensers, unbalanced expanders, and extractors. Combinatorica 27(2), 213–240 (2007)
6. Teutsch, J.: Short lists for shorter programs in short time, CORR Technical Report arXiv:1212.6104 (2012)

Author Index